MODERN COSMOLOGY

T0172863

Studies in High Energy Physics, Cosmology and Gravitation

Other books in the series

Electron–Positron Physics at the Z
M G Green, S L Lloyd, P N Ratoff and D R Ward

Non-accelerator Particle Physics
Paperback edition
H V Klapdor-Kleingrothaus and A Staudt

Ideas and Methods of Supersymmetry and Supergravity
or **A Walk Through Superspace**
Revised edition
I L Buchbinder and S M Kuzenko

Pulsars as Astrophysical Laboratories for Nuclear and Particle Physics
F Weber

Classical and Quantum Black Holes
Edited by P Fré, V Gorini, G Magli and U Moschella

Particle Astrophysics
Revised paperback edition
H V Klapdor-Kleingrothaus and K Zuber

The World in Eleven Dimensions
Supergravity, Supermembranes and M-Theory
Edited by M J Duff

Gravitational Waves
Edited by I Ciufolini, V Gorini, U Moschella and P Fré

MODERN COSMOLOGY

Edited by

Silvio Bonometto

Department of Physics,
University of Milan—Bicocca, Milan

Vittorio Gorini and Ugo Moschella

Department of Chemical, Mathematical and Physical Sciences,
University of Insubria at Como

INSTITUTE OF PHYSICS PUBLISHING
BRISTOL AND PHILADELPHIA

© IOP Publishing Ltd 2002

All rights reserved. No part of this publication may be reproduced, stored in a retrieval system or transmitted in any form or by any means, electronic, mechanical, photocopying, recording or otherwise, without the prior permission of the publisher. Multiple copying is permitted in accordance with the terms of licences issued by the Copyright Licensing Agency under the terms of its agreement with the Committee of Vice-Chancellors and Principals.

British Library Cataloguing-in-Publication Data

A catalogue record for this book is available from the British Library.

ISBN 0 7503 0810 9

Library of Congress Cataloging-in-Publication Data are available

Commissioning Editor: James Revill
Production Editor: Simon Laurenson
Production Control: Sarah Plenty
Cover Design: Victoria Le Billon
Marketing Executive: Laura Serratrice

Published by Institute of Physics Publishing, wholly owned by The Institute of Physics, London

Institute of Physics Publishing, Dirac House, Temple Back, Bristol BS1 6BE, UK

US Office: Institute of Physics Publishing, The Public Ledger Building, Suite 1035, 150 South Independence Mall West, Philadelphia, PA 19106, USA

Typeset in LaTeX 2_ε by Text 2 Text, Torquay, Devon
Printed in the UK by MPG Books Ltd, Bodmin, Cornwall

Contents

Preface

Cosmology is a new science, but cosmological questions are as old as mankind. Turning philosophical and metaphysical problems into problems that physics can treat and hopefully solve has been an achievement of the 20th century. The main contributions have come from the discovery of galaxies and the invention of a relativistic theory of gravitation. At the edge of the new millennium, in the spring of 2000, SIGRAV—Società Italiana di Relatività e Gravitazione (Italian Society of Relativity and Gravitation) and the University of Insubria sponsored a doctoral school on 'Relativistic Cosmology: Theory and Observation', which took place at the Centre for Scientific Culture 'Alessandro Volta', located in the beautiful environment of Villa Olmo in Como, Italy. This book brings together the reports of the courses held by a number of outstanding scientists currently working in various research fields in cosmology. Topics covered range over several different aspects of modern cosmology from observational matters to advanced theoretical speculations.

The main financial support for the school came from the University of Insubria at Como–Varese. Other contributors were the Department of Chemical, Physical and Mathematical Sciences of the same University, the National Institute of Nuclear Physics and the Physics Departments of the Universities of Milan, Turin, Rome La Sapienza and Rome Tor Vergata.

We are grateful to all the members of the scientific organizing committee and to the scientific coordinator of the Centro Volta, Professor Giulio Casati, for their invaluable help in the organization. We also acknowledge the essential support of the secretarial conference staff of the Centro Volta, in particular of Chiara Stefanetti.

S Bonometto, V Gorini and U Moschella
23 January 2001

Chapter 1

The physics of the early universe (an overview)

Silvio Bonometto
Department of Physics, University of Milan–Bicocca, Milan, Italy

1.1 The physics of the early universe: an overview

Modern cosmology has a precise birthdate, Hubble's discovery of Cepheids and ordinary stars in *Nebulae*. The nature of nebulae had been disputed for centuries. As early as 1755, in his *General History of Nature and Theory of the Sky*, Immanuel Kant suggested that nebulae could be galaxies. The main objection to this hypothesis has been supernovae. Today we know that, close to its peak, a supernova can exceed the luminosity of its host galaxy. But, while this remained unknown, single stars as luminous as whole nebulae were a severe objection to the claim that nebulae were made of as many as hundreds of billions stars. For instance, in 1893, the British astronomer Mary Clark reported the observation of two stellar bursts in a single nebula, one 25 years after the other. She wrote that: *The light of the nebula has been practically cancelled by the bursts, which... should have been of an order of magnitude so large, that even our imagination refuses in conceiving it.* Clark was not alone in having problems conceiving the energetics of supernovae.

After the recognition that most nebulae were galaxies, Hubble also claimed that they receded from one another, as fragments of a huge explosion. Such an expansive trend, currently named the *Hubble flow*, has been confirmed by the whole present data-set. Although there are no doubts that Hubble's intuition was great, the point is that his data-set did not show that much. At the distances where he pretended to see an expansive trend, the 'Hubble flow' is still dominated by peculiar motions of individual galaxies. Discovering the true nature of nebulae was, however, essential. It is the galactic scale which sets the boundary above which dynamical evolution is mostly due to pure gravity. Dissipative forces, of course, still play an essential role above such a scale. But even the huge x-ray

1

emission from galaxy clusters, now the principal tool for their detection, bears limited dynamical effects.

Galaxies, therefore, are the inhabitants of a super-world whose rules are set by relativistic gravitation. Their average distances are gradually increasing, within the Hubble flow. The Friedmann equations tell us the ensuing rate of matter density decrease and how such a rate varies with density itself. No doubts, then, that the early universe must have been very dense. The cosmic clock, telling us how long ago density was above a given level, is set by the Hubble constant $H = 100h$ km s^{-1} Mpc^{-1}. Here h conveys our residual ignorance, but it is likely that $0.6 < h < 0.8$, while almost no one suggests that h lies outside the interval 0.5–0.9. (One can appreciate how far from reality Hubble was, considering that he had estimated that $h \simeq 5$.)

A realistic measure of h came shortly before the discovery of the cosmic background radiation (CBR). The Friedmann equations could then also determine how temperature varies with time and it was soon clear that, besides being dense, the early universe was hot. This defined the early environment and, until the 1980s, modern cosmologists essentially used known physics within the frame of such exceptional environments. In a sense, this extended Newton's claim that the same gravity laws hold on Earth and in the skies. On the basis of spectroscopical analysis it had already become clear that such a claim could be extended beyond gravity to the laws governing all physical phenomena, thereby leading cosmologists to extend these laws back in time, besides far in space.

1.1.1   The middle-age cosmology

This program, essentially based on the use of general relativity, led to great results. It was shown that, during its early stages, the universe had been homogeneous and isotropic, apart from tiny fluctuations, seeds of the present inhomogeneities. Cosmic times (t) can be associated with redshifts (z), which relate the *scale factor* $a(t)$ to the present scale factor a_0, through the relation

$$1 + z = a_0/a(t).$$

The redshift z also tells us the temperature of the background radiation, which is $T_0(1 + z)$ ($T_0 \simeq 2.73$ K is today's temperature).

On average, linearity held for $z > 30$–100. For $z > 1000$, the high-energy tail of the black body (BB) distribution contained enough photons, with an energy exceeding $B_H = 13.6$ eV, to keep all baryonic matter ionized. Roughly above the same redshift, the radiation density exceeds the baryon density. This occurs above the so-called *equivalence* redshift $z_{eq} = 2.5 \times 10^4 \Omega_b h^2$. Here Ω_b is the ratio between the present density of baryon matter and the present critical density ρ_{cr}, setting the boundary between parabolic and hyperbolic models. It can be shown that $\rho_{cr} = 3H_0^2/8\pi G$.

The relativistic theory of fluctuation growth, developed by Lifshitz, also showed that, in their linear stages, inhomogeneities would grow proportionally to $(1 + z)^{-1}$, if the content of the universe were assumed to be a single fluid. This moderate growth rate tells us that the actual inhomogeneities could not arise from purely statistical fluctuations. When the Lifshitz result was generalized to any kind of matter contents, it also became clear that fluctuations compatible with observed anisotropies in the CBR were too small to turn into galaxies, unless another material component existed, already fully decoupled from radiation at $z \simeq 1000$, besides baryons.

Various hypotheses were then put forward, on the nature of such *dark* matter, whose density, today, is $\Omega_c \rho_{cr}$. (The world is then characterized by an overall *matter* density parameter $\Omega_m = \Omega_c + \Omega_b$.) But, as far as cosmology is concerned, only the redshift z_d when the quanta of dark matter become non-relativistic matters. Let M_d be the mass scale entering the horizon at z_d and let us also recall that the mass scale entering the horizon at $z_{eq} = 2.5 \times 10^4 \Omega_m h^2$ is $\sim 10^{16} M_\odot$. Early fluctuations, over scales $< M_d$, are fully erased by free-streaming, at the horizon entry. If one wants to preserve a fluctuation spectrum extending to quite small scales, it is therefore important for z_d to be large.

As far as cosmology is concerned, the nature of dark matter can therefore be classified according to the minimal size of fluctuations able to survive. If fluctuations are preserved down to scales well below the galactic scale ($M_g \sim 10^8$–$10^{12} M_\odot$), we say that dark matter is *cold*. If dark matter particles are too fast, and become non-relativistic only at late times, so that $M_d > M_g$, we say that dark matter is *hot*. In principle, in the latter case galaxies could also form, because of the fragmentation of greater structures in their nonlinear collapse, which, in general, is not spherically symmetric. But such *top–down* scenarios were soon shown not to fit observational data. This is why cold dark matter (CDM) became a basic ingredient of all cosmological models.

This argument is quite independent from the assumption that Ω_m has to approach unity, in order for the geometry of spatial world sections to be flat. However, once we accept that CDM exists, the temptation to imagine that $\Omega_m = 1$ is great. There is another class of arguments which prevents Ω_b from approaching unity by itself alone. These are related to the early formation of light elements, like ^2H, ^4He, ^7Li. The study of big-bang nucleosynthesis (BBNS) has shown that, in order to obtain the observed abundances of light nuclides, we ought to have $\Omega_b h^2 \simeq 0.02$. BBNS occurred when the temperature of the universe was between 900 and 60 keV (ν decoupling and the opening of the deuterium bottleneck, respectively). At even larger temperatures, strongly interacting matter had to be in the quark–hadron plasma form. Going backwards in time we reach T_{ew}, when the weak and electromagnetic interactions separated. To go still further backwards, we need to speculate on physical theories, as experimental data are lacking. The physics of cosmology, therefore, starts from hydrodynamics and reaches advanced particle physics. In this book, a review of the physics of cosmology is provided in the contribution by John Peacock.

All these ages, starting from the quark–hadron transition, through the era when lepton pairs were abundant, then through BBNS, to arrive at the moment when matter became denser than radiation and finally to matter–radiation decoupling and fluctuation growth, are the so-called *middle ages* of the world. Their study, until the 1980s, was the main duty of cosmologists. Not all problems, of course, were solved then. Moreover, as fresh data flowed in, theoretical questions evolved. In his contribution Piero Rosati reviews the present status of observational cosmology, in relation to the most recent data.

The world we observe today is the result of fluctuation growth through linear and nonlinear stages. The initial *simplicity* of the model has been heavily polluted by nonlinear and dissipative physics. Tracing back the initial conditions from data requires both a theoretical and a numerical effort. In his contribution Anatoly Klypin presents such numerical techniques, the role of which is becoming more and more important. Using recent parallel computing programs, it is now possible to try to reproduce the events leading to the shaping of the universe.

The point, however, is that, once this self-consistent scenario became clear, cosmology was ready for another leap. Since the 1980s, it has become a new paradigm within which very high-energy physics could be tested.

1.1.2 Inflationary theories

The world we observe is extremely complex and inhomogeneous. The level of inhomogeneity gradually decreases when we go to greater scales (on this subject, see the contribution by Luigi Guzzo; another less shared point of view is exposed by Marco Montuori and Luciano Pietronero). But only the observations of CBR show a 'substance' close to homogeneity. In spite of this, the driving scheme of the cosmological quest had been that the present complexity came from an initial simplicity and much effort has been spent in developing a framework able to show that this is what truly occurred. When this desire for unity was fulfilled, cosmologists realized that it had taken them to a deadlock: the conditions from which the observed world had evidently arisen, which so nicely fulfilled their intimate expectations, were so exceptional as to require an exceptional explanation.

This is the starting point of the next chapter of cosmological research, which started in the 1980s and was made possible by the great achievements of previous cosmological research. The new quest took two alternative directions. The most satisfactory possibility occurred if, starting from generic metric conditions, their eventual evolution necessarily created the exceptional 'initial conditions' needed to give a start to the observed world. An alternative, weaker requirement, was that, starting from a generic metric, its eventual evolution necessarily created *somewhere* the exceptional 'initial conditions' needed to give a start to the observed world.

The basic paradigm for implementing one of such requirement is set by inflationary theories. The paradoxes such theories are called to justify can be listed as follows:

(i) Homogeneity and isotropy: apart from tiny fluctuations, whose distribution is itself isotropic, the conditions holding in the universe, at $z > 1000$, are substantially identical anywhere we can observe them. The domain our observations reach has a size $\sim ct_0$ (c, the speed of light; t_0, the present cosmic time). This is the size of the regions causally connected today. At $z \sim 10^3$, the domain causally connected was smaller, just because the cosmic time was $\sim 10^{4.5}$ times smaller than t_0. Let us take a sphere whose radius is $\sim ct_0$. Its surface includes ~ 1000 regions which were then causally disconnected one from another. In spite of that, temperature, fluctuation spectrum, baryon content, etc, were equal anywhere. What made them so?

(ii) Flatness: According to observations, the present matter density parameter Ω_m cannot deviate from unity by more than a factor 10. (Recent observations on the CBR have reduced such a possible discrepancy further.) But, in order for $\Omega_m \sim 0.1$ today, we need to *fine-tune* the initial conditions, at the Planck time, by $1:10^{60}$. To avoid such tuning we can only assume that the spatial section of the metric is Euclidean. Then it remains as such forever.

(iii) Fluctuation spectrum: Let us assume that it reads:

$$P(k) = Ak^n.$$

Here $k = 2\pi/L$ and L are comoving length scales. This spectral shape, apparently depending on A and n only (spectral amplitude and spectral index, respectively), tries to minimize the scale dependence. But a fully scale-independent spectrum is obtained only if $n = 1$. It can then be shown that fluctuations on any scale have an identical amplitude when they enter the horizon. This fully scale-independent spectrum, first introduced by Harrison and Zel'dovich, approaches all features of the observed large-scale structure (LSS). How could such fluctuations arise and why did they have such a spectrum?

Apart from these basic requirements, there are a few other requests such as the absence of topological monsters that we shall not discuss here.

The scheme of inflationary theories amounts then to seeking a theory of fundamental interactions which eliminates these paradoxes. The essential ingredient in achieving such an aim is to prescribe a long period of cosmic expansion dominated by a false vacuum, rather than by any kind of *substance*. Early periods of vacuum dominance are indeed expected, within most elementary particle theories, and this sets the bridge between fundamental interaction theories and cosmological requirements.

In this book, inflationary theories and their framework are discussed in detail by Andrei Linde and George Ellis, and therefore we refrain from treating them further in this introduction. Let us rather outline what is the overall resulting scheme. One assumes that, around the Planck time, the universe emerges from quantum gravity in a *chaotic* status. Hence, anisotropies, inhomogeneities, discontinuities, etc, were dominant then.

However, such a variety of initial conditions has nothing to do with the present observed variety. The universe is indeed anisotropic, inhomogeneous, discontinuous, etc, today; and more and more so, as we go to smaller and smaller scales. But such *secondary* chaos has nothing to do with the *primeval* chaos. It is a kind of *moderate* chaos that we have reached after passing through intermediate highly symmetric conditions. The sequence *complex* → *simple* → *complex* had to run, so that today's world could arise.

1.1.3   Links between cosmology and particle physics

There are, therefore, at least two fields where the connections between particle physics and cosmology have grown strong. As we have just outlined, explaining why and how an inflationary era arose and runs is certainly a duty that cosmologists and particle physicists have to fulfill together.

In a sense, however, this is a more speculative domain, compared with the one opened by the need for a dark component. The first idea on the nature of dark matter was that neutrinos had mass. A neutrino background, similar to the CBR, must exist, if the universe ever had a temperature above ~1 MeV. Such a background would be made by ~100 neutrinos/cm^3, for each neutrino flavour. It is then sufficient to assume that neutrinos have a mass ~10–100 eV, to reach $\Omega_m \sim 1$.

Such an appealing picture, which needs no hypothetical new quanta, but refers to surely existing particles only, was, however, shown not to hold. Neutrinos could be *hot* dark matter, as they become non-relativistic around z_{eq}. As we have already stated, the *top–down* scenario, where structures on galactic scales form thanks to greater structure fragmentation, is widely contradicted by observations.

This does not mean that massive neutrinos may not have a role in shaping the present condition of the universe. Models with a mix of cold and hot dark matter were considered quite appealing until a couple of years ago. Their importance, today, has somehow faded, owing to recent data on dark energy. Recent data on the neutrino mass spectrum are reviewed by Gianluigi Fogli in his contribution.

Alternative ideas on the nature of dark matter then came from supersymmetries. The lightest neutral supersymmetric partner of existing bosons is likely to be stable. In current literature this particle is often called the *neutralino*. There are quite a few parameters, concerning supersymmetries, which are not deducible from known data and, after all, supersymmetries themselves have not yet been shown to be viable. However, well within observationally acceptable values, it is possible for neutralinos to have mass and abundance such as to yield $\Omega_m \sim 1$.

In their contribution Antonio Masiero and Silvia Pascoli focus on the interface between particle physics and cosmology, discussing in detail the nature of CDM. Andrea Giuliani's paper deals with current work aiming at detecting dark matter quanta in laboratories and the contribution by Rita Bernabei *et al* relates possible evidence for the detection of neutralinos. Various hypotheses were considered, about dark matter setting. Its distribution may differ from visible matter, on various scales. By definition, its main interaction, in the present epoch, occurs via gravity and gravitational lensing is the basic way to trace its presence. In his contribution Philippe Jetzer reviews the basic pattern to detect dark matter, over different scales, using the relativistic bending of light rays.

1.1.4 Basic questions and tentative answers

There can be little doubt that the last century has witnessed a change of the context within which the very word 'cosmology' is used. Man has always asked basic questions, concerning the origin of the world and the nature of things. The only answers to such questions, for ages, came from metaphysics or religious beliefs. During the last century, instead, a large number of such questions could be put into a scientific form and quite a significant number could be answered.

As an example, it is now clear that the universe is evolutionary. At the beginning of modern cosmology, models claiming a steady state (SS) of the universe had been put forward. They have been completely falsified, although it is now clear that the stationary expansion regime, introduced by SS models, is not so different from the inflationary expansion regime, needed to make bigbang models self-consistent. Furthermore, if recent measures of the deceleration parameter are confirmed, we seem to be living today in a phase of accelerated expansion, quite similar to inflation. It ought to be emphasized that the strength of the data, supporting this kind of expansion, is currently balanced by the theoretical prejudices of wise researchers. In fact, an accelerated expansion requires a desperate fine-tuning of the vacuum energy, which seems to spoil all the beauty of the inflationary paradigm.

Since Hubble's hazardous conclusion that the universe was expanding, the century which has just closed has seen a number of results, initially supported more by their elegance than by data. The Galilean scheme of experimental science is not being forgotten, but one must always remember that such a scheme is far from requiring pure experimental activity. The basic pattern to physical knowledge is set by the intricate network of observations, experiments and predictions that the researcher has to base on data, but goes well beyond them. With the growing complication of current research, the theoretical phase of scientific thought is acquiring greater and greater weight. During such a stage, the lead is taken by the same criteria which drove mathematical research to its extraordinary achievements.

Besides Hubble's findings, within the cosmological context, we may quote Peebles' discovery of the correlation length r_0, based on angular data, which have recently been shown to allow quite different interpretations. Outside cosmology, the main example is given by gauge theories, which are now the basic ingredient of the standard model of fundamental interactions, and were deepened, from 1954 to the early 1970s, only because they were *too beautiful not to be true*. At least two other fields of research in fundamental physics are now driven by similar criteria—supersymmetries and string theories (see the paper by Renata Kallosh).

While supersymmetries can soon be confirmed, either by the discovery of neutralinos by passive detectors or at CERN's new accelerator, string theories might only find confirmation if signals arriving from the Planck era can be observed. This might be possible if future analyses of CBR anisotropies and polarization show the presence of tensor modes. In this book a review of current procedures for CBR analysis is provided by Arthur Kosowsky.

Also within the cosmological domain, leading criteria linked to aesthetical categories are now being pursued. However, in this field, the concept of beauty is often directly connected with ideological prejudices. Questions such as '*can the universe tunnel from nothing*' have been asked and replied within precise physical contexts. It is, however, clear that the ideological charge of such research is dominant. Moreover, when theoretical results, in this field, are quoted by the media, the distinction between valid speculations and scientific acquisitions often fully fades.

But the main question, for physicists, is different. For at least two centuries, basic mathematics has developed without making reference to experimental reality. The criterion driving mathematicians to new acquisitions was the *mathematical beauty*. Only a tiny part of such mathematical developments then found a role in physics. Tensor calculus was developed well before Einstein found a role for it in special and general relativity. Hilbert spaces found a role in quantum mechanics. Lie groups found a role in gauge theories. But there are plenty of other chapters of beautiful advanced mathematics which are, as yet, unexplored by physicists and may remain so forever.

There is, however, no question about that. Mathematics is an *intellectual* construction and its advancement is based on *intellectual* criteria. The problem arises when physicists begin to use similar criteria to put order in the physical world. Let us emphasize that this is not new in the history of research. The Pythagorean school, in ancient Greece, centered its teaching on mathematical beauty. They also found important physical results, e.g. in acoustics, starting from their criterion that the world should be a reflection of mathematical purity. In the ancient world, the views of Pythagoreans were then taken up by the whole Platonic school, in opposition to the Aristoteleans who thought that the world was ugly and complicated, so that attempting a quantitative description was in vain.

Even though we now believe that the final word has to be provided by the experimental data, there is no doubt that theoretical developments, often long and articulate, are grounded on mathematical beauty. This is true for any field of physics, of course, but the impact of such criteria in the quest for the origin is intellectually disturbing. What seems implicit in all this is that the human mind, for some obscure reason, although in a confused form, owns in itself the basic categories enabling it to distinguish the truth and to assert what is adherent to physical reality.

It is not our intention to take a stand on such points. However, we believe that they should be very present in the mind of all readers, when considering recent developments in basic physics and modern cosmology.

Chapter 2

An introduction to the physics of cosmology

John A Peacock
Institute for Astronomy, University of Edinburgh, United
Kingdom

In asking me to write on 'The Physics of Cosmology', the editors of this book have placed no restrictions on the material, since the wonderful thing about modern cosmology is that it draws on just about every branch of physics. In practice, this chapter attempts to set the scene for some of the later more specialized topics by discussing the following subjects:

(1) some cosmological aspects of general relativity,
(2) basics of the Friedmann models,
(3) quantum fields and physics of the vacuum and
(4) dynamics of cosmological perturbations.

2.1 Aspects of general relativity

The aim of general relativity is to write down laws of physics that are valid descriptions of nature as seen from any viewpoint. Special relativity shares the same philosophy, but is restricted to inertial frames. The mathematical tool for the job is the 4-vector; this allows us to write equations that are valid for all observers because the quantities on either side of the equation will transform in the same way. We ensure that this is so by constructing physical 4-vectors out of the fundamental interval

$$\mathrm{d}x^{\mu} = (c\,\mathrm{d}t, \mathrm{d}x, \mathrm{d}y, \mathrm{d}z) \qquad \mu = 0, 1, 2, 3,$$

using relativistic invariants such as the the rest mass m and proper time $\mathrm{d}\tau$.

For example, defining the 4-momentum $P^{\mu} = m\,\mathrm{d}x^{\mu}/\mathrm{d}\tau$ allows an immediate relativistic generalization of conservation of mass and momentum, since the equation $\Delta P^\mu = 0$ reduces to these laws for an observer who sees a set of slowly-moving particles.

None of this seems to depend on whether or not observers move at constant velocity. We have in fact already dealt with the main principle of general relativity, which states that the only valid physical laws are those that equate two quantities that transform in the same way under any arbitrary change of coordinates. We may distinguish equations that are *covariant*—i.e. relate two tensors of the same rank—and *invariants*, where contraction of a tensor yields a number that is the same for all observers:

$$\Delta P^\mu = 0 \qquad \text{covariant}$$
$$P^\mu P_\mu = m^2 c^2 \qquad \text{invariant.}$$

The constancy of the speed of light is an example of this: with $dx_\mu = (c\, dt, -dx, -dy, -dz)$, we have $dx^\mu\, dx_\mu = 0$.

Before getting too pleased with ourselves, we should ask how we are going to construct general analogues of 4-vectors. We want general 4-vectors V^μ to transform like dx^μ under the adoption of a new set of coordinates x'^μ:

$$V'^\mu = \frac{\partial x'^\mu}{\partial x^\nu} V^\nu.$$

This relation applies for 4-velocity $U^\mu = dx^\mu/\tau$, but fails when we try to differentiate this equation to form the 4-acceleration $A^\mu = dU^\mu/d\tau$:

$$A'^\mu = \frac{\partial x'^\mu}{\partial x^\nu} A^\nu + \frac{\partial^2 x'^\mu}{\partial \tau \partial x^\nu} U^\nu.$$

The second term on the right-hand side is zero only when the transformation coefficients are constants. This is so for the Lorentz transformation, but not in general.

The need is therefore to be able to remove the effects of such *local* coordinate transformations from the laws of physics. Technically, we say that physics should be invariant under *Lorentz group symmetry*.

One difficulty with this programme is that general relativity makes no distinction between coordinate transformations associated with the motion of the observer and a simple change of variable. For example, we might decide that henceforth we will write down coordinates in the order (x, y, z, ct) rather than (ct, x, y, z). General relativity can cope with these changes automatically. Indeed, this flexibility of the theory is something of a problem: it can sometimes be hard to see when some feature of a problem is 'real', or just an artifact of the coordinates adopted. People attempt to distinguish this second type of coordinate change by distinguishing between 'active' and 'passive' Lorentz transformations; a more common term for the latter class is *gauge transformations*.

2.1.1 The equivalence principle

The problem of how to generalize the laboratory laws of special relativity is solved by using the equivalence principle, in which the physics in the vicinity of freely falling observers is assumed to be equivalent to special relativity. We can in fact obtain the full equations of general relativity in this way, in an approach pioneered by Weinberg (1972). In what follows, Greek indices run from 0 to 3 (spacetime), Roman from 1 to 3 (spatial). The summation convention on repeated indices of either type is assumed.

Consider freely falling observers, who erect a special-relativity coordinate frame ξ^μ in their neighbourhood. The equation of motion for nearby particles is simple:

$$\frac{d^2\xi^\mu}{d\tau^2} = 0; \qquad \xi^\mu = (ct, x, y, z),$$

i.e. they have zero acceleration, and we have Minkowski spacetime

$$c^2 \, d\tau^2 = \eta_{\alpha\beta} \, d\xi^\alpha \, d\xi^\beta,$$

where $\eta_{\alpha\beta}$ is just a diagonal matrix $\eta_{\alpha\beta} = \text{diag}(1, -1, -1, -1)$. Now suppose the observers make a transformation to some other set of coordinates x^μ. What results is the perfectly general relation

$$d\xi^\mu = \frac{\partial \xi^\mu}{\partial x^\nu} \, dx^\nu,$$

which, on substitution, leads to the two principal equations of dynamics in general relativity:

$$\frac{d^2 x^\mu}{d\tau^2} + \Gamma^\mu_{\alpha\beta} \frac{dx^\alpha}{d\tau} \frac{dx^\beta}{d\tau} = 0$$
$$c^2 \, d\tau^2 = g_{\alpha\beta} \, dx^\alpha \, dx^\beta.$$

At this stage, the new quantities appearing in these equations are defined only in terms of our transformation coefficients:

$$\Gamma^\mu_{\alpha\beta} = \frac{\partial x^\mu}{\partial \xi^\nu} \frac{\partial^2 \xi^\nu}{\partial x^\alpha \partial x^\beta}$$
$$g_{\mu\nu} = \frac{\partial \xi^\alpha}{\partial x^\mu} \frac{\partial \xi^\beta}{\partial x^\nu} \eta_{\alpha\beta}.$$

This tremendously neat argument effectively uses the equivalence principle to prove what is often merely assumed as a starting point in discussions of relativity: that spacetime is governed by Riemannian geometry. There is a metric tensor, and the gravitational force is to be interpreted as arising from non-zero derivatives of this tensor.

The most well-known example of the power of the equivalence principle is the thought experiment that leads to gravitational time dilation. Consider an accelerating frame, which is conventionally a rocket of height h, with a clock mounted on the roof that regularly disgorges photons towards the floor. If the rocket accelerates upwards at g, the floor acquires a speed $v = gh/c$ in the time taken for a photon to travel from roof to floor. There will thus be a blueshift in the frequency of received photons, given by $\Delta v/v = gh/c^2$, and it is easy to see that the rate of reception of photons will increase by the same factor.

Now, since the rocket can be kept accelerating for as long as we like, and since photons cannot be stockpiled anywhere, the conclusion of an observer on the floor of the rocket is that in a real sense the clock on the roof is running fast. When the rocket stops accelerating, the clock on the roof will have gained a time Δt by comparison with an identical clock kept on the floor. Finally, the equivalence principle can be brought in to conclude that gravity must cause the same effect. Noting that $\Delta \phi = gh$ is the difference in potential between roof and floor, it is simple to generalize this to

$$\frac{\Delta t}{t} = \frac{\Delta \phi}{c^2}.$$

The same thought experiment can also be used to show that light must be deflected in a gravitational field: consider a ray that crosses the rocket cabin horizontally when stationary. This track will appear curved when the rocket accelerates.

2.1.2 Applications of gravitational time dilation

For many purposes, the effects of weak gravitational fields can be dealt with by bolting gravitational time dilation onto Newtonian physics. One good example is in resolving the twin paradox (see p 8 of Peacock 1999).

Another nice paradox is the following: Why do distant stars suffer no time dilation due to their apparently high transverse velocities as viewed from the frame of the rotating Earth? At cylindrical radius r, a star appears to move at $v = r\omega$, implying time dilation by a factor $\Gamma \simeq 1 + r^2\omega^2/2c^2$; this is not observed. However, in order to maintain the stars in circular orbits, a centripetal acceleration $a = v^2/r$ is needed. This is supplied by an apparent gravitational acceleration in the rotating frame (a 'non-inertial' force). The necessary potential is $\Phi = r^2\omega^2/2$, so gravitational blueshift of the radiation cancels the kinematic redshift (at least to order r^2). This example captures very well the main philosophy of general relativity: correct laws of physics should allow us to explain what we see, whatever our viewpoint.

For a more important practical application of gravitational time dilation, consider the *Sachs–Wolfe effect*. This is the dominant source of large-scale anisotropies in the cosmic microwave background (CMB), which arise from potential perturbations at last scattering. These have two effects:

(i)   they redshift the photons we see, so that an overdensity *cools* the background as the photons climb out, $\delta T/T = \delta\Phi/c^2$;

(ii) they cause time dilation at the last-scattering surface, so that we seem to be looking at a younger (and hence *hotter*) universe where there is an overdensity.

The time dilation is $\delta t/t = \delta\Phi/c^2$; since the time dependence of the scale factor is $a \propto t^{2/3}$ and $T \propto 1/a$, this produces the counterterm $\delta T/T = -(2/3)\delta\Phi/c^2$. The net effect is thus one-third of the gravitational redshift:

$$\frac{\delta T}{T} = \frac{\delta\Phi}{3c^2}.$$

This effect was originally derived by Sachs and Wolfe (1967) and bears their name. It is common to see the first argument alone, with the factor $1/3$ attributed to some additional complicated effect of general relativity. However, in weak fields, general relativistic effects should already be incorporated within the concept of gravitational time dilation; the previous argument shows that this is indeed all that is required to explain the full result.

2.2   The energy–momentum tensor

The only ingredient now missing from a classical theory of relativistic gravitation is a field equation: the presence of mass must determine the gravitational field. To obtain some insight into how this can be achieved, it is helpful to consider first the weak-field limit and the analogy with electromagnetism. Suppose we guess that the weak-field form of gravitation will look like electromagnetism, i.e. that we will end up working with both a scalar potential ϕ and a vector potential A that together give a velocity-dependent acceleration $a = -\nabla\phi - \dot{A} + v\wedge(\nabla\wedge A)$. Making the usual $e/4\pi\epsilon_0 \to Gm$ substitution would suggest the field equation

$$\partial^\nu\partial_\nu A^\mu \equiv \Box A^\mu = \frac{4\pi G}{c^2}J^\mu,$$

where $\Box$ is the d'Alembertian wave operator, $A^\mu = (\phi/c, A)$ is the 4-potential and $J^\mu = (\rho c, j)$ is a quantity that resembles a 4-current, whose components are a mass density and mass flux density. The solution to this equation is well known:

$$A^\mu(r) = \frac{G}{c^2}\int\frac{[J^\mu(x)]}{|r - x|}\,d^3x,$$

where the square brackets denote retarded values.

Now, in fact this analogy can be discarded immediately as a theory of gravitation in the weak-field limit. The problem lies in the vector J^μ: what would the meaning of such a quantity be? In electromagnetism, it describes conservation of charge via

$$\partial_\mu J^\mu = \dot{\rho} + \nabla\cdot j = 0$$

(notice how neatly such a conservation law can be expressed in 4-vector form). When dealing with mechanics, on the other hand, we have not one conserved quantity, but *four*: energy and vector momentum.

The electromagnetic analogy is nevertheless useful, as it suggests that the source of gravitation might still be mass and momentum: what we need first is to find the object that will correctly express conservation of 4-momentum. Informally, what is needed is a way of writing four conservation laws for each component of P^μ. We can clearly write four equations of the previous type in matrix form:

$$\partial_\nu T^{\mu\nu} = 0.$$

Now, if this equation is to be covariant, $T^{\mu\nu}$ must be a tensor and is known as the *energy–momentum tensor* (or sometimes as the stress–energy tensor). The meanings of its components in words are: $T^{00} = c^2 \times$ (mass density) = energy density; $T^{12} = x$-component of current of y-momentum etc. From these definitions, the tensor is readily seen to be symmetric. Both momentum density and energy flux density are the product of a mass density and a net velocity, so $T^{0\mu} = T^{\mu 0}$. The spatial stress tensor T^{ij} is also symmetric because any small volume element would otherwise suffer infinite angular acceleration: any asymmetric stress acting on a cube of side L gives a couple $\propto L^3$, whereas the moment of inertia is $\propto L^5$.

An important special case is the energy–momentum tensor for a perfect fluid. In matrix form, the rest-frame $T^{\mu\nu}$ is given by just $\mathrm{diag}(c^2\rho, p, p, p)$ (using the fact that the meaning of the pressure p is just the flux density of x-momentum in the x direction etc.). We can bypass the step of carrying out an explicit Lorentz transformation (which would be rather cumbersome in this case) by the powerful technique of manifest covariance. The following expression is clearly a tensor and reduces to the previous rest-frame answer in special relativity:

$$T^{\mu\nu} = (\rho + p/c^2)U^\mu U^\nu - pg^{\mu\nu}.$$

Thus it must be the general expression for the energy–momentum tensor of a perfect fluid.

2.2.1 Relativistic fluid mechanics

A nice application of the energy–momentum tensor is to show how it generates the equations of relativistic fluid mechanics. Given $T^{\mu\nu}$ for a perfect fluid, all that needs to be done is to insert the specific components $U^\mu = \gamma(c, \boldsymbol{v})$ into the fundamental conservation laws: $\partial T^{\mu\nu}/\partial x^\nu = 0$. The manipulation of the resulting equations is a straightforward exercise. Note that it is immediately clear that the results will involve the total or *convective derivative*:

$$\frac{\mathrm{d}}{\mathrm{d}t} \equiv \frac{\partial}{\partial t} + \boldsymbol{v} \cdot \nabla = \gamma^{-1} U^\mu \partial_\mu.$$

The idea here is that the changes experienced by an observer moving with the fluid are inevitably a mixture of temporal and spatial changes. This two-part derivative arises automatically in the relativistic formulation through the 4-vector dot product $U^\mu \partial_\mu$, which arises from the 4-divergence of an energy–momentum tensor containing a term $\propto U^\mu U^\nu$.

The equations that result from unpacking $T^{\mu\nu}{}_{,\nu} = 0$ in this way have a familiar physical interpretation. The $\mu = 1, 2, 3$ components of $T^{\mu\nu}_{,\nu} = 0$ give the relativistic generalization of *Euler's equation* for momentum conservation in fluid mechanics (not to be confused with Euler's equation in variational calculus):

$$\frac{d}{dt} v = -\frac{1}{\gamma^2 (\rho + p/c^2)} (\nabla p + \dot{p} v/c^2),$$

and the $\mu = 0$ component gives a generalization of conservation of energy:

$$\frac{d}{dt}[\gamma^2(\rho + p/c^2)] = \dot{p}/c^2 - \gamma^2(\rho + p/c^2)\nabla \cdot v,$$

where $\dot{p} \equiv \partial p/\partial t$. The meaning of this equation may be made clearer by introducing one further conservation law: particle number. This is governed by a 4-current having zero 4-divergence:

$$\frac{d}{dx^\mu} J^\mu = 0, \qquad J^\mu \equiv n U^\mu = \gamma n(c, v).$$

If we now introduce the *relativistic enthalpy* $w = \rho + p/c^2$, then energy conservation becomes

$$\frac{d}{dt}\left(\frac{\gamma w}{n}\right) = \frac{\dot{p}}{\gamma n c^2}.$$

Thus, in steady flow, $\gamma \times$ (enthalpy per particle) is constant.

A very useful general procedure can be illustrated by *linearizing* the fluid equations. Consider a small perturbation about each quantity ($\rho \to \rho + \delta\rho$ etc) and subtract the unperturbed equations to yield equations for the perturbations valid to first order. This means that any higher-order term such as $\delta v \cdot \nabla \delta\rho$ is set equal to zero. If we take the initial state to have constant density and pressure and zero velocity, then the resulting equations are simple:

$$\frac{\partial}{\partial t}\delta v = -\frac{1}{\rho + p/c^2}\nabla\delta p$$

$$\frac{\partial}{\partial t}\delta\rho = -(\rho + p/c^2)\nabla \cdot \delta v.$$

Now eliminate the perturbed velocity (via the divergence of the first of these equations minus the time derivative of the second) to yield the wave equation:

$$\nabla^2 \delta\rho - \left(\frac{\partial\rho}{\partial p}\right)\frac{\partial^2\delta\rho}{\partial t^2} = 0.$$

This defines the speed of sound to be $c_S^2 = \partial p / \partial \rho$. Notice that, by a fortunate coincidence, this is exactly the same as is derived from the non-relativistic equations, although we could not have relied upon this in advance. Thus, the speed of sound in a radiation-dominated fluid is just $c / \sqrt{3}$.

2.3   The field equations

The energy–momentum tensor plausibly plays the role that the charge 4-current J^μ plays in the electromagnetic field equations, $\Box A^\mu = \mu_0 J^\mu$. The tensor on the left-hand side of the gravitational field equations is rather more complicated. Weinberg (1972) showed that it is only possible to make one tensor that is linear in second derivatives of the metric, which is the *Riemann tensor*:

$$R^\mu{}_{\alpha\beta\gamma} = \frac{\partial \Gamma^\mu_{\alpha\gamma}}{\partial x^\beta} - \frac{\partial \Gamma^\mu_{\alpha\beta}}{\partial x^\gamma} + \Gamma^\mu_{\sigma\beta}\Gamma^\sigma_{\gamma\alpha} - \Gamma^\mu_{\sigma\gamma}\Gamma^\sigma_{\beta\alpha}.$$

This tensor gives a covariant description of spacetime curvature. For the field equations, we need a second-rank tensor to match $T^{\mu\nu}$, and the Riemann tensor may be contracted to the Ricci tensor $R^{\mu\nu}$, or further to the *curvature scalar R*:

$$R_{\alpha\beta} = R^\mu{}_{\alpha\beta\mu}$$
$$R = R_\mu{}^\mu = g^{\mu\nu} R_{\mu\nu}.$$

Unfortunately, these definitions are not universally agreed, All authors, however, agree on the definition of the Einstein tensor $G^{\mu\nu}$:

$$G^{\mu\nu} = R^{\mu\nu} - \tfrac{1}{2} g^{\mu\nu} R.$$

This tensor is what is needed, because it has zero covariant divergence. Since $T^{\mu\nu}$ also has zero covariant divergence by virtue of the conservation laws it expresses, it therefore seems reasonable to guess that the two are proportional:

$$G^{\mu\nu} = -\frac{8\pi G}{c^4} T^{\mu\nu}.$$

These are Einstein's gravitational field equations, where the correct constant of proportionality has been inserted. This is obtained by considering the weak-field limit.

2.3.1   Newtonian limit

The relation between Einstein's and Newton's descriptions of gravity involves taking the limit of weak gravitational fields ($\phi / c^2 \ll 1$). We also need to consider a classical source of gravity, with $p \ll \rho c^2$, so that the only non-zero component of $T^{\mu\nu}$ is $T^{00} = c^2 \rho$. Thus, the spatial parts of $R^{\mu\nu}$ must be given by

$$R^{ij} = \tfrac{1}{2} g^{ij} R.$$

Converting this to an equation for R^i_j, it follows that $R = R^{00} + \frac{3}{2}R$ and hence that

$$G^{00} = G_{00} = 2R_{00}.$$

Discarding nonlinear terms in the definition of the Riemann tensor leaves

$$R_{\alpha\beta} = \frac{\partial \Gamma^\mu_{\alpha\mu}}{\partial x^\beta} - \frac{\partial \Gamma^\mu_{\alpha\beta}}{\partial x^\mu} \Rightarrow R_{00} = -\Gamma^i_{00,i}$$

for the case of a stationary field. We have already seen that $c^2\Gamma^i_{00}$ plays the role of the Newtonian acceleration, so the required limiting expression for G^{00} is

$$G^{00} = -\frac{2}{c^2}\nabla^2\phi,$$

and comparison with Poisson's equation gives us the constant of proportionality in the field equations.

2.3.2 Pressure as a source of gravity

Newtonian gravitation is modified in the case of a relativistic fluid (i.e. where we cannot assume $p \ll \rho c^2$). It helps to begin by recasting the field equations (this would also have simplified the previous discussion). Contract the equation using $g^\mu_\mu = 4$ to obtain $R = (8\pi G/c^4)T$. This allows us to write an equation for $R^{\mu\nu}$ directly:

$$R^{\mu\nu} = -\frac{8\pi G}{c^4}(T^{\mu\nu} - \frac{1}{2}g^{\mu\nu}T).$$

Since $T = c^2\rho - 3p$, we get a modified Poisson equation:

$$\nabla^2\phi = 4\pi G(\rho + 3p/c^2).$$

What does this mean? For a gas of particles all moving at the same speed u, the effective gravitational mass density is $\rho(1 + u^2/c^2)$; thus a radiation-dominated fluid generates a gravitational attraction twice as strong as one would expect from Newtonian arguments. In fact, this factor applies also to individual particles and leads to an interesting consequence. One can turn the argument round by going to the rest frame of the gravitating mass. We will then conclude that a passing test particle will exhibit an acceleration transverse to its path greater by a factor $(1 + u^2/c^2)$ than that of a slowly moving particle. This gives an extra factor of two deflection in the trajectories of photons, which is of critical importance in gravitational lensing.

2.3.3 Energy density of the vacuum

One consequence of the gravitational effects of pressure that may seem of mathematical interest only is that a negative-pressure equation of state that achieved $\rho c^2 + 3p < 0$ would produce *antigravity*. Although such a possibility may seem physically nonsensical, it is in fact one of the most important concepts in contemporary cosmology. The origin of the idea goes back to the time when Einstein was first thinking about the cosmological consequences of general relativity. At that time, the universe was believed to be static—although this was simply a prejudice, rather than being founded on any observational facts. The problem of how a uniform distribution of matter could remain static was one that had faced Newton, and Einstein gave a very simple Newtonian solution. He reasoned that a static homogeneous universe required both the density, ρ, and the gravitational potential, Φ, to be constants. This does not solve Poisson's equation, $\nabla^2 \Phi = 4\pi G\rho$, so he suggested that the equation should be changed to $(\nabla^2 + \lambda)\Phi = 4\pi G\rho$, where λ is a new constant of nature: the *cosmological constant*. Almost as an afterthought, Einstein pointed out that this equation has the natural relativistic generalization of

$$G^{\mu\nu} + \Lambda g^{\mu\nu} = -\frac{8\pi G}{c^4} T^{\mu\nu}.$$

What is the physical meaning of Λ? In the current form, it represents the curvature of empty space. The modern approach is to move the Λ term to the right-hand side of the field equations. It now looks like the energy–momentum tensor of the vacuum:

$$T_{\text{vac}}^{\mu\nu} = \frac{\Lambda c^4}{8\pi G} g^{\mu\nu}.$$

How can a vacuum have a non-zero energy density and pressure? Surely these are zero by definition in a vacuum? What we can be sure of is that the absence of a preferred frame means that $T_{\text{vac}}^{\mu\nu}$ must be the same for all observers in special relativity . Now, apart from zero, there is only one isotropic tensor of rank 2: $\eta^{\mu\nu}$. Thus, in order for $T_{\text{vac}}^{\mu\nu}$ to be unaltered by Lorentz transformations, the only requirement we can have is that it must be proportional to the metric tensor. Therefore, it is inevitable that the vacuum (at least in special relativity) will have a negative-pressure equation of state:

$$p_{\text{vac}} = -\rho_{\text{vac}} c^2.$$

In this case, $\rho c^2 + 3p$ is indeed negative: a positive Λ will act to cause a large-scale repulsion. The vacuum energy density can thus play a crucial part in the dynamics of the early universe.

It may seem odd to have an energy density that does not change as the universe expands. What saves us is the peculiar equation of state of the vacuum: the work done by the pressure is just sufficient to maintain the energy density constant (see figure 2.1). In effect, the vacuum acts as a reservoir of unlimited energy, which can supply as much as is required to inflate a given region to any required size at constant energy density. This supply of energy is what is used in 'inflationary' theories of cosmology to create the whole universe out of almost nothing.

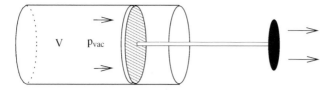

Figure 2.1. A thought experiment to illustrate the application of conservation of energy to the vacuum. If the vacuum density is ρ_{vac} then the energy created by withdrawing the piston by a volume dV is $\rho_{vac}c^2\,dV$. This must be supplied by work done by the vacuum pressure $p_{vac}\,dV$, and so $p_{vac} = -\rho_{vac}c^2$, as required.

2.4 The Friedmann models

Many of the chapters in this book discuss observational cosmology, assuming a body of material on standard homogeneous cosmological models. This section attempts to set the scene by summarizing the key basic features of relativistic cosmology.

2.4.1 Cosmological coordinates

The simplest possible mass distribution is one whose properties are *homogeneous* (constant density) and *isotropic* (the same in all directions). The first point to note is that something suspiciously like a universal time exists in an isotropic universe. Consider a set of observers in different locations, all of whom are at rest with respect to the matter in their vicinity (these characters are usually termed *fundamental observers*). We can envisage them as each sitting on a different galaxy, and so receding from each other with the general expansion. We can define a global time coordinate t, which is the time measured by the clocks of these observers—i.e. t is the proper time measured by an observer at rest with respect to the local matter distribution. The coordinate is useful globally rather than locally because the clocks can be synchronized by the exchange of light signals between observers, who agree to set their clocks to a standard time when, e.g., the universal homogeneous density reaches some given value. Using this time coordinate plus isotropy, we already have enough information to conclude that the metric must take the following form:

$$c^2\,d\tau^2 = c^2\,dt^2 - R^2(t)[f^2(r)\,dr^2 + g^2(r)\,d\psi^2].$$

Here, we have used the equivalence principle to say that the proper time interval between two distant events would look locally like special relativity to a fundamental observer on the spot: for them, $c^2\,d\tau^2 = c^2\,dt^2 - dx^2 - dy^2 - dz^2$. Since we use the same time coordinate as they do, our only difficulty is in the spatial part of the metric: relating their dx etc to spatial coordinates centred on us.

Because of spherical symmetry, the spatial part of the metric can be decomposed into a radial and a transverse part (in spherical polars, $d\psi^2 = d\theta^2 + \sin^2\theta\, d\phi^2$). Distances have been decomposed into a product of a time-dependent *scale factor* $R(t)$ and a time-independent *comoving coordinate r*. The functions f and g are arbitrary; however, we can choose our radial coordinate such that either $f = 1$ or $g = r^2$, to make things look as much like Euclidean space as possible. Furthermore, we can determine the form of the remaining function from symmetry arguments.

To get some feeling for the general answer, it should help to think first about a simpler case: the metric on the surface of a sphere. A balloon being inflated is a common popular analogy for the expanding universe, and it will serve as a two-dimensional example of a space of constant curvature. If we call the polar angle in spherical polars r instead of the more usual θ, then the element of length on the surface of a sphere of radius R is

$$d\sigma^2 = R^2(dr^2 + \sin^2 r\, d\phi^2).$$

It is possible to convert this to the metric for a 2-space of constant by the device of considering an imaginary radius of curvature, $R \to iR$. If we simultaneously let $r \to ir$, we obtain

$$d\sigma^2 = R^2(dr^2 + \sinh^2 r\, d\phi^2).$$

These two forms can be combined by defining a new radial coordinate that makes the transverse part of the metric look Euclidean:

$$d\sigma^2 = R^2\left(\frac{dr^2}{1 - kr^2} + r^2\, d\phi^2\right),$$

where $k = +1$ for positive curvature and $k = -1$ for negative curvature.

An isotropic universe has the same form for the comoving spatial part of its metric as the surface of a sphere. This is no accident, since it it possible to define the equivalent of a sphere in higher numbers of dimensions, and the form of the metric is always the same. For example, a *3-sphere* embedded in four-dimensional Euclidean space would be defined as the coordinate relation $x^2 + y^2 + z^2 + w^2 = R^2$. Now define the equivalent of spherical polars and write $w = R\cos\alpha$, $z = R\sin\alpha\cos\beta$, $y = R\sin\alpha\sin\beta\cos\gamma$, $x = R\sin\alpha\sin\beta\sin\gamma$, where α, β and γ are three arbitrary angles. Differentiating with respect to the angles gives a four-dimensional vector (dx, dy, dz, dw), and it is a straightforward exercise to show that the squared length of this vector is

$$|(dx, dy, dz, dw)|^2 = R^2[d\alpha^2 + \sin^2\alpha(d\beta^2 + \sin^2\beta\, d\gamma^2)],$$

which is the Robertson–Walker metric for the case of positive spatial curvature. This $k = +1$ metric describes a closed universe, in which a traveller who sets off along a trajectory of fixed β and γ will eventually return to their starting point (when $\alpha = 2\pi$). In this respect, the positively curved 3D universe is identical to the case of the surface of a sphere: it is finite, but unbounded. By contrast, the $k = -1$ metric describes an open universe of infinite extent.

The Robertson–Walker metric (which we shall often write in the shorthand *RW metric*) may be written in a number of different ways. The most compact forms are those where the comoving coordinates are *dimensionless*. Define the very useful function

$$S_k(r) = \begin{cases} \sin r & (k = 1) \\ \sinh r & (k = -1) \\ r & (k = 0) \end{cases}$$

and its cosine-like analogue, $C_k(r) \equiv \sqrt{1 - kS_k^2(r)}$. The metric can now be written in the preferred form that we shall use throughout:

$$c^2\,d\tau^2 = c^2\,dt^2 - R^2(t)[dr^2 + S_k^2(r)\,d\psi^2].$$

The most common alternative is to use a different definition of comoving distance, $S_k(r) \to r$, so that the metric becomes

$$c^2\,d\tau^2 = c^2\,dt^2 - R^2(t)\left(\frac{dr^2}{1 - kr^2} + r^2\,d\psi^2\right).$$

There should of course be two different symbols for the different comoving radii, but each is often called r in the literature, so we have to learn to live with this ambiguity; the presence of terms like $S_k(r)$ or $1 - kr^2$ will usually indicate which convention is being used. Alternatively, one can make the scale factor dimensionless, defining

$$a(t) \equiv \frac{R(t)}{R_0},$$

so that $a = 1$ at the present.

2.4.2 The redshift

At small separations, where things are Euclidean, the proper separation of two fundamental observers is just $R(t)\,dr$, so that we obtain Hubble's law, $v = Hd$, with

$$H = \frac{\dot{R}}{R}.$$

At large separations where spatial curvature becomes important, the concept of radial velocity becomes a little more slippery—but in any case how could one measure it directly in practice? At small separations, the recessional velocity gives the Doppler shift

$$\frac{\nu_{\text{emit}}}{\nu_{\text{obs}}} \equiv 1 + z \simeq 1 + \frac{v}{c}.$$

This defines the *redshift z* in terms of the shift of spectral lines. What is the equivalent of this relation at larger distances? Since photons travel on null geodesics of zero proper time, we see directly from the metric that

$$r = \int \frac{c \, dt}{R(t)}.$$

The comoving distance is constant, whereas the domain of integration in time extends from t_{emit} to t_{obs}; these are the times of emission and reception of a photon. Photons that are emitted at later times will be received at later times, but these changes in t_{emit} and t_{obs} cannot alter the integral, since r is a comoving quantity. This requires the condition $dt_{emit}/dt_{obs} = R(t_{emit})/R(t_{obs})$, which means that events on distant galaxies time dilate according to how much the universe has expanded since the photons we see now were emitted. Clearly (think of events separated by one period), this dilation also applies to frequency, and we therefore get

$$\frac{\nu_{emit}}{\nu_{obs}} \equiv 1 + z = \frac{R(t_{obs})}{R(t_{emit})}.$$

In terms of the normalized scale factor $a(t)$ we have simply $a(t) = (1 + z)^{-1}$. Photon wavelengths therefore stretch with the universe, as is intuitively reasonable.

2.4.3 Dynamics of the expansion

The equation of motion for the scale factor can be obtained in a quasi-Newtonian fashion. Consider a sphere about some arbitrary point, and let the radius be $R(t)r$, where r is arbitrary. The motion of a point at the edge of the sphere will, in Newtonian gravity, be influenced only by the interior mass. We can therefore write down immediately a differential equation (Friedmann's equation) that expresses conservation of energy: $(\dot{R}r)^2/2 - GM/(Rr) = $ constant. The Newtonian result that the gravitational field inside a uniform shell is zero does still hold in general relativity, and is known as *Birkhoff's theorem*. General relativity becomes even more vital in giving us the constant of integration in Friedmann's equation:

$$\dot{R}^2 - \frac{8\pi G}{3} \rho R^2 = -kc^2.$$

Note that this equation covers all contributions to ρ, i.e. those from matter, radiation and vacuum; it is independent of the equation of state.

For a given rate of expansion, there is thus a critical density that will yield $k = 0$, making the comoving part of the metric look Euclidean:

$$\rho_c = \frac{3H^2}{8\pi G}.$$

A universe with a density above this critical value will be *spatially closed*, whereas a lower-density universe will be *spatially open*.

The 'flat' universe with $k = 0$ arises for a particular critical density. We are therefore led to define a density parameter as the ratio of density to critical density:

$$\Omega \equiv \frac{\rho}{\rho_c} = \frac{8\pi G\rho}{3H^2}.$$

Since ρ and H change with time, this defines an epoch-dependent density parameter. The current value of the parameter should strictly be denoted by Ω_0. Because this is such a common symbol, we shall keep the formulae uncluttered by normally dropping the subscript; the density parameter at other epochs will be denoted by $\Omega(z)$. The critical density therefore just depends on the rate at which the universe is expanding. If we now also define a dimensionless (current) Hubble parameter as

$$h \equiv \frac{H_0}{100 \text{ km s}^{-1} \text{ Mpc}^{-1}},$$

then the current density of the universe may be expressed as

$$\rho_0 = 1.88 \times 10^{-26} \Omega h^2 \text{ kg m}^{-3}$$
$$= 2.78 \times 10^{11} \Omega h^2 M_\odot \text{ Mpc}^{-3}.$$

A powerful approximate model for the energy content of the universe is to divide it into pressureless matter ($\rho \propto R^{-3}$), radiation ($\rho \propto R^{-4}$) and vacuum energy (ρ constant). The first two relations just say that the number density of particles is diluted by the expansion, with photons also having their energy reduced by the redshift; the third relation applies for Einstein's cosmological constant. In terms of observables, this means that the density is written as

$$\frac{8\pi G\rho}{3} = H_0^2(\Omega_v + \Omega_m a^{-3} + \Omega_r a^{-4})$$

(introducing the normalized scale factor $a = R/R_0$). For some purposes, this separation is unnecessary, since the Friedmann equation treats all contributions to the density parameter equally:

$$\frac{kc^2}{H^2R^2} = \Omega_m(a) + \Omega_r(a) + \Omega_v(a) - 1.$$

Thus, a flat $k = 0$ universe requires $\sum \Omega_i = 1$ at all times, whatever the form of the contributions to the density, even if the equation of state cannot be decomposed in this simple way.

Lastly, it is often necessary to know the present value of the scale factor, which may be read directly from the Friedmann equation:

$$R_0 = \frac{c}{H_0}[(\Omega - 1)/k]^{-1/2}.$$

The Hubble constant thus sets the *curvature length*, which becomes infinitely large as Ω approaches unity from either direction.

2.4.4 Solutions to the Friedmann equation

The Friedmann equation may be solved most simply in 'parametric' form, by recasting it in terms of the conformal time $d\eta = c\,dt/R$ (denoting derivatives with respect to η by primes):

$$R'^2 = \frac{8\pi G}{3c^2}\rho R^4 - kR^2.$$

Because $H_0^2 R_0^2 = kc^2/(\Omega - 1)$, the Friedmann equation becomes

$$a'^2 = \frac{k}{(\Omega - 1)}[\Omega_r + \Omega_m a - (\Omega - 1)a^2 + \Omega_v a^4],$$

which is straightforward to integrate provided $\Omega_v = 0$.

To the observer, the evolution of the scale factor is most directly characterized by the change with redshift of the Hubble parameter and the density parameter; the evolution of $H(z)$ and $\Omega(z)$ is given immediately by the Friedmann equation in the form $H^2 = 8\pi G\rho/3 - kc^2/R^2$. Inserting this dependence of ρ on a gives

$$H^2(a) = H_0^2[\Omega_v + \Omega_m a^{-3} + \Omega_r a^{-4} - (\Omega - 1)a^{-2}].$$

This is a crucial equation, which can be used to obtain the relation between redshift and comoving distance. The radial equation of motion for a photon is $R\,dr = c\,dt = c\,dR/\dot{R} = c\,dR/(RH)$. With $R = R_0/(1+z)$, this gives

$$R_0\,dr = \frac{c}{H(z)}\,dz = \frac{c}{H_0}\,dz[(1-\Omega)(1+z)^2 + \Omega_v + \Omega_m(1+z)^3 + \Omega_r(1+z)^4]^{-1/2}.$$

This relation is arguably the single most important equation in cosmology, since it shows how to relate comoving distance to the observables of redshift, the Hubble constant and density parameters.

Lastly, using the expression for $H(z)$ with $\Omega(a) - 1 = kc^2/(H^2 R^2)$ gives the redshift dependence of the total density parameter:

$$\Omega(z) - 1 = \frac{\Omega - 1}{1 - \Omega + \Omega_v a^2 + \Omega_m a^{-1} + \Omega_r a^{-2}}.$$

This last equation is very important. It tells us that, at high redshift, all model universes apart from those with only vacuum energy will tend to look like the $\Omega = 1$ model. If $\Omega \neq 1$, then in the distant past $\Omega(z)$ must have differed from unity by a tiny amount: the density and rate of expansion needed to have been finely balanced for the universe to expand to the present. This tuning of the initial conditions is called the *flatness problem*.

The solution of the Friedmann equation becomes more complicated if we allow a significant contribution from vacuum energy—i.e. a non-zero cosmological constant. Detailed discussions of the problem are given by Felten and Isaacman (1986) and Carroll *et al* (1992); the most important features are outlined later.

The Friedmann equation itself is independent of the equation of state, and just says $H^2 R^2 = kc^2/(\Omega - 1)$, whatever the form of the contributions to Ω. In terms of the cosmological constant itself, we have

$$\Omega_v = \frac{8\pi G \rho_v}{3H^2} = \frac{\Lambda c^2}{3H^2}.$$

With the addition of Λ, the Friedmann equation can only in general be solved numerically. However, we can find the conditions for the different behaviours described earlier analytically, at least if we simplify things by ignoring radiation. The equation in the form of the time-dependent Hubble parameter looks like

$$\frac{H^2}{H_0^2} = \Omega_v (1 - a^{-2}) + \Omega_m (a^{-3} - a^{-2}) + a^{-2}.$$

This equation allows the left-hand side to vanish, defining a turning point in the expansion. Vacuum energy can thus remove the possibility of a big bang in which the scale factor goes to zero. Setting the right-hand side to zero yields a cubic equation, and it is possible to give the conditions under which this has a solution (see Felten and Isaacman 1986). The main results of this analysis are summed up in figure 2.2. Since the radiation density is very small today, the main task of relativistic cosmology is to work out where on the Ω_{matter}–Ω_{vacuum} plane the real universe lies. The existence of high-redshift objects rules out the bounce models, so that the idea of a hot big bang cannot be evaded.

The most important model in cosmological research is that with $k = 0 \Rightarrow \Omega_{total} = 1$; when dominated by matter, this is often termed the *Einstein–de Sitter* model. Paradoxically, this importance arises because it is an unstable state: as we have seen earlier, the universe will evolve away from $\Omega = 1$, given a slight perturbation. For the universe to have expanded by so many *e-foldings* (factors of e expansion) and yet still have $\Omega \sim 1$ implies that it was very close to being spatially flat at early times.

It now makes more sense to work throughout in terms of the normalized scale factor $a(t)$, so that the Friedmann equation for a matter–radiation mix is

$$\dot{a}^2 = H_0^2 (\Omega_m a^{-1} + \Omega_r a^{-2}),$$

which may be integrated to give the time as a function of scale factor:

$$H_0 t = \frac{2}{3\Omega_m^2} \left[\sqrt{\Omega_r + \Omega_m a} (\Omega_m a - 2\Omega_r) + 2\Omega_r^{3/2} \right];$$

this goes to $\frac{2}{3} a^{3/2}$ for a matter-only model, and to $\frac{1}{2} a^2$ for radiation only.

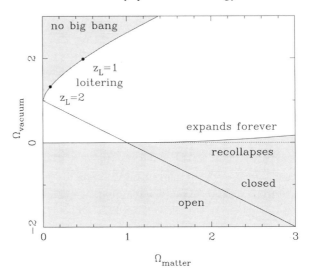

Figure 2.2. This plot shows the different possibilities for the cosmological expansion as a function of matter density and vacuum energy. Models with total $\Omega > 1$ are always spatially closed (open for $\Omega < 1$), although closed models can still expand to infinity if $\Omega_v \neq 0$. If the cosmological constant is negative, recollapse always occurs; recollapse is also possible with a positive Ω_v if $\Omega_m \gg \Omega_v$. If $\Omega_v > 1$ and Ω_m is small, there is the possibility of a 'loitering' solution with some maximum redshift and infinite age (top left); for even larger values of vacuum energy, there is no big bang singularity.

One further way of presenting the model's dependence on time is via the density. Following this, it is easy to show that

$$t = \sqrt{\frac{1}{6\pi G\rho}} \qquad \text{(matter domination)}$$

$$t = \sqrt{\frac{3}{32\pi G\rho}} \qquad \text{(radiation domination)}.$$

An alternative $k = 0$ model of greater observational interest has a significant cosmological constant, so that $\Omega_m + \Omega_v = 1$ (radiation being neglected for simplicity). The advantage of this model is that it is the only way of retaining the theoretical attractiveness of $k = 0$ while changing the age of the universe from the relation $H_0 t_0 = 2/3$, which characterizes the Einstein–de Sitter model. Since much observational evidence indicates that $H_0 t_0 \simeq 1$, this model has received a good deal of interest in recent years. To keep things simple we shall neglect radiation, so that the Friedmann equation is

$$\dot{a}^2 = H_0^2 [\Omega_m a^{-1} + (1 - \Omega_m)a^2],$$

and the $t(a)$ relation is

$$H_0 t(a) = \int_0^a \frac{x \, dx}{\sqrt{\Omega_m x + (1 - \Omega_m)x^4}}.$$

The x^4 on the bottom looks like trouble, but it can be rendered tractable by the substitution $y = \sqrt{x^3 |\Omega_m - 1| / \Omega_m}$, which turns the integral into

$$H_0 t(a) = \frac{2}{3} \frac{S_k^{-1} \left(\sqrt{a^3 |\Omega_m - 1| / \Omega_m} \right)}{\sqrt{|\Omega_m - 1|}}.$$

Here, k in S_k is used to mean sin if $\Omega_m > 1$, otherwise sinh; these are still $k = 0$ models. Since there is nothing special about the current era, we can clearly also rewrite this expression as

$$H(a)t(a) = \frac{2}{3} \frac{S_k^{-1} \left(\sqrt{|\Omega_m(a) - 1| / \Omega_m(a)} \right)}{\sqrt{|\Omega_m(a) - 1|}} \simeq \frac{2}{3} \Omega_m(a)^{-0.3},$$

where we include a simple approximation that is accurate to a few per cent over the region of interest ($\Omega_m \gtrsim 0.1$). In the general case of significant Λ but $k \neq 0$, this expression still gives a very good approximation to the exact result, provided Ω_m is replaced by $0.7\Omega_m - 0.3\Omega_v + 0.3$ (Carroll *et al* 1992).

2.4.5 Horizons

For photons, the radial equation of motion is just $c \, dt = R \, dr$. How far can a photon get in a given time? The answer is clearly

$$\Delta r = \int_{t_0}^{t_1} \frac{c \, dt}{R(t)} = \Delta \eta,$$

i.e. just the interval of conformal time. What happens as $t_0 \to 0$ in this expression? We can replace dt by $dR/\dot{R}$, which the Friedmann equation says is proportional to $dR/\sqrt{\rho R^2}$ at early times. Thus, this integral converges if $\rho R^2 \to \infty$ as $t_0 \to 0$, otherwise it diverges. Provided the equation of state is such that ρ changes faster than R^{-2}, light signals can only propagate a finite distance between the big bang and the present; there is then said to be a *particle horizon*. Such a horizon therefore exists in conventional big-bang models, which are dominated by radiation at early times.

2.4.6 Observations in cosmology

We can now assemble some essential formulae for interpreting cosmological observations. Our observables are the redshift, z, and the angular difference between two points on the sky, $d\psi$. We write the metric in the form

$$c^2 \, d\tau^2 = c^2 \, dt^2 - R^2(t)[dr^2 + S_k^2(r) \, d\psi^2],$$

so that the *comoving* volume element is

$$dV = 4\pi [R_0 S_k(r)]^2 R_0 \, dr.$$

The *proper* transverse size of an object seen by us is its comoving size $d\psi \, S_k(r)$ times the scale factor at the time of emission:

$$d\ell = d\psi \, R_0 S_k(r)/(1+z).$$

Probably the most important relation for observational cosmology is that between monochromatic flux density and luminosity. Start by assuming isotropic emission, so that the photons emitted by the source pass with a uniform flux density through any sphere surrounding the source. We can now make a shift of origin, and consider the RW metric as being centred on the source; however, because of homogeneity, the comoving distance between the source and the observer is the same as we would calculate when we place the origin at our location. The photons from the source are therefore passing through a sphere, on which we sit, of proper surface area $4\pi [R_0 S_k(r)]^2$. But redshift still affects the flux density in four further ways: photon energies and arrival rates are redshifted, reducing the flux density by a factor $(1+z)^2$; opposing this, the bandwidth dv is reduced by a factor $1+z$, so the energy flux per unit bandwidth goes down by one power of $1+z$; finally, the observed photons at frequency v_0 were emitted at frequency $v_0(1+z)$, so the flux density is the luminosity at this frequency, divided by the total area, divided by $1+z$:

$$S_v(v_0) = \frac{L_v([1+z]v_0)}{4\pi R_0^2 S_k^2(r)(1+z)}.$$

The flux density received by a given observer can be expressed by definition as the product of the *specific intensity* I_v (the flux density received from unit solid angle of the sky) and the solid angle subtended by the source: $S_v = I_v \, d\Omega$. Combining the angular size and flux–density relations thus gives the relativistic version of surface-brightness conservation. This is independent of cosmology:

$$I_v(v_0) = \frac{B_v([1+z]v_0)}{(1+z)^3},$$

where B_v is *surface brightness* (luminosity emitted into unit solid angle per unit area of source). We can integrate over v_0 to obtain the corresponding total or *bolometric* formulae, which are needed, for example, for spectral-line emission:

$$S_{tot} = \frac{L_{tot}}{4\pi R_0^2 S_k^2(r)(1+z)^2};$$

$$I_{tot} = \frac{B_{tot}}{(1+z)^4}.$$

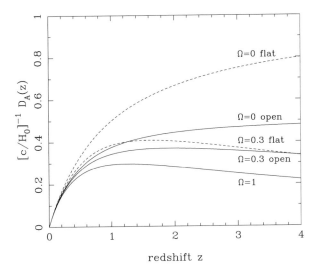

Figure 2.3. A plot of dimensionless angular-diameter distance versus redshift for various cosmologies. Full curves show models with zero vacuum energy; broken curves show flat models with $\Omega_m + \Omega_v = 1$. In both cases, results for $\Omega_m = 1, 0.3, 0$ are shown; higher density results in lower distance at high z, due to gravitational focusing of light rays.

The form of these relations lead to the following definitions for particular kinds of distances:

$$\text{angular-diameter distance:} \quad D_A = (1+z)^{-1} R_0 S_k(r)$$
$$\text{luminosity distance:} \quad D_L = (1+z) R_0 S_k(r).$$

The angular-diameter distance is plotted against redshift for various models in figure 2.3.

The last element needed for the analysis of observations is a relation between redshift and age for the object being studied. This brings in our earlier relation between time and comoving radius (consider a null geodesic traversed by a photon that arrives at the present):

$$c\, dt = R_0\, dr/(1+z).$$

The general relation between comoving distance and redshift was given earlier as

$$R_0\, dr = \frac{c}{H(z)}\, dz = \frac{c}{H_0}\, dz[(1-\Omega)(1+z)^2 + \Omega_v + \Omega_m(1+z)^3 + \Omega_r(1+z)^4]^{-1/2}.$$

2.4.7 The meaning of an expanding universe

Finally, having dealt with some of the formal apparatus of cosmology, it may be interesting to step back and ask what all this means. The idea of an expanding universe can easily lead to confusion, and this section tries to counter some of the more tenacious misconceptions.

The worst of these is the 'expanding space' fallacy. The RW metric written in comoving coordinates emphasizes that one can think of any given fundamental observer as fixed at the centre of their local coordinate system. A common interpretation of this algebra is to say that the galaxies separate 'because the space between them expands' or some such phrase. This suggests some completely new physical effect that is not covered by Newtonian concepts. However, on scales much smaller than the current horizon, we should be able to ignore curvature and treat galaxy dynamics as occurring in Minkowski spacetime; this approach works in deriving the Friedmann equation. How do we relate this to 'expanding space'? It should be clear that Minkowski spacetime does not expand – indeed, the very idea that the motion of distant galaxies could affect local dynamics is profoundly anti-relativistic: the equivalence principle says that we can always find a tangent frame in which physics is locally special relativity.

To clarify the issues here, it should help to consider an explicit example, which makes quite a neat paradox. Suppose we take a nearby low-redshift galaxy and give it a velocity boost such that its redshift becomes zero. At a later time, will the expansion of the universe have cause the galaxy to recede from us, so that it once again acquires a positive redshift? To idealize the problem, imagine that the galaxy is a massless test particle in a homogeneous universe.

The 'expanding space' idea would suggest that the test particle should indeed start to recede from us, and it appears that one can prove this formally, as follows. Consider the peculiar velocity with respect to the Hubble flow, δv. A completely general result is that this declines in magnitude as the universe expands:

$$\delta v \propto \frac{1}{a(t)}.$$

This is the same law that applies to photon energies, and the common link is that it is particle momentum in general that declines as $1/a$, just through the accumulated Lorentz transforms required to overtake successively more distant particles that are moving with the Hubble flow. So, at $t \to \infty$, the peculiar velocity tends to zero, leaving the particle moving with the Hubble flow, however it started out: 'expanding space' has apparently done its job.

Now look at the same situation in a completely different way. If the particle is nearby compared with the cosmological horizon, a Newtonian analysis should be valid: in an isotropic universe, Birkhoff's theorem assures us that we can neglect the effect of all matter at distances greater than that of the test particle, and all that counts is the mass between the particle and us. Call the proper separation of the particle from the origin r. Our initial conditions are that $\dot{r} = 0$ at $t = t_0$, when $r = r_0$. The equation of motion is just

$$\ddot{r} = \frac{-GM(\langle r|t)}{r^2},$$

and the mass internal to r is just

$$M(\langle r|t) = \frac{4\pi}{3}\rho r^3 = \frac{4\pi}{3}\rho_0 a^{-3} r^3,$$

where we assume $a_0 = 1$ and a matter-dominated universe. The equation of motion can now be re-expressed as

$$\ddot{r} = -\frac{\Omega_0 H_0^2}{2a^3} r.$$

Adding vacuum energy is easy enough:

$$\ddot{r} = -\frac{H_0^2}{2} r (\Omega_m a^{-3} - 2\Omega_v).$$

The -2 in front of the vacuum contribution comes from the effective mass density $\rho + 3p/c^2$.

We now show that this Newtonian equation is identical to what is obtained from $\delta v \propto 1/a$. In our present notation, this becomes

$$\dot{r} - H(t)r = -H_0 r_0/a;$$

the initial peculiar velocity is just $-Hr$, cancelling the Hubble flow. We can differentiate this equation to obtain $\ddot{r}$, which involves $\dot{H}$. This can be obtained from the standard relation

$$H^2(t) = H_0^2[\Omega_v + \Omega_m a^{-3} + (1 - \Omega_m - \Omega_v)a^{-2}].$$

It is then a straightforward exercise to show that the equation for $\ddot{r}$ is the same as obtained previously (remembering $H = \dot{a}/a$).

Now for the paradox. It will suffice at first to solve the equation for the case of the Einstein–de Sitter model, choosing time units such that $t_0 = 1$, with $H_0 t_0 = 2/3$:

$$\ddot{r} = -2r/9t^2.$$

The acceleration is negative, so the particle moves *inwards*, in complete apparent contradiction to our 'expanding space' conclusion that the particle would tend with time to pick up the Hubble expansion. The resolution of this contradiction comes from the full solution of the equation. The differential equation clearly has power-law solutions $r \propto t^{1/3}$ or $t^{2/3}$, and the combination with the correct boundary conditions is

$$r(t) = r_0(2t^{1/3} - t^{2/3}).$$

At large t, this becomes $r = -r_0 t^{2/3}$. This is indeed the equation of motion of a particle moving with the Hubble flow, but it arises because the particle has fallen right through the origin and emerged on the other side. In no sense, therefore, can 'expanding space' be said to have operated: in an Einstein–de Sitter model, a particle initially at rest with respect to the origin falls towards the origin, passes through it, and asymptotically regains its initial comoving radius on the opposite side of the sky. This behaviour can be understood quantitatively using only Newtonian dynamics.

Two further cases are worth considering. In an empty universe, the equation of motion is $\ddot{r} = 0$, so the particle remains at $r = r_0$, while the universe expands linearly with $a \propto t$. In this case, $H = 1/t$, so that $\delta v = -H r_0$, which declines as $1/a$, as required. Finally, models with vacuum energy are of more interest. Provided $\Omega_v > \Omega_m/2$, $\ddot{r}$ is initially positive, and the particle does move away from the origin. This is the criterion for $q_0 < 0$ and an accelerating expansion. In this case, there is a tendency for the particle to expand away from the origin, and this is caused by the repulsive effects of vacuum energy. In the limiting case of pure de Sitter space ($\Omega_m = 0$, $\Omega_v = 1$), the particle's trajectory is

$$r = r_0 \cosh H_0(t - t_0),$$

which asymptotically approaches half the $r = r_0 \exp H_0(t - t_0)$ that would have applied if we had never perturbed the particle in the first place. In the case of vacuum-dominated models, then, the repulsive effects of vacuum energy cause all pairs of particles to separate at large times, whatever their initial kinematics; this behaviour could perhaps legitimately be called 'expanding space'. Nevertheless, the effect stems from the clear physical cause of vacuum repulsion, and there is no new physical influence that arises purely from the fact that the universe expands. The earlier examples have proved that 'expanding space' is in general a fundamentally flawed way of thinking about an expanding universe.

2.5   Inflationary cosmology

We now turn from classical cosmology to aspects of cosmology in which quantum processes are important. This is necessary in order to solve the major problems of the simple big bang:

(1) The expansion problem. Why is the universe expanding at $t = 0$? This appears as an initial condition, but surely a mechanism is required to lauch the expansion?

(2) The flatness problem. Furthermore, the expansion needs to be launched at just the correct rate, so that is is very close to the critical density, and can thus expand from perhaps near the Planck era to the present (a factor of over 10^{30}).

(3) The horizon problem. Models in which the universe is radiation dominated (with $a \propto t^{1/2}$ at early times) have a finite horizon. There is apparently no causal means for different parts of the universe to agree on the mean density or rate of expansion.

The list of problems with conventional cosmology provides a strong hint that the equation of state of the universe may have been very different at very early times. To solve the horizon problem and allow causal contact over the whole of the region observed at last scattering requires a universe that expands 'faster than light' near $t = 0$: $R \propto t^\alpha$, with $\alpha > 1$. If such a phase had existed, the integral for the comoving horizon would have diverged, and there would be no difficulty in understanding the overall homogeneity of the universe—this could then be established by causal processes. Indeed, it is tempting to assert that the observed homogeneity *proves* that such causal contact must once have occurred.

What condition does this place on the equation of state? In the integral for r_H, we can replace dt by $dR/\dot{R}$, which the Friedmann equation says is proportional to $dR/\sqrt{\rho R^2}$ at early times. Thus, the horizon diverges provided the equation of state is such that ρR^2 vanishes or is finite as $R \rightarrow 0$. For a perfect fluid with $p \equiv (\Gamma - 1)\epsilon$ as the relation between pressure and energy density, we have the adiabatic dependence $p \propto R^{-3\Gamma}$, and the same dependence for ρ if the rest-mass density is negligible. A period of inflation therefore needs

$$\Gamma < 2/3 \Rightarrow \rho c^2 + 3p < 0.$$

Such a criterion can also solve the flatness problem. Consider the Friedmann equation,

$$\dot{R}^2 = \frac{8\pi G \rho R^2}{3} - kc^2.$$

As we have seen, the density term on the right-hand side must exceed the curvature term by a factor of at least 10^{60} at the Planck time, and yet a more natural initial condition might be to have the matter and curvature terms being of comparable order of magnitude. However, an inflationary phase in which ρR^2 increases as the universe expands can clearly make the curvature term relatively as small as required, provided inflation persists for sufficiently long.

We have seen that inflation will require an equation of state with negative pressure, and the only familiar example of this is the $p = -\rho c^2$ relation that applies for vacuum energy; in other words, we are led to consider inflation as happening in a universe dominated by a cosmological constant. As usual, any initial expansion will redshift away matter and radiation contributions to the density, leading to increasing dominance by the vacuum term. If the radiation and vacuum densities are initially of comparable magnitude, we quickly reach a state where the vacuum term dominates. The Friedmann equation in the vacuum-dominated case has three solutions:

$$R \propto \begin{cases} \sinh Ht & (k = -1) \\ \cosh Ht & (k = +1) \\ \exp Ht & (k = 0), \end{cases}$$

where $H = \sqrt{\Lambda c^2/3} = \sqrt{8\pi G \rho_{vac}/3}$; all solutions evolve towards the exponential $k = 0$ solution, known as *de Sitter spacetime*. Note that H is not the Hubble parameter at an arbitrary time (unless $k = 0$), but it becomes so exponentially fast as the hyperbolic trigonometric functions tend to the exponential.

Because de Sitter space clearly has H^2 and ρ in the right ratio for $\Omega = 1$ (obvious, since $k = 0$), the density parameter in all models tends to unity as the Hubble parameter tends to H. If we assume that the initial conditions are not fine tuned (i.e. $\Omega = O(1)$ initially), then maintaining the expansion for a factor f produces

$$\Omega = 1 + O(f^{-2}).$$

This can solve the flatness problem, provided f is large enough. To obtain Ω of order unity today requires $|\Omega - 1| \lesssim 10^{-52}$ at the Grand Unified Theory (GUT) epoch, and so $\ln f \gtrsim 60$ *e*-foldings of expansion are needed; it will be proved later that this is also exactly the number needed to solve the horizon problem. It then seems almost inevitable that the process should go to completion and yield $\Omega = 1$ to measurable accuracy today.

2.5.1 Inflation field dynamics

The general concept of inflation rests on being able to achieve a negative-pressure equation of state. This can be realized in a natural way by quantum fields in the early universe.

The critical fact we shall need from quantum field theory is that quantum fields can produce an energy density that mimics a cosmological constant. The discussion will be restricted to the case of a scalar field ϕ (complex in general, but often illustrated using the case of a single real field). The restriction to scalar fields is not simply for reasons of simplicity, but because the scalar sector of particle physics is relatively unexplored. While vector fields such as electromagnetism are well understood, it is expected in many theories of unification that additional scalar fields such as the Higgs field will exist. We now need to look at what these can do for cosmology.

The Lagrangian density for a scalar field is as usual of the form of a kinetic minus a potential term:

$$\mathcal{L} = \tfrac{1}{2} \partial_\mu \phi \partial^\mu \phi - V(\phi).$$

In familiar examples of quantum fields, the potential would be

$$V(\phi) = \tfrac{1}{2} m^2 \phi^2,$$

where m is the mass of the field in natural units. However, it will be better to keep the potential function general at this stage. As usual, Noether's theorem gives the energy–momentum tensor for the field as

$$T^{\mu\nu} = \partial^\mu \phi \partial^\nu \phi - g^{\mu\nu} \mathcal{L}.$$

From this, we can read off the energy density and pressure:

$$\rho = \tfrac{1}{2} \dot{\phi}^2 + V(\phi) + \tfrac{1}{2} (\nabla\phi)^2$$
$$p = \tfrac{1}{2} \dot{\phi}^2 - V(\phi) - \tfrac{1}{6} (\nabla\phi)^2.$$

If the field is constant both spatially and temporally, the equation of state is then $p = -\rho$, as required if the scalar field is to act as a cosmological constant; note that derivatives of the field spoil this identification.

Treating the field classically (i.e. considering the expectation value $\langle\phi\rangle$, we get from energy–momentum conservation ($T^{\mu\nu}_{;\nu} = 0$) the equation of motion

$$\ddot{\phi} + 3H\dot{\phi} - \nabla^2\phi + dV/d\phi = 0.$$

This can also be derived more easily by the direct route of writing down the action $S = \int \mathcal{L}\sqrt{-g}\,d^4x$ and applying the Euler–Lagrange equation that arises from a stationary action ($\sqrt{-g} = R^3(t)$ for an FRW model, which is the origin of the Hubble drag term $3H\dot{\phi}$).

The solution of the equation of motion becomes tractable if we both ignore spatial inhomogeneities in ϕ and make the *slow-rolling approximation* that $|\ddot{\phi}|$ is negligible in comparison with $|3H\dot{\phi}|$ and $|dV/d\phi|$. Both these steps are required in order that inflation can happen; we have shown earlier that the vacuum equation of state only holds if in some sense ϕ changes slowly both spatially and temporally. Suppose there are characteristic temporal and spatial scales T and X for the scalar field; the conditions for inflation are that the negative-pressure equation of state from $V(\phi)$ must dominate the normal-pressure effects of time and space derivatives:

$$V \gg \phi^2/T^2, \qquad V \gg \phi^2/X^2,$$

hence $|dV/d\phi| \sim V/\phi$ must be $\gg \phi/T^2 \sim \ddot{\phi}$. The $\ddot{\phi}$ term can therefore be neglected in the equation of motion, which then takes the slow-rolling form for homogeneous fields:

$$3H\dot{\phi} = -dV/d\phi.$$

The conditions for inflation can be cast into useful dimensionless forms. The basic condition $V \gg \dot{\phi}^2$ can now be rewritten using the slow-roll relation as

$$\epsilon \equiv \frac{m_P^2}{16\pi}(V'/V)^2 \ll 1.$$

Also, we can differentiate this expression to obtain the criterion $V'' \ll V'/m_P$. Using slow-roll once more gives $3H\dot{\phi}/m_P$ for the right-hand side, which is in turn $\ll 3H\sqrt{V}/m_P$ because $\dot{\phi}^2 \ll V$, giving finally

$$\eta \equiv \frac{m_P^2}{8\pi}(V''/V) \ll 1$$

(recall that for de Sitter space $H = \sqrt{8\pi G V(\phi)/3} \sim \sqrt{V}/m_P$ in natural units). These two criteria make perfect intuitive sense: the potential must be flat in the sense of having small derivatives if the field is to roll slowly enough for inflation to be possible.

Similar arguments can be made for the spatial parts. However, they are less critical: what matters is the value of $\nabla\phi = \nabla_{\text{comoving}}\phi/R$. Since R increases exponentially, these perturbations are damped away: assuming V is large enough for inflation to start in the first place, inhomogeneities rapidly become negligible. This 'stretching' of field gradients as we increase the cosmological horizon beyond the value predicted in classical cosmology also solves a related problem that was historically important in motivating the invention of inflation—the *monopole problem*. Monopoles are point-like topological defects that would be expected to arise in any phase transition at around the GUT scale ($t \sim 10^{-35}$ s). If they form at approximately one per horizon volume at this time, then it follows that the present universe would contain $\Omega \gg 1$ in monopoles. This unpleasant conclusion is avoided if the horizon can be made much larger than the classical one at the end of inflation; the GUT fields have then been aligned over a vast scale, so that topological-defect formation becomes extremely rare.

2.5.2   Ending inflation

Although spatial derivatives of the scalar field can thus be neglected, the same is not always true for time derivatives. Although they may be negligible initially, the relative importance of time derivatives increases as ϕ rolls down the potential and V approaches zero (leaving aside the subtle question of how we know that the minimum is indeed at zero energy). Even if the potential does not steepen, sooner or later we will have $\epsilon \simeq 1$ or $|\eta| \simeq 1$ and the inflationary phase will cease. Instead of rolling slowly 'downhill', the field will oscillate about the bottom of the potential, with the oscillations becoming damped by the $3H\dot{\phi}$ friction term. Eventually, we will be left with a stationary field that either continues to inflate without end, if $V(\phi = 0) > 0$, or which simply has zero density. This would be a most boring universe to inhabit, but fortunately there is a more realistic way in which inflation can end. We have neglected so far the couplings of the scalar field to matter fields. Such couplings will cause the rapid oscillatory phase to produce particles, leading to *reheating*. Thus, even if the minimum of $V(\phi)$ is at $V = 0$, the universe is left containing roughly the same energy density as it started with, but now in the form of normal matter and radiation—-which starts the usual FRW phase, albeit with the desired special 'initial' conditions.

As well as being of interest for completing the picture of inflation, it is essential to realize that these closing stages of inflation are the *only* ones of observational relevance. Inflation might well continue for a huge number of *e*-foldings, all but the last few satisfying $\epsilon, \eta \ll 1$. However, the scales that left the de Sitter horizon at these early times are now vastly greater than our observable horizon, c/H_0, which exceeds the de Sitter horizon by only a finite factor. If inflation was terminated by reheating to the GUT temperature, then the expansion factor required to reach the present epoch is

$$a_{\text{GUT}}^{-1} \simeq E_{\text{GUT}}/E_\gamma.$$

The comoving horizon size at the end of inflation was therefore

$$d_H(t_{GUT}) \simeq a_{GUT}^{-1}[c/H_{GUT}] \simeq [E_P/E_\gamma]E_{GUT}^{-1},$$

where the last expression in natural units uses $H \simeq \sqrt{V}/E_P \simeq E_{GUT}^2/E_P$. For a GUT energy of 10^{15} GeV, this is about 10 m. This is a sobering illustration of the magnitude of the horizon problem; if we relied on causal processes at the GUT era to produce homogeneity, then the universe would only be smooth in patches a few comoving metres across. To solve the problem, we need enough e-foldings of inflation to have stretched this GUT-scale horizon to the present horizon size

$$N_{obs} = \ln\left[\frac{3000h^{-1}\,\text{Mpc}}{(E_P/E_\gamma)E_{GUT}^{-1}}\right] \simeq 60.$$

By construction, this is enough to solve the horizon problem, and it is also the number of e-foldings needed to solve the flatness problem. This is no coincidence, since we saw earlier that the criterion in this case was

$$N \gtrsim \frac{1}{2}\ln\left(\frac{a_{eq}}{a_{GUT}^2}\right).$$

Now, $a_{eq} = \rho_\gamma/\rho$, and $\rho = 3H^2\Omega/(8\pi G)$. In natural units, this translates to $\rho \sim E_P^2(c/H_0)^{-2}$, or $a_{eq}^{-1} \sim E_P^2(c/H_0)^{-2}/E_\gamma^4$. The expression for N is then identical to that in the case of the horizon problem: the same number of e-folds will always solve both.

Successful inflation in any of these models requires > 60 e-foldings of the expansion. The implications of this are easily calculated using the slow-roll equation, which gives the number of e-foldings between ϕ_1 and ϕ_2 as

$$N = \int H\,dt = -\frac{8\pi}{m_P^2}\int_{\phi_1}^{\phi_2}\frac{V}{V'}\,d\phi.$$

For any potential that is relatively smooth, $V' \sim V/\phi$, and so we get $N \sim (\phi_{start}/m_P)^2$, assuming that inflation terminates at a value of ϕ rather smaller than at the start. The criterion for successful inflation is thus that the initial value of the field exceeds the Planck scale:

$$\phi_{start} \gg m_P.$$

By the same argument, it is easily seen that this is also the criterion needed to make the slow-roll parameters ϵ and $\eta \ll 1$. To summarize, any model in which the potential is sufficiently flat that slow-roll inflation can commence will probably achieve the critical 60 e-foldings. Counterexamples can of course be constructed, but they have to be somewhat special cases.

It is interesting to review this conclusion for some of the specific inflation models listed earlier. Consider a mass-like potential $V = m^2\phi^2$. If inflation starts near the Planck scale, the fluctuations in V are $\sim m_P^4$ and these will drive ϕ_{start} to $\phi_{start} \gg m_P$ provided $m \ll m_P$; similarly, for $V = \lambda\phi^4$, the condition is weak coupling: $\lambda \ll 1$. Any field with a rather flat potential will thus tend to inflate, just because typical fluctuations leave it a long way from home in the form of the potential minimum. In a sense, inflation is realized by means of 'inertial confinement': there is nothing to prevent the scalar field from reaching the minimum of the potential—-but it takes a long time to do so, and the universe has meanwhile inflated by a large factor.

2.5.3   Relic fluctuations from inflation

The idea of launching a flat and causally connected expanding universe, using only vacuum-energy antigravity, is attractive. What makes the package of inflationary ideas especially compelling is that there it is an inevitable outcome of this process that the post-inflation universe will be inhomogeneous to some extent. There is not time to go into much detail on this here, but we summarize some of the key aspects, in order to make a bridge to the following material on structure formation.

The key idea is to appreciate that the inflaton field cannot be a classical object, but must display quantum fluctuations. Well inside the horizon of de Sitter space, these must be calculable by normal flat-space quantum field theory. If we can calculate how these fluctuations evolve as the universe expands, we have a mechanism for seeding inhomogeneities in the expanding universe—which can then grow under gravity to make structure.

To anticipate the detailed treatment, the inflationary prediction is of a horizon-scale fractional perturbation to the density

$$\delta_H = \frac{H^2}{2\pi\dot{\phi}}$$

which can be understood as follows. Imagine that the main effect of fluctuations is to make different parts of the universe have fields that are perturbed by an amount $\delta\phi$. In other words, we are dealing with various copies of the same rolling behaviour $\phi(t)$, but viewed at different times

$$\delta t = \frac{\delta\phi}{\dot{\phi}}.$$

These universes will then finish inflation at different times, leading to a spread in energy densities (figure 2.4). The horizon-scale density amplitude is given by the different amounts that the universes have expanded following the end of inflation:

$$\delta_H \simeq H\delta t = \frac{H^2}{2\pi\dot{\phi}},$$

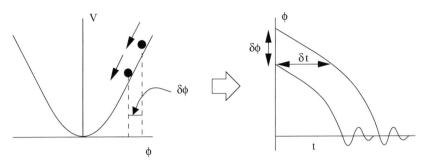

Figure 2.4. This plot shows how fluctuations in the scalar field transform themselves into density fluctuations at the end of inflation. Different points of the universe inflate from points on the potential perturbed by a fluctuation $\delta\phi$, like two balls rolling from different starting points. Inflation finishes at times separated by δt in time for these two points, inducing a density fluctuation $\delta = H\delta t$.

where the last step uses the crucial input of quantum field theory, which says that the rms $\delta\phi$ is given by $H/2\pi$. This is the classical amplitude that results from the stretching of sub-horizon flat-space quantum fluctuations. We will not attempt to prove this key result here (see chapter 12 of Peacock 1999, or Liddle and Lyth 1993, 2000).

Because the de Sitter expansion is invariant under time translation, the inflationary process produces a universe that is fractal-like in the sense that scale-invariant fluctuations correspond to a metric that has the same 'wrinkliness' per log length-scale. It then suffices to calculate that amplitude on one scale—i.e. the perturbations that are just leaving the horizon at the end of inflation, so that super-horizon evolution is not an issue. It is possible to alter this prediction of scale invariance only if the expansion is non-exponential; we have seen that such deviations plausibly do exist towards the end of inflation, so it is clear that exact scale invariance is not to be expected. This is discussed further later.

In summary, we have the following three key equations for basic inflationary model building. The fluctuation amplitude can be thought of as supplying the variance per $\ln k$ in potential perturbations, which we show later does not evolve with time:

$$\delta_{\mathrm{H}}^2 \equiv \Delta_\Phi^2(k) = \frac{H^4}{(2\pi\dot\phi)^2}$$

$$H^2 = \frac{8\pi}{3}\frac{V}{m_{\mathrm{P}}^2}$$

$$3H\dot\phi = -V'.$$

We have also written once again the exact relation between H and V and the slow-roll condition, since manipulation of these three equations is often required in derivations.

2.5.4 Gravity waves and tilt

The density perturbations left behind as a residue of the quantum fluctuations in the inflaton field during inflation are an important relic of that epoch, but are not the only one. In principle, a further important test of the inflationary model is that it also predicts a background of gravitational waves, whose properties couple with those of the density fluctuations.

It is easy to see in principle how such waves arise. In linear theory, any quantum field is expanded in a similar way into a sum of oscillators with the usual creation and annihilation operators; this analysis of quantum fluctuations in a scalar field is thus readily adapted to show that analogous fluctuations will be generated in other fields during inflation. In fact, the linearized contribution of a gravity wave, $h_{\mu\nu}$, to the Lagrangian looks like a scalar field $\phi = (m_P/4\sqrt{\pi})h_{\mu\nu}$, the expected rms gravity-wave amplitude is

$$h_{\rm rms} \sim H/m_P.$$

The fluctuations in ϕ are transmuted into density fluctuations, but gravity waves will survive to the present day, albeit redshifted.

This redshifting produces a break in the spectrum of waves. Prior to horizon entry, the gravity waves produce a scale-invariant spectrum of metric distortions, with amplitude $h_{\rm rms}$ per $\ln k$. These distortions are observable via the large-scale CMB anisotropies, where the tensor modes produce a spectrum with the same scale dependence as the Sachs–Wolfe gravitational redshift from scalar metric perturbations. In the scalar case, we have $\delta T/T \sim \phi/3c^2$, i.e. of order the Newtonian metric perturbation; similarly, the tensor effect is

$$\left(\frac{\delta T}{T}\right)_{\rm GW} \sim h_{\rm rms} \lesssim \delta_H \sim 10^{-5},$$

where the second step follows because the tensor modes can constitute no more than 100% of the observed CMB anisotropy.

A detailed estimate of the ratio between the tensor effect of gravity waves and the normal scalar Sachs–Wolfe effect was first analysed in a prescient paper by Starobinsky (1985). Denote the fractional temperature variance per natural logarithm of angular wavenumber by Δ^2 (constant for a scale-invariant spectrum). The tensor and scalar contributions are, respectively,

$$\Delta_T^2 \sim h_{\rm rms}^2 \sim (H^2/m_P^2) \sim V/m_P^4$$
$$\Delta_S^2 \sim \delta_H^2 \sim \frac{H^2}{\dot{\phi}} \sim \frac{H^6}{(V')^2} \sim \frac{V^3}{m_P^6 V'^2}.$$

The ratio of the tensor and scalar contributions to the variance of microwave background anisotropies is therefore proportional to the inflationary parameter ϵ:

$$\frac{\Delta_T^2}{\Delta_S^2} \simeq 12.4\epsilon,$$

inserting the exact coefficient from Starobinsky (1985). If it could be measured, the gravity-wave contribution to CMB anisotropies would therefore give a measure of ϵ, one of the dimensionless inflation parameters. The less 'de Sitter-like' the inflationary behaviour is, the larger the relative gravitational-wave contribution is.

Since deviations from exact exponential expansion also manifest themselves as density fluctuations with spectra that deviate from scale invariance, this suggests a potential test of inflation. Define the *tilt* of the fluctuation spectrum as follows:

$$\text{tilt} \equiv 1 - n \equiv -\frac{d \ln \delta_H^2}{d \ln k}.$$

We then want to express the tilt in terms of parameters of the inflationary potential, ϵ and η. These are of order unity when inflation terminates; ϵ and η must therefore be evaluated when the observed universe left the horizon, recalling that we only observe the last 60-odd e-foldings of inflation. The way to introduce scale dependence is to write the condition for a mode of given comoving wavenumber to cross the de Sitter horizon,

$$a/k = H^{-1}.$$

Since H is nearly constant during the inflationary evolution, we can replace $d/d \ln k$ by $d \ln a$, and use the slow-roll condition to obtain

$$\frac{d}{d \ln k} = a \frac{d}{da} = \frac{\dot{\phi}}{H} \frac{d}{d\phi} = -\frac{m_P^2}{8\pi} \frac{V'}{V} \frac{d}{d\phi}.$$

We can now work out the tilt, since the horizon-scale amplitude is

$$\delta_H^2 = \frac{H^4}{(2\pi \dot{\phi})^2} = \frac{128\pi}{3} \left(\frac{V^3}{m_P^6 V'^2} \right),$$

and derivatives of V can be expressed in terms of the dimensionless parameters ϵ and η. The tilt of the density perturbation spectrum is thus predicted to be

$$1 - n = 6\epsilon - 2\eta.$$

In section 2.8.5 on CMB anisotropies, we discuss whether this relation is observationally testable.

2.5.5 Evidence for vacuum energy at late times

The idea of inflation is audacious, but undeniably speculative. However, once we accept the idea that quantum fields can generate an equation of state resembling a cosmological constant, we need not confine this mechanism to GUT-scale energies. There is no known mechanism that requires the minimum of $V(\phi)$ to lie exactly at zero energy, so it is quite plausible that there remains in the universe today some non-zero vacuum energy.

The most direct way of detecting vacuum energy has been the immense recent progress in the use of supernovae as standard candles. Type Ia SNe have been used as standard objects for around two decades, with an rms scatter in luminosity of 40%, and so a distance error of 20%. The big breakthrough came when it was realized that the intrinsic timescale of the SNe correlates with luminosity (a brighter SNe lasts longer). Taking out this effect produces corrected standard candles that are capable of measuring distances to about 5% accuracy. Large search campaigns have made it possible to find of the order of 100 SNe over the range $0.1 \lesssim z \lesssim 1$, and two teams have used this strategy to make an empirical estimate of the cosmological distance–redshift relation.

The results of the *Supernova Cosmology Project* (e.g. Perlmutter *et al* 1998) and the *High-z Supernova Search* (e.g. Riess *et al* 1998) are highly consistent. Figure 2.5 shows the Hubble diagram from the latter team. The SNe magnitudes are K-corrected, so that their variation with redshift should be a direct measure of luminosity distance as a function of redshift.

We have seen earlier that this is written as the following integral, which must usually be evaluated numerically:

$$D_L(z) = (1 + z) R_0 S_k(r) = (1 + z) \frac{c}{H_0} |1 - \Omega|^{-1/2}$$
$$\times S_k \left[\int_0^z \frac{|1 - \Omega|^{1/2} \, \mathrm{d}z'}{\sqrt{(1 - \Omega)(1 + z')^2 + \Omega_v + \Omega_m(1 + z')^3}} \right],$$

where $\Omega = \Omega_m + \Omega_v$, and S_k is sinh if $\Omega < 1$, otherwise sin. It is clear from figure 2.5 that the empirical distance–redshift relation is very different from the simplest inflationary prediction, which is the $\Omega = 1$ Einstein–de Sitter model; by redshift 0.6, the SNe are fainter than expected in this model by about 0.5 magnitudes. If this model fails, we can try adjusting Ω_m and Ω_v in an attempt to do better. Comparing each such model to the data yields the likelihood contours shown in figure 2.6, which can be used in the standard way to set confidence limits on the cosmological parameters. The results very clearly require a low-density universe. For $\Lambda = 0$, a very low density is just barely acceptable, with $\Omega_m \lesssim 0.1$. However, the discussion of the CMB later shows that such a heavily open model is hard to sustain. The preferred model has $\Omega_v \simeq 1$; if we restrict ourselves to the inflationary $k = 0$, then the required parameters are very close to $(\Omega_m, \Omega_v) = (0.3, 0.7)$.

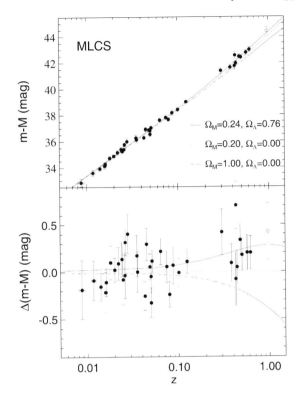

Figure 2.5. The Hubble diagram produced by the High-z Supernova search team (Riess *et al* 1998). The lower panel shows the data divided by a default model ($\Omega_m = 0.2$, $\Omega_v = 0$). The results lie clearly above this model, favouring a non-zero Λ. The lowest line is the Einstein–de Sitter model, which is in gross disagreement with observation.

2.5.6 Cosmic coincidence

This is an astonishing result—an observational detection of the physical reality of vacuum energy. The error bars continue to shrink, and no convincing systematic error has been suggested that could yield this result spuriously; this is one of the most important achievements of 20th century physics.

And yet, accepting the reality of vacuum energy raises a difficult question. If the universe contains a constant vacuum density and normal matter with $\rho \propto a^{-3}$, there is a unique epoch at which these two contributions cross over, and we seem to be living near to that time. This coincidence calls for some explanation. One might think of appealing to anthropic ideas, and these can limit Λ to some extent: if the universe became vacuum-dominated at $z > 1000$, gravitational instability as discussed in the next section would have been impossible—so that galaxies, stars and observers would not have been possible. However, Weinberg (1989) argues

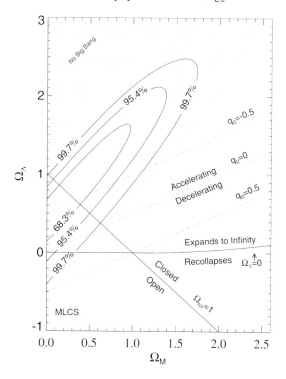

Figure 2.6. Confidence contours on the Ω_v–Ω_m plane, according to Riess *et al* (1998). Open models of all but the lowest densities are apparently ruled out, and non-zero Λ is strongly preferred. If we restrict ourselves to $k = 0$, then $\Omega_m \simeq 0.3$ is required. The constraints perpendicular to the $k = 0$ line are not very tight, but CMB data can help here in limiting the allowed degree of curvature.

that Λ could have been much larger than its actual value without making observers impossible. Efstathiou (1995) attempted to construct a probability distribution for Λ by taking this to be proportional to the number density of galaxies that result in a given model. However, there is no general agreement on how to set a probability measure for this problem.

It would be more satisfactory if we had some physical mechanism that guaranteed the coincidence, and one possibility has been suggested. We already have one coincidence, in that we live relatively close in time to the era of matter–radiation equality ($z \sim 10^3$, as opposed to $z \sim 10^{80}$ for the GUT era). What is required is a cosmological 'constant' that switches on around the equality era. Zlatev *et al* (1999) have suggested how this might happen. The idea is to use the vacuum properties of a homogeneous scalar field as the physical origin of the negative-pressure term detected via SNe. This idea of a 'rolling' Λ was first explored by Ratra and Peebles (1988), and there has recently been a tendency towards use of the fanciful term 'quintessence'. In any case, it is important to appreciate that the idea uses exactly the same physical elements that we discussed in the context of inflation: there is some $V(\phi)$, causing the expectation value of ϕ to obey the damped oscillator equation of motion, so the energy density and pressure are

$$\rho_\phi = \dot\phi^2/2 + V$$
$$p_\phi = \dot\phi^2/2 - V.$$

This gives us two extreme equations of state:

(i) vacuum-dominated, with $V \gg \dot\phi^2/2$, so that $p = -\rho$;
(ii) kinetic-dominated, with $V \ll \dot\phi^2/2$, so that $p = \rho$.

In the first case, we know that ρ does not alter as the universe expands, so the vacuum rapidly tends to dominate over normal matter. In the second case, the equation of state is the unusual $\Gamma = 2$, so we get the rapid behaviour $\rho \propto a^{-6}$. If a quintessence-dominated universe starts off with a large kinetic term relative to the potential, it may seem that things should always evolve in the direction of being potential-dominated. However, this ignores the detailed dynamics of the situation: for a suitable choice of potential, it is possible to have a *tracker field*, in which the kinetic and potential terms remain in a constant proportion, so that we can have $\rho \propto a^{-\alpha}$, where α can be anything we choose.

Putting this condition in the equation of motion shows that the potential is required to be exponential in form. More importantly, we can generalize to the case where the universe contains scalar field and ordinary matter. Suppose the latter dominates, and obeys $\rho_m \propto a^{-\alpha}$. It is then possible to have the scalar-field density obeying the same $\rho \propto a^{-\alpha}$ law, provided

$$V(\phi) = \frac{2}{\lambda^2}(6/\alpha - 1)\exp[-\lambda\phi].$$

The scalar-field density is $\rho_\phi = (\alpha/\lambda^2)\rho_{\text{total}}$ (see, e.g., Liddle and Scherrer 1999). The impressive thing about this solution is that the quintessence density stays a fixed fraction of the total, whatever the overall equation of state: it automatically scales as a^{-4} at early times, switching to a^{-3} after the matter–radiation equality.

This is not quite what we need, but it shows how the effect of the overall equation of state can affect the rolling field. Because of the $3H\dot\phi$ term in the equation of motion, ϕ 'knows' whether or not the universe is matter dominated. This suggests that a more complicated potential than the exponential may allow the arrival of matter domination to trigger the desired Λ-like behaviour. Zlatev *et al* suggest two potentials which might achieve this:

$$V(\phi) = M^{4+\beta}\phi^{-\beta} \qquad \text{or} \qquad V(\phi) = M^4[\exp(m_P/\phi) - 1].$$

The evolution in these potentials may be described by $w(t)$, where $w = p/\rho$. We need $w \simeq 1/3$ in the radiation era, changing to $w \simeq -1$ today. The evolution

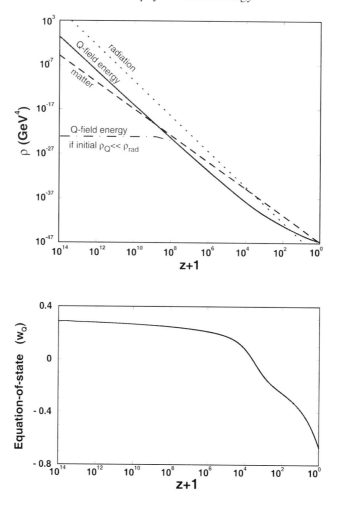

Figure 2.7. This figure, taken from Zlatev *et al* (1999), shows the evolution of the density in the 'quintessence' field (top panel), together with the effective equation of state of the quintessence vacuum (bottom panel), for the case of the inverse exponential potential. This allows vacuum energy to lurk at a few per cent of the total throughout the radiation era, but switching on a cosmological constant after the universe becomes matter dominated.

in the inverse exponential potential is shown in figure 2.7, demonstrating that the required behaviour can be found. However, a slight fine-tuning is still required, in that the trick only works for $M \sim 1$ meV, so there has to be an energy coincidence with the energy scale of matter–radiation equality.

So, the idea of tracker fields does not remove completely the puzzle concerning the level of the present-day vacuum energy. In a sense, relegating the solution to a potential of unexplained form may seem a retrograde step. However, it is at least a testable step: the prediction of figure 2.7 is that $w \simeq -0.8$ today, so that the quintessence density scales as $\rho \propto a^{-0.6}$. This is a significant difference from the classical $w = -1$ vacuum energy, and it should be detectable as the SNe data improve. The existing data already require approximately $w < -0.5$, so there is the entrancing prospect that the equation of state for the vacuum will soon become the subject of experimental study.

2.6   Dynamics of structure formation

The overall properties of the universe are very close to being homogeneous; and yet telescopes reveal a wealth of detail on scales varying from single galaxies to *large-scale structures* of size exceeding 100 Mpc. This section summarizes some of the key results concerning how such structure can arise via gravitational instability.

2.6.1   Linear perturbations

The study of cosmological perturbations can be presented as a complicated exercise in linearized general relativity; fortunately, much of the essential physics can be extracted from a Newtonian approach. We start by writing down the fundamental equations governing fluid motion (non-relativistic for now):

$$\text{Euler:} \qquad \frac{D\mathbf{v}}{Dt} = -\frac{\nabla p}{\rho} - \nabla \Phi$$

$$\text{energy:} \qquad \frac{D\rho}{Dt} = -\rho \nabla \cdot \mathbf{v}$$

$$\text{Poisson:} \qquad \nabla^2 \Phi = 4\pi G \rho,$$

where $D/Dt = \partial/\partial t + \mathbf{v} \cdot \nabla$ is the usual convective derivative. We now produce the *linearized equations of motion* by collecting terms of first order in perturbations about a homogeneous background: $\rho = \rho_0 + \delta\rho$ etc. As an example, consider the energy equation:

$$[\partial/\partial t + (\mathbf{v}_0 + \delta\mathbf{v}) \cdot \nabla](\rho_0 + \delta\rho) = -(\rho_0 + \delta\rho)\nabla \cdot (\mathbf{v}_0 + \delta\mathbf{v}).$$

For no perturbation, the zero-order equation is

$$(\partial/\partial t + \mathbf{v}_0 \cdot \nabla)\rho_0 = -\rho_0 \nabla \cdot \mathbf{v}_0;$$

since ρ_0 is homogeneous and $\mathbf{v}_0 = H\mathbf{x}$ is the Hubble expansion, this just says $\dot{\rho}_0 = -3H\rho_0$. Expanding the full equation and subtracting the zeroth-order equation gives the equation for the perturbation:

$$(\partial/\partial t + \mathbf{v}_0 \cdot \nabla)\delta\rho + \delta\mathbf{v} \cdot \nabla(\rho_0 + \delta_\rho) = -(\rho_0 + \delta\rho)\nabla \cdot \delta\mathbf{v} - \delta\rho \nabla \cdot \mathbf{v}_0.$$

Now, for sufficiently small perturbations, terms containing a product of perturbations such as $\delta v \cdot \nabla \delta_\rho$ must be negligible in comparison with the first-order terms. Remembering that ρ_0 is homogeneous leaves the linearized equation

$$[\partial/\partial t + v_0 \cdot \nabla]\delta\rho = -\rho_0 \nabla \cdot \delta v - \delta\rho \nabla \cdot v_0.$$

It is straightforward to perform the same steps with the other equations; the results look simpler if we define the fractional density perturbation

$$\delta \equiv \frac{\delta\rho}{\rho_0}.$$

As before, when dealing with time derivatives of perturbed quantities, the full convective time derivative D/Dt can always be replaced by d/d$t \equiv \partial/\partial t + v_0 \cdot \nabla$, which is the time derivative for an observer comoving with the unperturbed expansion of the universe. We then can write

$$\frac{\mathrm{d}}{\mathrm{d}t}\delta v = -\frac{\nabla \delta p}{\rho_0} - \nabla\delta\Phi - (\delta v \cdot \nabla)v_0$$
$$\frac{\mathrm{d}}{\mathrm{d}t}\delta = -\nabla \cdot \delta v$$
$$\nabla^2\delta\Phi = 4\pi G\rho_0\delta.$$

There is now only one complicated term to be dealt with: $(\delta v \cdot \nabla)v_0$ on the right-hand side of the perturbed Euler equation. This is best attacked by writing it in components:

$$[(\delta v \cdot \nabla)v_0]_j = [\delta v]_i \nabla_i [v_0]_j = H[\delta v]_j,$$

where the last step follows because $v_0 = Hx_0 \Rightarrow \nabla_i[v_0]_j = H\delta_{ij}$. This leaves a set of equations of motion that have no explicit dependence on the global expansion speed v_0; this is only present implicitly through the use of convective time derivatives d/dt.

These equations of motion are written in Eulerian coordinates: proper length units are used, and the Hubble expansion is explicitly present through the velocity v_0. The alternative approach is to use the comoving coordinates formed by dividing the Eulerian coordinates by the scale factor $a(t)$:

$$x(t) = a(t)r(t)$$
$$\delta v(t) = a(t)u(t).$$

The next step is to translate spatial derivatives into comoving coordinates:

$$\nabla_x = \frac{1}{a}\nabla_r.$$

To keep the notation simple, subscripts on ∇ will normally be omitted hereafter, and spatial derivatives will be with respect to comoving coordinates.   The linearized equations for conservation of momentum and matter as experienced by fundamental observers moving with the Hubble flow then take the following simple forms in comoving units:

$$\dot{u} + 2\frac{\dot{a}}{a}u = \frac{g}{a} - \frac{\nabla \delta p}{\rho_0}$$
$$\dot{\delta} = -\nabla \cdot u,$$

where dots stand for d/dt. The peculiar gravitational acceleration $\nabla \delta \Phi / a$ is denoted by g.

Before going on, it is useful to give an alternative derivation of these equations, this time working in comoving length units right from the start. First note that the comoving peculiar velocity u is just the time derivative of the comoving coordinate r:

$$\dot{x} = \dot{a}r + a\dot{r} = Hx + a\dot{r},$$

where the right-hand side must be equal to the Hubble flow Hx, plus the peculiar velocity $\delta v = au$. In this equation, dots stand for exact convective time derivatives—i.e. time derivatives measured by an observer who follows a particle's trajectory—rather than partial time derivatives $\partial/\partial t$. This allows us to apply the continuity equation immediately in comoving coordinates, since this equation is simply a statement that particles are conserved, independent of the coordinates used. The exact equation is

$$\frac{D}{Dt}\rho_0(1 + \delta) = -\rho_0(1 + \delta)\nabla \cdot u,$$

and this is easy to linearize because the background density ρ_0 is independent of time when comoving length units are used. This gives the first-order equation $\dot{\delta} = -\nabla \cdot u$ immediately. The equation of motion follows from writing the Eulerian equation of motion as $\ddot{x} = g_0 + g$, where $g = \nabla \delta \Phi / a$ is the peculiar acceleration defined earlier, and g_0 is the acceleration that acts on a particle in a homogeneous universe (neglecting pressure forces, for simplicity). Differentiating $x = ar$ twice gives

$$\ddot{x} = a\dot{u} + 2\dot{a}u + \frac{\ddot{a}}{a}x = g_0 + g.$$

The unperturbed equation corresponds to zero peculiar velocity and zero peculiar acceleration: $(\ddot{a}/a)x = g_0$; subtracting this gives the perturbed equation of motion $\dot{u} + 2(\dot{a}/a)u = g$, as before. This derivation is rather more direct than the previous route of working in Eulerian space. Also, it emphasizes that the equation of motion is exact, even though it happens to be linear in the perturbed quantities.

After doing all this, we still have three equations in the four variables δ, u, $\delta \Phi$ and δp. The system needs an equation of state in order to be closed; this may be specified in terms of the sound speed

$$c_s^2 \equiv \frac{\partial p}{\partial \rho}.$$

Now think of a plane-wave disturbance $\delta \propto e^{-i\mathbf{k} \cdot \mathbf{r}}$, where $\mathbf{k}$ is a comoving wavevector; in other words, suppose that the wavelength of a single Fourier mode stretches with the universe. All time dependence is carried by the amplitude of the wave, and so the spatial dependence can be factored out of time derivatives in these equations (which would not be true with a constant comoving wavenumber k/a). An equation for the amplitude of δ can then be obtained by eliminating $\mathbf{u}$:

$$\ddot{\delta} + 2\frac{\dot{a}}{a}\dot{\delta} = \delta(4\pi G\rho_0 - c_s^2 k^2/a^2).$$

This equation is the one that governs the gravitational amplification of density perturbations.

There is a critical proper wavelength, known as the *Jeans length*, at which we switch from the possibility of exponential growth for long-wavelength modes to standing sound waves at short wavelengths. This critical length is

$$\lambda_J = c_s\sqrt{\frac{\pi}{G\rho}},$$

and clearly delineates the scale at which sound waves can cross an object in about the time needed for gravitational free-fall collapse. When considering perturbations in an expanding background, things are more complex. Qualitatively, we expect to have no growth when the 'driving term' on the right-hand side is negative. However, owing to the expansion, λ_J will change with time, and so a given perturbation may switch between periods of growth and stasis.

2.6.2 Dynamical effects of radiation

At early enough times, the universe was radiation dominated ($c_s = c/\sqrt{3}$) and the analysis so far does not apply. It is common to resort to general relativity perturbation theory at this point. However, the fields are still weak, and so it is possible to generate the results we need by using special relativity fluid mechanics and Newtonian gravity with a relativistic source term. For simplicity, assume that accelerations due to pressure gradients are negligible in comparison with gravitational accelerations (i.e. restrict the analysis to $\lambda \gg \lambda_J$ from the start). The basic equations are then a simplified Euler equation and the full energy and gravitational equations:

$$\text{Euler:} \quad \frac{D\mathbf{v}}{Dt} = -\nabla\Phi$$

$$\text{energy:} \quad \frac{D}{Dt}(\rho + p/c^2) = \frac{\partial}{\partial t}(p/c^2) - (\rho + p/c^2)\nabla \cdot \mathbf{v}$$

$$\text{Poisson:} \quad \nabla^2\Phi = 4\pi G(\rho + 3p/c^2).$$

For total radiation domination, $p = \rho c^2/3$, and it is easy to linearize these equations as before. The main differences come from factors of 2 and 4/3 due to the non-negligible contribution of the pressure. The result is a continuity equation $\nabla \cdot \mathbf{u} = -(3/4)\dot{\delta}$, and the evolution equation for δ:

$$\ddot{\delta} + 2\frac{\dot{a}}{a}\dot{\delta} = \frac{32\pi}{3}G\rho_0\delta,$$

so the net result of all the relativistic corrections is a driving term on the right-hand side that is a factor 8/3 higher than in the matter-dominated case.

In both matter- and radiation-dominated universes with $\Omega = 1$, we have $\rho_0 \propto 1/t^2$:

$$\text{matter domination } (a \propto t^{2/3}): \qquad 4\pi G\rho_0 = \frac{2}{3t^2}$$

$$\text{radiation domination } (a \propto t^{1/2}): \qquad 32\pi G\rho_0/3 = \frac{1}{t^2}.$$

Every term in the equation for δ is thus the product of derivatives of δ and powers of t, and a power-law solution is obviously possible. If we try $\delta \propto t^n$, then the result is $n = 2/3$ or -1 for matter domination; for radiation domination, this becomes $n = \pm 1$. For the growing mode, these can be combined rather conveniently using the *conformal time* $\eta \equiv \int dt/a$:

$$\delta \propto \eta^2.$$

Recall that η is proportional to the comoving size of the horizon.

It is also interesting to think about the growth of matter perturbations in universes with non-zero vacuum energy, or even possibly some other exotic background with a peculiar equation of state. The differential equation for δ is as before, but $a(t)$ is altered. The way to deal with this is to treat a spherical perturbation as a small universe. Consider the Friedmann equation in the form

$$(\dot{a})^2 = \Omega_0^{\text{tot}} H_0^2 a^2 + K,$$

where $K = -kc^2/R_0^2$; this emphasizes that K is a constant of integration. A second constant of integration arises in the expression for time:

$$t = \int_0^a \dot{a}^{-1} \, da + C.$$

This lets us argue as before in the case of decaying modes: if a solution to the Friedmann equation is $a(t, K, C)$, then valid density perturbations are

$$\delta \propto \left(\frac{\partial \ln a}{\partial K}\right)_t \qquad \text{or} \qquad \left(\frac{\partial \ln a}{\partial C}\right)_t.$$

Since $\partial(\dot{a}^2)/\partial K = 1$, this gives the growing and decaying modes as

$$\delta \propto \begin{cases} (\dot{a}/a) \int_0^a (\dot{a})^{-3}\, da & \text{(growing mode)} \\ (\dot{a}/a) & \text{(decaying mode)}. \end{cases}$$

(Heath 1977, see also section 10 of Peebles 1980).

The equation for the growing mode requires numerical integration in general, with $\dot{a}(a)$ given by the Friedmann equation. A very good approximation to the answer is given by Carroll *et al* (1992):

$$\frac{\delta(z=0, \Omega)}{\delta(z=0, \Omega=1)} \simeq \frac{5}{2}\Omega_m \left[\Omega_m^{4/7} - \Omega_v + \left(1 + \frac{1}{2}\Omega_m\right)\left(1 + \frac{1}{70}\Omega_v\right) \right]^{-1}.$$

This fitting formula for the growth suppression in low-density universes is an invaluable practical tool. For flat models with $\Omega_m + \Omega_v = 1$, it says that the growth suppression is less marked than for an open universe—approximately $\Omega^{0.23}$ as against $\Omega^{0.65}$ if $\Lambda = 0$. This reflects the more rapid variation of Ω_v with redshift; if the cosmological constant is important dynamically, this only became so very recently, and the universe spent more of its history in a nearly Einstein–de Sitter state by comparison with an open universe of the same Ω_m.

What about the case of collisionless matter in a radiation background? The fluid treatment is not appropriate here, since the two species of particles can interpenetrate. A particularly interesting limit is for perturbations well inside the horizon: the radiation can then be treated as a smooth, unclustered background that affects only the overall expansion rate. This is analogous to the effect of Λ, but an analytical solution does exist in this case. The perturbation equation is as before

$$\ddot{\delta} + 2\frac{\dot{a}}{a}\dot{\delta} = 4\pi G\rho_m\delta,$$

but now $H^2 = 8\pi G(\rho_m + \rho_r)/3$. If we change variable to $y \equiv \rho_m/\rho_r = a/a_{eq}$, and use the Friedmann equation, then the growth equation becomes

$$\delta'' + \frac{2 + 3y}{2y(1+y)}\delta' - \frac{3}{2y(1+y)}\delta = 0$$

(for $k = 0$, as appropriate for early times). It may be seen by inspection that a growing solution exists with $\delta'' = 0$:

$$\delta \propto y + 2/3.$$

It is also possible to derive the decaying mode. This is simple in the radiation-dominated case ($y \ll 1$): $\delta \propto -\ln y$ is easily seen to be an approximate solution in this limit.

What this says is that, at early times, the dominant energy of radiation drives the universe to expand so fast that the matter has no time to respond, and δ is frozen at a constant value. At late times, the radiation becomes negligible, and the growth increases smoothly to the Einstein–de Sitter $\delta \propto a$ behaviour (Mészáros 1974). The overall behaviour is therefore similar to the effects of pressure on a coupled fluid: for scales greater than the horizon, perturbations in matter and radiation can grow together, but this growth ceases once the perturbations enter the horizon. However, the explanations of these two phenomena are completely different.

2.6.3 The peculiar velocity field

The foregoing analysis shows that gravitational collapse inevitably generates deviations from the Hubble expansion, which are interesting to study in detail.

Consider first a galaxy that moves with some peculiar velocity in an otherwise uniform universe. Even though there is no peculiar gravitational acceleration acting, its velocity will decrease with time as the galaxy attempts to catch up with successively more distant (and therefore more rapidly receding) neighbours. If the proper peculiar velocity is v, then after time dt the galaxy will have moved a proper distance $x = v \, dt$ from its original location. Its near neighbours will now be galaxies with recessional velocities $Hx = Hv \, dt$, relative to which the peculiar velocity will have fallen to $v - Hx$. The equation of motion is therefore just

$$\dot{v} = -Hv = -\frac{\dot{a}}{a}v,$$

with the solution $v \propto a^{-1}$: peculiar velocities of non-relativistic objects suffer redshifting by exactly the same factor as photon momenta. It is often convenient to express the peculiar velocity in terms of its comoving equivalent, $v \equiv au$, for which the equation of motion becomes $\dot{u} = -2Hu$. Thus, in the absence of peculiar accelerations and pressure forces, comoving peculiar velocities redshift away through the Hubble drag term $2Hu$.

If we now include the effects of peculiar acceleration, this simply adds the acceleration g on the right-hand side. This gives the equation of motion

$$\dot{u} + \frac{2\dot{a}}{a}u = -\frac{g}{a},$$

where $g = \nabla \delta\Phi/a$ is the peculiar gravitational acceleration. Pressure terms have been neglected, so $\lambda \gg \lambda_{\mathrm{J}}$. Remember that throughout we are using comoving length units, so that $\nabla_{\mathrm{proper}} = \nabla/a$. This equation is the exact equation of motion for a single galaxy, so that the time derivative is $d/dt = \partial/\partial t + u \cdot \nabla$. In linear theory, the second part of the time derivative can be neglected, and the equation then turns into one that describes the evolution of the linear peculiar velocity field at a fixed point in comoving coordinates.

The solutions for the peculiar velocity field can be decomposed into modes either parallel to g or independent of g (these are the homogeneous and inhomogeneous solutions to the equation of motion). The interpretation of these solutions is aided by knowing that the velocity field satisfies the *continuity equation*: $\dot{\rho} = -\nabla \cdot (\rho \boldsymbol{v})$ in proper units, which obviously takes the same form $\dot{\rho} = -\nabla \cdot (\rho \boldsymbol{u})$ if lengths and densities are in comoving units. If we express the density as $\rho = \rho_0(1+\delta)$ (where in comoving units ρ_0 is just a number independent of time), the continuity equation takes the form

$$\dot{\delta} = -\nabla \cdot [(1 + \delta)\boldsymbol{u}],$$

which becomes just

$$\nabla \cdot \boldsymbol{u} = -\dot{\delta}$$

in linear theory when both δ and $\boldsymbol{u}$ are small. This states that it is possible to have vorticity modes with $\nabla \cdot \boldsymbol{u} = 0$, for which $\dot{\delta}$ vanishes. We have already seen that δ either grows or decays as a power of time, so these modes require zero density perturbation, in which case the associated peculiar gravity also vanishes. These vorticity modes are thus the required homogeneous solutions, and they decay as $v = au \propto a^{-1}$, as with the kinematic analysis for a single particle. For any gravitational-instability theory, in which structure forms via the collapse of small perturbations laid down at very early times, it should therefore be a very good approximation to say that the linear velocity field must be curl-free.

For the growing modes, we want to try looking for a solution $\boldsymbol{u} = F(t)\boldsymbol{g}$. Then using continuity plus Gauss's theorem, $\nabla \cdot \boldsymbol{g} = 4\pi G a \rho \delta$, gives us

$$\delta \boldsymbol{v} = \frac{2f(\Omega)}{3H\Omega}\boldsymbol{g},$$

where the function $f(\Omega) \equiv (a/\delta)\,\mathrm{d}\delta/\mathrm{d}a$. A very good approximation to this (Peebles 1980) is $g \simeq \Omega^{0.6}$ (a result that is almost independent of Λ; Lahav *et al* 1991). Alternatively, we can work in Fourier terms. This is easy, as $\boldsymbol{g}$ and $\boldsymbol{k}$ are parallel, so that $\nabla \cdot \boldsymbol{u} = -\mathrm{i}\boldsymbol{k} \cdot \boldsymbol{u} = -\mathrm{i}ku$. Thus, directly from the continuity equation,

$$\delta \boldsymbol{v}_k = -\frac{\mathrm{i}Hf(\Omega)a}{k}\delta_k \hat{\boldsymbol{k}}.$$

The $1/k$ factor shows clearly that peculiar velocities are much more sensitive probes of large-scale inhomogeneities than are density fluctuations. The existence of large-scale homogeneity in density requires $n > -3$, whereas peculiar velocities will diverge unless $n > -1$ on large scales.

2.6.4 Transfer functions

We have seen that power spectra at late times result from modifications of any primordial power by a variety of processes: growth under self-gravitation; the effects of pressure; dissipative processes. We now summarize the two main ways in which the power spectrum that exists at early times may differ from that which emerges at the present, both of which correspond to a reduction of small-scale fluctuations (at least, for adiabatic fluctuations; we shall not consider isocurvature modes here):

(1) Radiation effects. Prior to matter–radiation equality, we have already seen that perturbations inside the horizon are prevented from growing by radiation pressure. Once z_{eq} is reached, if collisionless dark matter dominates, perturbations on all scales can grow. We therefore expect a feature in the transfer function around $k \sim 1/r_H(z_{eq})$. In the matter-dominated approximation, we get

$$d_H = \frac{2c}{H_0}(\Omega z)^{-1/2} \Rightarrow d_{eq} = 39(\Omega h^2)^{-1} \text{ Mpc}.$$

The exact distance–redshift relation is

$$R_0 \, dr = \frac{c}{H_0} \frac{dz}{(1+z)\sqrt{1+\Omega_m z + (1+z)^2 \Omega_r}},$$

from which it follows that the correct answer for the horizon size including radiation is a factor $\sqrt{2} - 1$ smaller: $d_{eq} = 16.0(\Omega h^2)^{-1}$ Mpc.

(2) Damping. In addition to having their growth retarded, very small-scale perturbations will be erased entirely, which can happen in one of two ways. For collisionless dark matter, perturbations are erased simply by *free-streaming*: random particle velocities cause blobs to disperse. At early times ($kT > mc^2$), the particles will travel at c, and so any perturbation that has entered the horizon will be damped. This process switches off when the particles become non-relativistic; for massive particles, this happens long before z_{eq} (resulting in cold dark matter (CDM)). For massive neutrinos, however, it happens *at* z_{eq}: only perturbations on very large scales survive in the case of hot dark matter (HDM). In a purely baryonic universe, the corresponding process is called *Silk damping*: the mean free path of photons due to scattering by the plasma is non-zero, and so radiation can diffuse out of a perturbation, convecting the plasma with it.

The overall effect is encapsulated in the *transfer function*, which gives the ratio of the late-time amplitude of a mode to its initial value:

$$T_k \equiv \frac{\delta_k(z=0)}{\delta_k(z)D(z)},$$

where $D(z)$ is the linear growth factor between redshift z and the present. The normalization redshift is arbitrary, so long as it refers to a time before any scale of interest has entered the horizon.

It is invaluable in practice to have some accurate analytic formulae that fit the numerical results for transfer functions. We give below results for some common models of particular interest (illustrated in figure 2.8, along with other cases where a fitting formula is impractical). For the models with collisionless dark matter, $\Omega_B \ll \Omega$ is assumed, so that all lengths scale with the horizon size at matter–radiation equality, leading to the definition

$$q \equiv \frac{k}{\Omega h^2 \text{ Mpc}^{-1}}.$$

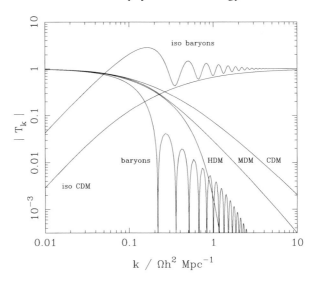

Figure 2.8. A plot of transfer functions for various models. For adiabatic models, $T_k \to 1$ at small k, whereas the opposite is true for isocurvature models. A number of possible matter contents are illustrated: pure baryons; pure CDM; pure HDM; MDM (30% HDM, 70% CDM). For dark-matter models, the characteristic wavenumber scales proportional to Ωh^2. The scaling for baryonic models does not obey this exactly; the plotted cases correspond to $\Omega = 1, h = 0.5$.

We consider the following cases:

(1) adiabatic CDM;
(2) adiabatic massive neutrinos (one massive, two massless); and
(3) isocurvature CDM; these expressions come from Bardeen *et al* (1986; BBKS).

Since the characteristic length-scale in the transfer function depends on the horizon size at matter–radiation equality, the temperature of the CMB enters. In these formulae, it is assumed to be exactly 2.7 K; for other values, the characteristic wavenumbers scale $\propto T^{-2}$. For these purposes massless neutrinos count as radiation, and three species of these contribute a total density that is 0.68 that of the photons.

(1) $T_k = \dfrac{\ln(1 + 2.34q)}{2.34q}[1 + 3.89q + (16.1q)^2 + (5.46q)^3 + (6.71q)^4]^{-1/4}$
(2) $T_k = \exp(-3.9q - 2.1q^2)$
(3) $T_k = (5.6q)^2(1 + [15.0q + (0.9q)^{3/2} + (5.6q)^2]^{1.24})^{-1/1.24}$.

The case of mixed dark matter (MDM: a mixture of massive neutrinos and CDM) is more complex. See Pogosyan and Starobinksy (1995) for a fit in this case.

These expressions assume pure dark matter, which is unrealistic. At least for CDM models, a non-zero baryonic density lowers the apparent dark-matter density parameter. We can define an apparent shape parameter Γ for the transfer function:

$$q \equiv (k/h \text{ Mpc}^{-1})/\Gamma,$$

and $\Gamma = \Omega h$ in a model with zero baryon content. This parameter was originally defined by Efstathiou *et al* (1992), in terms of a CDM model with $\Omega_B = 0.03$. Peacock and Dodds (1994) showed that the effect of increasing Ω_B was to preserve the CDM-style spectrum shape, but to shift to lower values of Γ. This shift was generalized to models with $\Omega \neq 1$ by Sugiyama (1995):

$$\Gamma = \Omega h \exp[-\Omega_B(1 + \sqrt{2h}/\Omega)].$$

Note the oscillations in $T(k)$ for high baryon content; these can be significant even in CDM-dominated models when working with high-precision data. Eisenstein and Hu (1998) are to be congratulated for their impressive persistence in finding an accurate fitting formula that describes these wiggles. This is invaluable for carrying out a search of a large parameter space. An interesting question is whether these 'wiggles' survive evolution into the nonlinear regime: Meiksin *et al* (1999) showed that most do not, but that observable signatures of baryons remain on large scales.

2.6.5 The spherical model

An overdense sphere is a very useful nonlinear model, as it behaves in exactly the same way as a closed sub-universe. The density perturbation needs not be a uniform sphere: any spherically symmetric perturbation will clearly evolve at a given radius in the same way as a uniform sphere containing the same amount of mass. In what follows, therefore, density refers to the *mean* density inside a given sphere. The equations of motion are the same as for the scale factor, and we can therefore write down the cycloid solution immediately. For a matter-dominated universe, the relation between the proper radius of the sphere and time is

$$r = A(1 - \cos\theta)$$
$$t = B(\theta - \sin\theta),$$

and $A^3 = GMB^2$, just from $\ddot{r} = -GM/r^2$. Expanding these relations up to order θ^5 gives $r(t)$ for small t:

$$r \simeq \frac{A}{2}\left(\frac{6t}{B}\right)^{2/3}\left[1 - \frac{1}{20}\left(\frac{6t}{B}\right)^{2/3}\right],$$

and we can identify the density perturbation within the sphere:

$$\delta \simeq \frac{3}{20}\left(\frac{6t}{B}\right)^{2/3}.$$

This all agrees with what we knew already: at early times the sphere expands with the $a \propto t^{2/3}$ Hubble flow and density perturbations grow proportional to a.

We can now see how linear theory breaks down as the perturbation evolves. There are three interesting epochs in the final stages of its development, which we can read directly from the above solutions. Here, to keep things simple, we compare only with linear theory for an $\Omega = 1$ background.

(1) *Turnround.* The sphere breaks away from the general expansion and reaches a maximum radius at $\theta = \pi$, $t = \pi B$. At this point, the true density enhancement with respect to the background is just $[A(6t/B)^{2/3}/2]^3/r^3 = 9\pi^2/16 \simeq 5.55$.

(2) *Collapse.* If only gravity operates, then the sphere will collapse to a singularity at $\theta = 2\pi$. This occurs when $\delta_{\mathrm{lin}} = (3/20)(12\pi)^{2/3} \simeq 1.69$.

(3) *Virialization.* Consider the time at which the sphere has collapsed by a factor 2 from maximum expansion. At this point, it has kinetic energy K related to potential energy V by $V = -2K$. This is the condition for equilibrium, according to the *virial theorem*. For this reason, many workers take this epoch as indicating the sort of density contrast to be expected as the endpoint of gravitational collapse. This occurs at $\theta = 3\pi/2$, and the corresponding density enhancement is $(9\pi + 6)^2/8 \simeq 147$, with $\delta_{\mathrm{lin}} \simeq 1.58$. Some authors prefer to assume that this virialized size is eventually achieved only at collapse, in which case the contrast becomes $(6\pi)^2/2 \simeq 178$.

These calculations are the basis for a common 'rule of thumb', whereby one assumes that linear theory applies until δ_{lin} is equal to some δ_c a little greater than unity, at which point virialization is deemed to have occurred. Although this only applies for $\Omega = 1$, analogous results can be worked out from the full $\delta_{\mathrm{lin}}(z, \Omega)$ and $t(z, \Omega)$ relations; $\delta_{\mathrm{lin}} \simeq 1$ is a good criterion for collapse for any value of Ω likely to be of practical relevance. The full density contrast at virialization may be approximated by

$$1 + \delta_{\mathrm{vir}} \simeq 178\Omega^{-0.7}$$

(although flat Λ-dominated models show less dependence on Ω; Eke *et al* 1996).

2.7 Quantifying large-scale structure

The next step is to see how these theoretical ideas can be confronted with statistical measures of the observed matter distribution, and to summarize what is known about the dimensionless density perturbation field

$$\delta(\boldsymbol{x}) \equiv \frac{\rho(\boldsymbol{x}) - \langle \rho \rangle}{\langle \rho \rangle}.$$

A critical feature of the δ field is that it inhabits a universe that is isotropic and homogeneous in its large-scale properties. This suggests that the statistical properties of δ should also be homogeneous, even though it is a field that describes inhomogeneities.

We will often need to use the $\langle \cdots \rangle$ symbol, that denotes averaging over an ensemble of realizations of the statistical δ process. In practice, this will usually be equated to the spatial average over a sufficiently large volume. Fields that satisfy this property, whereby

$$\text{volume average} \leftrightarrow \text{ensemble average}$$

are termed *ergodic*.

2.7.1   Fourier analysis of density fluctuations

It is often convenient to consider building up a general field by the superposition of many modes. For a flat comoving geometry, the natural tool for achieving this is via Fourier analysis. How do we make a Fourier expansion of the density field in an infinite universe? If the field were periodic within some box of side L, then we would just have a sum over wave modes:

$$F(x) = \sum F_k e^{-ik \cdot x}.$$

The requirement of periodicity restricts the allowed wavenumbers to harmonic!boundary conditions

$$k_x = n \frac{2\pi}{L}, \qquad n = 1, 2 \ldots,$$

with similar expressions for k_y and k_z. Now, if we let the box become arbitrarily large, then the sum will go over to an integral that incorporates the density of states in k-space, exactly as in statistical mechanics. The Fourier relations in n dimensions are thus

$$F(x) = \left(\frac{L}{2\pi} \right)^n \int F_k(k) \exp(-ik \cdot x) \, d^n k$$

$$F_k(k) = \left(\frac{1}{L} \right)^n \int F(x) \exp(ik \cdot x) \, d^n x.$$

As an immediate example of the Fourier machinery in action, consider the important quantity

$$\xi(r) \equiv \langle \delta(x)\delta(x + r)\rangle,$$

which is the autocorrelation function of the density field—usually referred to simply as the *correlation function*. The angle brackets indicate an averaging over the normalization volume V. Now express δ as a sum and note that δ is real, so that we can replace one of the two δ's by its complex conjugate, obtaining

$$\xi = \left\langle \sum_k \sum_{k'} \delta_k \delta_{k'}^* e^{i(k'-k) \cdot x} e^{-ik \cdot r} \right\rangle.$$

Alternatively, this sum can be obtained without replacing $\langle \delta\delta \rangle$ by $\langle \delta\delta^* \rangle$, from the relation between modes with opposite wavevectors that holds for any real field: $\delta_k(-k) = \delta_k^*(k)$. Now, by the periodic boundary conditions, all the cross terms with $k' \neq k$ average to zero. Expressing the remaining sum as an integral, we have

$$\xi(r) = \frac{V}{(2\pi)^3} \int |\delta_k|^2 e^{-ik \cdot r} \, d^3k.$$

In short, the correlation function is the Fourier transform of the *power spectrum*. This relation has been obtained by volume averaging, so it applies to the specific mode amplitudes and correlation function measured in any given realization of the density field. Taking ensemble averages of each side, the relation clearly also holds for the ensemble average power and correlations—which are really the quantities that cosmological studies aim to measure. We shall hereafter often use the alternative notation

$$P(k) \equiv \langle |\delta_k|^2 \rangle$$

for the ensemble-average power.

In an isotropic universe, the density perturbation spectrum cannot contain a preferred direction, and so we must have an isotropic power spectrum: $\langle |\delta_k|^2(k) \rangle = |\delta_k|^2(k)$. The angular part of the k-space integral can therefore be performed immediately: introduce spherical polars with the polar axis along k, and use the reality of ξ so that $e^{-ik \cdot x} \rightarrow \cos(kr \cos\theta)$. In three dimensions, this yields

$$\xi(r) = \frac{V}{(2\pi)^3} \int P(k) \frac{\sin kr}{kr} 4\pi k^2 \, dk.$$

We shall usually express the power spectrum in dimensionless form, as the variance per $\ln k$ ($\Delta^2(k) = d\langle \delta^2 \rangle / d \ln k \propto k^3 P[k]$):

$$\Delta^2(k) \equiv \frac{V}{(2\pi)^3} 4\pi k^3 P(k) = \frac{2}{\pi} k^3 \int_0^\infty \xi(r) \frac{\sin kr}{kr} r^2 \, dr.$$

This gives a more easily visualizable meaning to the power spectrum than does the quantity $V P(k)$, which has dimensions of volume: $\Delta^2(k) = 1$ means that there are order-unity density fluctuations from modes in the logarithmic bin around wavenumber k. $\Delta^2(k)$ is therefore the natural choice for a Fourier-space counterpart to the dimensionless quantity $\xi(r)$.

This shows that the power spectrum is a central quantity in cosmology, but how can we predict its functional form? For decades, this was thought to be impossible, and so a minimal set of assumptions was investigated. In the absence of a physical theory, we should not assume that the spectrum contains any preferred length scale, otherwise we should then be compelled to explain this feature. Consequently, the spectrum must be a featureless power law:

$$\langle |\delta_k|^2 \rangle \propto k^n.$$

The index n governs the balance between large-and small-scale power.

A power-law spectrum implies a power-law correlation function. If $\xi(r) = (r/r_0)^{-\gamma}$, with $\gamma = n + 3$, the corresponding 3D power spectrum is

$$\Delta^2(k) = \frac{2}{\pi}(kr_0)^{\gamma}\Gamma(2-\gamma)\sin\frac{(2-\gamma)\pi}{2} \equiv \beta(kr_0)^{\gamma}$$

($= 0.903(kr_0)^{1.8}$ if $\gamma = 1.8$). This expression is only valid for $n < 0$ ($\gamma < 3$); for larger values of n, ξ must become negative at large r (because $P(0)$ must vanish, implying $\int_0^\infty \xi(r)r^2\,dr = 0$). A cut-off in the spectrum at large k is needed to obtain physically sensible results.

Most important of all is the scale-invariant spectrum, which corresponds to the value $n = 1$, i.e. $\Delta^2 \propto k^4$. To see how the name arises, consider a perturbation $\delta\Phi$ in the gravitational potential:

$$\nabla^2\delta\Phi = 4\pi G\rho_0\delta \Rightarrow \delta\Phi_k = -4\pi G\rho_0\delta_k/k^2.$$

The two powers of k pulled down by ∇^2 mean that, if $\Delta^2 \propto k^4$ for the power spectrum of density fluctuations, then Δ_Φ^2 is a constant. Since potential perturbations govern the flatness of spacetime, this says that the scale-invariant spectrum corresponds to a metric that is a *fractal*: spacetime has the same degree of 'wrinkliness' on each resolution scale.

2.7.2 The CDM model

The CDM model is the simplest model for structure formation, and it is worth examining in some detail. The CDM linear-theory spectrum modifications are illustrated in figure 2.9. The primordial power-law spectrum is reduced at large k, by an amount that depends on both the quantity of dark matter and its nature. Generally the bend in the spectrum occurs near $1/k$ of the order of the horizon size at matter–radiation equality, proportional to $(\Omega h^2)^{-1}$. For a pure CDM universe, with scale-invariant initial fluctuations ($n = 1$), the observed spectrum depends only on two parameters. One is the shape $\Gamma = \Omega h$, and the other is a normalization. On the shape front, a government health warning is needed, as follows. It has been quite common to take Γ-based fits to observations as indicating a *measurement* of Ωh, but there are three reasons why this may give incorrect answers:

(1) The dark matter may not be CDM. An admixture of HDM will damp the spectrum more, mimicking a lower CDM density.
(2) Even in a CDM-dominated universe, baryons can have a significant effect, making Γ lower than Ωh.
(3) The strongest (and most-ignored) effect is tilt: if $n \neq 1$, then even in a pure CDM universe a Γ-model fit to the spectrum will give a badly incorrect estimate of the density (the change in Ωh is roughly $0.3(n-1)$; Peacock and Dodds 1994).

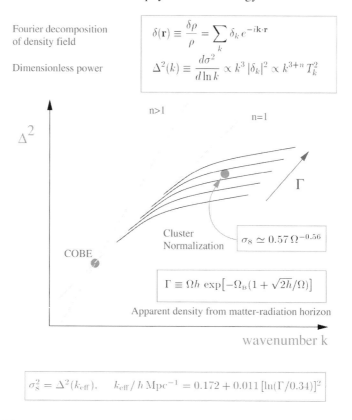

Fourier decomposition of density field

$$\delta(\mathbf{r}) \equiv \frac{\delta\rho}{\rho} = \sum_k \delta_k e^{-i\mathbf{k}\cdot\mathbf{r}}$$

Dimensionless power

$$\Delta^2(k) \equiv \frac{d\sigma^2}{d\ln k} \propto k^3 |\delta_k|^2 \propto k^{3+n} T_k^2$$

n>1

n=1

Δ^2

Γ

Cluster Normalization

$$\sigma_8 \simeq 0.57\,\Omega^{-0.56}$$

COBE

$$\Gamma \equiv \Omega h \, \exp\left[-\Omega_{\mathrm{B}}(1 + \sqrt{2h}/\Omega)\right]$$

Apparent density from matter-radiation horizon

wavenumber k

$$\sigma_8^2 = \Delta^2(k_{\mathrm{eff}}). \qquad k_{\mathrm{eff}}/h\,\mathrm{Mpc}^{-1} = 0.172 + 0.011\,[\ln(\Gamma/0.34)]^2$$

Figure 2.9. This figure illustrates how the primordial power spectrum is modified as a function of density in a CDM model. For a given tilt, it is always possible to choose a density that satisfies both the COBE and cluster normalizations.

The other parameter is the normalization. This can be set at a number of points. The COBE normalization comes from large-angle CMB anisotropies, and is sensitive to the power spectrum at $k \simeq 10^{-3} h$ Mpc^{-1}. The alternative is to set the normalization near the quasilinear scale, using the abundance of rich clusters. Many authors have tried this calculation, and there is good agreement on the answer:

$$\sigma_8 \simeq (0.5 - 0.6)\Omega_{\mathrm{m}}^{-0.6},$$

where σ_8 is the fractional rms variation in the linear density field, when convolved with a sphere of radium $8h^{-1}$ Mpc (White *et al* 1993, Eke *et al* 1996, Viana and Liddle 1996). In many ways, this is the most sensible normalization to use for LSS studies, since it does not rely on an extrapolation from larger scales.

2.7.3   Karhunen–Loève and all that

A key question for these statistical measures is how accurate they are—i.e. how much does the result for a given finite sample depart from the ideal statistic averaged over an infinite universe? Terminology here can be confusing, in that a distinction is sometimes made between *sampling variance* and *cosmic variance*. The former is to be understood as arising from probing a given volume only with a finite number of galaxies (e.g. just the bright ones), so that $\sqrt{N}$ statistics limit our knowledge of the mass distribution within that region. The second term concerns whether we have reached a fair sample of the universe, and depends on whether there is significant power in density perturbation modes with wavelengths larger than the sample depth. Clearly, these two aspects are closely related.

The quantitative analysis of these errors is most simply performed in Fourier space, and was given by Feldman *et al* (1994). The results can be understood most simply by comparison with an idealized complete and uniform survey of a volume L^3, with periodicity scale L. For an infinite survey, the arbitrariness of the spatial origin means that different modes are uncorrelated:

$$\langle \delta_k(\mathbf{k}_i)\delta_k^*(\mathbf{k}_j)\rangle = P(k)\delta_{ij}.$$

Each mode has an exponential distribution in power (because the complex coefficients δ_k are 2D Gaussian-distributed variables on the Argand plane), for which the mean and rms are identical. The fractional uncertainty in the mean power measured over some k-space volume is then just determined by the number of uncorrelated modes averaged over

$$\frac{\delta \bar{P}}{\bar{P}} = \frac{1}{N_{\mathrm{modes}}^{1/2}}; \qquad N_{\mathrm{modes}} = \left(\frac{L}{2\pi}\right)^3 \int d^3k.$$

The only subtlety is that, because the density field is real, modes at k and $-k$ are perfectly correlated. Thus, if the k-space volume is a shell, the effective number of uncorrelated modes is only half this expression.

Analogous results apply for an arbitrary survey selection function. In the continuum limit, the Kroneker delta in the expression for mode correlation would be replaced a term proportional to a delta-function, $\delta[\mathbf{k}_i - \mathbf{k}_j]$. Now, multiplying the infinite ideal survey by a survey window, $\rho(\mathbf{r})$, is equivalent to convolution in the Fourier domain, with the result that the power per mode is correlated over k-space separations of order $1/D$, where D is the survey depth.

Given this expression for the fractional power, it is clear that the precision of the estimate can be manipulated by appropriate weighting of the data: giving increased weight to the most distant galaxies increases the effective survey volume, boosting the number of modes. This sounds too good to be true, and of course it is: the previous expression for the fractional power error applies to the sum of true clustering power and shot noise. The latter arises because we transform a point process. Given a set of N galaxies, we would estimate Fourier coefficients via $\delta_k = (1/N) \sum_i \exp(-i\mathbf{k} \cdot x_i)$. From this, the expectation power is

$$\langle |\delta_k|^2 \rangle = P(k) + 1/N.$$

The existence of an additive discreteness correction is no problem, but the *fluctuations* on the shot noise hide the signal of interest. Introducing weights boosts the shot noise, so there is an optimum choice of weight that minimizes the uncertainty in the power after shot-noise subtraction. Feldman *et al* (1994) showed that this weight is

$$w = (1 + \bar{n} P)^{-1},$$

where $\bar{n}$ is the expected galaxy number density as a function of position in the survey.

Since the correlation of modes arises from the survey selection function, it is clear that weighting the data changes the degree of correlation in k space. Increasing the weight in low-density regions increases the effective survey volume, and so shrinks the k-space coherence scale. However, the coherence scale continues to shrink as distant regions of the survey are given greater weight, whereas the noise goes through a minimum. There is thus a trade-off between the competing desirable criteria of high k-space resolution and low noise. Tegmark (1996) shows how weights may be chosen to implement any given prejudice concerning the relative importance of these two criteria. See also Hamilton (1997a, b) for similar arguments.

Finally, we note that this discussion strictly applies only to the case of Gaussian density fluctuations—which cannot be an accurate model on nonlinear scales. In fact, the errors in the power spectrum are increased on nonlinear scales, and modes at all k have their amplitudes coupled to some degree by nonlinear evolution. These effects are not easy to predict analytically, and are best dealt with by running numerical simulations (see Meiksin and White 1999, Scoccimarro *et al* 1999).

Given these difficulties with correlated results, it is attractive to seek a method where the data can be decomposed into a set of statistics that are completely uncorrelated with each other. Such a method is provided by the Karhunen–Loève formalism. Vogeley and Szalay (1996) argued as follows. Define a column vector of data $\underline{d}$; this can be quite abstract in nature, and could be e.g. the numbers of galaxies in a set of cells, or a set of Fourier components of the transformed galaxy number counts. Similarly, for CMB studies, $\underline{d}$ could be $\delta T / T$ in a set of pixels, or spherical-harmonic coefficients $a_{\ell m}$. We assume that the mean can be identified and subtracted off, so that $\langle \underline{d} \rangle = 0$ in ensemble average. The statistical properties of the data are then described by the covariance matrix

$$C_{ij} \equiv \langle d_i d_j^* \rangle$$

(normally the data will be real, but it is convenient to keep things general and include the complex conjugate).

Suppose we seek to expand the datavector in terms of a set of new orthonormal vectors:

$$\underline{d} = \sum_i a_i \underline{\psi}_i; \quad \underline{\psi}_i^* \cdot \underline{\psi}_j = \delta_{ij}.$$

The expansion coefficients are extracted in the usual way: $a_j = \underline{d} \cdot \underline{\psi}_j^*$. Now require that these coefficients be statistically uncorrelated, $\langle a_i a_j^* \rangle = \lambda_i \delta_{ij}$ (no sum on i). This gives

$$\underline{\psi}_i^* \cdot \langle \underline{d}\,\underline{d}^* \rangle \cdot \underline{\psi}_j = \lambda_i \delta_{ij},$$

where the dyadic $\langle \underline{d}\,\underline{d}^* \rangle$ is $\underline{\underline{C}}$, the correlation matrix of the data vector: $(\underline{d}\,\underline{d}^*)_{ij} \equiv d_i d_j^*$. Now, the effect of operating this matrix on one of the $\underline{\psi}_i$ must be expandable in terms of the complete set, which shows that the $\underline{\psi}_j$ must be the eigenvectors of the correlation matrix:

$$\langle \underline{d}\,\underline{d}^* \rangle \cdot \underline{\psi}_j = \lambda_j \underline{\psi}_j.$$

Vogeley and Szalay (1996) further show that these uncorrelated modes are optimal for representing the data: if the modes are arranged in order of decreasing λ, and the series expansion truncated after n terms, the rms truncation error is minimized for this choice of eigenmodes. To prove this, consider the truncation error

$$\underline{\epsilon} = \underline{d} - \sum_{i=1}^{n} a_i \underline{\psi}_i = \sum_{i=n+1}^{\infty} a_i \underline{\psi}_i.$$

The square of this is

$$\langle \epsilon^2 \rangle = \sum_{i=n+1}^{\infty} \langle |a_i|^2 \rangle,$$

where $\langle |a_i|^2 \rangle = \underline{\psi}_i^* \cdot \underline{\underline{C}} \cdot \underline{\psi}_i$, as before. We want to minimize $\langle \epsilon^2 \rangle$ by varying the $\underline{\psi}_i$, but we need to do this in a way that preserves normalization. This is achieved by introducing a Lagrange multiplier, and minimizing

$$\sum_i \underline{\psi}_i^* \cdot \underline{\underline{C}} \cdot \underline{\psi}_i + \lambda(1 - \underline{\psi}_i^* \cdot \underline{\psi}_i).$$

This is easily solved if we consider the more general problem where $\underline{\psi}_i^*$ and $\underline{\psi}_i$ are independent vectors:

$$\underline{\underline{C}} \cdot \underline{\psi}_i = \lambda \psi_i.$$

In short, the eigenvectors of $\underline{\underline{C}}$ are optimal in a least-squares sense for expanding the data. The process of truncating the expansion is a form of lossy *data compression*, since the size of the data vector can be greatly reduced without significantly affecting the fidelity of the resulting representation of the universe.

The process of diagonalizing the covariance matrix of a set of data also goes by the more familiar name of *principal components analysis* (PCA), so what is the difference between the KL approach and PCA? In the previous discussion, they are identical, but the idea of choosing an optimal eigenbasis is more general than PCA. Consider the case where the covariance matrix can be decomposed into a 'signal' and a 'noise' term:

$$\underline{\underline{C}} = \underline{\underline{S}} + \underline{\underline{N}},$$

where $\underline{\underline{S}}$ depends on cosmological parameters that we might wish to estimate, whereas $\underline{\underline{N}}$ is some fixed property of the experiment under consideration. In the simplest imaginable case, $\underline{\underline{N}}$ might be a diagonal matrix, so PCA diagonalizes both $\underline{\underline{S}}$ and $\underline{\underline{N}}$. In this case, ranking the PCA modes by eigenvalue would correspond to ordering the modes according to signal-to-noise ratio. Data compression by truncating the mode expansion then does the sensible thing: it rejects all modes of low signal-to-noise ratio.

However, in general these matrices will not commute, and there will not be a single set of eigenfunctions that are common to the $\underline{\underline{S}}$ and $\underline{\underline{N}}$ matrices. Normally, this would be taken to mean that it is impossible to find a set of coordinates in which both are diagonal. This conclusion can however be evaded, as follows. When considering the effect of coordinate transformations on vectors and matrices, we are normally forced to consider only rotation-like transformations that preserve the norm of a vector (e.g. in quantum mechanics, so that states stay normalized). Thus, we write $\underline{d}' = \underline{\underline{R}} \cdot \underline{d}$, where $\underline{\underline{R}}$ is unitary, so that $\underline{\underline{R}} \cdot \underline{\underline{R}}^\dagger = \underline{\underline{I}}$. If $\underline{\underline{R}}$ is chosen so that its columns are the eigenvalues of $\underline{\underline{N}}$, then the transformed noise matrix, $\underline{\underline{R}} \cdot \underline{\underline{N}} \cdot \underline{\underline{R}}^\dagger$, is diagonal. Nevertheless, if the transformed $\underline{\underline{S}}$ is not diagonal, the two will not commute. This apparently insuperable problem can be solved by using the fact that the data vectors are entirely abstract at this stage. There is therefore no reason not to consider the further transformation of scaling the data, so that $\underline{\underline{N}}$ becomes proportional to the identity matrix. This means that the transformation is no longer unitary – but there is no physical reason to object to a change in the normalization of the data vectors.

Suppose we therefore make a further transformation

$$\underline{d}'' = \underline{\underline{W}} \cdot \underline{d}'.$$

The matrix $\underline{\underline{W}}$ is related to the rotated noise matrix:

$$\underline{\underline{N}}' = \text{diag}(n_1, n_2, \ldots) \Rightarrow \underline{\underline{W}} = \text{diag}(1/\sqrt{n_1}, 1/\sqrt{n_2}, \ldots).$$

This transformation is termed *prewhitening* by Vogeley and Szalay (1996), since it converts the noise matrix to white noise, in which each pixel has a unit noise that is uncorrelated with other pixels. The effect of this transformation on the full covariance matrix is

$$C_{ij}'' \equiv \langle d_i'' d_j''^* \rangle \Rightarrow \underline{\underline{C}}'' = (\underline{\underline{W}} \cdot \underline{\underline{R}}) \cdot \underline{\underline{C}} \cdot (\underline{\underline{W}} \cdot \underline{\underline{R}})^\dagger.$$

After this transformation, the noise and signal matrices certainly do commute, and the optimal modes for expanding the new data are once again the PCA eigenmodes in the new coordinates:

$$\underline{\underline{C}}'' \cdot \underline{\psi}''_i = \lambda \underline{\psi}''_i.$$

These eigenmodes must be expressible in terms of some modes in the original coordinates, $\underline{e}_i$:

$$\underline{\psi}''_i = (\underline{\underline{W}} \cdot \underline{\underline{R}}) \cdot \underline{e}_i.$$

In these terms, the eigenproblem is

$$(\underline{\underline{W}} \cdot \underline{\underline{R}}) \cdot \underline{\underline{C}} \cdot (\underline{\underline{W}} \cdot \underline{\underline{R}})^\dagger \cdot (\underline{\underline{W}} \cdot \underline{\underline{R}}) \cdot \underline{e}_i = \lambda (\underline{\underline{W}} \cdot \underline{\underline{R}}) \cdot \underline{e}_i.$$

This can be simplified using $\underline{\underline{W}}^\dagger \cdot \underline{\underline{W}} = \underline{\underline{N}}'^{-1}$ and $\underline{\underline{N}}'^{-1} = \underline{\underline{R}} \cdot \underline{\underline{N}}^{-1} \underline{\underline{R}}^\dagger$, to give

$$\underline{\underline{C}} \cdot \underline{\underline{N}}^{-1} \cdot \underline{e}_i = \lambda \underline{e}_i,$$

so the required modes are eigenmodes of $\underline{\underline{C}} \cdot \underline{\underline{N}}^{-1}$. However, care is required when considering the orthonormality of the $\underline{e}_i$: $\underline{\psi}^\dagger_i \cdot \underline{\psi}_j = \underline{e}^\dagger_i \cdot \underline{\underline{N}}^{-1} \cdot \underline{e}_j$, so the $\underline{e}_i$ are not orthonormal. If we write $\underline{d} = \sum_i a_i \underline{e}_i$, then

$$a_i = (\underline{\underline{N}}^{-1} \cdot \underline{e}_i)^\dagger \cdot \underline{d} \equiv \underline{\psi}^\dagger_i \cdot \underline{d}.$$

Thus, the modes used to extract the compressed data by dot product satisfy $\underline{\underline{C}} \cdot \underline{\psi} = \lambda \underline{\underline{N}} \cdot \underline{\psi}$, or finally

$$\underline{\underline{S}} \cdot \underline{\psi} = \lambda \underline{\underline{N}} \cdot \underline{\psi},$$

given a redefinition of λ. The optimal modes are thus eigenmodes of $\underline{\underline{N}}^{-1} \cdot \underline{\underline{S}}$, hence the name *signal-to-noise eigenmodes* (Bond 1995, Bunn 1995).

It is interesting to appreciate that the set of KL modes just discussed is also the 'best' set of modes to choose from a completely different point of view: they are the modes that are optimal for estimation of a parameter via maximum likelihood. Suppose we write the compressed data vector, $\underline{x}$, in terms of a non-square matrix $\underline{\underline{A}}$ (whose rows are the basis vectors $\underline{\psi}^*_i$):

$$\underline{x} = \underline{\underline{A}} \cdot \underline{d}.$$

The transformed covariance matrix is

$$\underline{\underline{D}} \equiv \langle \underline{x} \underline{x}^\dagger \rangle = \underline{\underline{A}} \cdot \underline{\underline{C}} \cdot \underline{\underline{A}}^\dagger.$$

For the case where the original data obeyed Gaussian statistics, this is true for the compressed data also, so the likelihood is

$$-2 \ln \mathcal{L} = \ln \det \underline{\underline{D}} + \underline{x}^* \cdot \underline{\underline{D}}^{-1} \cdot \underline{x} + \text{constant}.$$

The normal variance on some parameter p (on which the covariance matrix depends) is

$$\frac{1}{\sigma_p^2} = \frac{d^2[-2 \ln \mathcal{L}]}{dq^2}.$$

Without data, we do not know this, so it is common to use the expectation value of the right-hand side as an estimate (recently, there has been a tendency to dub this the 'Fisher matrix').

We desire to optimize σ_p by an appropriate choice of data-compression vectors, $\underline{\psi}_i$. By writing σ_p in terms of $\underline{\underline{A}}$, $\underline{\underline{C}}$ and $\underline{d}$, it may eventually be shown that the desired optimal modes satisfy

$$\left(\frac{\mathrm{d}}{\mathrm{d}p}\underline{\underline{C}}\right) \cdot \underline{\psi} = \lambda\,\underline{\underline{C}} \cdot \underline{\psi}.$$

For the case where the parameter of interest is the cosmological power, the matrix on the left-hand side is just proportional to $\underline{\underline{S}}$, so we have to solve the eigenproblem

$$\underline{\underline{S}} \cdot \underline{\psi} = \lambda\underline{\underline{C}} \cdot \underline{\psi}.$$

With a redefinition of λ, this becomes

$$\underline{\underline{S}} \cdot \underline{\psi} = \lambda\underline{\underline{N}} \cdot \underline{\psi}.$$

The optimal modes for parameter estimation in the linear case are thus identical to the PCA modes of the prewhitened data discussed earlier. The more general expression was given by Tegmark *et al* (1997), and it is only in this case, where the covariance matrix is not necessarily linear in the parameter of interest, that the KL method actually differs from PCA.

The reason for going to all this trouble is that the likelihood can now be evaluated much more rapidly, using the compressed data. This allows extensive model searches over large parameter spaces that would be infeasible with the original data (since inversion of an $N \times N$ covariance matrix takes a time proportional to N^3). Note, however, that the price paid for this efficiency is that a different set of modes need to be chosen depending on the model of interest, and that these modes will not in general be optimal for expanding the dataset itself. Nevertheless, it may be expected that application of these methods will inevitably grow as datasets increase in size. Present applications mainly prove that the techniques work: see Matsubara *et al* (2000) for application to the LCRS (Las Campanas Redshift Survey), or Padmanabhan *et al* (1999) for the UZC (Updated Zwicky Catalog) survey. The next generation of experiments will probably be forced to resort to data compression of this sort, rather than using it as an elegant alternative method of analysis.

2.7.4   Projection on the sky

A more common situation is where we lack any distance data; we then deal with a projection on the sky of a magnitude-limited set of galaxies at different depths. The statistic that is observable is the angular correlation function, $w(\theta)$, or its angular power spectrum Δ_θ^2. If the sky were flat, the relation between these would be the usual *Hankel transform* pair:

$$w(\theta) = \int_0^\infty \Delta_\theta^2 J_0(K\theta)\, dK/K,$$

$$\Delta_\theta^2 = K^2 \int_0^\infty w(\theta) J_0(K\theta)\theta \, d\theta.$$

For power-law clustering, $w(\theta) = (\theta/\theta_0)^{-\epsilon}$, this gives

$$\Delta_\theta^2(K) = (K\theta_0)^\epsilon 2^{1-\epsilon} \frac{\Gamma(1-\epsilon/2)}{\Gamma(\epsilon/2)},$$

which is equal to $0.77(K\theta_0)^\epsilon$ for $\epsilon = 0.8$. At large angles, these relations are not quite correct. We should really expand the sky distribution in *spherical harmonics*:

$$\delta(\hat{q}) = \sum a_\ell^m Y_{\ell m}(\hat{q}),$$

where $\hat{q}$ is a unit vector that specifies direction on the sky. The functions $Y_{\ell m}$ are the eigenfunctions of the angular part of the ∇^2 operator: $Y_{\ell m}(\theta, \phi) \propto \exp(im\phi) P_\ell^m(\cos\theta)$, where P_ℓ^m are the *associated Legendre polynomials* (see e.g. section 6.8 of Press *et al* 1992). Since the spherical harmonics satisfy the orthonormality relation $\int Y_{\ell m} Y_{\ell' m'}^* \, d^2q = \delta_{\ell\ell'}\delta_{mm'}$, the inverse relation is

$$a_\ell^m = \int \delta(\hat{q}) Y_{\ell m}^* \, d^2q.$$

The analogues of the Fourier relations for the correlation function and power spectrum are

$$w(\theta) = \frac{1}{4\pi} \sum_\ell \sum_{m=-\ell}^{m=+\ell} |a_\ell^m|^2 P_\ell(\cos\theta)$$

$$|a_\ell^m|^2 = 2\pi \int_{-1}^1 w(\theta) P_\ell(\cos\theta) \, d\cos\theta.$$

For small θ and large ℓ, these go over to a form that looks like a flat sky, as follows. Consider the asymptotic forms for the Legendre polynomials and the J_0 Bessel function:

$$P_\ell(\cos\theta) \simeq \sqrt{\frac{2}{\pi\ell\sin\theta}} \cos\left[\left(\ell+\frac{1}{2}\right)\theta - \frac{1}{4}\pi\right]$$

$$J_0(z) \simeq \sqrt{\frac{2}{\pi z}} \cos\left[z - \frac{1}{4}\pi\right],$$

for respectively $\ell \to \infty$, $z \to \infty$; see chapters 8 and 9 of Abramowitz and Stegun (1965). This shows that, for $\ell \gg 1$, we can approximate the small-angle correlation function in the usual way in terms of an angular power spectrum Δ_θ^2 and angular wavenumber K:

$$w(\theta) = \int_0^\infty \Delta_\theta^2(K) J_0(K\theta) \frac{dK}{K}$$

$$\Delta_\theta^2(K = \ell + \tfrac{1}{2}) = \frac{2\ell+1}{8\pi} \sum_m |a_\ell^m|^2.$$

An important relation is that between the angular and spatial power spectra. In outline, this is derived as follows. The perturbation seen on the sky is

$$\delta(\hat{q}) = \int_0^\infty \delta(y) y^2 \phi(y) \, dy,$$

where $\phi(y)$ is the *selection function*, normalized such that $\int y^2 \phi(y) \, dy = 1$, and y is comoving distance. The function ϕ is the comoving density of objects in the survey, which is given by the integrated luminosity function down to the luminosity limit corresponding to the limiting flux of the survey seen at different redshifts; a flat universe ($\Omega = 1$) is assumed for now. Now write down the Fourier expansion of δ. The plane waves may be related to spherical harmonics via the expansion of a plane wave in *spherical Bessel functions* $j_\ell(x) = (\pi/2x)^{1/2} J_{n+1/2}(x)$ (see chapter 10 of Abramowitz and Stegun (1965) or section 6.7 of Press *et al* (1992)):

$$e^{ikr\cos\theta} = \sum_0^\infty (2\ell+1) i^\ell P_\ell(\cos\theta) j_\ell(kr),$$

plus the spherical harmonic addition theorem

$$P_\ell(\cos\theta) = \frac{4\pi}{2\ell+1} \sum_{m=-\ell}^{m=+\ell} Y_{\ell m}^*(\hat{q}) Y_{\ell m}(\hat{q}'); \qquad \hat{q} \cdot \hat{q}' = \cos\theta.$$

These relations allow us to take the angular correlation function $w(\theta) = \langle \delta(\hat{q}) \delta(\hat{q}') \rangle$ and transform it to give the angular power spectrum coefficients. The actual manipulations involved are not as intimidating as they may appear, but they are left as an exercise and we simply quote the final result (Peebles 1973):

$$\langle |a_\ell^m|^2 \rangle = 4\pi \int \Delta^2(k) \frac{dk}{k} \left[\int y^2 \phi(y) j_\ell(ky) \, dy \right]^2.$$

What is the analogue of this formula for small angles? Rather than manipulating large-ℓ Bessel functions, it is easier to start again from the correlation function. By writing as before the overdensity observed at a particular direction on the sky as a radial integral over the spatial overdensity, with a weighting of $y^2\phi(y)$, we see that the angular correlation function is

$$\langle\delta(\hat{q}_1)\delta(\hat{q}_2)\rangle = \int\int \langle\delta(y_1)\delta(y_2)\rangle y_1^2 y_2^2 \phi(y_1)\phi(y_2)\,dy_1\,dy_2.$$

We now change variables to the mean and difference of the radii, $y \equiv (y_1+y_2)/2$; $x \equiv (y_1 - y_2)$. If the depth of the survey is larger than any correlation length, we only get a signal when $y_1 \simeq y_2 \simeq y$. If the selection function is a slowly varying function, so that the thickness of the shell being observed is also of order of the depth, the integration range on x may be taken as being infinite. For small angles, we then obtain *Limber's equation*:

$$w(\theta) = \int_0^\infty y^4\phi^2\,dy \int_{-\infty}^\infty \xi\left(\sqrt{x^2 + y^2\theta^2}\right)dx$$

(see sections 51 and 56 of Peebles 1980). Theory usually supplies a prediction about the linear density field in the form of the power spectrum, and so it is convenient to recast Limber's equation:

$$w(\theta) = \int_0^\infty y^4\phi^2\,dy \int_0^\infty \pi\Delta^2(k)J_0(ky\theta)\,dk/k^2.$$

This power-spectrum version of Limber's equation is already in the form required for the relation to the angular power spectrum, and so we obtain the direct small-angle relation between spatial and angular power spectra:

$$\Delta_\theta^2 = \frac{\pi}{K}\int \Delta^2(K/y)y^5\phi^2(y)\,dy.$$

This is just a convolution in log space, and is considerably simpler to evaluate and interpret than the w–ξ version of Limber's equation.

Finally, note that it is not difficult to make allowance for spatial curvature in this discussion. Write the RW metric in the form

$$c^2\,d\tau^2 = c^2\,dt^2 - R^2\left[\frac{dr^2}{1 - kr^2} + r^2\theta^2\right];$$

for $k = 0$, the notation $y = R_0 r$ was used for comoving distance, where $R_0 = (c/H_0)|1 - \Omega|^{-1/2}$. The radial increment of comoving distance was $dx = R_0\,dr$, and the comoving distance between two objects was $(dx^2+y^2\theta^2)^{1/2}$. To maintain this version of Pythagoras's theorem, we clearly need to keep the definition of y and redefine radial distance: $dx = R_0\,dr\,C(y)$, where $C(y) = [1 - k(y/R_0)^2]^{-1/2}$. The factor $C(y)$ appears in the non-Euclidean comoving volume element, $dV \propto y^2 C(y)\,dy$, so that we now require the normalization equation for ϕ to be

$$\int_0^\infty y^2\phi(y)C(y)\,dy = 1.$$

The full version of Limber's equation therefore gains two powers of $C(y)$, but one of these is lost in converting between $R_0 \, dr$ and dx:

$$w(\theta) = \int_0^\infty [C(y)]^2 y^4 \phi^2 \, dy \int_{-\infty}^\infty \xi \left(\sqrt{x^2 + y^2\theta^2} \right) \frac{dx}{C(y)}.$$

The net effect is therefore to replace $\phi^2(y)$ by $C(y)\phi^2(y)$, so that the full power-spectrum equation is

$$\Delta_\theta^2 = \frac{\pi}{K} \int \Delta^2(K/y) C(y) y^5 \phi^2(y) \, dy.$$

It is also straightforward to allow for evolution. The power version of Limber's equation is really just telling us that the angular power from a number of different radial shells adds incoherently, so we just need to use the actual evolved power at that redshift. These integral equations can be inverted numerically to obtain the real-space 3D clustering results from observations of 2D clustering; see Baugh and Efstathiou (1993, 1994).

2.7.5 Nonlinear clustering: a problem for CDM?

Observations of galaxy clustering extend into the highly nonlinear regime, $\xi \lesssim 10^4$, so it is essential to understand how this nonlinear clustering relates to the linear-theory initial conditions. A useful trick for dealing with this problem is to think of the density field under full nonlinear evolution as consisting of a set of collapsed, virialized clusters. What is the density profile of one of these objects? At least at separations smaller than the clump separation, the density profile of the clusters is directly related to the correlation function, since this just measures the number density of neighbours of a given galaxy. For a very steep cluster profile, $\rho \propto r^{-\epsilon}$, most galaxies will lie near the centres of clusters, and the correlation function will be a power law, $\xi(r) \propto r^{-\gamma}$, with $\gamma = \epsilon$. In general, because the correlation function is the convolution of the density field with itself, the two slopes differ. In the limit that clusters do not overlap, the relation is $\gamma = 2\epsilon - 3$ (for $3/2 < \epsilon < 3$; see Peebles 1974 or McClelland and Silk 1977). In any case, the critical point is that the correlation function may be be thought of as arising directly from the density profiles of clumps in the density field.

In this picture, it is easy to see how ξ will evolve with redshift, since clusters are virialized objects that do not expand. The hypothesis of *stable clustering* states that, although the separation of clusters will alter as the universe expands, their internal density structure will stay constant with time. This hypothesis clearly breaks down in the outer regions of clusters, where the density contrast is small and linear theory applies, but it should be applicable to small-scale clustering. Regarding ξ as a density profile, its small-scale shape should therefore be fixed in *proper* coordinates, and its amplitude should scale as $(1 + z)^{-3}$ owing to the changing mean density of unclustered galaxies, which dilute the clustering at high redshift. Thus, with $\xi \propto r^{-\gamma}$, we obtain the comoving evolution

$$\xi(r, z) \propto (1+z)^{\gamma-3} \qquad \text{(nonlinear)}.$$

Since the observed $\gamma \simeq 1.8$, this implies slower evolution than is expected in the linear regime:

$$\xi(r, z) \propto (1+z)^{-2}g(\Omega) \qquad \text{(linear)}.$$

This argument does not so far give a relation between the nonlinear slope γ and the index n of the linear spectrum. However, the linear and nonlinear regimes match at the scale of quasilinearity, i.e. $\xi(r_0) = 1$; each regime must make the same prediction for how this break point evolves. The linear and nonlinear predictions for the evolution of r_0 are, respectively, $r_0 \propto (1+z)^{-2/(n+3)}$ and $r_0 \propto (1+z)^{-(3-\gamma)/\gamma}$, so that $\gamma = (3n+9)/(n+5)$. In terms of an effective index $\gamma = 3 + n_{\text{NL}}$, this becomes

$$n_{\text{NL}} = -\frac{6}{5+n}.$$

The power spectrum resulting from power-law initial conditions will evolve self-similarly with this index. Note the narrow range predicted: $-2 < n_{\text{NL}} < -1$ for $-2 < n < +1$, with an $n = -2$ spectrum having the same shape in both linear and nonlinear regimes.

For many years it was thought that only these limiting cases of extreme linearity or nonlinearity could be dealt with analytically, but in a marvelous piece of alchemy, Hamilton *et al* (1991; HKLM) suggested a general way of understanding the linear ↔ nonlinear mapping. This initial idea was extended into a workable practical scheme by Peacock and Dodds (1996), allowing the effects of nonlinear evolution to be calculated to a few per cent accuracy for a wide class of spectra.

Indications from the angular clustering of faint galaxies (Efstathiou *et al* 1991) and directly from redshift surveys (Le Fèvre *et al* 1996) are that the observed clustering of galaxies evolves at about the linear-theory rate for $z \lesssim 0.5$, rather more rapidly than the scaling solution would indicate. However, any interpretation of such data needs to assume that galaxies are unbiased tracers of the mass, whereas the observed high amplitude of clustering of quasars at $z \simeq 1$ ($r_0 \simeq 7h^{-1}$ Mpc; see Shanks *et al* 1987, Shanks and Boyle 1994) were an early warning that some high-redshift objects had clustering that is apparently not due to gravity alone. When it eventually became possible to measure correlations of normal galaxies at $z \gtrsim 1$ directly, a similar effect was found, with the comoving strength of clustering being comparable to its value at $z = 0$ (e.g. Adelberger *et al* 1998, Carlberg *et al* 2000). This presumably states that the increasing degree of bias due to high-redshift galaxies being rare objects swamps the gravitational evolution of density fluctuations.

A number of authors have pointed out that the detailed spectral shape inferred from galaxy data appears to be inconsistent with that of nonlinear evolution from CDM initial conditions. (e.g. Efstathiou *et al* 1990, Klypin *et al* 1996, Peacock 1997). Perhaps the most detailed work was carried out by the Virgo consortium, who carried out $N = 256^3$ simulations of a number of CDM models (Jenkins *et al* 1998). Their results are shown in figure 2.10, which gives the nonlinear power spectrum at various times (cluster normalization is chosen for $z = 0$) and contrasts this with the APM data. The lower small panels are the scale-dependent bias that would be required if the model did, in fact, describe the real universe, defined as

$$b(k) \equiv \left(\frac{\Delta^2_{\text{gals}}(k)}{\Delta^2_{\text{mass}}} \right)^{1/2}.$$

In all cases, the required bias is non-monotonic; it rises at $k \gtrsim 5h^{-1}$ Mpc, but also displays a bump around $k \simeq 0.1h^{-1}$ Mpc. If real, this feature seems impossible to understand as a genuine feature of the mass power spectrum; certainly, it is not at a scale where the effects of even a large baryon fraction would be expected to act (Eisenstein *et al* 1998, Meiksin *et al* 1999).

2.7.6 Real-space and redshift-space clustering

Peculiar velocity fields are responsible for the distortion of the clustering pattern in redshift space, as first clearly articulated by Kaiser (1987). For a survey that subtends a small angle (i.e. in the *distant-observer approximation*), a good approximation to the anisotropic redshift-space Fourier spectrum is given by the Kaiser function together with a damping term from nonlinear effects:

$$\delta^s_k = \delta^r_k (1 + \beta \mu^2) D(k\sigma \mu),$$

where $\beta = \Omega_m^{0.6}/b$, b being the linear bias parameter of the galaxies under study, and $\mu = \hat{\boldsymbol{k}} \cdot \hat{\boldsymbol{r}}$. For an exponential distribution of relative small-scale peculiar velocities (as seen empirically), the damping function is $D(y) \simeq (1 + y^2/2)^{-1/2}$, and $\sigma \simeq 400$ km s^{-1} is a reasonable estimate for the pairwise velocity dispersion of galaxies (e.g. Ballinger *et al* 1996).

In principle, this distortion should be a robust way to determine Ω (or at least β). In practice, the effect has not been easy to see with past datasets. This is mainly a question of depth: a large survey is needed in order to beat down the shot noise, but this tends to favour bright spectroscopic limits. This limits the result both because relatively few modes in the linear regime are sampled, and also because local survey volumes will tend to violate the small-angle approximation. Strauss and Willick (1995) and Hamilton (1998) review the practical application of redshift-space distortions. In the next section, preliminary results are presented from the 2dF Galaxy Redshift Survey, which shows the distortion effect clearly for the first time.

Peculiar velocities may be dealt with by using the correlation function evaluated explicitly as a 2D function of transverse ($r_\perp$) and radial ($r_\parallel$) separation.

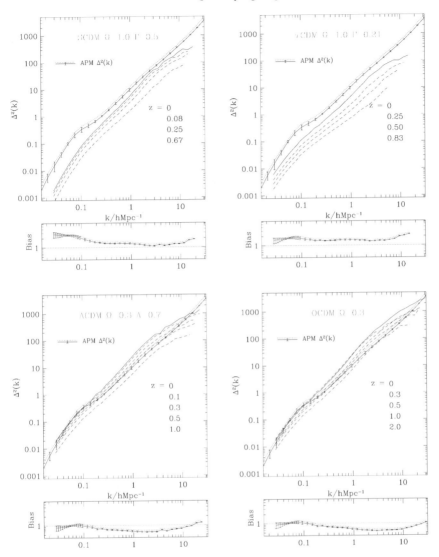

Figure 2.10. The nonlinear evolution of various CDM power spectra, as determined by the Virgo consortium (Jenkins *et al* 1998). The broken curves show the evolving spectra for the mass, which at no time match the shape of the APM data. This is expressed in the lower small panels as a scale-dependent bias at $z = 0$: $b^2(k) = P_{\text{APM}}/P_{\text{mass}}$.

Integrating along the redshift axis then gives the *projected correlation function*, which is independent of the velocities

$$w_{\text{p}}(r_\perp) \equiv \int_{-\infty}^{\infty} \xi(r_\perp, r_\parallel)\, dr_\parallel = 2 \int_{r_\perp}^{\infty} \xi(r) \frac{r\, dr}{(r^2 - r_\perp^2)^{1/2}}.$$

In principle, this statistic can be used to recover the real-space correlation function by using the inverse relation for the *Abel integral equation*:

$$\xi(r) = -\frac{1}{\pi} \int_r^{\infty} w_p'(y) \frac{\mathrm{d}y}{(y^2 - r^2)^{1/2}}.$$

An alternative notation for the projected correlation function is $\Xi(r_\perp)$ (Saunders *et al* 1992). Note that the projected correlation function is not dimensionless, but has dimensions of length. The quantity $\Xi(r_\perp)/r_\perp$ is more convenient to use in practice as the projected analogue of $\xi(r)$.

2.7.7   The state of the art in LSS

We now consider the confrontation of some of these tools with observations. In the past few years, much attention has been attracted by the estimate of the galaxy power spectrum from the automatic plate measuring (APM) survey (Baugh and Efstathiou 1993, 1994, Maddox *et al* 1996). The APM result was generated from a catalogue of $\sim 10^6$ galaxies derived from UK Schmidt Telescope photographic plates scanned with the Cambridge APM machine; because it is based on a *deprojection* of angular clustering, it is immune to the complicating effects of redshift-space distortions. The difficulty, of course, is in ensuring that any low-level systematics from e.g. spatial variations in magnitude zero point are sufficiently well controlled that they do not mask the cosmological signal, which is of order $w(\theta) \lesssim 0.01$ at separations of a few degrees.

   The best evidence that the APM survey has the desired uniformity is the *scaling test*, where the correlations in fainter magnitude slices are expected to move to smaller scales and be reduced in amplitude. If we increase the depth of the survey by some factor D, the new angular correlation function will be

$$w'(\theta) = \frac{1}{D} w(D\theta).$$

The APM survey passes this test well; once the overall redshift distribution is known, it is possible to obtain the spatial power spectrum by inverting a convolution integral:

$$w(\theta) = \int_0^{\infty} y^4 \phi^2 \, \mathrm{d}y \int_0^{\infty} \pi \Delta^2(k) J_0(ky\theta) \, \mathrm{d}k / k^2$$

(where zero spatial curvature is assumed). Here, $\phi(y)$ is the comoving density at comoving distance y, normalized so that $\int y^2 \phi(y) \, \mathrm{d}y = 1$.

   This integral was inverted numerically by Baugh and Efstathiou (1993), and gives an impressively accurate determination of the power spectrum. The error estimates are derived empirically from the scatter between independent regions of the sky, and so should be realistic. If there are no undetected systematics, these error bars state that the power is very accurately determined. The APM result has been investigated in detail by a number of authors (e.g. Gaztañaga and Baugh 1998, Eisenstein and Zaldarriaga 1999) and found to be robust; this has significant implications if true.

Because of the sheer number of galaxies, plus the large volume surveyed, the APM survey outperforms redshift surveys of the past, at least for the purpose of determining the power spectrum. The largest surveys of recent years (CfA: Huchra *et al* 1990, LCRS: Shectman *et al* 1996, PSCz: Saunders *et al* 2000) contain of the order of 10^4 galaxy redshifts, and their statistical errors are considerably larger than those of the APM. On the other hand, it is of great importance to compare the results of deprojection with clustering measured directly in 3D.

This comparison was carried out by Peacock and Dodds (1994; PD94). The exercise is not straightforward, because the 3D results are affected by redshift-space distortions; also, different galaxy tracers can be biased to different extents. The approach taken was to use each dataset to reconstruct an estimate of the linear spectrum, allowing the relative bias factors to float in order to make these estimates agree as well as possible (figure 2.11). To within a scatter of perhaps a factor 1.5 in power, the results were consistent with a $\Gamma \simeq 0.25$ CDM model. Even though the subsequent sections will discuss some possible disagreements with the CDM models at a higher level of precision, the general existence of CDM-like curvature in the spectrum is likely to be an important clue to the nature of the dark matter.

An important general lesson can be drawn from the lack of large-amplitude features in the power spectrum. This is a strong indication that collisionless matter is deeply implicated in forming large-scale structure. Purely baryonic models contain large bumps in the power spectrum around the Jeans' length prior to recombination ($k \sim 0.03\Omega h^2$ Mpc^{-1}), whether the initial conditions are isocurvature or adiabatic. It is hard to see how such features can be reconciled with the data, beyond a 'visibility' in the region of 20%.

The proper resolution of many of the observational questions regarding the large-scale distribution of galaxies requires new generations of redshift survey that push beyond the $N = 10^5$ barrier. Two groups are pursuing this goal. The Sloan survey (e.g. Margon 1999) is using a dedicated 2.5-m telescope to measure redshifts for approximately 700 000 galaxies to $r = 18.2$ in the North Galactic Cap. The 2dF Galaxy Redshift Survey (e.g. Colless 1999) is using a fraction of the time on the 3.9-m Anglo-Australian Telescope plus Two-Degree Field spectrograph to measure 250 000 galaxies from the APM survey to $B_J = 19.45$ in the South Galactic Cap. At the time of writing, the Sloan spectroscopic survey has yet to commence. However, the 2dFGRS project has measured in excess of 100 000 redshifts, and some preliminary clustering results are given here. For more details of the survey, particularly the team members whose hard work has made all this possible, see http://www.mso.anu.edu.au/2dFGRS/.

One of the advantages of 2dFGRS is that it is a fully sampled survey, so that the space density out to the depth imposed by the magnitude limit

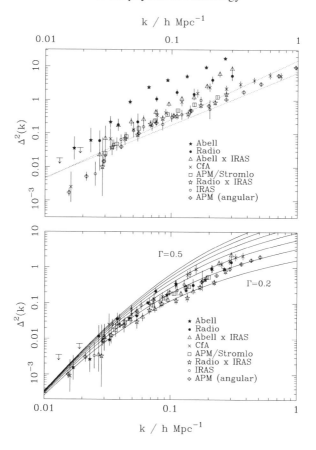

Figure 2.11. The PD94 compilation of power-spectrum measurements. The upper panel shows raw power measurements; the lower shows these data corrected for relative bias, nonlinear effects and redshift-space effects.

(median $z = 0.12$) is as high as nature allows: apart from a tail of low surface brightness galaxies (inevitably omitted from any spectroscopic survey), the 2dFGRS measure all the galaxies that exist over a cosmologically representative volume. It is the first to achieve this goal. The fidelity of the resulting map of the galaxy distribution can be seen in figure 2.12, which shows a small subset of the data: a slice of thickness 4 degrees, centred at declination $-27°$.

An issue with using the 2dFGRS data in their current form is that the sky has to be divided into circular 'tiles' each two degrees in diameter ('2dF' = 'two-degree field', within which the AAT is able to measure 400 spectra simultaneously; see http://www.aao.gov.au/2df/ for details of the instrument). The tiles are positioned adaptively, so that larger overlaps occur in regions of high galaxy density. In this way, it is possible to place a fibre on >95% of all galaxies.

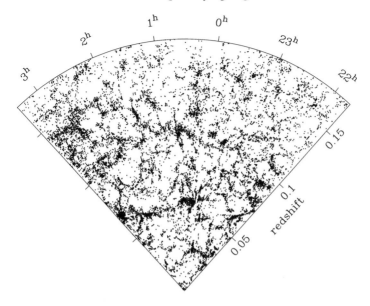

Figure 2.12. A four-degree thick slice of the southern strip of the 2dF Galaxy Redshift Survey. This restricted region alone contains 16 419 galaxies.

However, while the survey is in progress, there exist parts of the sky where the overlapping tiles have not yet been observed, and so the effective sampling fraction is only $\simeq 50\%$. These effects can be allowed for in two different ways. In clustering analyses, we compare the counts of pairs (or n-tuplets) of galaxies in the data to the corresponding counts involving an unclustered random catalogue. The effects of variable sampling can therefore be dealt with either by making the density of random points fluctuate according to the sampling, or by weighting observed galaxies by the reciprocal of the sampling factor for the zone in which they lie. The former approach is better from the point of view of shot noise, but the latter may be safer if there is any suspicion that the sampling fluctuations are correlated with real structure on the sky. In practice, both strategies give identical answers for the results below.

At the two-point level, the most direct quantity to compute is the *redshift–space correlation function*. This is an anisotropic function of the orientation of a galaxy pair, owing to peculiar velocities. We therefore evaluate ξ as a function of 2D separation in terms of coordinates both parallel and perpendicular to the line of sight. If the comoving radii of two galaxies are y_1 and y_2 and their total separation is r, then we define coordinates

$$\pi \equiv |y_1 - y_2|; \qquad \sigma = \sqrt{r^2 - \pi^2}.$$

The correlation function measured in these coordinates is shown in figure 2.13. In

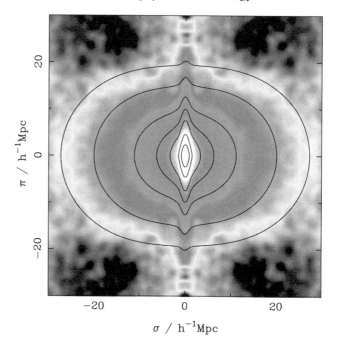

Figure 2.13. The redshift–space correlation function for the 2dFGRS, $\xi(\sigma, \pi)$, plotted as a function of transverse (σ) and radial (π) pair separation. The function was estimated by counting pairs in boxes of side $0.2h^{-1}$ Mpc, and then smoothing with a Gaussian of rms width $0.5h^{-1}$ Mpc. This plot clearly displays redshift distortions, with 'fingers of God' elongations at small scales and the coherent Kaiser flattening at large radii. The overplotted contours show model predictions with flattening parameter $\beta \equiv \Omega^{0.6}/b = 0.4$ and a pairwise dispersion of $\sigma_p = 4h^{-1}$ Mpc. Contours are plotted at $\xi = 10, 5, 2, 1, 0.5, 0.2, 0.1$.

evaluating $\xi(\sigma, \pi)$, the optimal radial weight discussed earlier has been applied, so that the noise at large r should be representative of true cosmic scatter.

The 2dFGRS results for the redshift-space correlation function results are shown in figure 2.13, and display very clearly the two signatures of redshift-space distortions discussed earlier. The *fingers of God* from small-scale random velocities are very clear, as indeed has been the case from the first redshift surveys (e.g. Davis and Peebles 1983). However, this is arguably the first time that the large-scale flattening from coherent infall has been really obvious in the data.

A good way to quantify the flattening is to analyse the clustering as a function of angle into Legendre polynomials:

$$\xi_\ell(r) = \frac{2\ell + 1}{2} \int_{-1}^{1} \xi(\sigma = r \sin\theta, \pi = r \cos\theta) P_\ell(\cos\theta) \, \mathrm{d}\cos\theta.$$

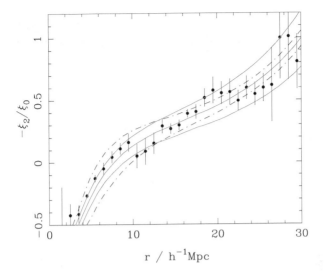

Figure 2.14. The flattening of the redshift–space correlation function is quantified by the quadrupole-to-monopole ratio, ξ_2/ξ_0. This quantity is positive where fingers-of-God distortion dominates, and is negative where coherent infall dominates. The full curves show model predictions for $\beta = 0.3, 0.4$ and 0.5, with $\sigma_p = 4h^{-1}$ Mpc (full), plus $\beta = 0.4$ with $\sigma_p = 3, 4$ and $5h^{-1}$ Mpc (chain). At large radii, the effects of fingers-of-God are becoming relatively small, and values of $\beta \simeq 0.4$ are clearly appropriate.

The quadrupole-to-monopole ratio should be a clear indicator of coherent infall. In linear theory, it is given by

$$\frac{\xi_2}{\xi_0} = f(n)\frac{4\beta/3 + 4\beta^2/7}{1 + 2\beta/3 + \beta^2/5},$$

where $f(n) = (3+n)/n$ (Hamilton 1992). On small and intermediate scales, the effective spectral index is negative, so the quadrupole-to-monopole ratio should be negative, as observed.

However, it is clear that the results on the largest scales are still significantly affected by finger-of-God smearing. The best way to interpret the observed effects is to calculate the same quantities for a model. To achieve this, we use the observed APM 3D power spectrum, plus the distortion model discussed earlier. This gives the plots shown in figure 2.14. The free parameter is β, and this has a best-fit value close to 0.4, approximately consistent with other arguments for a universe with $\Omega = 0.3$ and a small degree of large-scale galaxy bias.

2.7.8 Galaxy formation and biased clustering

We now come to the difficult question of the relation between the galaxy distribution and the large-scale density field. The formation of galaxies must be a non-local process to some extent, and the modern paradigm was introduced by White and Rees (1978): galaxies form through the cooling of baryonic material in virialized halos of dark matter. The virial radii of these systems are in excess of 0.1 Mpc, so there is the potential for large differences in the correlation properties of galaxies and dark matter on these scales.

A number of studies have indicated that the observed galaxy correlations may indeed be reproduced by CDM models. The most direct approach is a numerical simulation that includes gas, and relevant dissipative processes. This is challenging, but just starting to be feasible with current computing power Pearce *et al* 1999). The alternative is 'semi-analytic' modelling, in which the merging history of dark-matter halos is treated via the extended Press–Schechter theory (Bond *et al* 1991), and the location of galaxies within halos is estimated using dynamical-friction arguments (e.g. Kauffmann *et al* 1993, 1999, Cole *et al* 1994, Somerville and Primack 1999, van Kampen *et al* 1999, Benson *et al* 2000a, b). Both these approaches have yielded similar conclusions, and shown how CDM models can match the galaxy data: specifically, the low-density flat ΛCDM model that is favoured on other grounds can yield a correlation function that is close to a single power law over $1000 \gtrsim \xi \gtrsim 1$, even though the mass correlations show a marked curvature over this range (Pearce *et al* 1999, Benson *et al* 2000a; see figure 2.15). These results are impressive, yet it is frustrating to have a result of such fundamental importance emerge from a complicated calculational apparatus. There is thus some motivation for constructing a simpler heuristic model that captures the main processes at work in the full semi-analytic models. The following section describes an approach of this sort (Peacock and Smith 2000; see also Seljak 2000).

An early model for galaxy clustering was suggested by Neyman *et al* (1953), in which the nonlinear density field was taken to be a superposition of randomly placed clumps. With our present knowledge about the evolution of CDM universes, we can make this idealized model considerably more realistic: hierarchical models are expected to contain a distribution of masses of clumps, which have density profiles that are more complicated than isothermal spheres. These issues are well studied in N-body simulations, and highly accurate fitting formulae exist, both for the mass function and for the density profiles. Briefly, we use the mass function of Sheth and Tormen (1999; ST) and the halo profiles of Moore *et al* (1999; M99).

$$f(\nu) = 0.216\,17[1 + (\sqrt{2}/\nu^2)^{0.3}]\exp[-\nu^2/(2\sqrt{2})]$$
$$\Rightarrow F(> \nu) = 0.322\,18[1 - \mathrm{erf}(\nu/2^{3/4})]$$
$$+ 0.147\,65\Gamma[0.2, \nu^2/(2\sqrt{2})],$$

where Γ is the incomplete gamma function.

Recently, it has been claimed by Moore *et al* (1999; M99) that the commonly adopted density profile of Navarro *et al* (1996; NFW) is in error at small r. M99

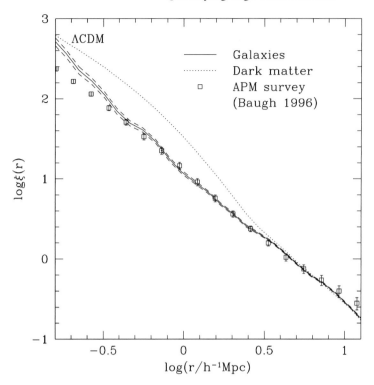

Figure 2.15. The correlation function of galaxies in the semi-analytical simulation of an LCDM universe by Benson *et al* (2000a).

proposed the alternative form

$$\rho/\rho_b = \frac{\Delta_c}{y^{3/2}(1 + y^{3/2})} \qquad (r < r_{\mathrm{vir}}); \qquad y \equiv r/r_c.$$

Using this model, it is then possible to calculate the correlations of the nonlinear density field, neglecting only the large-scale correlations in halo positions. The power spectrum determined in this way is shown in figure 2.16, and turns out to agree very well with the exact nonlinear result on small and intermediate scales. The lesson here is that a good deal of the nonlinear correlations of the dark matter field can be understood as a distribution of random clumps, provided these are given the correct distribution of masses and mass-dependent density profiles.

How can we extend this model to understand how the clustering of galaxies can differ from that of the mass? There are two distinct ways in which a degree of bias is inevitable:

(1) Halo occupation numbers. For low-mass halos, the probability of obtaining an L^* galaxy must fall to zero. For halos with mass above this lower limit,

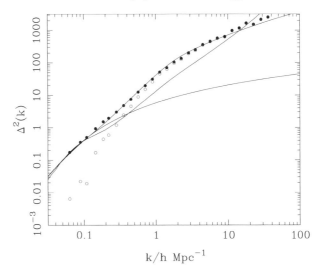

Figure 2.16. The power spectrum for the ΛCDM model. The full lines contrast the linear spectrum with the nonlinear spectrum, calculated according to the approximation of PD96. The spectrum according to randomly placed halos is denoted by open circles; if the linear power spectrum is added, the main features of the nonlinear spectrum are well reproduced.

the number of galaxies will in general not scale with halo mass.

(2) Non-locality. Galaxies can orbit within their host halos, so the probability of forming a galaxy depends on the overall halo properties, not just the density at a point. Also, the galaxies will end up at special places within the halos: for a halo containing only one galaxy, the galaxy will clearly mark the halo centre. In general, we expect one central galaxy and a number of satellites.

The numbers of galaxies that form in a halo of a given mass is the prime quantity that numerical models of galaxy formation aim to calculate. However, for a given assumed background cosmology, the answer may be determined empirically. Galaxy redshift surveys have been analysed via grouping algorithms similar to the 'friends-of-friends' method widely employed to find virialized clumps in N-body simulations. With an appropriate correction for the survey limiting magnitude, the observed number of galaxies in a group can be converted to an estimate of the total stellar luminosity in a group. This allows a determination of the All Galaxy System (AGS) luminosity function: the distribution of virialized clumps of galaxies as a function of their total luminosity, from small systems like the Local Group to rich Abell clusters.

The AGS function for the CfA survey was investigated by Moore *et al* (1993), who found that the result in blue light was well described by

$$\mathrm{d}\phi = \phi^*[(L/L^*)^\beta + (L/L^*)^\gamma]^{-1}\,\mathrm{d}L/L^*,$$

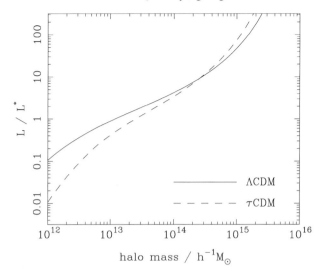

Figure 2.17. The empirical luminosity–mass relation required to reconcile the observed AGS luminosity function with two variants of CDM. L^* is the characteristic luminosity in the AGS luminosity function ($L^* = 7.6 \times 10^{10} h^{-2} L_\odot$). Note the rather flat slope around $M = 10^{13}$–$10^{14} h^{-1} M_\odot$, especially for ΛCDM.

where $\phi^* = 0.001\,26 h^3$ Mpc^{-3}, $\beta = 1.34$, $\gamma = 2.89$; the characteristic luminosity is $M^* = -21.42 + 5\log_{10} h$ in Zwicky magnitudes, corresponding to $M_B^* = -21.71 + 5\log_{10} h$, or $L^* = 7.6 \times 10^{10} h^{-2} L_\odot$, assuming $M_B^\odot = 5.48$. One notable feature of this function is that it is rather flat at low luminosities, in contrast to the mass function of dark-matter halos (see Sheth and Tormen 1999). It is therefore clear that any fictitious galaxy catalogue generated by randomly sampling the mass is unlikely to be a good match to observation. The simplest cure for this deficiency is to assume that the stellar luminosity per virialized halo is a monotonic, but nonlinear, function of halo mass. The required luminosity–mass relation is then easily deduced by finding the luminosity at which the integrated AGS density $\Phi(> L)$ matches the integrated number density of halos with mass $> M$. The result is shown in figure 2.17.

We can now return to the halo-based galaxy power spectrum and use the correct occupation number, N, as a function of mass. This needs a little care at small numbers, however, since the number of halos with occupation number unity affects the correlation properties strongly. These halos contribute no correlated pairs, so they simply dilute the signal from the halos with $N \geq 2$. The existence of antibias on intermediate scales can probably be traced to the fact that a large fraction of galaxy groups contain only one $> L_*$ galaxy. Finally, we need to put the galaxies in the correct location, as discussed before. If one galaxy always occupies the halo centre, with others acting as satellites, the small-scale

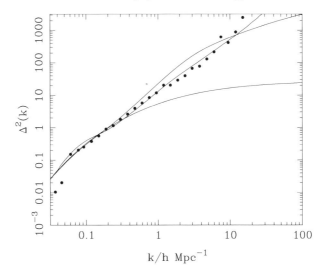

Figure 2.18. The power spectrum for a galaxy catalogue constructed from the ΛCDM model. A reasonable agreement with the APM data (full line) is achieved by simple empirical adjustment of the occupation number of galaxies as a function of halo mass, plus a scheme for placing the halos non-randomly within the halos. In contrast, the galaxy power spectrum differs significantly in shape from that of the dark matter (linear and nonlinear theory shown as in figure 2.16).

correlations automatically follow the slope of the halo density profile, which keeps them steep. The results of this exercise are shown in figure 2.18.

The results of this simple model are encouragingly similar to the scale-dependent bias found in the detailed calculations of Benson *et al* (2000a), shown in figure 2.15. There are thus grounds for optimism that we may be starting to attain a physical understanding of the origin of galaxy bias.

2.8   Cosmic background fluctuations

2.8.1   The hot big bang and the microwave background

What was the state of matter in the early phases of the big bang? Since the present-day expansion will cause the density to decline in the future, conditions in the past must have corresponded to high density—and thus to high temperature. We can deal with this quantitatively by looking at the thermodynamics of the fluids that make up a uniform cosmological model.

The expansion is clearly *adiathermal*, since the symmetry means that there can be no net heat flow through any surface. If the expansion is also reversible, then we can go one step further, because entropy change is defined in terms of the heat that flows during a reversible change. If no heat flows during a reversible change, then entropy must be conserved, and the expansion will be *adiabatic*. This can only be an approximation, since there will exist irreversible microscopic processes. In practice, however, it will be shown later that the effects of these processes are overwhelmed by the entropy of thermal background radiation in the universe. It will therefore be an excellent approximation to treat the universe as if the matter content were a simple dissipationless fluid undergoing a reversible expansion. This means that, for a ratio of specific heats Γ, we get the usual adiabatic behaviour

$$T \propto R^{-3(\Gamma-1)}.$$

For radiation, $\Gamma = 4/3$ and we get just $T \propto 1/R$. A simple model for the energy content of the universe is to distinguish pressureless 'dust-like' matter (in the sense that $p \ll \rho c^2$) from relativistic 'radiation-like' matter (photons plus neutrinos). If these are assumed not to interact, then the energy densities scale as

$$\rho_m \propto R^{-3} \qquad \rho_r \propto R^{-4}$$

The universe must therefore have been *radiation-dominated* at some time in the past, where the densities of matter and radiation cross over. To anticipate, we know that the current radiation density corresponds to thermal radiation with $T \simeq 2.73$ K. In addition to this CMB, we also expect a background in neutrinos. This arises in the same way as the CMB: both photons and neutrinos are in thermal equilibrium at high redshift, but eventually fall out of equilibrium as the universe expands and reaction timescales lengthen. Subsequently, the number density of frozen-out background particles scales as $n \propto a^{-3}$, exactly as expected for a thermal background with $T \propto 1/a$. The background appears to stay in thermal equilibrium even though it has frozen out. If the neutrinos are massless and therefore relativistic, they contribute an energy density comparable to that of the photons (to be exact, a factor 0.68 times the photon density—see p 280 of Peacock (1999)). If there are no other contributions to the energy density from relativistic particles, then the total effective radiation density is $\Omega_r h^2 \simeq 4.2 \times 10^{-5}$ and the redshift of *matter–radiation equality* is

$$1 + z_{eq} = 23\,900\Omega h^2 (T/2.73 \text{ K})^{-4}.$$

The time of this change in the global equation of state is one of the key epochs in determining the appearance of the present-day universe. By a coincidence, this epoch is close to another important event in cosmological history: *recombination*. Once the temperature falls below $\simeq 10^4$ K, ionized material can form neutral hydrogen. Observational astronomy is only possible from this point on, since Thomson scattering from electrons in ionized material prevents photon propagation. In practice, this limits the maximum redshift of observational interest to about 1000; unless Ω is very low or vacuum energy is important, a matter-dominated model is therefore a good approximation to reality.

In a famous piece of serendipity, the redshifted radiation from the last-scattering photosphere was detected as a 2.73 K microwave background by Penzias and Wilson (1965). Since the initial detection of the microwave background at $\lambda = 7.3$ cm, measurements of the spectrum have been made over an enormous range of wavelengths, from the depths of the Rayleigh–Jeans regime at 74 cm to well into the Wien tail at 0.5 mm. The most accurate measurements come from *COBE*—the NASA cosmic background explorer satellite. Early data showed the spectrum to be very close to a pure Planck function (Mather *et al* 1990), and the final result verifies the lack of any distortion with breathtaking precision. The COBE temperature measurement and 95% confidence range of

$$T = 2.728 \pm 0.004 \text{ K}$$

improves significantly on the ground-based experiments. The lack of distortion in the shape of the spectrum is astonishing, and limits the chemical potential to $|\mu| < 9 \times 10^{-5}$ (Fixsen *et al* 1996). These results also allow the limit $y \lesssim 1.5 \times 10^{-5}$ to be set on the Compton-scattering distortion parameter. These limits are so stringent that many competing cosmological models can be eliminated.

2.8.2 Mechanisms for primary fluctuations

At the last-scattering redshift ($z \simeq 1000$), gravitational instability theory says that fractional density perturbations $\delta \gtrsim 10^{-3}$ must have existed in order for galaxies and clusters to have formed by the present. A long-standing challenge in cosmology has been to detect the corresponding fluctuations in brightness temperature of the CMB radiation, and it took over 25 years of ever more stringent upper limits before the first detections were obtained, in 1992. The study of CMB fluctuations has subsequently blossomed into a critical tool for pinning down cosmological models.

This can be a difficult subject; the treatment given here is intended to be the simplest possible. For technical details see, e.g., Bond (1997), Efstathiou (1990), Hu and Sugiyama (1995), Seljak and Zaldarriaga (1996); for a more general overview, see White *et al* (1994) or Partridge (1995). The exact calculation of CMB anisotropies is complicated because of the increasing photon mean free path at recombination: a fluid treatment is no longer fully adequate. For full accuracy, the Boltzmann equation must be solved to follow the evolution of the photon distribution function. A convenient means for achieving this is provided by the public domain *CMBFAST* code (Seljak and Zaldarriaga 1996). Fortunately, these exact results can usually be understood via a more intuitive treatment, which is quantitatively correct on large and intermediate scales. This is effectively what would be called local thermodynamic equilibrium in stellar structure: imagine that the photons we see each originated in a region of space in which the radiation field was a Planck function of a given characteristic temperature. The observed

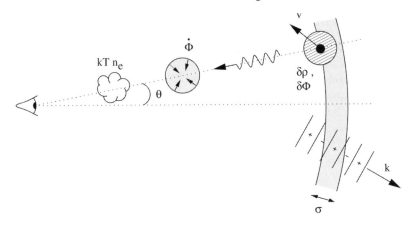

Figure 2.19. Illustrating the physical mechanisms that cause CMB anisotropies. The shaded arc on the right represents the last-scattering shell; an inhomogeneity on this shell affects the CMB through its potential, adiabatic and Doppler perturbations. Further perturbations are added along the line of sight by time-varying potentials (the Rees–Sciama effect) and by electron scattering from hot gas (the Sunyaev–Zeldovich effect). The density field at last scattering can be Fourier analysed into modes of wavevector k. These spatial perturbation modes have a contribution that is, in general, damped by averaging over the shell of last scattering. Short-wavelength modes are more heavily affected (1) because more of them fit inside the scattering shell and (2) because their wavevectors point more nearly radially for a given projected wavelength.

brightness temperature field can then be thought of as arising from a superposition of these fluctuations in thermodynamic temperature.

   We distinguish *primary anisotropies* (those that arise due to effects at the time of recombination) from *secondary anisotropies*, which are generated by scattering along the line of sight. There are three basic primary effects, illustrated in figure 2.19, which are important on respectively large, intermediate and small angular scales:

(1) Gravitational (Sachs–Wolfe) perturbations. Photons from high-density regions at last scattering have to climb out of potential wells, and are thus redshifted.
(2) Intrinsic (adiabatic) perturbations. In high-density regions, the coupling of matter and radiation can compress the radiation also, giving a higher temperature.
(3) Velocity (Doppler) perturbations. The plasma has a non-zero velocity at recombination, which leads to Doppler shifts in frequency and hence brightness temperature.

To make quantitative progress, the next step is to see how to predict the size of these effects in terms of the spectrum of mass fluctuations.

2.8.3   The temperature power spectrum

The statistical treatment of CMB fluctuations is very similar to that of spatial density fluctuations. We have a 2D field of random fluctuations in brightness temperature, and this can be analysed by the same tools that are used in the case of 2D galaxy clustering.

Suppose that the fractional temperature perturbations on a patch of sky of side L are Fourier expanded:

$$\frac{\delta T}{T}(X) = \frac{L^2}{(2\pi)^2} \int T_K \exp(-i\mathbf{K} \cdot \mathbf{X})\, \mathrm{d}^2 K$$

$$T_K(\mathbf{K}) = \frac{1}{L^2} \int \frac{\delta T}{T}(X) \exp(i\mathbf{K} \cdot \mathbf{X})\, \mathrm{d}^2 X,$$

where X is a 2D position vector on the sky, and K is a 2D wavevector. This is only a valid procedure if the patch of sky under consideration is small enough to be considered flat; we give the full machinery later. We will normally take the units of length to be angle on the sky, although they could also in principle be h^{-1} Mpc at a given redshift. The relation between angle and comoving distance on the last-scattering sphere requires the comoving angular-diameter distance to the last-scattering sphere; because of its high redshift, this is effectively identical to the horizon size at the present epoch, R_H:

$$R_\mathrm{H} = \frac{2c}{\Omega_\mathrm{m} H_0} \qquad \text{(open)}$$

$$R_\mathrm{H} \simeq \frac{2c}{\Omega_\mathrm{m}^{0.4} H_0} \qquad \text{(flat)};$$

the latter approximation for models with $\Omega_\mathrm{m} + \Omega_\mathrm{v} = 1$ is due to Vittorio and Silk (1991).

As with the density field, it is convenient to define a dimensionless power spectrum of fractional temperature fluctuations,

$$\mathcal{T}^2 \equiv \frac{L^2}{(2\pi)^2} 2\pi K^2 |T_K|^2,$$

so that $\mathcal{T}^2$ is the fractional variance in temperature from modes in a unit range of $\ln K$. The corresponding dimensionless spatial statistic is the two-point correlation function

$$C(\theta) = \left\langle \frac{\delta T}{T}(\psi) \frac{\delta T}{T}(\psi + \theta) \right\rangle,$$

which is the Fourier transform of the power spectrum, as usual:

$$C(\theta) = \int \mathcal{T}^2(K) J_0(K\theta) \frac{\mathrm{d}K}{K}.$$

Here, the Bessel function comes from the angular part of the Fourier transform:

$$\int \exp(ix\cos\phi)\,d\phi = 2\pi J_0(x).$$

Now, in order to predict the observed anisotropy of the microwave background, the problem we must solve is to integrate the temperature perturbation field through the *last-scattering shell*. In order to do this, we assume that the sky is flat; we also neglect curvature of the 3-space, although this is only strictly valid for flat models with $k = 0$. Both these restrictions mean that the results are not valid for very large angles. Now, introducing the Fourier expansion of the 3D temperature perturbation field (with coefficients T_k^{3D}) we can construct the observed 2D temperature perturbation field by integrating over k space and optical depth:

$$\frac{\delta T}{T} = \frac{V}{(2\pi)^3}\int\int T_k^{3D}e^{-i\mathbf{k}\cdot\mathbf{r}}\,d^3k\,e^{-\tau}\,d\tau.$$

A further simplification is possible if we approximate $e^{-\tau}\,d\tau$ by a Gaussian in comoving radius:

$$\exp(-\tau)\,d\tau \propto \exp[-(r - r_{LS})^2/2\sigma_r^2]\,dr.$$

This says that we observe radiation from a last-scattering shell centred at comoving distance r_{LS} (which is very nearly identical to r_H, since the redshift is so high). The thickness of this shell is of the order of the mean free path to Compton scattering at recombination, which is approximately

$$\sigma_r = 7(\Omega h^2)^{-1/2}\,\text{Mpc}$$

(see p 287 of Peacock 1999).

The 2D power spectrum is thus a smeared version of the 3D one: any feature that appears at a particular wavenumber in 3D will cause a corresponding feature at the same wavenumber in 2D. A particularly simple converse to this rule arises when there are *no* features: the 3D power spectrum is scale-invariant ($T_{3D}^2 = $ constant). In this case, for scales large enough that we can neglect the radial smearing from the last-scattering shell,

$$T_{2D}^2 = T_{3D}^2$$

so that the pattern on the CMB sky is also scale invariant. To apply this machinery for a general spectrum, we now need quantitative expressions for the spatial temperature anisotropies.

Sachs–Wolfe effect. To relate to density perturbations, use Poisson's equation $\nabla^2\delta\Phi_k = 4\pi G\rho\delta_k$. The effect of ∇^2 is to pull down a factor of $-k^2/a^2$ (a^2 because k is a comoving wavenumber). Eliminating ρ in terms of Ω and z_{LS} gives

$$T_k = -\frac{\Omega(1 + z_{LS})}{2}\left(\frac{H_0}{c}\right)^2\frac{\delta_k(z_{LS})}{k^2}.$$

Doppler source term. The effect here is just the Doppler effect from the scattering of photons by moving plasma:

$$\frac{\delta T}{T} = \frac{\delta \boldsymbol{v} \cdot \hat{\boldsymbol{r}}}{c}.$$

Using the standard expression for the linear peculiar velocity, the corresponding k-space result is

$$T_k = -i\sqrt{\Omega(1+z_{\mathrm{LS}})} \left(\frac{H_0}{c}\right) \frac{\delta_k(z_{\mathrm{LS}})}{k} \hat{\boldsymbol{k}} \cdot \hat{\boldsymbol{r}}.$$

Adiabatic source term. This is the simplest of the three effects mentioned earlier:

$$T_k = \frac{\delta_k(z_{\mathrm{LS}})}{3},$$

because $\delta n_\gamma / n_\gamma = \delta\rho/\rho$ and $n_\gamma \propto T^3$. However, this simplicity conceals a paradox. Last scattering occurs only when the universe recombines, which occurs at roughly a fixed temperature: $kT \sim \chi$, the ionization potential of hydrogen. Surely, then, we should just be looking back to a surface of constant temperature? Hot and cold spots should normalize themselves away, so that the last-scattering sphere appears uniform. The solution is that a denser spot recombines *later*: it is therefore less redshifted and appears hotter. In algebraic terms, the observed temperature perturbation is

$$\left(\frac{\delta T}{T}\right)_{\mathrm{obs}} = -\frac{\delta z}{1+z} = \frac{\delta\rho}{\rho},$$

where the last expression assumes linear growth, $\delta \propto (1+z)^{-1}$. Thus, even though a more correct picture for the temperature anisotropies seen on the sky is of a crinkled surface at constant temperature, thinking of hot and cold spots gives the right answer. Any observable cross-talk between density perturbations and delayed recombination is confined to effects of order higher than linear.

We now draw these results together to form the spatial power spectrum of CMB fluctuations in terms of the power spectrum of mass fluctuations at last scattering:

$$T_{\mathrm{3D}}^2 = [(f_{\mathrm{A}} + f_{\mathrm{SW}})^2(k) + f_{\mathrm{V}}^2(k)\mu^2]\Delta_k^2(z_{\mathrm{LS}}),$$

where $\mu \equiv \hat{\boldsymbol{k}} \cdot \hat{\boldsymbol{r}}$. The dimensionless factors can be written most simply as

$$f_{\mathrm{SW}} = -\frac{2}{(kD_{\mathrm{LS}})^2}$$

$$f_{\mathrm{V}} = \frac{2}{kD_{\mathrm{LS}}}$$

$$f_{\mathrm{A}} = 1/3,$$

where

$$D_{LS} = \frac{2c}{\Omega_m^{1/2} H_0}(1 + z_{LS})^{-1/2} = 184(\Omega h^2)^{-1/2} \text{ Mpc}$$

is the comoving horizon size at last scattering (a result that is independent of whether there is a cosmological constant).

We can see immediately from these expressions the relative importance of the various effects on different scales. The Sachs–Wolfe effect dominates for wavelengths $\gtrsim 1h^{-1}$ Gpc; Doppler effects then take over but are almost immediately dominated by adiabatic effects on the smallest scales.

These expressions apply to perturbations for which only gravity has been important up until last scattering, i.e. those larger than the horizon at z_{eq}. For smaller wavelengths, a variety of additional physical processes act on the radiation perturbations, generally reducing the predicted anisotropies. An accurate treatment of these effects is not really possible without a more complicated analysis, as is easily seen by considering the thickness of the last-scattering shell, $\sigma_r = 7(\Omega h^2)^{-1/2}$ Mpc. This clearly has to be of the same order of magnitude as the photon mean free path at this time; on any smaller scales, a fluid approximation for the radiation is inadequate and a proper solution of the Boltzmann equation is needed. Nevertheless, some qualitative insight into the small-scale processes is possible. The radiation fluctuations will be damped relative to the baryon fluid by photon diffusion, characterized by the Silk-damping scale, $\lambda_S = 2.7(\Omega \Omega_B h^6)^{-1/4}$ Mpc. Below the horizon scale at z_{eq}, $16(\Omega h^2)^{-1}$ Mpc, there is also the possibility that dark-matter perturbations can grow while the baryon fluid is still held back by radiation pressure, which results in adiabatic radiation fluctuations that are less than would be predicted from the dark-matter spectrum alone. In principle, this suggests a suppression factor of $(1 + z_{eq})/(1 + z_{LS})$ or roughly a factor 10. In detail, the effect is an oscillating function of scale, since we have seen that baryonic perturbations oscillate as sound waves when they come inside the horizon:

$$\delta_b \propto (3c_S)^{1/4} \exp\left(\pm i \int k c_S \, d\tau\right);$$

here, τ stands for conformal time. There is thus an oscillating signal in the CMB, depending on the exact phase of these waves at the time of last scattering. These oscillations in the fluid of baryons plus radiation cause a set of acoustic peaks in the small-scale power spectrum of the CMB fluctuations (see later).

2.8.4 Large-scale fluctuations and CMB power spectrum

The flat-space formalism becomes inadequate for very large angles; the proper basis functions to use are the spherical harmonics:

$$\frac{\delta T}{T}(\hat{q}) = \sum a_\ell^m Y_{\ell m}(\hat{q}),$$

where $\hat{q}$ is a unit vector that specifies direction on the sky. Since the spherical harmonics satisfy the orthonormality relation $\int Y_{\ell m} Y^*_{\ell' m'} \, d^2 q = \delta_{\ell \ell'} \delta_{mm'}$, the inverse relation is

$$a^m_\ell = \int \frac{\delta T}{T} Y^*_{\ell m} \, d^2 q.$$

The analogues of the Fourier relations for the correlation function and power spectrum are

$$C(\theta) = \frac{1}{4\pi} \sum_\ell \sum_{m=-\ell}^{m=+\ell} |a^m_\ell|^2 P_\ell(\cos\theta)$$

$$|a^m_\ell|^2 = 2\pi \int_{-1}^{1} C(\theta) P_\ell(\cos\theta) \, d\cos\theta.$$

These are exact relations, governing the actual correlation structure of the observed sky. However, the sky we see is only one of infinitely many possible realizations of the statistical process that yields the temperature perturbations; as with the density field, we are more interested in the *ensemble average power*. A common notation is to define C_ℓ as the expectation value of $|a^m_\ell|^2$:

$$C(\theta) = \frac{1}{4\pi} \sum_\ell (2\ell + 1) C_\ell P_\ell(\cos\theta), \qquad C_\ell \equiv \langle |a^m_\ell|^2 \rangle,$$

where now $C(\theta)$ is the ensemble-averaged correlation. For small θ and large ℓ, the exact form reduces to a Fourier expansion:

$$C(\theta) = \int_0^\infty \mathcal{T}^2(K) J_0(K\theta) \frac{dK}{K},$$

$$\mathcal{T}^2(K = \ell + \tfrac{1}{2}) = \frac{(\ell + \tfrac{1}{2})(2\ell + 1)}{4\pi} C_\ell.$$

The effect of filtering the microwave sky with the beam of a telescope may be expressed as a multiplication of the C_ℓ, as with convolution in Fourier space:

$$C_S(\theta) = \frac{1}{4\pi} \sum_\ell (2\ell + 1) W_\ell^2 C_\ell P_\ell(\cos\theta).$$

When the telescope beam is narrow in angular terms, the Fourier limit can be used to deduce the appropriate ℓ-dependent filter function. For example, for a Gaussian beam of *FWHM* (full-width to half maximum) 2.35σ, the filter function is $W_\ell = \exp(-\ell^2 \sigma^2 / 2)$.

For the large-scale temperature anisotropy, we have already seen that what matters is the Sachs–Wolfe effect, for which we have derived the spatial anisotropy power spectrum. The spherical harmonic coefficients for a spherical slice through such a field can be deduced using the results for large-angle galaxy clustering, in the limit of a selection function that goes to a delta function in radius:

$$C_\ell^{\text{SW}} = 16\pi \int (k D_{\text{LS}})^{-4} \Delta_k^2(z_{\text{LS}}) j_\ell^2(k R_{\text{H}}) \frac{dk}{k},$$

where the j_ℓ are *spherical Bessel functions* (see chapter 10 of Abramowitz and Stegun 1965). This formula, derived by Peebles (1982), strictly applies only to spatially flat models, since the Fourier expansion of the density field is invalid in an open model. Nevertheless, since the curvature radius R_0 subtends an angle of $\Omega/[2(1 - \Omega)^{1/2}]$, even the lowest few multipoles are not seriously affected by this point, provided $\Omega \gtrsim 0.1$.

For simple mass spectra, the integral for the C_ℓ can be performed analytically. The case of most practical interest is a scale-invariant spectrum ($\Delta_k^2 \propto k^4$), for which the integral scales as

$$C_\ell = \frac{6}{\ell(\ell + 1)} C_2$$

(see equation (6.574.2) of Gradshteyn and Ryzhik 1980). The direct relation between the mass fluctuation spectrum and the multipole coefficients of CMB fluctuations mean that either can be used as a measure of the normalization of the spectrum.

2.8.5 Predictions of CMB anisotropies

We are now in a position to understand the characteristic angular structure of CMB fluctuations. The change-over from scale-invariant Sachs–Wolfe fluctuations to fluctuations dominated by Doppler scattering has been shown to occur at $k \simeq D_{\text{LS}}$. This is one critical angle (call it θ_1); its definition is $\theta_1 = D_{\text{LS}}/R_{\text{H}}$, and for a matter-only model it takes the value

$$\theta_1 = 1.8\Omega^{1/2} \text{ degrees.}$$

For flat low-density models with significant vacuum density, R_{H} is smaller; θ_1 and all subsequent angles would then be larger by about a factor $\Omega^{-0.6}$ (i.e. θ_1 is roughly independent of Ω in flat Λ-dominated models).

The second dominant scale is the scale of last-scattering smearing set by $\sigma_{\text{r}} = 7(\Omega h^2)^{-1/2}$ Mpc. This subtends an angle

$$\theta_2 = 4\Omega^{1/2} \text{ arcmin.}$$

Finally, a characteristic scale in many density power spectra is set by the horizon at z_{eq}. This is $16(\Omega h^2)^{-1}$ Mpc and subtends

$$\theta_3 = 9h^{-1} \text{ arcmin,}$$

independent of Ω. This is quite close to θ_2, so that alterations in the transfer function are an effect of secondary importance in most models.

We therefore expect that all scale-invariant models will have similar CMB power spectra: a flat Sachs–Wolfe portion down to $K \simeq 1 \text{ deg}^{-1}$, followed by a bump where Doppler and adiabatic effects come in, which turns over on arcminute scales through damping and smearing. This is illustrated well in figure 2.22, which shows some detailed calculations of 2D power spectra, generated with the CMBFAST package. From these plots, the key feature of the anisotropy spectrum is clearly the peak at $\ell \sim 100$. This is often referred to as the *Doppler peak*, but it is not so clear that this name is accurate. Our simplified analysis suggests that Sachs–Wolfe anisotropy should dominate for $\theta > \theta_1$, with Doppler and adiabatic terms becoming of comparable importance at θ_1, and adiabatic effects dominating at smaller scales. There are various effects that cause the simple estimate of adiabatic effects to be too large, but they clearly cannot be neglected for $\theta < \theta_1$. A better name, which is starting to gain currency, is the *acoustic peak*. In any case, it is clear that the peak is the key diagnostic feature of the CMB anisotropy spectrum: its height above the SW 'plateau' is sensitive to Ω_B and its angular location depends on Ω and Λ. It is therefore no surprise that many experiments are currently attempting accurate measurements of this feature. Furthermore, it is apparent that sufficiently accurate experiments will be able to detect higher 'harmonics' of the peak, in the form of smaller oscillations of amplitude perhaps 20% in power, around $\ell \simeq 500$–1000. These features arise because the matter–radiation fluid undergoes small-scale oscillations, the phase of which at last scattering depends on wavelength, since the density oscillation varies roughly as $\delta \propto \exp(ic_S k \tau)$. Accurate measurement of these oscillations would pin down the sound speed at last scattering, and help give an independent measurement of the baryon density.

Since large-scale CMB fluctuations are expected to be dominated by gravitational potential fluctuations, it was possible to make relatively clear predictions of the likely level of CMB anisotropies, even in advance of the first detections. What was required was a measurement of the typical depth of large-scale potential wells in the universe, and many lines of argument pointed inevitably to numbers of order 10^{-5}. This was already clear from the existence of massive clusters of galaxies with velocity dispersions of up to 1000 km s^{-1}:

$$v^2 \sim \frac{GM}{r} \Rightarrow \frac{\Phi}{c^2} \sim \frac{v^2}{c^2},$$

so the potential well of a cluster is of order 10^{-5} deep. More exactly, the abundance of rich clusters is determined by the amplitude σ_8, which measures $[\Delta^2(k)]^{1/2}$ at an effective wavenumber of very nearly $0.17h \text{ Mpc}^{-1}$. If we assume that this is a large enough scale so that what we are measuring is the amplitude of any scale-invariant spectrum, then the earlier expression for the temperature power spectrum gives

$$\sqrt{\mathcal{T}_{\text{SW}}^2} \simeq 10^{-5.7} \Omega \sigma_8 [g(\Omega)]^{-1}.$$

There were thus strong grounds to expect that large-scale fluctuations would be present at about the 10^{-5} level, and it was a significant boost to the credibility of the gravitational-instability model that such fluctuations were eventually seen.

In more detail, it is possible to relate the COBE anisotropy to the large-scale portion of the power spectrum. Górski *et al* (1995), Bunn *et al* (1995), and White and Bunn (1995) discuss the large-scale normalization from the two-year COBE data in the context of CDM-like models. The final four-year COBE data favour very slightly lower results, and we scale to these in what follows. For scale-invariant spectra and $\Omega = 1$, the best normalization is

$$\text{COBE} \Rightarrow \Delta^2(k) = \left(\frac{k}{0.0737h \text{ Mpc}^{-1}}\right)^4.$$

Translated into other common notation for the normalization, this is equivalent to $Q_{\text{rms-ps}} = 18.0 \ \mu\text{K}$, or $\delta_H = 2.05 \times 10^{-5}$ (see e.g. Peacock and Dodds 1994).

For low-density models, the earlier discussion suggests that the power spectrum should depend on Ω and the growth factor g as $P \propto g^2/\Omega^2$. Because of the time dependence of the gravitational potential (integrated Sachs–Wolfe effect) and because of spatial curvature, this expression is not exact, although it captures the main effect. From the data of White and Bunn (1995), a better approximation is

$$\Delta^2(k) \propto \frac{g^2}{\Omega^2} g^{0.7}.$$

This applies for low-Ω models both with and without vacuum energy, with a maximum error of 2% in density fluctuation provided $\Omega > 0.2$. Since the rough power-law dependence of g is $g(\Omega) \simeq \Omega^{0.65}$ and $\Omega^{0.23}$ for open and flat models respectively, we see that the implied density fluctuation amplitude scales approximately as $\Omega^{-0.12}$ and $\Omega^{-0.69}$ respectively for these two cases. The dependence is weak for open models, but vacuum energy implies much larger fluctuations for low Ω.

Within the CDM model, it is always possible to satisfy both the large-scale COBE normalization and the small-scale σ_8 constraint, by appropriate choice of Γ and n. This is illustrated in figure 2.20. Note that the vacuum energy affects the answer; for reasonable values of h and reasonable baryon content, flat models require $\Omega_m \simeq 0.3$, whereas open models require $\Omega_m \simeq 0.5$ in order to be consistent with scale-invariant primordial fluctuations.

2.8.6 Geometrical degeneracy

The statistics of CMB fluctuations depend on a large number of parameters, and it can be difficult to understand what the effect of changing each one will be. Furthermore, the effects of some parameters tend to change things in opposite directions, so that there are degenerate directions in the parameter space, along which changes leave the CMB unaffected. These were analysed comprehensively by Efstathiou and Bond (1999), and we now summarize the main results.

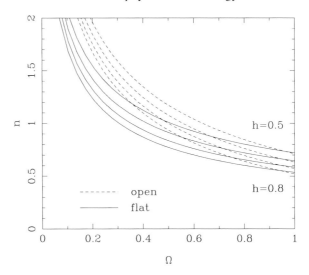

Figure 2.20. For 10% baryons, the value of n needed to reconcile COBE and the cluster normalization in CDM models.

The usual expression for the comoving angular-diameter distance is

$$R_0 S_k(r) = \frac{c}{H_0}|1 - \Omega|^{-1/2} S_k\left[\int_0^z \frac{|1 - \Omega|^{1/2}\,dz'}{\sqrt{(1 - \Omega)(1 + z')^2 + \Omega_v + \Omega_m(1 + z')^3}}\right],$$

where $\Omega = \Omega_m + \Omega_v$. Defining $\omega_i \equiv \Omega_i h^2$, this can be rewritten in a way that has no explicit h dependence:

$$R_0 S_k(r) = \frac{3000\,\text{Mpc}}{|\omega_k|^{1/2}} S_k\left[\int_0^z \frac{|\omega_k|^{1/2}\,dz'}{\sqrt{\omega_k(1 + z')^2 + \omega_v + \omega_m(1 + z')^3}}\right],$$

where $\omega_k \equiv (1 - \Omega_m - \Omega_v)h^2$. This parameter describes the curvature of the universe, treating it effectively as a physical density that scales as $\rho \propto a^{-2}$. This is convenient for the present formalism, but it is important to appreciate that curvature differs fundamentally from a hypothetical fluid with such an equation of state: the value of ω_k also sets the curvature index k.

The horizon distance at last scattering is $184\omega_m^{-1/2}$ Mpc. Similarly, other critical length scales such as the sound horizon are governed by the relevant physical density, ω_b. Thus, if ω_m and ω_b are given, the shape of the spatial power spectrum is determined. The translation of this into an angular spectrum depends on the angular-diameter distance, which is a function of these parameters, plus ω_k and ω_v. Models in which $\omega_m^{1/2} R_0 S_k(r)$ is a constant have the same angular horizon size. There is therefore a degeneracy between curvature (ω_k) and vacuum (ω_v): these two parameters can be varied simultaneously to keep the same apparent distance, as illustrated in figure 2.21.

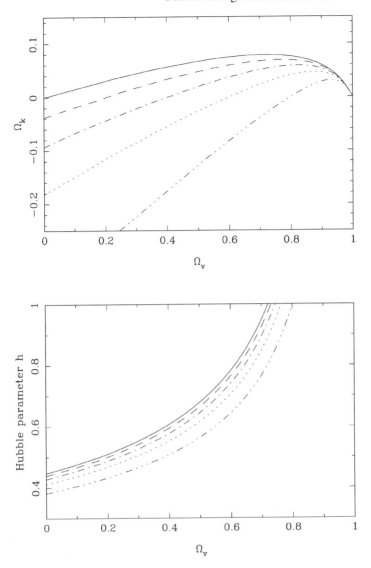

Figure 2.21. The geometrical degeneracy in the CMB means that models with fixed $\Omega_m h^2$ and $\Omega_b h^2$ can be made to look identical by varying the curvature against vacuum energy, while also varying the Hubble parameter. This degeneracy is illustrated here for the case $\omega_m \equiv \Omega_m h^2 = 0.2$. Models along a given line are equivalent from a CMB point of view; corresponding lines in the upper and lower panels have the same line style. The sensitivity to curvature is strong: if the universe appears to be flat, then it really must be so, unless it is very heavily vacuum dominated. Note that supplying external information about h breaks the degeneracy. This figure assumes scalar fluctuations only; allowing tensor modes introduces additional degeneracies—mainly between the tensor fraction and tilt.

The physical degree of freedom here can be thought of as the Hubble constant. This is involved via the relation

$$h^2 = \omega_m + \omega_v + \omega_k,$$

so specifying h in addition to the physical matter density fixes $\omega_v + \omega_k$ and removes the degeneracy.

2.8.7   Small-scale data and outlook

The study of large-scale CMB anisotropies had a huge impact on cosmology in the 1990s, and the field seems likely to be of increasing importance over the next decade. This school was held at a particularly exciting time, as major new data on the CMB power spectrum arrived during the lectures (de Bernardis *et al* 2000, Hanany *et al* 2000). Although these developments are very recent, the situation already seems a good deal clearer than previously, and it is interesting to try to guess where the field is heading.

One immediate conclusion is that it increasingly seems that the relevant models are ones in which the primordial fluctuations were close to being adiabatic and Gaussian. Isocurvature models suffer from the high amplitude of the large-scale perturbations, and do not become any more attractive when modelled in detail (Hu *et al* 1995). Topological defects were for a long time hard to assess, since accurate predictions of their CMB properties were difficult to make. Recent progress does, however, indicate that these theories may have difficulty matching the main details of CMB anisotropies, even as they are presently known (Pen *et al* 1997).

We shall therefore concentrate on interpreting the data in terms of the simplest gravitational-instability models. Many of the features of these models are generic, although they are often spoken of as 'the inflationary predictions'. This statement needs to be examined carefully, since one of the possible prizes from a study of the CMB may be a test of inflation. CMB anisotropies in theories where structure forms via gravitational collapse were calculated in largely the modern way well before inflation was ever considered, by Peebles and Yu (1970). The difficulty in these calculations is the issue of super-horizon fluctuations. In a conventional hot big bang, these must be generated by some acausal process—indeed, an acausal origin is required even for large-scale homogeneity. Inflation is so far the only theory that generates such superhorizon modes at all naturally. Nevertheless, it is not acceptable to claim that detection of super-horizon modes amounts to a proof of inflation. Rather, we need some more characteristic signature of the specific process used by inflation: amplified quantum fluctuations.

We should thus review the predictions that simple models of inflation make for CMB anisotropies (see, e.g., chapter 11 of Peacock 1999 or Liddle and Lyth 2000 for more details). Inflation is driven by a scalar field ϕ, with a potential

$V(\phi)$. As well as the characteristic energy density of inflation, V, this can be characterized by two dimensionless parameters

$$\epsilon \equiv \frac{m_P^2}{16\pi}(V'/V)^2$$

$$\eta \equiv \frac{m_P^2}{8\pi}(V''/V),$$

where m_P is the Planck mass, $V' = dV/d\phi$, and all quantities are evaluated towards the end of inflation, when the present large-scale structure modes were comparable in size to the inflationary horizon. Prior to transfer-function effects, the primordial fluctuation spectrum is specified by a horizon-scale amplitude (extrapolated to the present) δ_H and a slope n:

$$\Delta^2(k) = \delta_H^2 \left(\frac{ck}{H_0}\right)^{3+n}.$$

The inflationary predictions for these numbers are

$$\delta_H \sim \frac{V^{1/2}}{m_P^2 \epsilon^{1/2}}$$

$$n = 1 - 6\epsilon + 2\eta,$$

which leaves us in the unsatisfactory position of having two observables and three parameters.

The critical ingredient for testing inflation by making further predictions is the possibility that, in addition to scalar modes, the CMB could also be affected by gravitational waves (following the original insight of Starobinsky 1985). We therefore distinguish explicitly between scalar and tensor contributions to the CMB fluctuations by using appropriate subscripts. The former category are those described by the Sachs–Wolfe effect, and are gravitational potential fluctuations that relate directly to mass fluctuations. The relative amplitude of tensor and scalar contributions depended on the inflationary parameter ϵ alone:

$$\frac{C_\ell^T}{C_\ell^S} \simeq 12.4\epsilon \simeq 6(1-n).$$

The second relation to the *tilt* (which is defined to be $1 - n$) is less general, as it assumes a polynomial-like potential, so that η is related to ϵ. If we make this assumption, inflation can be tested by measuring the tilt and the tensor contribution. For simple models, this test should be feasible: $V = \lambda\phi^4$ implies $n \simeq 0.95$ and $C_\ell^T/C_\ell^S \simeq 0.3$. To be safe, we need one further observation, and this is potentially provided by the spectrum of C_ℓ^T. Suppose we write separate power-law index definitions for the scalar and tensor anisotropies:

$$C_\ell^S \propto \ell^{n_S - 3}, \qquad C_\ell^T \propto \ell^{n_T - 3}.$$

From the discussion of the Sachs–Wolfe effect, we know that, on large scales, the scalar index is the same as index in the matter power spectrum: $n_S = n = 1 - 6\epsilon + 2\eta$. By the same method, it is easily shown that $n_T = 1 - 2\epsilon$ (although different definitions of n_T are in use in the literature; the convention here is that $n = 1$ always corresponds to a constant $\mathcal{T}^2(\ell)$). Finally, then, we can write the *inflationary consistency equation*:

$$\frac{C_\ell^T}{C_\ell^S} = 6.2(1 - n_T).$$

The slope of the scalar perturbation spectrum is the only quantity that contains η, and so n_S is not involved in a consistency equation, since there is no independent measure of η with which to compare it.

From the point of view of an inflationary purist, the scalar spectrum is therefore an annoying distraction from the important business of measuring the tensor contribution to the CMB anisotropies. A certain degree of degeneracy exists here (see Bond *et al* 1994), since the tensor contribution has no acoustic peak; C_ℓ^T is roughly constant up to the horizon scale and then falls. A spectrum with a large tensor contribution therefore closely resembles a scalar-only spectrum with smaller Ω_b (and hence a relatively lower peak). One way in which this degeneracy may be lifted is through polarization of the CMB fluctuations. A non-zero polarization is inevitable because the electrons at last scattering experience an anisotropic radiation field. Thomson scattering from an anisotropic source will yield polarization, and the practical size of the fractional polarization P is of the order of the quadrupole radiation anisotropy at last scattering: $P \gtrsim 1\%$. Furthermore, the polarization signature of tensor perturbations differs from that of scalar perturbations (e.g. Seljak 1997, Hu and White 1997); the different contributions to the total unpolarized C_ℓ can in principle be disentangled, allowing the inflationary test to be carried out.

How do these theoretical expectations match with the recent data, shown in figure 2.22? In many ways, the match to prediction is startlingly good: there is a very clear acoustic peak at $\ell \simeq 220$, which has very much the height and width expected for the principal peak in adiabatic models. As we have seen, the location of this peak is sensitive to Ω, since it measures directly the angular size of the horizon at last scattering, which scales as $\ell \propto \Omega^{-1/2}$ for open models. The cut-off at $\ell \simeq 1000$ caused by last-scattering smearing also moves to higher ℓ for low Ω; if Ω were small enough, the smearing cut-off would be carried to large ℓ, where it would be inconsistent with the upper limits to anisotropies on 10-arcminute scales. This tendency for open models to violate the upper limits to arcminute-scale anisotropies is in fact a long-standing problem, which allowed Bond and Efstathiou (1984) to deduce the following limit on CDM universes:

$$\Omega \gtrsim 0.3 h^{-4/3}.$$

The known lack of a CMB peak at high ℓ was thus already a very strong argument for a flat universe (with the caveats expressed in the earlier section on geometrical

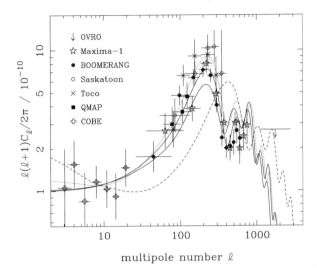

multipole number ℓ

Figure 2.22. Angular power spectra $T^2(\ell) = \ell(\ell + 1)C_\ell/2\pi$ for the CMB, plotted against angular wavenumber ℓ in rad^{-1}. The experimental data are an updated version of the compilation described in White *et al* (1994), communicated by M White; see also Hancock *et al* (1997) and Jaffe *et al* (2000). Various model predictions for adiabatic scale-invariant CDM fluctuations are shown. The two full curves correspond to $(\Omega, \Omega_B, h) = (1, 0.05, 0.5)$ and $(1,0.1,0.5)$, with the higher Ω_B increasing power by about 20% at the peak. The dotted line shows a flat Λ-dominated model with $(\Omega, \Omega_B, h) = (0.3, 0.05, 0.65)$; the broken curve shows an open model with the same parameters. Note the very similar shapes of all the curves. The normalization has been set to the large-scale amplitude, and so any dependence on Ω is quite modest. The main effects are that open models shift the peak to the right, and that the height of the peak increases with Ω_B and h.

degeneracy). Now that we have a direct detection of a peak at low ℓ, this argument for a flat universe is even stronger.

If the basic adiabatic CDM paradigm is adopted, then we can move beyond generic statements about flatness to attempt to use the CMB to measure cosmological parameters. In a recent analysis (Jaffe *et al* 2000), the following best-fitting values for the densities in collisionless matter (c), baryons (b) and vacuum (v) were obtained, together with tight constraints on the power-spectrum index:

$$\Omega_c + \Omega_b + \Omega_v = 1.11 \pm 0.07$$
$$\Omega_c h^2 = 0.14 \pm 0.06$$
$$\Omega_b h^2 = 0.032 \pm 0.005$$

$$n = 1.01 \pm 0.09.$$

The only parameter left undetermined by the CMB is the Hubble constant, h. Recent work (e.g. Mould *et al* 2000) suggests that this is now determined to an rms accuracy of 10%, and we adopt a central value of $h = 0.70$. This completes the cosmological model, requiring a total matter density parameter $\Omega_c + \Omega_b = 0.35 \pm 0.14$, very nicely consistent with what is required to be consistent with σ_8 for exactly scale-invariant fluctuations. The predicted fluctuation shape is also very sensible for this model: $\Gamma = 0.18$.

The fact that such a 'vanilla' model matches the main cosmological data so well is a striking achievement, but it raises a number of issues. One is that the baryon density inferred from the data exceeds that determined via primordial nucleosynthesis by about a factor 1.5. This may sound like good agreement, but both the CMB and nucleosynthesis are now impressively precise areas of science, and neither can easily accommodate the other's figure. The boring solution is that small systematics will eventually be identified that allow a compromise figure. Alternatively, this inconsistency could be the first sign that something is rotten in the basic framework. However, it is too early to make strong claims in this direction.

Of potentially greater significance is the fact that this successful fit has been achieved using scalar fluctuations alone; indeed, the tensor modes are not even mentioned by Jaffe *et al* (2000). To a certain extent, the presence of tensor modes can be hidden by adjustments in the other parameters. There can be no acoustic peak in the tensor contribution, so that the addition of tensors would require larger peak in the scalar component to compensate, pushing in the direction of universes that are of lower density, with larger baryon fractions. However, this would make it harder to keep higher harmonics of the acoustic oscillations low – and it is the lack of detection of any second and third peaks that forces the high baryon density in this solution. There would also be the danger of spoiling the very good agreement with other constraints, such as the σ_8 normalization. We therefore have to face the unpalatable fact that there is as yet no sign of the two generic signatures expected from inflationary models: tilt and a significant tensor contribution. It may be that the next generation of CMB experiments will detect such features at a low level. If they do not, and the initial conditions for structure formation remain as they presently seem to be (scale-invariant adiabatic scalar modes), then the heroic vision of using cosmology to probe physics near the Planck scale may not be achieved. The stakes are high.

References

Abramowitz M and Stegun I A 1965 *Handbook of Mathematical Functions* (New York: Dover)

Adelberger K, Steidel C, Giavalisco M, Dickinson M, Pettini M and Kellogg M 1998 *Astrophys. J.* **505** 18

Ballinger W E, Peacock J A and Heavens A F 1996 *Mon. Not. R. Astron. Soc.* **282** 877

Bardeen J M, Bond J R, Kaiser N and Szalay A S 1986 *Astrophys. J.* **304** 15

Baugh C M and Efstathiou G 1993 *Mon. Not. R. Astron. Soc.* **265** 145

——1994 *Mon. Not. R. Astron. Soc.* **267** 323

Benson A J, Cole S, Frenk C S, Baugh C M and Lacey C G 2000a *Mon. Not. R. Astron. Soc.* **311** 793

Benson A J, Baugh C M, Cole S, Frenk C S, Lacey C G 2000b *Mon. Not. R. Astron. Soc.* **316** 107

Bond J R 1995 *Phys. Rev. Lett.* **74** 4369

——1997 Cosmology and large-scale structure *Proc. 60th Les Houches School* ed R Schaeffer *et al* (Amsterdam: Elsevier) p 469

Bond J R, Cole S, Efstathiou G and Kaiser N 1991 *Astrophys. J.* **379** 440

Bond J R, Crittenden R, Davis R L, Efstathiou G and Steinhardt P J 1994 *Phys. Rev. Lett.* **72** 13

Bond J R and Efstathiou G 1984 *Astrophys. J.* **285** L45

Bunn E F 1995 *PhD Thesis* University of California, Berkeley

Bunn E F, Scott D and White M 1995 *Astrophys. J.* **441** 9

Carlberg R G, Yee H K C, Morris S L, Lin H, Hall P B, Patton D, Sawicki M and Shepherd C W 2000 *Astrophys. J.* **542** 57

Carroll S M, Press W H and Turner E L 1992 *Annu. Rev. Astron. Astrophys.* **30** 499

Cole S, Aragón-Salamanca A, Frenk C S, Navarro J F and Zepf S E 1994 *Mon. Not. R. Astron. Soc.* **271** 781

Colless M 1999 *Phil. Trans. R. Soc.* A **357** 105

Davis M and Peebles P J E 1983 *Astrophys. J.* **267** 465

de Bernardis P *et al* 2000 *Nature* **404** 955

Efstathiou G 1990 Physics of the early Universe *Proc. 36th Scottish Universities Summer School in Physics* ed J A Peacock, A F Heavens and A T Davies (Bristol: Adam Hilger) p 361

——1995 *Mon. Not. R. Astron. Soc.* **274** L73

Efstathiou G, Bernstein G, Katz N, Tyson T and Guhathakurta P 1991 *Astrophys. J.* **380** 47

Efstathiou G and Bond J R 1999 *Mon. Not. R. Astron. Soc.* **304** 75

Efstathiou G, Bond J R and White S D M 1992 *Mon. Not. R. Astron. Soc.* **258** 1P

Efstathiou G, Sutherland W and Maddox S J 1990 *Nature* **348** 705

Eisenstein D J and Hu W 1998 *Astrophys. J.* **496** 605

Eisenstein D J and Zaldarriaga M 1999 *Astrophys. J.* **546** 2

Eke V R, Cole S and Frenk C S 1996 *Mon. Not. R. Astron. Soc.* **282** 263

Feldman H A, Kaiser N and Peacock J A 1994 *Astrophys. J.* **426** 23

Felten J E and Isaacman R 1986 *Rev. Mod. Phys.* **58** 689

Fixsen D J, Cheng E S, Gales J M, Mather J C, Shafer R A and Wright E L 1996 *Astrophys. J.* **473** 576

Gaztañaga E and Baugh C M 1998 *Mon. Not. R. Astron. Soc.* **294** 229

Górski K M, Ratra B, Sugiyama N and Banday A J 1995 *Astrophys. J.* **444** L65

Gradshteyn I S and Ryzhik I M 1980 *Table of Integrals, Series and Products* (New York: Academic Press)

Hamilton A J S 1992 *Astrophys. J.* **385** L5

——1997a *Mon. Not. R. Astron. Soc.* **289** 285

——1997b *Mon. Not. R. Astron. Soc.* **289** 295

——1998 *The Evolving Universe* ed D Hamilton (Dordrecht: Kluwer) pp 185–275

Hamilton A J S, Kumar P, Lu E and Matthews A 1991 *Astrophys. J.* **374** L1

Hanany S *et al* 2000 *Astrophys. J.* **545** L5

Hancock S *et al* 1997 *Mon. Not. R. Astron. Soc.* **289** 505

Heath D 1977 *Mon. Not. R. Astron. Soc.* **179** 351

Hu W, Bunn E F and Sugiyama N 1995 *Astrophys. J.* **447** L59

Hu W and Sugiyama N 1995 *Astrophys. J.* **444** 489

Hu W and White M 1997 *New Astronomy* **2** 323

Huchra J P, Geller M J, de Lapparant V and Corwin H G 1990 *Astrophys. J. Suppl.* **72** 433

Jaffe A *et al* 2000 *Preprint* astro-ph/0007333

Jenkins A, Frenk C S, Pearce F R, Thomas P A, Colberg J M, White S D M, Couchman H M P, Peacock J A, Efstathiou G and Nelson A H 1998 *Astrophys. J.* **499** 20

Kaiser N 1987 *Mon. Not. R. Astron. Soc.* **227** 1

Kauffmann G, Colberg J M, Diaferio A and White S D M 1999 *Mon. Not. R. Astron. Soc.* **303** 188

Kauffmann G, White S D M and Guiderdoni B 1993 *Mon. Not. R. Astron. Soc.* **264** 201

Klypin A, Primak J and Holtzman J 1996 *Astrophys. J.* **466** 13

Lahav O, Lilje P B, Primack J R and Rees M J 1991 *Mon. Not. R. Astron. Soc.* **251** 128

Le Fèvre O *et al* 1996 *Astrophys. J.* **461** 534

Liddle A R and Lyth D 1993 *Phys. Rep.* **231** 1

——2000 *Cosmological Inflation & Large-Scale Structure* (Cambridge: Cambridge University Press)

Liddle A R and Scherrer R J 1999 *Phys. Rev.* D **59** 023509 (astro-ph/9809272)

Mészáros P 1974 *Astron. Astrophys.* **37** 225

Maddox S J, Efstathiou G, Sutherland W J 1996 *Mon. Not. R. Astron. Soc.* **283** 1227

Margon B 1999 *Phil. Trans. R. Soc.* A **357** 93

Mather J C *et al* 1990 *Astrophys. J.* **354** L37

Matsubara T, Szalay A S and Landy S D 2000 *Astrophys. J.* **535** 1

McClelland J and Silk J 1977 *Astrophys. J.* **217** 331

Meiksin A A, White M 1999 *Mon. Not. R. Astron. Soc.* **308** 1179

Meiksin A A, White M and Peacock J A 1999 *Mon. Not. R. Astron. Soc.* **304** 851

Moore B, Frenk C S and White S D M 1993 *Mon. Not. R. Astron. Soc.* **261** 827

Moore B, Quinn T, Governato F, Stadel J and Lake G 1999 *Mon. Not. R. Astron. Soc.* **310** 1147

Mould J R *et al* 2000 *Astrophys. J.* **529** 786

Navarro J F, Frenk C S and White S D M 1996 *Astrophys. J.* **462** 563

Neyman J, Scott E L and Shane C D 1953 *Astrophys. J.* **117** 92

Padmanabhan N, Tegmark M and Hamilton A J S 1999 *Astrophys. J.* **550** 52

Partridge R B 1995 *3K: The Cosmic Microwave Background* (Cambridge: Cambridge University Press)

Peacock J A 1997 *Mon. Not. R. Astron. Soc.* **284** 885

——1999 *Cosmological Physics* (Cambridge: Cambridge University Press)

Peacock J A and Dodds S J 1994 *Mon. Not. R. Astron. Soc.* **267** 1020

——1996 *Mon. Not. R. Astron. Soc.* **280** L19

Peacock J A and Smith R E 2000 *Mon. Not. R. Astron. Soc.* **318** 1144

Pearce F R *et al* 1999 *Astrophys. J.* **521** L99

Peebles P J E 1973 *Astrophys. J.* **185** 413

——1974 *Astrophys. J.* **32** 197

——1980 *The Large-Scale Structure of the Universe* (Princeton, NJ: Princeton University Press)

——1982 *Astrophys. J.* **263** L1

Peebles P J E and Yu J T 1970 *Astrophys. J.* **162** 815

Pen U-L, Seljak U and Turok N 1997 *Phys. Rev. Lett.* **79** 1611

Penzias A A and Wilson R W 1965 *Astrophys. J.* **142** 419

Perlmutter S *et al* 1998 *Astrophys. J.* **517** 565

Pogosyan D and Starobinsky A A 1995 *Astrophys. J.* **447** 465

Press W H, Teukolsky S A, Vetterling W T and Flannery B P 1992 *Numerical Recipes* 2nd edn (Cambridge: Cambridge University Press)

Ratra B and Peebles P J E 1988 *Phys. Rev.* D **37** 3406

Riess A G *et al* 1998 *Astron. J.* **116** 1009

Sachs R K and Wolfe A M 1967 *Astrophys. J.* **147** 73

Saunders W *et al* 2000 *Mon. Not. R. Astron. Soc.* **317** 55

Saunders W, Rowan-Robinson M and Lawrence A 1992 *Mon. Not. R. Astron. Soc.* **258** 134

Scoccimarro R, Zaldarriaga M and Hui L 1999 *Astrophys. J.* **527** 1

Seljak U 1997 *Astrophys. J.* **482** 6

——2000 *Mon. Not. R. Astron. Soc.* **318** 203

Seljak U and Zaldarriaga M 1996 *Astrophys. J.* **469** 437

Shanks T and Boyle B J 1994 *Mon. Not. R. Astron. Soc.* **271** 753

Shanks T, Fong R, Boyle B J and Peterson B A 1987 *Mon. Not. R. Astron. Soc.* **227** 739

Shectman S A, Landy S D, Oemler A, Tucker D L, Lin H, Kirshner R P and Schechter P L 1996 *Astrophys. J.* **470** 172

Sheth R K and Tormen G 1999 *Mon. Not. R. Astron. Soc.* **308** 11

Somerville R S and Primack J R 1999 *Mon. Not. R. Astron. Soc.* **310** 1087

Starobinsky A A 1985 *Sov. Astron. Lett.* **11** 133

Strauss M A and Willick J A 1995 *Phys. Rep.* **261** 271

Sugiyama N 1995 *Astrophys. J. Suppl.* **100** 281

Tegmark M 1996 *Mon. Not. R. Astron. Soc.* **280** 299

Tegmark M, Taylor A N and Heavens A F 1997 *Astrophys. J.* **480** 22

van Kampen E, Jimenez R and Peacock J A 1999 *Mon. Not. R. Astron. Soc.* **310** 43

Viana P T and Liddle A R 1996 *Mon. Not. R. Astron. Soc.* **281** 323

Vittorio N and Silk J 1991 *Astrophys. J.* **385** L9

Vogeley M S and Szalay A S 1996 *Astrophys. J.* **465** 34

Weinberg S 1972 *Gravitation & Cosmology* (New York: Wiley)

——1989 *Rev. Mod. Phys.* **61** 1

White M and Bunn E F 1995 *Astrophys. J.* **450** 477

White M, Scott D and Silk J 1994 *Annu. Rev. Astron. Astrophys.* **32** 319

White S D M, Efstathiou G and Frenk C S 1993 *Mon. Not. R. Astron. Soc.* **262** 1023

White S D M and Rees M 1978 *Mon. Not. R. Astron. Soc.* **183** 341

Zlatev I, Wang L and Steinhardt P J 1999 *Phys. Rev. Lett.* **82** 896

Chapter 3

Cosmological models

George F R Ellis
Mathematics Department, University of Cape Town, South Africa

3.1 Introduction

The current standard models of the universe are the Friedmann–Lemaître (FL) family of models, based on the Robertson–Walker (RW) spatially homogeneous and isotropic geometries but with a much more complex set of matter constituents than originally envisaged by Friedmann and Lemaître. It is appropriate then to ask whether the universe is indeed well described by an RW geometry. There is reasonable evidence supporting these models on the largest observable scales, but at smaller scales they are clearly a bad description. Thus a better form of the question is: *On what scales and in what domains is the universe's geometry nearly RW? What are the best-fit RW parameters in the observable domain?*

Given that the universe is apparently well described by the RW geometry on the largest scales in the observable domain, the next question is: *Why is it RW? How did the universe come to have such an improbable geometry?* The predominant answer to this question at present is that it results from a very early epoch when inflation took place (a period of accelerating expansion through many e-folds of the scale of the universe). It is important to consider how good an answer this is. One can only do so by considering alternatives to RW geometries, as well as the models based on those geometries.

The third question is: *How did astronomical structure come to exist on smaller scales? Given a smooth structure on the largest scales, how was that smoothness broken on smaller scales?* Again, inflationary theory applied to perturbed FL models gives a general answer to that question: quantum fluctuations in the very early universe formed the seeds of inhomogeneities that could then grow, on scales bigger than the (time-dependent) Jeans' scale, by gravitational attraction. It is important to note, however, that not only do structure-formation effects depend in important ways on the background model, but also (and indeed, in consequence of this remark) many of the ways of estimating the model parameters depend on models of structure formation. Thus the previous questions and this one interact in a number of ways.

This review will look at the first two questions in some depth, and only briefly consider the third (which is covered in depth in Peacock's chapter). To examine these questions, we need to consider the family of cosmological solutions with observational properties like those of the real universe at some stage of their histories. Thus we are interested in the *full state space of solutions*, allowing us to see how realistic (lumpy) models are related to each other and to higher symmetry models, including, in particular, the FL models. This chapter develops general techniques for examining this family of models, and describes some specific models of interest. The first part looks at exact general relations valid in all cosmological models, the second part examines exact cosmological solutions of the field equations and the third part looks at the observational properties of these models and then returns to considering the previous questions. The chapter concludes by emphasizing some of the fundamental issues that make it difficult to obtain definitive answers if one tries to pursue the chain of cause and effect to extremely early times.

3.1.1 Spacetime

We will make the standard assumption that on large scales, physics is dominated by gravity, which is well described by general relativity (see, e.g. d'Inverno [19], Wald [129], Hawking and Ellis [68] or Stephani [117]), with gravitational effects resulting from spacetime curvature. The starting point for describing a spacetime is an atlas of local coordinates $\{x^i\}$ covering the four-dimensional spacetime manifold $\mathcal{M}$, and a Lorentzian metric tensor $g_{ij}(x^k)$ at each point of $\mathcal{M}$, representing the spacetime geometry near the point on a particular scale. This then determines the connection components $\Gamma^i_{jk}(x^s)$, and, hence, the spacetime curvature tensor R_{ijkl}, at that scale. The curvature tensor can be decomposed into its trace-free part (the Weyl tensor $C_{ijkl} : C^i{}_{jil} = 0$) and its trace (the Ricci tensor $R_{ik} \equiv R^s{}_{isk}$) by the relation

$$R_{ijkl} = C_{ijkl} - \tfrac{1}{2}(R_{ik}g_{jl} + R_{jl}g_{ik} - R_{il}g_{jk} - R_{jk}g_{il}) + \tfrac{1}{6}R(g_{ik}g_{jl} - g_{il}g_{jk}), \quad (3.1)$$

where $R \equiv R^a{}_a$ is the Ricci scalar. The coordinates may be chosen arbitrarily in each neighbourhood in $\mathcal{M}$. To be useful in an explanatory role, a cosmological model must be easy to describe—this means they have symmetries or special properties of some kind or other.

3.1.2 Field equations

The metric tensor is determined, at the relevant averaging scale, by the *Einstein gravitational field equations* ('EFEs')

$$(R_{ij} - \tfrac{1}{2}Rg_{ij}) + \lambda g_{ij} = \kappa T_{ij} \Leftrightarrow R_{ij} = \lambda g_{ij} + \kappa(T_{ij} - \tfrac{1}{2}Tg_{ij}) \quad (3.2)$$

where λ is the cosmological constant and κ the gravitational constant. Here T_{ij} (with trace $T = T^a_a$) is the total energy–momentum–stress tensor for all the matter and fields present, described at the relevant averaging scale. This covariant equation (a set of second-order nonlinear equations for the metric tensor components) shows that the Ricci tensor is determined pointwise by the matter present at each point, but the Weyl tensor is not so determined; rather it is fixed by suitable boundary conditions, together with the Bianchi identities for the curvature tensor:

$$\nabla_{[e} R_{ab]cd} = 0 \Leftrightarrow \nabla_{[e} R^e_{ab]cd} = 0 \qquad (3.3)$$

(the equivalence of the full equations on the left with the first contracted equations on the right holding only for four dimensions or less). Consequently it is this tensor that enables gravitational 'action at a distance' (gravitational radiation, tidal forces, and so on). Contracting the right-hand of equation (3.3) and substituting into the divergence of equation (3.2) shows T_{ij} necessarily obeys the energy–momentum conservation equations

$$\nabla_j T^{ij} = 0 \qquad (3.4)$$

(the divergence of λg_{ij} vanishes provided λ is indeed constant, as we assume). Thus matter determines the geometry which, in turn, determines the motion of the matter (see e.g. [132]). We can look for exact solutions of these equations, or approximate solutions obtained by suitable linearization of the equations; and one can also consider how the solutions relate to Newtonian theory solutions. Care must be exercised in the latter two cases, both because of the nonlinearity of the theory, and because there is no fixed background spacetime available in general relativity theory. This makes it essentially different from both Newtonian theory and special relativity.

3.1.3   Matter description

The total stress tensor T_{ij} is the sum of the N stress tensors T_{nij} for the various matter components labelled by index n (baryons, radiation, neutrinos, etc):

$$T_{ij} = \Sigma_n T_{nij} \qquad (3.5)$$

each component being described by suitable equations of state which encapsulate their physics. The most common forms of matter in the cosmological context will often to a good approximation, each have a 'perfect fluid' stress tensor;

$$T_{nij} = (\mu_n + p_n)u_{ni}u_{nj} + p_n g_{ij} \qquad (3.6)$$

with unit 4-velocity u^i_n ($u_{ni} u^i_n = -1$), energy density μ_n and pressure p_n, with suitable equations of state relating μ_n and p_n. In simple cases, they will be related by a barotropic relation $p_n = p_n(\mu_n)$; for example, for baryons, $p_b = 0$ and for radiation, e.g. the cosmic background radiation ('CBR'), $p_r = \mu_r/3$,. However, in more complex cases there will be further variables determining p_n and μ_n; for example, in the case of a massless scalar field ϕ with potential $V(\phi)$, on choosing u^i as the unit vector normal to spacelike surfaces $\phi = $ constant, the stress tensor takes the form (3.6) with

$$4\pi p_\phi = \tfrac{1}{2}\dot{\phi}^2 - V(\phi), \qquad 4\pi\mu_\phi = \tfrac{1}{2}\dot{\phi}^2 + V(\phi). \qquad (3.7)$$

It must be noted that, in general, different matter components will each have a different 4-velocity u_n^i, and the total stress tensor (3.5) of perfect fluid stress tensors (3.6) itself has the perfect fluid form if and only if the 4-velocities of all contributing matter components are the same, i.e. $u_n^i = u^i$ for all n; in that case,

$$T_{ij} = (\mu + p)u_i u_j + pg_{ij}, \qquad \mu \equiv \Sigma_n \mu_n, \qquad p \equiv \Sigma_n p_n \qquad (3.8)$$

where μ is the total energy density and p the total pressure.

The individual matter components will each separately satisfy the conservation equation (3.4) if they are non-interacting with the other components; however this will no longer be the case if interactions lead to exchanges of energy and momentum between the different components. The key to a physically realistic cosmological model is the representation of suitable matter components, with realistic equations of state for each matter component and equations describing the interactions between the components. For reasonable behaviour of matter, irrespective of its constitution we require the 'energy condition'

$$\mu + p > 0 \qquad (3.9)$$

on cosmological averaging scales (the vacuum case $\mu + p = 0$ can apply only to regions described on averaging scales less than or equal to that of clusters of galaxies).

3.1.4 Cosmology

A key feature of cosmological models, as contrasted with general solutions of the EFEs, is that in them, at each point a unique 4-velocity u^a is defined representing the preferred motion of matter there on a cosmological scale. Whenever the matter present is well described by the perfect fluid stress tensor (3.8), because of (3.9) there will be a unique timelike eigenvector of this tensor that can be used to define the vector u, representing the average motion of the matter, and conventionally referred to as defining the *fundamental world-lines* of the cosmology. Unless stated otherwise, we will assume that observers move with this 4-velocity. At late times, a unique frame is defined by choosing a 4-velocity such that the CBR anisotropy dipole vanishes; the usual assumption is that this is the same frame as defined locally by the average motion of matter [26]; indeed this assumption is what underlies studies of large-scale motions and the 'Great Attractor'.

The description of matter and radiation in a cosmological model must be sufficiently complete to determine the *observational relations* predicted by the model for both discrete sources and the background radiation, implying a well-developed theory of *structure growth* for very small and for very large physical scales (i.e. for light atomic nuclei and for galaxies and clusters of galaxies), and of *radiation absorbtion and emission*. Clearly an essential requirement for a viable cosmological model is that it should be able to reproduce current large-scale astronomical observations accurately.

I will deal with both the $1 + 3$ covariant approach [21, 26, 28, 91] and the orthonormal tetrad approach, which serves as a completion to the $1 + 3$ covariant approach [41].

3.2 $1 + 3$ covariant description: variables

3.2.1 Average 4-velocity of matter

The preferred 4-velocity is

$$u^a = \frac{dx^a}{d\tau}, \qquad u_a u^a = -1, \tag{3.10}$$

where τ is the proper time measured along the fundamental world-lines. Given u^a, unique *projection tensors* can be defined:

$$U^a{}_b = -u^a u_b \Rightarrow U^a{}_c U^c{}_b = U^a{}_b, U^a{}_a = 1, U_{ab} u^b = u_a,$$
$$h_{ab} = g_{ab} + u_a u_b \Rightarrow h^a{}_c h^c{}_b = h^a{}_b, h^a{}_a = 3, h_{ab} u^b = 0. \tag{3.11}$$

The first projects parallel to the velocity vector u^a, and the second determines the metric properties of the (orthogonal) instantaneous rest-spaces of observers moving with 4-velocity u^a. A *volume element* for the rest spaces:

$$\eta_{abc} = u^d \eta_{dabc} \Rightarrow \eta_{abc} = \eta_{[abc]}, \eta_{abc} u^c = 0, \tag{3.12}$$

where η_{abcd} is the four-dimensional volume element ($\eta_{abcd} = \eta_{[abcd]}$, $\eta_{0123} = \sqrt{|\det g_{ab}|}$) is also defined.

Furthermore, two derivatives are defined: the covariant time derivative "˙" along the fundamental world-lines, where for any tensor $T^{ab}{}_{cd}$

$$\dot{T}^{ab}{}_{cd} = u^e \nabla_e T^{ab}{}_{cd}, \tag{3.13}$$

and the fully orthogonally projected covariant derivative $\tilde{\nabla}$ where, for any tensor $T^{ab}{}_{cd}$,

$$\tilde{\nabla}_e T^{ab}{}_{cd} = h^a{}_f h^b{}_g h^p{}_c h^q{}_d h^r{}_e \nabla_r T^{fg}{}_{pq}, \tag{3.14}$$

with total projection on all free indices. The tilde serves as a reminder that if u^a has *non-zero* vorticity, $\tilde{\nabla}$ is *not* a proper three-dimensional covariant derivative (see equation (3.20)). The projected time and space derivatives of U_{ab}, h_{ab} and η_{abc} all vanish. Finally, following [91] we use angle brackets to denote orthogonal projections of vectors and the orthogonally projected symmetric trace-free part of tensors:

$$v^{\langle a \rangle} = h^a{}_b v^b, \qquad T^{\langle ab \rangle} = [h^{(a}{}_c h^{b)}{}_d - \tfrac{1}{3} h^{ab} h_{cd}] T^{cd}; \qquad (3.15)$$

for convenience the angle brackets are also used to denote othogonal projections of covariant time derivatives along u^a ('*Fermi derivatives*'):

$$\dot{v}^{\langle a \rangle} = h^a{}_b \dot{v}^b, \qquad \dot{T}^{\langle ab \rangle} = [h^{(a}{}_c h^{b)}{}_d - \tfrac{1}{3} h^{ab} h_{cd}] \dot{T}^{cd}. \qquad (3.16)$$

3.2.2 Kinematic quantities

The orthogonal vector $\dot{u}^a = u^b \nabla_b u^a$ is the *acceleration vector*, representing the degree to which the matter moves under forces other than gravity plus inertia (which cannot be covariantly separated from each other in general relativity). The acceleration vanishes for matter in free fall (i.e. moving under gravity plus inertia alone).

We split the first covariant derivative of u_a into its irreducible parts, defined by their symmetry properties:

$$\nabla_a u_b = -u_a \dot{u}_b + \tilde{\nabla}_a u_b = -u_a \dot{u}_b + \tfrac{1}{3} \Theta h_{ab} + \sigma_{ab} + \omega_{ab} \qquad (3.17)$$

where the trace $\Theta = \tilde{\nabla}_a u^a$ is the *(volume) rate of expansion* of the fluid (with $H = \Theta/3$ the Hubble parameter); $\sigma_{ab} = \tilde{\nabla}_{\langle a} u_{b \rangle}$ is the trace-free symmetric *shear* tensor ($\sigma_{ab} = \sigma_{(ab)}$, $\sigma_{ab} u^b = 0$, $\sigma^a{}_a = 0$), describing the rate of distortion of the matter flow; and $\omega_{ab} = \tilde{\nabla}_{[a} u_{b]}$ is the skew-symmetric *vorticity* tensor ($\omega_{ab} = \omega_{[ab]}$, $\omega_{ab} u^b = 0$), describing the rotation of the matter relative to a non-rotating (Fermi-propagated) frame. The meaning of these quantities follows from the evolution equation for a relative position vector $\eta^a_\perp = h^a{}_b \eta^b$, where η^a is a deviation vector for the family of fundamental world-lines, i.e. $u^b \nabla_b \eta^a = \eta^b \nabla_b u^a$. Writing $\eta^a_\perp = \delta\ell e^a$, $e_a e^a = 1$, we find the relative distance $\delta\ell$ obeys the propagation equation

$$\frac{(\delta\ell)^{\cdot}}{\delta\ell} = \tfrac{1}{3} \Theta + (\sigma_{ab} e^a e^b), \qquad (3.18)$$

(the generalized Hubble law), and the relative direction vector e^a the propagation equation

$$\dot{e}^{\langle a \rangle} = (\sigma^a{}_b - (\sigma_{cd} e^c e^d) h^a{}_b - \omega^a{}_b) e^b, \qquad (3.19)$$

giving the observed rate of change of position in the sky of distant galaxies [21, 26].

Each function f satisfies the important *commutation relation* for the $\tilde{\nabla}$-derivative [40]

$$\tilde{\nabla}_{[a} \tilde{\nabla}_{b]} f = \eta_{abc} \omega^c \dot{f}. \qquad (3.20)$$

Applying this to the energy density μ shows that if $\omega^a \dot\mu \neq 0$ in an open set then $\tilde\nabla_a \mu \neq 0$ there, so non-zero vorticity implies anisotropic number counts in an expanding universe [61] (this is because there are then no 3-surfaces orthogonal to the fluid flow; see [21, 26]).

3.2.2.1 *Auxiliary quantities*

It is useful to define some associated kinematical quantities:

- the *vorticity vector* $\omega^a = \frac{1}{2}\eta^{abc}\omega_{bc} \Rightarrow \omega_a u^a = 0,\ \omega_{ab}\omega^b = 0$,
- the magnitudes $\omega^2 = \frac{1}{2}(\omega_{ab}\omega^{ab}) \geq 0,\ \sigma^2 = \frac{1}{2}(\sigma_{ab}\sigma^{ab}) \geq 0$, and
- the *average length scale* S determined by $\frac{\dot S}{S} = \frac{1}{3}\Theta$, so the volume of a fluid element varies along the fluid flow lines as S^3.

3.2.3 Matter tensor

Both the total matter *energy–momentum tensor* T_{ab} and each of its components can be decomposed relative to u^a in the form

$$T_{ab} = \mu u_a u_b + q_a u_b + u_a q_b + p h_{ab} + \pi_{ab}, \tag{3.21}$$

where $\mu = (T_{ab}u^a u^b)$ is the *relativistic energy density* relative to u^a, $q^a = -T_{bc}u^b h^{ca}$ is the *relativistic momentum density* ($q_a u^a = 0$), which is also the energy flux relative to u^a, $p = \frac{1}{3}(T_{ab}h^{ab})$ is the *isotropic pressure*, and $\pi_{ab} = T_{cd}h^c{}_{\langle a}h^d{}_{b\rangle}$ is the trace-free *anisotropic pressure* ($\pi^a{}_a = 0,\ \pi_{ab} = \pi_{(ab)}$, $\pi_{ab}u^b = 0$). A different choice of u^a will result in a different splitting. The physics of the situation is in the equations of state relating these quantities; for example, the commonly imposed restrictions

$$q^a = \pi_{ab} = 0 \Leftrightarrow T_{ab} = \mu u_a u_b + p h_{ab} \tag{3.22}$$

characterize a 'perfect fluid' moving with the chosen 4-velocity u_a as in equation (3.8) with, in general, an equation of state $p = p(\mu, s)$ where s is the entropy [21, 26].

3.2.4 Electromagnetic field

The *Maxwell field tensor* F_{ab} of an electromagnetic field is split relative to u^a into electric and magnetic parts by the relations (see [28])

$$E_a = F_{ab}u^b \Rightarrow E_a u^a = 0, \tag{3.23}$$
$$H_a = \frac{1}{2}\eta_{abc}F^{bc} \Rightarrow H_a u^a = 0. \tag{3.24}$$

Again, a different choice of u^a will result in a different split.

3.2.5 Weyl tensor

In analogy to F_{ab}, the *Weyl conformal curvature tensor* C_{abcd} defined by equation (3.1) is split relative to u^a into 'electric' and 'magnetic' *Weyl curvature* parts according to

$$E_{ab} = C_{acbd}u^c u^d \Rightarrow E^a{}_a = 0, \ E_{ab} = E_{(ab)}, \ E_{ab}u^b = 0, \quad (3.25)$$

$$H_{ab} = \tfrac{1}{2}\eta_{ade}C^{de}{}_{bc}u^c \Rightarrow H^a{}_a = 0, \ H_{ab} = H_{(ab)}, \ H_{ab}u^b = 0. \quad (3.26)$$

These influence the motion of matter and radiation through the *geodesic deviation equation* for timelike and null vectors, see, respectively, [107] and [120].

3.3 1 + 3 Covariant description: equations

There are three sets of equations to be considered, resulting from EFE (3.2) and its associated integrability conditions.

3.3.1 Energy–momentum conservation equations

We obtain from the conservation equations (3.4), on projecting parallel and perpendicular to u^a and using (3.21), the propagation equations

$$\dot{\mu} + \tilde{\nabla}_a q^a = -\Theta(\mu + p) - 2(\dot{u}_a q^a) - (\sigma^a{}_b \pi^b{}_a), \quad (3.27)$$

$$\dot{q}^{\langle a\rangle} + \tilde{\nabla}^a p + \tilde{\nabla}_b \pi^{ab} = -\tfrac{4}{3}\Theta q^a - \sigma^a{}_b q^b - (\mu + p)\dot{u}^a - \dot{u}_b \pi^{ab} - \eta^{abc}\omega_b q_c. \quad (3.28)$$

For perfect fluids, characterized by equation (3.8), these reduce to

$$\dot{\mu} = -\Theta(\mu + p), \quad (3.29)$$

the *energy conservation equation*, and the *momentum conservation equation*

$$0 = \tilde{\nabla}_a p + (\mu + p)\dot{u}_a \quad (3.30)$$

(which because of the perfect fluid assumption, has changed from a time-derivative equation for q^a to an algebraic equation for $\dot{u}_a$, and thus a time-derivative equation for u^a). These equations show that $(\mu + p)$ is both the inertial mass density and that it governs the conservation of energy. It is clear that if this quantity is zero (the case of an effective cosmological constant) or negative, the behaviour of matter will be anomalous; in particular velocities will be unstable if $\mu + p \to 0$, because the acceleration generated by a given force will diverge in this limit. If we assume a perfect fluid with a (linear) γ-law equation of state, then (3.29) shows that

$$p = (\gamma - 1)\mu, \ \dot{\gamma} = 0 \Rightarrow \mu = M/S^{3\gamma}, \ \dot{M} = 0. \quad (3.31)$$

One can approximate ordinary matter in this way, with $1 \leq \gamma \leq 2$ in order that the causality and energy conditions are valid. Radiation corresponds to $\gamma = \frac{4}{3} \Rightarrow \mu = M/S^4$, so from Stefan's law ($\mu \propto T^4$) we find that $T \propto 1/S$. Another useful case is *pressure-free matter* (often described as 'baryonic' or 'cold dark matter (CDM)'); the momentum conservation: (3.30) shows that such matter moves geodesically (as expected from the equivalence principle):

$$\gamma = 1 \Leftrightarrow p = 0 \Rightarrow \dot{u}_a = 0, \mu = M/S^3. \tag{3.32}$$

This is the case of *pure gravitation*, without fluid dynamical effects. Another important case is that of a scalar field, see (3.7).

3.3.2   Ricci identities

The second set of equations arise from the *Ricci identities* for the vector field u^a, i.e.

$$2\nabla_{[a}\nabla_{b]}u^c = R_{ab}{}^c{}_d u^d. \tag{3.33}$$

On substituting from (3.17), using (3.2), and separating out the parallelly and orthogonally projected parts into a trace, symmetric trace-free and skew symmetric part, we obtain three propagation equations and three constraint equations. The *propagation equations* are the Raychaudhuri equation, the vorticity propagation equation and the shear propagation equation.

3.3.2.1   *The Raychaudhuri equation*

This equation

$$\dot{\Theta} = -\tfrac{1}{3}\Theta^2 + \nabla_a \dot{u}^a - 2\sigma^2 + 2\omega^2 - \tfrac{1}{2}(\mu + 3p) + \lambda, \tag{3.34}$$

the *basic equation of gravitational attraction* [21, 26, 28], shows the repulsive nature of a positive cosmological constant and leads to the identification of $(\mu + 3p)$ as the active gravitational mass density. Rewriting it in terms of the average scale factor S, this equation can be rewritten in the form

$$3\frac{\ddot{S}}{S} = -2(\sigma^2 - \omega^2) + \nabla_a \dot{u}^a - \frac{1}{2}(\mu + 3p) + \lambda, \tag{3.35}$$

showing how the curvature of the curve $S(\tau)$ along each world-line (in terms of proper time τ along that world-line) is determined by the shear, vorticity and acceleration; the total energy density and pressure in terms of the combination $(\mu + 3p)$—the *active gravitational mass*; and the cosmological constant λ. This gives the basic singularity theorem.

Singularity theorem.   *[21, 26, 28] In a universe where the active gravitational mass is positive at all times,*

$$(\mu + 3p) > 0, \tag{3.36}$$

the cosmological constant vanishes (or is negative); $\lambda \leq 0$, and the vorticity and acceleration vanish; $\dot{u}^a = \omega^a = 0$ at all times, at any instant when $H_0 = \frac{1}{3}\Theta_0 > 0$, there must have been a time $t_0 < 1/H_0$ ago such that $S \to 0$ as $t \to t_0$; a spacetime singularity occurs there, where $\mu \to \infty$ and $T \to \infty$.

The further singularity theorems of Hawking and Penrose [68,69,124] utilize this result or its null version as an essential part of their proofs.

Closely related to this are three other results:

(1) a static universe model containing ordinary matter requires $\lambda > 0$ (Einstein's discovery of 1917);
(2) the Einstein static universe is unstable (Eddington's discovery of 1930);
(3) in a universe satisfying the requirements of the singularity theorem, at each instant t the age of the universe is less that $1/H(t)$, so for example the hot early stage of the universe takes place extremely rapidly.

Proofs follow directly from (3.35). The energy condition $(\mu + 3p) > 0$ will be satisfied by all ordinary matter but will not, in general, be satisfied by a scalar field, see (3.7).

3.3.2.2   *The vorticity propagation equation*

$$\dot{\omega}^{\langle a \rangle} - \frac{1}{2}\eta^{abc}\tilde{\nabla}_b\dot{u}_c = -\frac{2}{3}\Theta\omega^a + \sigma^a{}_b\omega^b. \tag{3.37}$$

If we have a barotropic perfect fluid:

$$q^a = \pi_{ab} = 0, \; p = p(\mu) \Rightarrow \eta^{abc}\tilde{\nabla}_b\dot{u}_c = 0, \tag{3.38}$$

then $\omega^a = 0$ is involutive: i.e. the statement

$$\omega^a = 0 \text{ initially} \Rightarrow \dot{\omega}^{\langle a \rangle} = 0 \Rightarrow \omega^a = 0 \text{ at later times}$$

follows from the vorticity conservation equation (3.37) (and it is also true in the special case $p = 0$). Thus non-trivial entropy dependence or an imperfect fluid is required to create vorticity.

When the vorticity vanishes $\Leftrightarrow \omega = 0$:

(1) The fluid flow is hypersurface-orthogonal, and there exists a cosmic time function t such that $u_a = -g(x^b)\nabla_a t$, allowing synchronization of the clocks of fundamental observers. If, in addition, the acceleration vanishes, we can set $g = 1$ and the time function can be proper time for all of them (whereas if the acceleration is non-zero, the coordinate time t will necessarily correspond to different proper times along different world-lines).

(2) The metric of the orthogonal 3-spaces $t = $ constant formed by meshing together the tangent spaces orthogonal to u_a is h_{ab}.

(3) From the Gauss equation and the Ricci identities for u^a, the Ricci tensor of these 3-spaces is given by [21, 26]

$$^3R_{ab} = -\dot{\sigma}_{\langle ab \rangle} - \Theta\sigma_{ab} + \tilde{\nabla}_{\langle a}\dot{u}_{b \rangle} + \dot{u}_{\langle a}\dot{u}_{b \rangle} + \pi_{ab} + \tfrac{1}{3}h_{ab}{}^3R, \quad (3.39)$$

and their Ricci scalar is given by

$$^3R = 2\mu - \tfrac{2}{3}\Theta^2 + 2\sigma^2 + 2\lambda, \quad (3.40)$$

which is a generalized Friedmann equation, showing how the matter tensor determines the 3-space average curvature. These equations fully determine the curvature tensor $^3R_{abcd}$ of the orthogonal 3-spaces, and so show how the EFEs result in *spatial* curvature (as well as spacetime curvature) [21, 26].

3.3.2.3 *The shear propagation equation*

$$\dot{\sigma}^{\langle ab \rangle} - \tilde{\nabla}^{\langle a}\dot{u}^{b \rangle} = -\tfrac{2}{3}\Theta\sigma^{ab} + \dot{u}^{\langle a}\dot{u}^{b \rangle} - \sigma^{\langle a}{}_c\sigma^{b \rangle c} - \omega^{\langle a}\omega^{b \rangle} - (E^{ab} - \tfrac{1}{2}\pi^{ab}). \quad (3.41)$$

This shows how the tidal gravitational field E_{ab} directly induces shear (which then feeds into the Raychaudhuri and vorticity propagation equations, thereby changing the nature of the fluid flow), and that the anisotropic pressure term π_{ab} also generates shear in an imperfect fluid situation. Shear-free solutions are very special solutions, because (in contrast to the case of vorticity) a conspiracy of terms is required to maintain the shear zero if it is zero at any initial time (see later for a specific example).

The *constraint equations* are as follows:

(1) The (0α)-*equation*

$$0 = (C_1)^a = \tilde{\nabla}_b\sigma^{ab} - \tfrac{2}{3}\tilde{\nabla}^a\Theta + \eta^{abc}[\tilde{\nabla}_b\omega_c + 2\dot{u}_b\omega_c] + q^a, \quad (3.42)$$

shows how the momentum flux q^a (zero for a comoving perfect fluid) relates to the spatial inhomogeneity of the expansion.

(2) The *vorticity divergence identity*

$$0 = (C_2) = \tilde{\nabla}_a\omega^a - (\dot{u}_a\omega^a), \quad (3.43)$$

follows because ω^a is a curl.

(3) The H_{ab}-*equation*

$$0 = (C_3)^{ab} = H^{ab} + 2\dot{u}^{\langle a}\omega^{b \rangle} + (\text{curl}\,\sigma)^{ab} - (\text{curl}\,\sigma)^{ab}, \quad (3.44)$$

characterizes the magnetic part of the Weyl tensor as being constructed from the 'curls' of the vorticity and shear tensors: $(\text{curl}\,\omega)^{ab} = \eta^{cd\langle a}\tilde{\nabla}_c\omega^{b \rangle}{}_d$, $(\text{curl}\,\sigma)^{ab} = \eta^{cd\langle a}\tilde{\nabla}_c\sigma^{b \rangle}{}_d$.

3.3.3 Bianchi identities

The third set of equations arises from the *Bianchi identities* (3.3). On using the splitting of R_{abcd} into R_{ab} and C_{abcd}, the $1 + 3$ splitting, (3.21),(3.25) of those quantities, and the EFE (3.2), these identities give two further propagation equations and two further constraint equations, which are similar in form to the Maxwell field equations for the electromagnetic field in an expanding universe (see [28]).

The *propagation equations* are:

$$(\dot{E}^{\langle ab \rangle} + \tfrac{1}{2}\dot{\pi}^{\langle ab \rangle}) = (\text{curl } H)^{ab} - \tfrac{1}{2}\tilde{\nabla}^{\langle a}q^{b \rangle} - \tfrac{1}{2}(\mu + p)\sigma^{ab} - \Theta(E^{ab} + \tfrac{1}{6}\pi^{ab})$$
$$+ 3\sigma^{\langle a}{}_c(E^{b \rangle c} - \tfrac{1}{6}\pi^{b \rangle c}) - \dot{u}^{\langle a}q^{b \rangle}$$
$$+ \eta^{cd\langle a}[2\dot{u}_c H^{b \rangle}{}_d + \omega_c(E^{b \rangle}{}_d + \tfrac{1}{2}\pi^{b \rangle}{}_d)], \tag{3.45}$$

the $\dot{E}$-*equation*, and

$$\dot{H}^{\langle ab \rangle} = - (\text{curl } E)^{ab} + \tfrac{1}{2}(\text{curl } \pi)^{ab} - \Theta H^{ab} + 3\sigma^{\langle a}{}_c H^{b \rangle c}$$
$$+ \tfrac{3}{2}\omega^{\langle a}q^{b \rangle} - \eta^{cd\langle a}[2\dot{u}_c E^{b \rangle}{}_d - \tfrac{1}{2}\sigma^{b \rangle}{}_c q_d - \omega_c H^{b \rangle}{}_d], \tag{3.46}$$

the $\dot{H}$-*equation*, where we have defined the 'curls':

$$(\text{curl } H)^{ab} = \eta^{cd\langle a}\tilde{\nabla}_c H^{b \rangle}{}_d, \qquad (\text{curl } E)^{ab} = \eta^{cd\langle a}\tilde{\nabla}_c E^{b \rangle}{}_d. \tag{3.47}$$

These equations show how gravitational radiation arises: as in the electromagnetic case, taking the time derivative of the $\dot{E}$-equation gives a term of the form $(\text{curl } H)$; commuting the derivatives and substituting from the $\dot{H}$-equation eliminates H, and results in a term in $\ddot{E}$ and a term of the form $(\text{curl curl } E)$, which together give the wave operator acting on E [20, 66]. Similarly the time derivative of the $\dot{H}$-equation gives a wave equation for H, and associated with these is a wave equation for the shear σ.

The *constraint equations* are

$$0 = (C_4)^a = \tilde{\nabla}_b(E^{ab} + \tfrac{1}{2}\pi^{ab}) - \tfrac{1}{3}\tilde{\nabla}^a\mu + \tfrac{1}{3}\Theta q^a$$
$$- \tfrac{1}{2}\sigma^a{}_b q^b - 3\omega_b H^{ab} - \eta^{abc}[\sigma_{bd} H^d{}_c - \tfrac{3}{2}\omega_b q_c], \tag{3.48}$$

the $(\text{div } E)$-*equation* with its source the spatial gradient of the energy density and

$$0 = (C_5)^a = \tilde{\nabla}_b H^{ab} + (\mu + p)\omega^a + 3\omega_b(E^{ab} - \tfrac{1}{6}\pi^{ab})$$
$$+ \eta^{abc}[\tfrac{1}{2}\tilde{\nabla}_b q_c + \sigma_{bd}(E^d{}_c + \tfrac{1}{2}\pi^d{}_c)], \tag{3.49}$$

the $(\text{div } H)$-*equation*, with its source the fluid vorticity. The $(\text{div } E)$-equation can be regarded as a (vector) analogue of the Newtonian Poisson equation [52], leading to the Newtonian limit and enabling tidal action at a distance. These equations respectively show that, generically, scalar modes will result in a non-zero divergence of E_{ab} (and hence a non-zero E-field) and vector modes in a non-zero divergence of H_{ab} (and hence a non-zero H-field).

3.3.4 Implications

Altogether, we have six propagation equations and six constraint equations; considered as a set of evolution equations for the $1+3$ covariant variables, they are a first-order system of equations. This set is determinate once the fluid equations of state are given; together they then form a dynamical system (the set closes up, but is essentially an infinite dimensional dynamical system because of the spatial derivatives that occur).

The *key issue* that arises is consistency of the constraints with the evolution equations. It is believed that they are *generally consistent* for physically reasonable and well-defined equations of state, i.e. they are consistent if no restrictions are placed on their evolution other than those implied by the constraint equations and the equations of state (this has been shown for irrotational dust [91]). It is this that makes consistent the overall hyperbolic nature of the equations with the 'instantaneous' action at a distance implicit in the Gauss-like equations (specifically, the (div E)-equation), the point being that the 'action at a distance' nature of the solutions to these equations is built into the initial data, which must be chosen so that the constraints are satisfied initially, and they then remain satisfied thereafter because the time evolution preserves these constraints (cf [49]).

3.3.5 Shear-free dust

One must be very cautious with imposing simplifying assumptions in order to obtain solutions: this can lead to major restrictions on the possible flows, and one can be badly misled if their consistency is not investigated carefully. A case of particular interest is *shear-free dust*, that is perfect-fluid solutions for which $\sigma_{ab} = 0$, $p = 0 \Rightarrow \dot{u}^a = 0$. In this case, careful study of the consistency conditions between all the equations [25] shows that necessarily $\omega\Theta = 0$: the solutions either do not rotate, or do not expand. This conclusion is of considerable importance, because if it were not true, there would be shear-free expanding and rotating solutions which would violate the Hawking–Penrose singularity theorems for cosmology [68,69] (integrating the vorticity equation along the fluid flow lines (3.37) gives $\omega = \omega_0/S^2$; substituting in the Raychaudhuri equation (3.34) and integrating, using the conservation equation (3.29), gives a first integral which is a generalized Friedmann equation, in which vorticity dominates expansion at early times and allows a bounce and singularity avoidance). The interesting point then is that *this result does not hold in Newtonian theory* [113], in which case there do indeed exist such solutions when suitable boundary conditions are imposed. If one uses these solutions as an argument against the singularity theorems, the argument is invalid; what they really do is point out the dangers of the Newtonian limit of cosmological equations.

3.4 Tetrad description

The 1+3 covariant equations are immediately transparent in terms of representing relations between $1 + 3$ covariantly defined quantities with clear geometrical and/or physical significance. However, they do not form a complete set of equations guaranteeing the existence of a corresponding metric and connection. For that we need to use a full tetrad description. The equations determined will then form a complete set, which will contain as a subset all the $1 + 3$ covariant equations just derived (albeit presented in a slightly different form) [53, 55]. First we summarize a generic tetrad formalism, and then describe its application to cosmological models (cf [25, 92]).

3.4.1 General tetrad formalism

A *tetrad* is a set of four linearly independent vector fields $\{e_a\}$, $a = 0, 1, 2, 3$, which serves as a basis for spacetime vectors and tensors. It can be written in terms of a local coordinate basis by means of the *tetrad components* $e_a{}^i(x^j)$:

$$e_a = e_a{}^i(x^j)\frac{\partial}{\partial x^i} \Leftrightarrow e_a(f) = e_a{}^i(x^j)\frac{\partial f}{\partial x^i}, \qquad e_a{}^i \equiv e_a(x^i), \qquad (3.50)$$

(the latter stating that the ith component of the ath tetrad vector is just the directional derivative of the ith coordinate x^i in the direction e_a). This relation can be thought of as just a change of vector basis, leading to a change of tensor components of the standard tensorial form:

$$T^{ab}{}_{cd} = e^a{}_i e^b{}_j e_c{}^k e_d{}^l T^{ij}{}_{kl}$$

with an obvious inverse, where the inverse components $e^a{}_i(x^j)$ (note the placing of the indices!) are defined by

$$e_a{}^i e^a{}_j = \delta^i{}_j \Leftrightarrow e_a{}^i e^b{}_i = \delta^b{}_a. \qquad (3.51)$$

However, this is a change from an integrable basis to a non-integrable one, so the non-tensorial relations (specifically the form of the metric and connection components) differ slightly from when coordinate bases are used. A change of one tetrad basis to another will also lead to transformations of the standard tensor form for all tensorial quantities: if $e_a = \lambda_a{}^{a'}(x^i)e_{a'}$ is a change of tetrad basis with inverse $e_{a'} = \lambda_{a'}{}^a(x^i)e_a$ (in the case of orthonormal bases, each of these matrices representing a Lorentz transformation), then

$$T^{ab}{}_{cd} = \lambda_{a'}{}^a \lambda_{b'}{}^b \lambda_c{}^{c'} \lambda_d{}^{d'} T^{a'b'}{}_{c'd'}.$$

Again the inverse is obvious. The *commutation functions* related to the tetrad are the quantities $\gamma^a{}_{bc}(x^i)$ defined by the *commutators* $[e_a, e_b]$ of the basis vectors:

$$[e_a, e_b] = \gamma^c{}_{ab}(x^i)e_c \Rightarrow \gamma^a{}_{bc}(x^i) = -\gamma^a{}_{cb}(x^i). \qquad (3.52)$$

It follows (apply this relation to the coordinate x^i) that in terms of the tetrad components,

$$\gamma^a{}_{bc}(x^i) = e^a{}_i(e_b{}^j \partial_j e_c{}^i - e_c{}^j \partial_j e_b{}^i) = -2e_b{}^i e_c{}^j \nabla_{[i} e^a{}_{j]}. \tag{3.53}$$

These quantities vanish iff the basis $\{e_a\}$ is a coordinate basis: that is, there exist coordinates x^i such that $e_a = \delta_a{}^i \partial/\partial x^i$, iff

$$[e_a, e_b] = 0 \Leftrightarrow \gamma^a{}_{bc} = 0.$$

The *metric tensor components* in the tetrad form are given by

$$g_{ab} = g_{ij} e_a{}^i e_b{}^j = e_a \cdot e_b. \tag{3.54}$$

The inverse equation

$$g_{ij}(x^k) = g_{ab} e^a{}_i(x^k) e^b{}_j(x^k) \tag{3.55}$$

explicitly constructs the coordinate components of the metric from the (inverse) tetrad components $e^a{}_i(x^j)$. We can raise and lower tetrad indices by use of the metric g_{ab} and its inverse g^{ab}. In the case of an orthonormal tetrad,

$$g_{ab} = \text{diag}(-1, +1, +1, +1) = g^{ab}, \tag{3.56}$$

showing by (3.54) that the basis vectors are unit vectors orthogonal to each other. Such a tetrad is defined up to an arbitrary position-dependent Lorentz transformation.

The *connection components* $\Gamma^a{}_{bc}$ for the tetrad are defined by the relations

$$\nabla_{e_b} e_a = \Gamma^c{}_{ab} e_c \Leftrightarrow \Gamma^c{}_{ab} = e^c{}_i e_b{}^j \nabla_j e_a{}^i, \tag{3.57}$$

i.e. it is the c-component of the covariant derivative in the b-direction of the a-vector. It follows that all covariant derivatives can be written out in tetrad components in a way completely analogous to the usual tensor form, for example

$$\nabla_a T_{bc} = e_a(T_{bc}) - \Gamma^d{}_{ba} T_{dc} - \Gamma^d{}_{ca} T_{bd},$$

where for any function f, $e_a(f) = e_a{}^i \partial f/\partial x^i$ is the derivative of f in the direction e_a. In the case of an orthonormal tetrad, (3.56) shows that $e_a(g_{bc}) = 0$; hence applying this relation to the metric tensor,

$$\nabla_a g_{bc} = 0 \Leftrightarrow \Gamma_{(ab)c} = 0, \tag{3.58}$$

—the connection components are skew in their first two indices, when we use the metric to raise and lower the first indices only, and are called 'Ricci rotation coefficients' or just *rotation coefficients*. We obtain from this and the assumption of vanishing torsion the relations for an orthonormal tetrad that are the analogues of the usual Christoffel relation:

$$\gamma^a{}_{bc} = -(\Gamma^a{}_{bc} - \Gamma^a{}_{cb}), \qquad \Gamma_{abc} = \tfrac{1}{2}(g_{ad}\gamma^d{}_{cb} - g_{bd}\gamma^d{}_{ca} + g_{cd}\gamma^d{}_{ab}). \quad (3.59)$$

This shows that the rotation coefficients and the commutation functions are each just linear combinations of the other.

Any set of vectors however must satisfy the *Jacobi identities*:

$$[X, [Y, Z]] + [Y, [Z, X]] + [Z, [X, Y]] = 0,$$

which follows from the definition of a commutator. Applying this to the basis vectors e_a, e_b and e_c gives the identities

$$e_{[a}(\gamma^d{}_{bc]}) + \gamma^e{}_{[ab}\gamma^d{}_{c]e} = 0, \qquad (3.60)$$

which are the integrability conditions that the $\gamma^a{}_{bc}(x^i)$ are the commutation functions for the set of vectors e_a.

If we apply the Ricci identities to the tetrad basis vectors e_a, we obtain the Riemann curvature tensor components in the form

$$R^a{}_{bcd} = \partial_c(\Gamma^a{}_{bd}) - \partial_d(\Gamma^a{}_{bc}) + \Gamma^a{}_{ec}\Gamma^e{}_{bd} - \Gamma^a{}_{ed}\Gamma^e{}_{bc} - \Gamma^a{}_{be}\gamma^e{}_{cd}. \quad (3.61)$$

Contracting this on a and c, one obtains the EFE in the form

$$R_{bd} = \partial_a(\Gamma^a{}_{bd}) - \partial_d(\Gamma^a{}_{ba}) + \Gamma^a{}_{ea}\Gamma^e{}_{bd} - \Gamma^a{}_{de}\Gamma^e{}_{ba} = T_{bd} - \tfrac{1}{2}T g_{bd} + \lambda g_{bd}. \tag{3.62}$$

It is not immediately obvious that this is symmetric, but this follows because (3.60) implies $R_{a[bcd]} = 0 \Rightarrow R_{ab} = R_{(ab)}$.

3.4.2 Tetrad formalism in cosmology

In detailed studies of families of exact non-vacuum solutions, it will usually be advantageous to use an orthonormal tetrad basis, because the tetrad vectors can be chosen in physically preferred directions. For a cosmological model we choose an orthonormal tetrad with the timelike vector e_0 chosen to be either the fundamental 4-velocity field u^a, or the normals n^a to surfaces of homogeneity when they exist. This fixing implies that the initial six-parameter freedom of using Lorentz transformations has been reduced to a three-parameter freedom of rotations of the spatial frame $\{e_\alpha\}$. The 24 algebraically independent rotation coefficients can then be split into (see [25, 45, 53]):

$$\Gamma_{\alpha 00} = \dot{u}_\alpha, \qquad \Gamma_{\alpha 0\beta} = \tfrac{1}{3}\Theta\delta_{\alpha\beta} + \sigma_{\alpha\beta} - \epsilon_{\alpha\beta\gamma}\omega^\gamma, \qquad \Gamma_{\alpha\beta 0} = \epsilon_{\alpha\beta\gamma}\Omega^\gamma \quad (3.63)$$

$$\Gamma_{\alpha\beta\gamma} = 2a_{[\alpha}\delta_{\beta]\gamma} + \epsilon_{\gamma\delta[\alpha}n^\delta{}_{\beta]} + \tfrac{1}{2}\epsilon_{\alpha\beta\delta}n^\delta{}_\gamma. \qquad (3.64)$$

The first two sets contain the kinematical variables for the chosen vector field. The third is the rate of rotation Ω^α of the spatial frame $\{e_\alpha\}$ with respect to a Fermi-propagated (physically non-rotating) basis along the fundamental flow lines. Finally, the quantities a^α and $n_{\alpha\beta} = n_{(\alpha\beta)}$ determine the nine spatial rotation coefficients. In terms of these quantities, the commutator equations (3.52) applied to any function f take the form

$$[\mathbf{e}_0, \mathbf{e}_\alpha](f) = \dot{u}_\alpha \mathbf{e}_0(f) - [\tfrac{1}{3}\Theta\delta_\alpha{}^\beta + \sigma_\alpha{}^\beta + \epsilon_\alpha{}^\beta{}_\gamma(\omega^\gamma + \Omega^\gamma)]\mathbf{e}_\beta(f), \quad (3.65)$$

$$[\mathbf{e}_\alpha, \mathbf{e}_\beta](f) = 2\epsilon_{\alpha\beta\gamma}\omega^\gamma \mathbf{e}_0(f) + [2a_{[\alpha}\delta^\gamma{}_{\beta]} + \epsilon_{\alpha\beta\delta}n^{\delta\gamma}]\mathbf{e}_\gamma(f). \quad (3.66)$$

3.4.3   Complete set

The full set of equations for a gravitating fluid can be written in tetrad form, using the matter variables, the rotation coefficients (3.57) and the tetrad components (3.50) as the primary variables. The equations needed are the conservation equations (3.27), (3.28) and all the Ricci equations (3.61) and Jacobi identities (3.60) for the tetrad basis vectors, together with the tetrad equations (3.50) and the commutator equations (3.53). This gives a set of constraints and a set of first-order evolution equations, which include the tetrad form of all the $1 + 3$ covariant equations given earlier, based on the chosen vector field. For a prescribed set of equations of state, this gives the complete set of relations needed to determine the spacetime structure. One has the option of including or not including the tetrad components of the Weyl tensor as variables in this set; whether it is better to include them or not depends on the problem to be solved (if they are included, there will be more equations in the corresponding complete set, for we must then include the full Bianchi identities). The full set of equations is given in [41, 55], and see [25, 118] for the use of tetrads to study locally rotationally symmetric spacetimes, and [45, 128] for the case of Bianchi universes.

Finally, when tetrad vectors are chosen uniquely in an invariant way (e.g. as eigenvectors of a non-degenerate shear tensor), then—because they are uniquely defined from $1 + 3$ covariant quantities—all the rotation coefficients are covariantly defined scalars, so these equations are all equations for scalar invariants. The only times when it is not possible to define unique tetrads in this way is when the spacetimes are isotropic or locally rotationally symmetric (these concepts are discussed later).

3.5   Models and symmetries

3.5.1   Symmetries of cosmologies

Symmetries of a space or a spacetime (generically, 'space') are transformations of the space into itself that leave the metric tensor and all physical and geometrical properties invariant. We deal here only with continuous symmetries, characterized by a continuous group of transformations and associated vector fields [24].

3.5.1.1 Killing vectors

A space or spacetime *symmetry*, or *isometry*, is a transformation that drags the metric along a certain congruence of curves into itself. The generating vector field ξ_i of such curves is called a *Killing vector (field)* (or 'KV'), and obeys Killing's equations,

$$(L_\xi g)_{ij} = 0 \Leftrightarrow \nabla_{(i}\xi_{j)} = 0 \Leftrightarrow \nabla_i \xi_j = -\nabla_j \xi_i, \tag{3.67}$$

where L_X is the *Lie derivative*. By the Ricci identities for a KV, this implies the curvature equation:

$$\nabla_i \nabla_j \xi_k = R^m{}_{ijk}\xi_m, \tag{3.68}$$

and hence the infinite series of further equations that follows by taking covariant derivatives of this one, e.g.

$$\nabla_l \nabla_i \nabla_j \xi_k = (\nabla_l R^m{}_{ijk})\xi_m + R^m{}_{ijk}\nabla_l \xi_m. \tag{3.69}$$

The set of all KVs forms a Lie algebra with a basis $\{\xi_a\}$, $a = 1, 2, \ldots, r$, of dimension $r \leq \frac{1}{2}n(n-1)$. ξ_a^i denotes the components with respect to a local coordinate basis, a, b and c label the KV basis and i, j and k the coordinate components. Any KV can be written in terms of this basis, with *constant coefficients*. Hence, if we take the commutator $[\xi_a, \xi_b]$ of two of the basis KVs, this is also a KV, and so can be written in terms of its components relative to the KV basis, which will be constants. We can write the constants as $C^c{}_{ab}$, obtaining

$$[\xi_a, \xi_b] = C^c{}_{ab}\xi_c, \qquad C^a{}_{bc} = C^a{}_{[bc]}. \tag{3.70}$$

By the Jacobi identities for the basis vectors, these structure constants must satisfy

$$C^a{}_{e[b}C^e{}_{cd]} = 0 \tag{3.71}$$

(which is just equation (3.60) specialized to a set of vectors with constant commutation functions). These are the integrability conditions that must be satisfied in order that the Lie algebra exist in a consistent way. The transformations generated by the Lie algebra form a Lie group of the same dimension (see Eisenhart [24] or Cohn [11]).

Arbitrariness of the basis: We can change the basis of KVs in the usual way;

$$\xi_{a'} = \lambda_{a'}{}^a \xi_a \Leftrightarrow \xi_{a'}^i = \lambda_{a'}{}^a \xi_a^i, \tag{3.72}$$

where the $\lambda_{a'}{}^a$ are constants with $\det(\lambda_{a'}{}^a) \neq 0$, so unique inverse matrices $\lambda^{a'}{}_a$ exist. Then the structure constants transform as tensors:

$$C^{c'}{}_{a'b'} = \lambda^{c'}{}_c \lambda_{a'}{}^a \lambda_{b'}{}^b C^c{}_{ab}. \tag{3.73}$$

Thus the possible equivalence of two Lie algebras is not obvious, as they may be given in different bases.

3.5.1.2   Groups of isometries

The isometries of a space of dimension n must be a group, as the identity is an isometry, the inverse of an isometry is an isometry, and the composition of two isometries is an isometry. Continuous isometries are generated by the Lie algebra of KVs. The group structure is determined locally by the Lie algebra, in turn characterized by the structure constants [11]. The action of the group is characterized by the nature of its orbits in space; this is only partially determined by the group structure (indeed the same group can act as a spacetime symmetry group in quite different ways).

3.5.1.3   Dimensionality of groups and orbits

Most spaces have no KVs, but special spaces (with symmetries) have some. The group action defines orbits in the space where it acts and the dimensionality of these orbits determines the kind of symmetry that is present.

The *orbit* of a point p is the set of all points into which p can be moved by the action of the isometries of a space. Orbits are necessarily homogeneous (all physical quantities are the same at each point). An *invariant variety* is a set of points moved into itself by the group. This will be bigger than (or equal to) all orbits it contains. The orbits are necessarily invariant varieties; indeed they are sometimes called *minimum invariant varieties*, because they are the smallest subspaces that are always moved into themselves by all the isometries in the group. *Fixed points* of a group of isometries are those points which are left invariant by the isometries (thus the orbit of such a point is just the point itself). These are the points where all KVs vanish (however, the derivatives of the KVs there are non-zero; the KVs generate isotropies about these points). *General points* are those where the dimension of the space spanned by the KVs (that is, the dimension of the orbit through the point) takes the value it has almost everywhere; *special points* are those where it has a lower dimension (e.g. fixed points). Consequently, the dimension of the orbits through special points is lower than that of orbits through general points. The dimension of the orbit and isotropy group is the same at each point of an orbit, because of the equivalence of the group action at all points on each orbit.

The group is *transitive on a surface S* (of whatever dimension) if it can move any point of S into any other point of S. Orbits are the largest surfaces through each point on which the group is transitive; they are therefore sometimes referred to as *surfaces of transitivity*. We define their dimension as follows, and determine limits from the maximal possible initial data for KVs: *dimension of the surface of transitivity* $= s$, where in a space of dimension n, $s \leq n$.

At each point we can also consider the dimension of the isotropy group (the group of isometries leaving that point fixed), generated by all those KVs that vanish at that point: *dimension of an isotropy group* $= q$, where $q \leq \frac{1}{2}n(n-1)$.

The *dimension r of the group of symmetries* of a space of dimension n is $r = s + q$ (translations plus rotations). The dimension q of the isotropy group can vary over the space (but not over an orbit): it can be greater at special points (e.g. an axis centre of symmetry) where the dimension s of the orbit is less, but r (the dimension of the total symmetry group) must stay the same everywhere. From these limits , $0 \leq r \leq n + \frac{1}{2}n(n-1) = \frac{1}{2}n(n+1)$ (the maximal number of translations and of rotations). This shows the Lie algebra of KVs is finite dimensional.

Maximal dimensions: If $r = \frac{1}{2}n(n+1)$, we have a space(time) of constant curvature (maximal symmetry for a space of dimension n). In this case,

$$R_{ijkl} = K(g_{ik}g_{jl} - g_{il}g_{jk}), \tag{3.74}$$

with K a constant. One cannot get $q = \frac{1}{2}n(n-1) - 1$ so $r \neq \frac{1}{2}n(n+1) - 1$.

A group is *simply transitive* if $r = s \Leftrightarrow q = 0$ (no redundancy: dimensionality of group of isometries is just sufficient to move each point in a surface of transitivity into each other point). There is no continuous isotropy group.

A group is *multiply transitive* if $r > s \Leftrightarrow q > 0$ (there is redundancy in that the dimension of the group of isometries is larger than is necessary to move each point in an orbit into each other point). There exist non-trivial isotropies.

3.5.2 Classification of cosmological symmetries

We consider non-empty perfect fluid models, i.e. (3.6) holds with $(\mu + p) > 0$, implying u^a is the uniquely defined timelike eigenvector of the Ricci tensor.

Spacetime is four-dimensional, so the possibilities for the dimension of the surface of transitivity are $s = 0, 1, 2, 3, 4$. Because u^a is invariant, the isotropy group at each point has to be a sub-group of the rotations $O(3)$ acting orthogonally to u^a, but there is no two-dimensional subgroup of $O(3)$. Thus the possibilities for isotropy at a general point are:

(1) *Isotropic*: $q = 3$, the matter is a perfect fluid, the Weyl tensor vanishes, all kinematical quantities vanish except Θ. All observations (at every point) are isotropic. This is the RW family of geometries.
(2) *Local rotational symmetry* ('LRS'): $q = 1$, the Weyl tensor is of algebraic Petrov type D, kinematical quantities are rotationally symmetric about a preferred spatial direction. All observations at every general point are rotationally symmetric about this direction. All metrics are known in the case of dust [25] and a perfect fluid [51, 118].
(3) *Anisotropic*: $q = 0$; there are no rotational symmetries. Observations in each direction are different from observations in each other direction.

Putting this together with the possibilities for the dimensions of the surfaces of transitivity, we have the following possibilities (see table 3.1).

Table 3.1. Classification of cosmological models (with $(\mu + p) > 0$) by isotropy and homogeneity.

Dimension, Isotropy group	Dim invariant variety		
	$s = 2$ Inhomogeneous	$s = 3$ Spatially homogeneous	$s = 4$ Spacetime homogeneous
$q = 0$ anisotropic	Generic metric form known. Spatially self-similar, Abelian G_2 on 2D spacelike surfaces, non-Abelian G_2	Bianchi: orthogonal, tilted	Osvath/Kerr
$q = 1$ LRS	Lemaître–Tolman– Bondi family	Kantowski–Sachs, LRS Bianchi	Gödel
$q = 3$ isotropic	None (cannot happen)	Friedmann	Einstein static
	Two non-ignorable coordinates	One non-ignorable coordinate	Algebraic EFE (no redshift)

Dimension Isotropy group	Dim invariant variety	
	$s = 0$ Inhomogeneous	$s = 1$ Inhomogeneous/ no isotropy group
$q = 0$	Szekeres–Szafron, Stephani–Barnes, Oleson type N	General metric form independent of one coord; KV h.s.o./not h.s.o.
	The real universe!	

3.5.2.1 *Spacetime homogeneous models*

These models with $s = 4$ are unchanging in space and time, hence μ is a constant, so by the energy conservation equation (3.29) they cannot expand: $\Theta = 0$. They cannot produce an almost isotropic redshift, and are not useful as models of the real universe. Nevertheless they are of some interest for their geometric properties.

The *isotropic case* $q = 3$ ($\Rightarrow r = 7$) is the Einstein static universe, the non-expanding FL model that was the first relativistic cosmological model found. It is not a viable cosmology because it has no redshifts, but it laid the foundation for the discovery of the expanding FLRW models.

The *LRS case q* = 1 ($\Rightarrow$ *r* = 5) is the Gödel stationary rotating universe [60], also with no redshifts. This model was important because of the new understanding it brought as to the nature of time in general relativity (see [68, 124]). It is a model in which causality is violated (there exist closed timelike lines through each spacetime point) and there exists no cosmic time function whatsoever.

The anisotropic models *q* = 0 ($\Rightarrow$ *r* = 4) are all known, but are interesting only for the light they shed on Mach's principle; see [101].

3.5.2.2 *Spatially homogeneous universes*

These models with *s* = 3 are the major models of theoretical cosmology, because they express mathematically the idea of the 'cosmological principle': all points of space at the same time are equivalent to each other [6].

The *isotropic case q* = 3 ($\Rightarrow$ *r* = 6) is the family of FL models, the standard models of cosmology, with the comoving RW metric form:

$$\mathrm{d}s^2 = -\mathrm{d}t^2 + S^2(t)(\mathrm{d}r^2 + f^2(r)(\mathrm{d}\theta^2 + \sin^2\theta\,\mathrm{d}\phi^2)), \qquad u^a = \delta^a_0. \quad (3.75)$$

Here the space sections are of constant curvature $K = k/S^2$ and

$$f(r) = \sin r, r, \sinh r \qquad (3.76)$$

if the normalized spatial curvature *k* is +1, 0, −1 respectively. The space sections are necessarily closed if *k* = +1.

The *LRS case q* = 1 ($\Rightarrow$ *r* = 4) is the family of Kantowski–Sachs universes [13,80] plus the LRS orthogonal [45] and tilted [77] Bianchi models. The simplest are the Kantowski–Sachs family, with comoving metric form

$$\mathrm{d}s^2 = -\mathrm{d}t^2 + A^2(t)\,\mathrm{d}r^2 + B^2(t)(\mathrm{d}\theta^2 + f^2(\theta)\,\mathrm{d}\phi^2), \qquad u^a = \delta^a_0, \quad (3.77)$$

where $f(\theta)$ is given by (3.76).

The *anisotropic case q* = 0 ($\Rightarrow$ *r* = 3) is the family of Bianchi universes with a group of isometries G_3 acting simply transitively on spacelike surfaces. They can be orthogonal or tilted. The simplest class is the Bianchi type I family, with an Abelian isometry group and metric form:

$$\mathrm{d}s^2 = -\mathrm{d}t^2 + A^2(t)\,\mathrm{d}x^2 + B^2(t)\,\mathrm{d}y^2 + C^2(t)\,\mathrm{d}z^2, \qquad u^a = \delta^a_0. \quad (3.78)$$

The family as a whole has quite complex properties; these models are discussed in the following section.

3.5.2.3 *Spatially inhomogeneous universes*

These models have $s \leq 2$. The *LRS cases* ($q = 1 \Rightarrow s = 2, r = 3$) are the spherically symmetric family with metric form:

$$\mathrm{d}s^2 = -C^2(t,r)\,\mathrm{d}t^2 + A^2(t,r)\,\mathrm{d}r^2 + B^2(t,r)(\mathrm{d}\theta^2 + f^2(\theta)\,\mathrm{d}\phi^2), \qquad u^a = \delta_0^a, \tag{3.79}$$

where $f(\theta)$ is given by (3.76). In the dust case, we can set $C(t,r) = 1$ and can integrate the EFE analytically; for $k = +1$, these are the ('LTB') spherically symmetric models [5,87]. They may have a centre of symmetry (a timelike worldline), and can even allow two such centres, but they cannot be isotropic about a general point (because isotropy everywhere implies spatial homogeneity).

Solutions with no symmetries at all have $r = 0 \Rightarrow s = 0, q = 0$. The real universe, of course, belongs to this class; all the others are intended as approximations to this unique universe. Remarkably, we know some exact solutions without any symmetries, specifically (a) the Szekeres quasi-spherical models [121, 122], (b) Stephani's conformally flat models [84, 116], and (c) Oleson's type-N solutions (for a discussion of these and all the other inhomogeneous models, see Krasiński [85] and Kramer *et al* [83]). One further interesting family without global symmetries are the 'Swiss-cheese' models, made by cutting and pasting segments of spherically symmetric models [23, 112].

Because of the nonlinearity of the equations, it is helpful to have exact solutions at hand as models of structure formation as well as studies of linearly perturbed FL models (briefly discussed later). The dust (Tolman–Bondi) and perfect fluid spherically symmetric models are useful here, in particular in terms of relating the time evolution to self-similar models. However, in the fully nonlinear regime numerical solutions of the full equations are needed.

3.6   Friedmann–Lemaître models

The FL models are discussed in detail in other chapters, so here I will only briefly mention some interesting properties of these solutions (and see also [33]). These models are perfect fluid solutions with metric form (3.75), characterized by

$$\dot{u} = 0 = \omega = \sigma = 0, \qquad \theta = 3\frac{\dot{S}}{S} \tag{3.80}$$

$$\Rightarrow \tilde{\nabla}_e \mu = \tilde{\nabla}_e p = \tilde{\nabla}_e \theta = 0, \qquad E_{ab} = H_{ab} = 0. \tag{3.81}$$

They are isotropic about every point ($q = 3$) and consequently are spatially homogeneous ($s = 3$). The equations that apply are the covariant equations (3.80), (3.83) with restrictions (3.80). The dynamical equations are the energy equation (3.29)

$$\dot{\mu} = -3\frac{\dot{S}}{S}(\mu + p), \tag{3.82}$$

the Raychaudhuri equation (3.34):

$$3\frac{\ddot{S}}{S} = -\tfrac{1}{2}(\mu + 3p) + \lambda \tag{3.83}$$

and the Friedmann equation (3.40), where $^3R = 6k/S^2$,

$$3\frac{\dot{S}^2}{S^2} - \kappa\mu - \lambda = -\frac{3k}{S^2}. \tag{3.84}$$

The Friedmann equation is a first integral of the other two when $\dot{S} \neq 0$. The solutions, of course, depend on the equation of state; for the currently favoured universe models, going backward in time there will be

(1) a cosmological-constant-dominated phase,
(2) a matter-dominated phase,
(3) a radiation-dominated phase,
(4) a scalar-field-dominated inflationary phase and
(5) a pre-inflationary phase where the physics is speculative (see the last section of this chapter). The normalized density parameter is $\Omega \equiv \kappa\mu/3H^2$, where as usual $H = \dot{S}/S$.

3.6.1 Phase planes and evolutionary paths

From these equations, one can obtain phase planes

(i) for the density parameter Ω against the deceleration parameter q, see [115];
(ii) for the density parameter Ω against the Hubble parameter H, see [128] for the case $\lambda = 0$; and
(iii) for the density parameter Ω against the scale parameter S, see [94], showing how Ω changes in inflationary and non-inflationary universes.

It is a consequence of the equations that the spatial curvature parameter k is a constant of the motion. In particular, flatness cannot change as the universe evolves: either $k = 0$ or not, depending on the initial conditions, and this is independent of any inflation that may take place. Thus while inflation can drive the spatial curvature $K = k/S^2$ very close indeed to zero, it cannot set $K = 0$.

If one has a scalar field matter source ϕ with potential $V(\phi)$, one can obtain essentially arbitrary functional forms for the scale function $S(t)$ by using the arbitrariness in the function $V(\phi)$ and running the field equations backwards, see [46].

3.6.2 Spatial topology

The Einstein field equations determine the time evolution of the metric and its spatial curvature, but they do not determine its spatial topology. Spatially closed

FL models can occur even if $k = 0$ or $k = -1$, for example with a toroidal topology [27]. These universes can be closed on a small enough spatial scale that we could have seen all the matter in the universe already, and indeed could have seen it many times over; see the discussion on 'small universes' later.

3.6.3 Growth of inhomogeneity

This is studied by looking at linear perturbations of the FL models, as well as by examining inhomogeneous models. The geometry and dynamics of perturbed FL models is described in detail in other talks, so I will again just make a few remarks. In dealing with perturbed FL models, one runs into the *gauge issue*: the background model is not uniquely defined by a realistic (lumpy) model and the definition of the perturbations depends on the choice of background model (the gauge chosen). Consequently it is advisable to use gauge-invariant variables, either coordinate-based [2] or covariant [39]. When dealing with multiple matter components, it is important to take carefully into account the separate velocities needed for each matter component, and their associated conservation equations. The CBR can best be described by kinetic theory, which again can be presented in a covariant and gauge invariant way [10, 59].

3.7 Bianchi universes ($s = 3$)

These are the models in which there is a group of isometries G_3 simply transitive on spacelike surfaces, so they are spatially homogeneous. There is only one essential dynamical coordinate (the time t) and the EFE reduce to ordinary differential equations, because the inhomogeneous degrees of freedom have been 'frozen out'. They are thus quite special in geometrical terms; nevertheless, they form a rich set of models where one can study the exact dynamics of the full nonlinear field equations. The solutions to the EFE will depend on the matter in the spacetime. In the case of a fluid (with uniquely defined flow lines), we have two different kinds of models:

(1) *Orthogonal models*, with the fluid flow lines orthogonal to the surfaces of homogeneity (Ellis and MacCallum [45], see also [128]). In this case the fluid 4-velocity u^a is *parallel* to the normal vectors n^a so the matter variables will be just the fluid density and pressure. The fluid flow is necessarily irrotational and geodesic.

(2) *Tilted models*, with the fluid flow lines not orthogonal to the surfaces of homogeneity. Thus the fluid 4-velocity is *not parallel* to the normals, and the components of the fluid peculiar velocity enter as further variables (King and Ellis [15, 77]). They determine the fluid energy–momentum tensor components relative to the normal vectors (a perfect fluid will appear as an imperfect fluid in that frame). Rotating models *must* be tilted, and are much more complex than non-rotating models.

3.7.1 Constructing Bianchi universes

The approach of Ellis and MacCallum [45]) uses an orthonormal tetrad based on the normals to the surfaces of homogeneity (i.e. $e_0 = n$, the unit normal vector to these surfaces). The tetrad is chosen to be invariant under the group of isometries, i.e. the tetrad vectors commute with the KVs. Then we have an orthonormal basis e_a, $a = 0, 1, 2, 3$, such that equation (3.52) becomes

$$[e_a, e_b] = \gamma^c{}_{ab}(t)e_c \qquad (3.85)$$

and all dynamic variables are function of time t only. The matter variables—$\mu(t)$, $p(t)$, and $u_\alpha(t)$ in the case of tilted models—and the commutation functions $\gamma^a{}_{bc}(t)$, which by (3.59) are equivalent to the rotation coefficients, are chosen to be these variables. The EFE (3.2) are first-order equations for these quantities, supplemented by the Jacobi identities for the $\gamma^a{}_{bc}(t)$, which are also first-order equations. Thus the equations needed are just the tetrad equations mentioned in section 3.3, for the case

$$\dot{u}^\alpha = \omega^\alpha = 0 = e_\alpha(\gamma^a{}_{bc}). \qquad (3.86)$$

The spatial commutation functions $\gamma^\alpha{}_{\beta\gamma}(t)$ can be decomposed into a time-dependent matrix $n_{\alpha\beta}(t)$ and vector $a^\alpha(t)$, see (3.66), and are equivalent to the structure constants $C^\alpha{}_{\beta\gamma}$ of the symmetry group at each point. In view of (3.86), the Jacobi identities (3.60) for the spatial vectors now take the simple form

$$n^{\alpha\beta}a_\beta = 0. \qquad (3.87)$$

The tetrad basis can be chosen to diagonalize $n_{\alpha\beta}$ at all times, to attain $n_{\alpha\beta} = \text{diag}(n_1, n_2, n_3)$, $a^\alpha = (a, 0, 0)$, so that the Jacobi identities are then simply $n_1 a = 0$. Consequently we define two major classes of structure constants (and so Lie algebras):

Class A: $a = 0$; and
Class B: $a \neq 0$.

Following Schücking's extension of Bianchi's work, the classification of G_3 group types used is as in table 3.2. Given a specific group type at one instant, this type will be preserved by the evolution equations for the quantities $n_\alpha(t)$ and $a(t)$. This is a consequence of a generic property of the EFE: they will always preserve symmetries in initial data (within the Cauchy development of that data); see Hawking and Ellis [68].

In some cases, the Bianchi groups allow higher symmetry subcases, i.e. they are compatible with isotropic (FL) or LRS models, see [45] for details. For us the interesting point is that $k = 0$ FL models are compatible with groups of type I and VII$_0$, $k = -1$ models with groups of types V and VII$_h$, and $k = +1$ models with groups of type IX.

Table 3.2. Canonical structure constants for different Bianchi types. The parameter $h = a^2/n_2 n_3$.

Class	Type	n_1	n_2	n_3	a	
A	I	0	0	0	0	Abelian
	II	+ve	0	0	0	
	VI_0	0	+ve	−ve	0	
	VII_0	0	+ve	+ve	0	
	VIII	−ve	+ve	+ve	0	
	IX	+ve	+ve	+ve	0	
B	V	0	0	0	+ve	
	IV	0	0	+ve	+ve	
	VI_h	0	+ve	−ve	+ve	$h < 0$
	III	0	+ve	−ve	$n_2 n_3$	same as VI_1
	VII_h	0	+ve	+ve	+ve	$h > 0$

The set of tetrad equations (section 3.3) with restrictions (3.86) will determine the evolution of all the commutation functions and matter variables and, hence, determine the metric and also the evolution of the Weyl tensor. One can relate these equations to variational principles and a Hamiltonian, thus expressing them in terms of a potential formalism that gives an intuitive feel for what the evolution will be like [92, 93]. They are also the basis of dynamical systems analyses.

3.7.2   Dynamical systems approach

The most illuminating description of the evolution of families of Bianchi models is a dynamical systems approach based on the use of orthonormal tetrads, presented in detail in Wainwright and Ellis [128]. The main variables used are essentially the commutation functions mentioned earlier, but rescaled by a common time-dependent factor.

3.7.2.1   Reduced differential equations

The basic idea [12, 126] is to write the EFE in a way that enables one to study the evolution of the various physical and geometrical quantities *relative to the overall rate of expansion of the universe*, as described by the rate of expansion scalar Θ or, equivalently, *the Hubble parameter* $H = \frac{1}{3}\Theta$. The remaining freedom in the choice of orthonormal tetrad needs to be eliminated by specifying the variables Ω^α implicitly or explicitly (for example by specifying the basis as eigenvectors of the $\sigma_{\alpha\beta}$). This also simplifies other quantities (for example the choice of a shear eigenframe will result in the tensor $\sigma_{\alpha\beta}$ being represented by two diagonal terms). One hence obtains a reduced set of variables, consisting of H and the remaining commutation functions, which we denote symbolically by $x = (\gamma^a{}_{bc}|_{\text{reduced}})$. The physical state of the model is thus described by the vector (H, x). The details of this reduction differ for classes A and B in the latter case, there is an algebraic constraint of the form $g(x) = 0$, where g is a homogeneous polynomial.

The idea is now to normalize x with the Hubble parameter H. Denoting the resulting variables by a vector $y \in R^n$, we write

$$y = \frac{x}{H}. \tag{3.88}$$

These new variables are *dimensionless*, and will be referred to as *expansion-normalized variables*. It is clear that each dimensionless state y determines a one-parameter family of physical states (x, H). The evolution equations for the $\gamma^a{}_{bc}$ lead to evolution equations for H and x and hence for y. In order that the evolution equations define a flow, it is necessary, in conjunction with the rescaling of the variables, to introduce a *dimensionless time variable* τ according to

$$S = S_0 e^\tau, \tag{3.89}$$

where S_0 is the value of the scale factor at some arbitrary reference time. Since S assumes values $0 < S < +\infty$ in an ever-expanding model, τ assumes all real values, with $\tau \to -\infty$ at the initial singularity and $\tau \to +\infty$ at late times. It follows that

$$\frac{dt}{d\tau} = \frac{1}{H} \tag{3.90}$$

and the evolution equation for H can be written

$$\frac{dH}{d\tau} = -(1+q)H, \tag{3.91}$$

where the *deceleration parameter* q is defined by $q = -\ddot{S}S/\dot{S}^2$, and is related to $\dot{H}$ by $\dot{H} = -(1+q)H^2$. Since the right-hand side of the evolution equations for the $\gamma^a{}_{bc}$ are homogeneous of degree 2 in the $\gamma^a{}_{bc}$, the change (3.90) of the time variable results in H cancelling out of the evolution equation for y, yielding an autonomous differential equation (DE):

$$\frac{dy}{d\tau} = f(y), \qquad y \in R^n. \tag{3.92}$$

The constraint $g(x) = 0$ translates into a constraint

$$g(y) = 0, \tag{3.93}$$

which is preserved by the DE. The functions $f : R^n \to R^n$ and $g : R^n \to R$ are polynomial functions in y. An essential feature of this process is that the evolution equation for H, namely (3.91), decouples from the remaining equations (3.92) and (3.93). Thus the DE (3.92) describes the evolution of the non-tilted Bianchi cosmologies, the transformation of variables essentially scaling away the effects of the overall expansion. An important consequence is that the new variables are bounded near the initial singularity.

3.7.2.2 *Equations and orbits*

Since τ assumes all real values (for models which expand indefinitely), the solutions of (3.92) are defined for all τ and hence define a *flow* $\{\phi_\tau\}$ on R^n. The evolution of the cosmological models can thus be analysed by studying the orbits of this flow in the physical region of state space, which is a subset of R^n defined by the requirement that the matter energy density μ be non-negative, i.e.

$$\Omega(y) = \frac{\kappa\mu}{3H^2} \geq 0, \tag{3.94}$$

where the density parameter Ω is a dimensionless measure of μ.

The *vacuum boundary*, defined by $\Omega(y) = 0$, describes the evolution of vacuum Bianchi models, and is an invariant set which plays an important role in the qualitative analysis because vacuum models can be asymptotic states for perfect fluid models near the big bang or at late times. There are other invariant sets which are also specified by simple restrictions on y which play a special role: the subsets representing each Bianchi type (table 3.2), and the subsets representing higher-symmetry models, specifically the FLRW models and the LRS Bianchi models (table 3.1).

It is desirable that the dimensionless state space D in R^n is a *compact set*. In this case each orbit will have non-empty future and past limit sets, and hence there will exist a past attractor and a future attractor in state space. When using expansion-normalized variables, compactness of the state space has a direct physical meaning for ever-expanding models: if the state space is compact, then at the big bang no physical or geometrical quantity diverges more rapidly than the appropriate power of H, and at late times no such quantity tends to zero less rapidly than the appropriate power of H. This will happen for many models; however, the state space for Bianchi type VII_0 and type VIII models is non-compact. This lack of compactness manifests itself in the behaviour of the Weyl tensor at late times.

3.7.2.3 *Equilibrium points and self-similar cosmologies*

Each ordinary orbit in the dimensionless state space corresponds to a one-parameter family of physical universes, which are conformally related by a constant rescaling of the metric. However, for an equilibrium point y^* of the DE (3.92), which satisfies $f(y^*) = 0$, the deceleration parameter q is a constant, i.e. $q(y^*) = q^*$, and we find

$$H(\tau) = H_0 e^{(1+q^*)\tau}.$$

In this case the parameter H_0 is no longer essential, since it can be set to unity by a translation of τ, $\tau \rightarrow \tau + \text{constant}$; then (3.90) implies that

$$Ht = \frac{1}{1+q^*},\qquad(3.95)$$

so that the commutation functions are of the form (constant) $\times t^{-1}$. It follows that the resulting cosmological model is *self-similar*. It thus turns out that *to each equilibrium point of the DE (3.92) there corresponds a unique self-similar cosmological model*. In such a model the physical states at different times differ only by an overall change in the length scale. Such models are expanding, but in such a way that their dimensionless state does not change. They include the flat FLRW model ($\Omega = 1$) and the Milne model ($\Omega = 0$). All vacuum and non-tilted perfect fluid self-similar Bianchi solutions have been given by Hsu and Wainwright [73]. The equilibrium points determine the asymptotic behaviour of other more general models.

3.7.2.4 Phase planes

Many phase planes can be constructed explicitly. The reader is referred to Wainright and Ellis [128] for a comprehensive presentation and survey of results. Several interesting points emerge

(1) *Variety of singularities.* Various types of singularity can occur in Bianchi universes: cigar, pancake and oscillatory in the orthogonal case. In the case of tilted models, one can, in addition get non-scalar singularities, associated with a change in the nature of the spacetime symmetries—a horizon occurs where the surfaces of homogeneity change from being timelike to being spacelike, so the model changes from being spatially homogeneous to spatially inhomogeneous [15, 42]. The fluid can then run into timelike singularities, quite unlike the spacelike singularities in FL models. Thus the singularity structure can be quite unlike that in a FL model, even in models that are arbitrarily similar to a FL model today and indeed since the time of decoupling.

(2) *Relation to lower dimensional spaces.* It seems that the lower dimensional spaces, delineating higher symmetry models, may be skeletons guiding the development of the higher dimensional spaces (the more generic models). This is one reason why study of the exact higher symmetry models is of significance.

(3) *Identification of models in state space.* The analysis of the phase planes for Bianchi models shows that the procedure sometimes adopted of identifying all points in state space corresponding to the same model, is not a good idea. For example the Kasner ring that serves as a framework for evolution of many other Bianchi models contains multiple realizations of the same Kasner model. To identify them as the same point in state space would make the

evolution patterns very difficult to follow. It is better to keep them separate, but to learn to identify where multiple realizations of the same model occur (which is just the *equivalence problem* for cosmological models).

3.7.3   Isotropization properties

An issue of importance is whether these models tend to isotropy at early or late times. An important paper by Collins and Hawking [16] shows that for ordinary matter, at late times, types I, V, VII, isotropize but other Bianchi models become anisotropic at very late times, even if they are very nearly isotropic at present. Thus isotropy is unstable in this case. However, a paper by Wald [130] showed that Bianchi models will tend to isotropize at late times if there is a positive cosmological constant present, implying that an inflationary era can cause anisotropies to die away. The latter work, however, while applicable to models with non-zero tilt angle, did not show this angle dies away, and indeed it does not do so in general (Goliath and Ellis [62]). Inflation also only occurs in Bianchi models if there is not too much anisotropy to begin with (Rothman and Ellis [111]), and it is not clear that shear and spatial curvature are in fact removed in all cases [109]. Hence, some Bianchi models isotropize due to inflation, but not all.

An important idea that arises out of this study is that of *intermediate isotropization*: namely, models that become very like a FLRW model for a period of their evolution but start and end quite unlike these models. It turns out that many Bianchi types allow intermediate isotropization, because the FLRW models are saddle points in the relevant phase planes. This leads to the following two interesting results:

Bianchi evolution theorem 1. *Consider a family of Bianchi models that allow intermediate isotropization. Define an ϵ-neighbourhood of a FLRW model as a region in state space where all geometrical and physical quantities are closer than ϵ to their values in a FLRW model. Choose a time scale L. Then no matter how small ϵ and how large L, there is an open set of Bianchi models in the state space such that each model spends longer than L within the corresponding ϵ-neighbourhood of the FLRW model.*

This follows because the saddle point is a fixed point of the phase flow; consequently the phase flow vector becomes arbitrarily close to zero at all points in a small enough open region around the FLRW point in state space. Consequently, although these models are quite unlike FLRW models at very early and very late times, there is an open set of them that are observationally indistinguishable from a FLRW model (choose L long enough to encompass from today to last coupling or nucleosynthesis, and ϵ to correspond to current observational bounds). Thus there exist many such models that are viable as models of the real universe in terms of compatibility with astronomical observations.

Bianchi evolution theorem 2. *In each set of Bianchi models of a type admitting intermediate isotropization, there will be spatially homogeneous models that are linearizations of these Bianchi models about FLRW models. These perturbation modes will occur in any almost-FLRW model that is generic rather than fine-tuned; however, the exact models approximated by these linearizations will be quite unlike FLRW models at very early and very late times.*

Proof is by linearizing the previous equations (see the following section) to obtain the Bianchi equations linearized about the FLRW models that occur at the saddle point leading to the intermediate isotropisation. These modes will be the solutions in a small neighbourhood about the saddle point permitted by the linearized equations (given existence of solutions to the nonlinear equations, linearization will not prevent corresponding linearized solutions existing).

The point is that these modes can exist as linearizations of the FLRW model; if they do not occur, then the initial data have been chosen to set these modes precisely to zero (rather than being made very small), which requires very special initial conditions. Thus these modes will occur in almost all almost-FLRW universes. Hence, if one believes in generality arguments, they will occur in the real universe. When they occur, they will, at early and late times grow until the model is very far from a FLRW geometry (while being arbitrarily close to an FLRW model for a very long time, as per the previous theorem).

3.8   Observations and horizons

The basic observational problem is that, because of the enormous scale of the universe, we can effectively only see it from one spacetime point, 'here and now' [26, 29]. Consequently what we are able to see is a projection onto a 2-sphere ('the sky') of all the objects in the universe, and our fundamental problem is determining the distances of the various objects we see in the images we obtain. In the standard universe models, redshift is a reliable zero-order distance indicator, but is unreliable at first order because of local velocity perturbations. Thus we need the array of other distance indicators (Tully–Fisher for example). Furthermore, to test cosmological models we need at least two reliable measurable properties of the objects we see, that we can plot against each other (magnitude and redshift, for example), and most of them are unreliable both because of intrinsic variation in source properties, and because of evolutionary effects associated with the inevitable lookback-time involved when we observe distant objects.

3.8.1   Observational variables and relations: FL models

The basic variables underlying direct observations of objects in the spatially homegenous and isotropic FL models are:

(1) the *redshift*, basically a time-dilation effect for all measurements of the source (determined by the radial component of velocity);

(2) the *area distance*, equivalent to the angular diameter distance in RW geometries, and also equivalent (up to a redshift factor) to the luminosity distance in all relativistic cosmological models, because of the reciprocity theorem [26]—this can best be calculated from the geodesic deviation equation [54]; and

(3) *number counts*, determined by (i) the number of objects in a given volume, (ii) the relation of that volume to increments in distance measures (determined by the spacetime geometry) and (iii) the selection and detection effects that determine which sources we can actually identify and measure (difficulties in detection being acute in the case of dark matter).

Thus to determine the spacetime geometry in these models, we need to correlate at least two of these variables against each other. Further observational data which must be consistent with the other observations comes from the following sources:

(4) *background radiation spectra* at all wavelengths particularly the 3K blackbody relic radiation ('CBR'); and

(5) *the 'relics' of processes taking place in the hot big-bang era*, for example the primeval element abundances resulting from baryosynthesis and nucleosynthesis in the hot early universe (and the CBR anisotropies and present large-scale structures can also be regarded in this light, for they are evidence about density fluctuations, which are one form of such relic).

The observational relations in FL models are covered in depth in other reports to this meeting (and see also [26, 33]), so I will just comment on two aspects here.

Selection/detection effects: The way we detect objects from available images depends on both their surface brightness and their apparent size. Thus we essentially need *two variables* to adequately characterize selection and detection effects; it simply is not adequate to discuss such effects on the basis of apparent magnitude or flux alone [47]. Hence one should regard with caution any catalogues that claim to be magnitude limited, for that cannot be an adequate criteria for detection limits; such catalogues may well be missing out many low surface-brightness objects.

Minimum angles and trapping surfaces: For ordinary matter, there is a redshift $z_\star$ such that apparent sizes of objects of fixed linear size reach a minimum at $z = z_\star$, and for larger redshift look larger again. What is happening here is that the universe as a whole is acting as a giant gravitational lens, refocusing our past light cone as a whole [26]; in an Einstein–de-Sitter universe, this happens at $z_\star = 5/4$; in a low density universe, it happens at about $z = 4$. This refocusing means that closed trapped surfaces occur in the universe, and hence via the Hawking–Penrose singularity theorems, leads to the prediction of the existence of a spacetime singularity in our past [68].

3.8.2   Particle horizons and visual horizons

For ordinary equations of state, because causal influences can travel at most at the speed of light, there is both a *particle horizon* [110, 124], limiting causal communication since the origin of the universe and a *visual horizon* [50], limiting visual communication since the decoupling of matter and radiation. The former depends on the equation of state of matter at early times, and can be changed drastically by an early period of inflation; however the latter depends only on the equation of state since decoupling, and is unaffected by whether inflation took place or not. From (3.75), at an arbitrary time of observation t_0, the radial comoving coordinate values corresponding to the particle and event horizons, respectively, of an observer at the origin of coordinates are:

$$u_{ph}(t_0) = \int_0^{t_0} \frac{dt}{S(t)}, \qquad u_{vh}(t_0) = \int_{t_d}^{t_0} \frac{dt}{S(t)}, \qquad (3.96)$$

where we have assumed the initial singularity occurred at $t = 0$ and decoupling at $t = t_d$. We cannot have had causal contact with objects lying at a coordinate value r greater than $u_{ph}(t_0)$, and cannot have received any type of electromagnetic radiation from objects lying at a coordinate value r greater than $u_{vh}(t_0)$.

It is fundamental to note, then, that no object can leave either of these horizons once it has entered it: *once two objects are in causal or visual contact, that contact cannot be broken, regardless of whether inflation or an accelerated expansion takes place or not.* This follows immediately from (3.96): $t_1 > t_0 \Rightarrow u_{ph}(t_1) > u_{ph}(t_0)$ (the integrand between t_0 and t_1 is positive, so $du_{ph}(t)/dt = 1/S(t) > 0$.) Furthermore the physical scales associated with these horizons cannot decrease while the universe is expanding. These are

$$D_{ph}(t) = S(t)u_{ph}(t), \qquad D_{vh}(t) = S(t)u_{vh}(t)$$

respectively, at time t; hence for example $d(D_{ph}(t))/dt = 1 + H(t)D_{ph}(t) > 0$. Much of the literature on inflation is misleading in this regard.

3.8.3   Small universes

The one case where visual horizons do not occur is when the universe has compact spatial sections whose physical size is less than the Hubble radius; consider, for example, the case of a $k = 0$ model universe of toroidal topology, with a length scale of identification of, say, 300 Mpc. In that case we can see right round the universe, with many images of each galaxy, and indeed many images of our own galaxy [48]. There are some philosophical advantages in such models [32], but they may or may not correspond to physical reality. If this is indeed the case, it would show up in multiple images of the same objects [48, 81], identical circles in the CBR anisotropy pattern across the sky [18], and altered CBR power spectra predictions [17]. A complete cosmological observational programme should test for the possibility of such small alternative universe topologies, as well as determining the fundamental cosmological parameters.

3.8.4   Observations in anisotropic and inhomogeneous models

In anisotropic models, new kinds of observations become possible. First, each of these relations will be anisotropic and so will vary with direction in the sky. In particular,

(6) *background radiation anisotropies* will occur and provide important information on the global spacetime geometry [100] as well as on local inhomogeneities [10, 59, 82] and gravitational waves [9];

(7) *image distortion effects* (strong and weak lensing) are caused by the Weyl tensor, which in turn is generated by local matter inhomogeneities through the 'div E' equation (3.48).

Finally, to fully determine the spacetime geometry [44, 86] we should also measure

(8) *transverse velocities*, corresponding to proper motions in the sky. However, these are so small as to be undetectable and so measurements only give weak upper limits in this case.

To evaluate the limits put on inhomogenity and anisotropy by observations, one must calculate observational relations predicted in anisotropic and inhomogenous models.

3.8.4.1   Bianchi observations

One can examine observational relations in the spatially homogeneous class of models, for example determining predicted Hubble expansion anisotropy, CBR anisotropy patterns, and nucleosynthesis results in Bianchi universes. These enable one to put strong limits on the anisotropy of these universe models since decoupling, and limits on the deviation from FL expansion rates during nucleosynthesis. However although these analyses put strong limits on the shear and vorticity in such models today, nevertheless they could have been very anisotropic at very early times—in particular, before nucleosynthesis—without violating the observational limits, and they could become anisotropic again at very late times. Also these limits are derived for specific spatially homogeneous models of particular Bianchi type, and there are others where they do not apply. For example, there exist Bianchi models in which rapid oscillations take place in the shear at late times, and these oscillations prevent a build up of CBR anisotropy, even though the universe is quite anisotropic at many times.

3.8.4.2 *Inhomogeneity and observations*

Similarly, one can examine observational relations in specific inhomogeneous models, for example the Tolman–Bondi spherically symmetric models and hierarchical Swiss-cheese models. We can then use these models to investigate the spatial homegenity of the universe (cf the next subsection).

The observational relations in linearly perturbed FL models, particularly (a) gravitational lensing properties and (b) CBR anisotropies have been the subject of intense theoretical study as well as observational exploration. A crucial issue that arises is on what scale we are representing the universe, for both its dynamic and observational properties may be quite different on small and large scales, and then the issue arises of how averaging over the small-scale behaviour can lead to the correct observational behaviour on large scales [32]. It seems that this will work out correctly, but really clear and compelling arguments that this is so are still lacking.

3.8.4.3 *Perturbed FL models and FL parameters*

As explained in detail in other chapters, the CBR anisotropies in perturbed FL models, in conjunction with studies of large-scale structure and models of the growth of inhomogeneities in such models, also using large-scale structure and supernovae observations, enables us to tie down the parameters of viable FL background models to a striking degree [8, 75].

3.8.5 Proof of almost-FL geometry

On a cosmological scale, observations appear almost isotropic about us (in particular number counts of many kinds of objects on the one hand, and the CBR temperature on the other). From this we may deduce that the observable region of the universe is, to a good approximation, also isotropic about us. A particular substantial issue, then, is how we can additionally prove the universe is spatially homogeneous, and so has an RW geometry, as is assumed in the standards models of cosmology.

3.8.5.1 *Direct proof*

Direct proof of spatial homogeneity would follow if we could show that the universe has precisely the relation between both area distance $r_0(z)$ and number counts $N(z)$ with redshift z that is predicted by the FL family of models. However, proving this observationally is not easily possible. Current supernova-based observations are indicate a non-zero cosmological constant rather than the relation predicted by the FL models with zero λ, and we are not able to test the $r_0(z)$ relationship accurately enough to show it takes a FL form with non-zero λ [95]. Furthermore number counts are only compatible with the FL models if we assume just the right source evolution takes place to make the observations compatible with spatial homogeneity; but once we take evolution into account, we can fit almost any observational relations by almost any spherically symmetric model (see [98] for exact theorems making this statement precise). Recent statistical observations of distant sources support spatial homogeneity on intermediate scales (between 30 and 400 Mpc [102]), but do not extend to larger scales because of sample limits.

3.8.5.2 Uniform thermal histories

A strong indication of spatial homogeneity is the fact that we see the same kinds of object, more or less, at high z as nearby. This suggests that they must have experienced more or less the same thermal history as nearby objects, as otherwise their structure would have come out different; and this, in turn, suggests that the spacetime geometry must have been rather similar near those objects as near to us, else (through the field equations) the thermal history would have come out different. This idea can be formulated in the *Postulate of Uniform Thermal Histories* (PUTH), stating that uniform thermal histories can occur only if the geometry is spatially homogeneous. Unfortunately, counter-examples to this conjecture have been found [7]. These are, however, probably exceptional cases and this remains a strong observationally-based argument for spatial homogeneity, indeed probably the most compelling at an intuitive level. However, relating the idea to observations also involves untangling the effects of time evolution, and it cannot be considered a formal proof of homogeneity.

3.8.5.3 Almost-EGS theorem

The most compelling precisely formulated argument is a based on our observations of the high degree of CBR anisotropy around us. If we assume we are not special observers, others will see the same high degree of anisotropy; and then that shows spatial homogeneity: exactly, in the case of exact isotropy (the Ehlers–Geren–Sachs (EGS) theorem [22]) and approximately in the case of almost-isotropy:

Almost-EGS-theorem. *[119]. If the Einstein–Liouville equations are satisfied in an expanding universe, where there is pressure-free matter with 4-velocity vector field u^a ($u_a u^a = -1$) such that (freely-propagating) background radiation is everywhere almost-isotropic relative to u^a in some domain U, then spacetime is almost-FLRW in U.*

This description is intended to represent the situation since decoupling to the present day. The pressure-free matter represents the galaxies on which fundamental observers live, who measure the radiation to be almost isotropic. This deduction is very plausible, particularly because of the argument just mentioned in the last subsection: conditions there *look* more or less the same, so there is no reason to think they are very different. Nevertheless, in the end this argument rests on an unproved philosophical assumption (that other observers see more or less what we do), and so is highly suggestive rather than a full observational proof. In addition, there is a technical issue of substance, namely what derivatives of the CBR temperature should be included in this formulation (remembering here that there are Bianchi models where the matter shear remains small but its time derivative can be large; these can have a large Weyl tensor but small CBR anisotropy [99]).

3.8.5.4 Theoretical arguments

Given the observational difficulties, one can propose theoretical rather than observational arguments for spatial homogeneity. Traditionally this was done by appeal to a *cosmological principle* [6,131]; however, this is no longer fashionable. Still some kinds of theoretical argument remain in vogue.

One can try to argue for spatial homogeneity on the basis of *probability*: this is more likely than the case of a spherically symmetric inhomogeneous universe, where we are near the centre (see [43] for detailed development of such a model). However, that argument is flawed [30], because spatially homogeneous universe models are intrinsically less likely than spherically symmetric inhomogeneous ones (as the latter have more degrees of freedom, and so are more general). In additionally, it is unclear that any probability arguments at all can be applied to the universe, because of its uniqueness [37].

Alternatively, one can argue that *inflation guarantees that the universe must be spatially homogeneous*. If we accept that argument, then the implication is that we are giving preference to a theoretically based analysis over what can, in fact, be established from observational data. In addition, it provides a partial rather than complete solution to the issues it addresses (see the discussion in the next section). Nevertheless it is an important argument that many find fully convincing.

Perhaps the most important argument in the end is that from *cumulative evidence*: none of these approaches by themselves proves spatial homogeneity, but taken together they give a sound cumulative argument that this is indeed the case—within the domain previously specified above.

3.8.5.5 Domains of plausibility

Accepting that argument, to what spacetime regions does it apply? We may take it as applying to the observable region of the universe V, that is, *the region both inside our visual horizon, and lying between the epoch of decoupling and the present day*. It will then also hold in some larger neighbourhood of this region, but there is no reason to believe it will hold elsewhere; specifically, it need not hold (i) very far out from us (say, 1000 Hubble radii away), hence chaotic inflation is a possibility; nor (ii) at very early times (say, before nucleosynthesis), so Bianchi anisotropic modes are possible at these early times; nor (iii) at very late times (say in another 50 Hubble times), so late-time anisotropic modes which are presently negligble could come to dominate (cf the discussion in the section on evolution of Bianchi models above). Thus we can observationally support the supposition of spatial homegeneity and isotropy within the domain $\mathcal{V}$, but not too far outside of it.

3.8.6   Importance of consistency checks

Because we have no knock-out observational proof of spatial homogeneity, it is important to consider all the possible observationally based consistency checks on the standard model geometry. The most important are as follows:

(1) *Ages.* This has been one of the oldest worries for expanding universe models: the requirement that the age of the universe must be greater than the ages of all objects in it. However with present estimates of the ages of stars on the one hand, and of the value of the Hubble constant on the other, this no longer seems problematic, particularly if current evidence for a positive cosmological constant turn out to be correct.

(2) *Anisotropic number counts.* If our interpretation of the CBR dipole as due to our motion relative to the FL model is correct, then this must also be accompanied by a dipole in all cosmological number counts at the 2% level [38]. Observationally verifying that this is so is a difficult task, but it is a crucial check on the validity of the standard model of cosmology.

(3) *High-z observations.* The best check on spatial homogeneity is to try to check the physical state of the universe at high redshifts and hence at great distances from us, and to compare the observations with theory. This can be done in particular (a) for the CBR, whose temperature can be measured via excited states of particular molecules; this can then be compared with the predicted temperature $T = T_0(1+z)$, where T_0 is the present day temperature of 2.75 K. It can also be done (b) for element abundances in distant objects, specifically helium abundances. This is particularly useful as it tests the thermal history of the universe at very early times of regions that are far out from us [34].

3.9   Explaining homogeneity and structure

This is the unique core business of physical cosmology: explaining both why the universe has the very improbable high-symmetry FL geometry on the largest scales, and how structures come into existence on all smaller scales. Clearly only cosmology itself can ask the first question; and it uniquely sets the initial conditions underlying the astrophysical and physical processes that are the key to the second, underlying all studies of origins.There is a creative tension between two aims: smoothing processes, on the one hand, and structure growth, on the other. Present day cosmology handles this tension by suggesting a change of equation of state: at early enough times, the equation of state was such as to cause smoothing on all scales; but at later times, it was such as to cause structure growth on particular scales. The inflationary scenario, and the models that build on it, are remarkably successful in this regard, particularly through predicting the CBR anisotropy patterns (the 'Doppler peaks') which seem to have been found now (but significant problems remain, particularly as regards compatibility with the well-established nucleosynthesis arguments).

Given these astrophysical and physical processes, explanation of the large-scale isotropy and homogeneity of the universe together with the creation of smaller-scale structures means determining the dynamical evolutionary trajectories relating initial to final conditions, and then essentially either (a) explaining initial conditions or (b) showing they are irrelevant.

3.9.1 Showing initial conditions are irrelevant

This can be attempted in a number of different ways.

3.9.1.1   *Initial conditions are irrelevant because they are forgotten*

Demonstrating minimal dependence of the large-scale final state on the initial conditions has been the aim of

- the *chaotic cosmology* programme of Misner, where physical processes such a viscosity wipe out memories of previous conditions [97]; and
- the *inflationary family of theories*, where the rapid exponential expansion driven by a scalar field smooths out the universe and so results in similar memory loss [79].

The (effective) scalar field is slow-rolling, so the energy condition (3.36) is violated and a period of accelerating expansion can take place through many e-foldings, until the scalar field decays into radiation at the end of inflation. This drives the universe model towards flatness, and is commonly believed to predict that the universe must be very close indeed to flatness today, even though this is an unstable situation, see the phase planes of Ω against S [94]. It can also damp out both anisotropy, as previously explained and inhomogeneity, if the initial situation is close enough to a FL model of that inflation can in fact start. In a chaotic inflationary scenario, with random initial conditions occurring at some initial time, inflation will not succeed in starting in most places, but those domains where it does start will expand so much that they will soon be the dominant feature of the universe: there will be many vast FL-like domains, each with different parameter values and perhaps even different physics, separated from each other by highly inhomogeneous transition regions (where physics may be very strange). In the almost-FL domains, quantum fluctuations are expanded to a very large scale in the inflationary era, and form the seeds for structure formation at later times. Inflation then goes on to provide a causal theory of initial structure formation from an essentially homogeneous early state (via amplification of initial quantum fluctuations)—a major success if the all the details can be sorted out.

This is an attractive scenario, particularly because it ties in the large-scale structure of the universe with high-energy physics. It works fine for those regions that start off close enough to FL models, and, as noted earlier this suffices to explain the existence of large FL-like domains, such as the one we inhabit. It does not necessarily rule out the early and late anisotropic modes that were discussed in the section on Bianchi models. It fits the observations provided one has enough auxiliary functions and parameters available to mediate between the basic theory and the observations (specifically, evolution functions, a bias parameter or function, a dark matter component, a cosmological constant or 'quintessence' component at late times). However, it is not at present a specific physical model, rather it is a family of models (see e.g. [78]), with many competing explanations for the origin of the inflaton, which is not yet identified with any specific matter component or field. It will become a well-defined physical theory when one or other of these competing possibilities is identified as the real physical driver of an inflationary early epoch.

There are three other issues to note here. First, the *issue of probability*: inflation is intended as a means of showing the observed region of the universe is in fact probable. But we have no proper measure of probability on the family of universe models, so this has not been demonstrated in a convincing way. Second, the *Trans-Planckian problem* [96]: inflation is generally very successful in generating a vast expansion of the universe. The consequence is that the spacetime region that has been expanded to macroscopic scales today is deep in the Planck (quantum-gravity) era, so the nature of what is predicted depends crucially on our assumptions about that era; but we do not know what conditions were like there, and indeed even lack proper tools to describe that epoch, which may have been of the nature of a spacetime foam, for example. Thus the results of inflation for large-scale structure depend on specific assumptions about the nature of spacetime in the strong quantum gravity regime, and we do not know what that nature is. Penrose suggests it was very inhomogeneous at that time, in which case inflation will amplify that inhomogeneous nature rather than creating spatial homgeneity. As in the previous case, whether or not the process succeeds will depend on the initial conditions for the expansion of the universe as it emerges from the Planck (quantum gravity) era. Thirdly, there are still unsolved problems regarding *the end of inflation*. These relate to the fact that if one has a very slow rolling field as is often claimed, then the inertial mass density is very close to zero so velocities are unstable.

It must be emphasized that in order to investigate this issue of isotropisation properly, *one must examine the dynamical behaviour of very anisotropic and inhomogeneous cosmologies*. This is seldom done—for example, almost all of the literature on inflation examines only its effects in RW geometries, which is precisely when there is no need for inflation take place in order to explain the smooth geometry—for then a smooth geometry has been assumed *a priori*.

When the full range of inhomogeneities and anisotropies is taken into account (e.g. [128]), it appears that *both approaches are partially successful*: with or without inflation one can explain a considerable degree of isotropization and homogenization of the physical universe (see e.g. [127]), but this will not work in all circumstances [105, 106]. It can only be guaranteed to work if initial conditions are somewhat restricted—so in order for the programme to succeed, we have to go back to the former issue of somehow explaining why it is probable for a restricted set of initial data to occur.

3.9.1.2   *Initial conditions are irrelevant because they never happened*

Some attempts involve avoiding a true beginning by going back to some form of eternal or cyclic state, so that the universe existed forever. Initial conditions are pushed back into the infinite past, and thus were never set. Examples are as follows.

- The original *steady state universe* proposal of Bondi [6], and its updated form as the *quasi-steady state universe* of Hoyle, Burbidge and Narlikar [71, 72].
- Linde's *eternal chaotic inflation*, where ever-forming new bubbles of expansion arising within old ones exist forever; this can prevent the universe from ever entering the quantum gravity regime [90].
- The Hartle–Hawking *'no-boundary'* proposal (cf [67]) avoids the initial singularity through a change of spacetime signature at very early times, thereby entering a positive-definite ('space–space') regime where the singularity theorems do not apply (the physical singularity of the big bang gets replaced by the coordinate singularity at the south pole of a sphere). There is no singularity and no boundary, and so there are no boundary conditions. This gets round the issue of a creation event in an ingenious way: there is no unique start to the universe, but there is a beginning of time.
- The Hawking–Turok *initial instanton proposal* is a variant of this scenario, where there is a weak singularity to start with, and one is then able to enter a low-density inflationary phase.
- Gott and Liu's *causality violation in the early universe* does the same kind of thing in a different way: causality violation takes place in the early universe, enabling the universe to 'create itself' [63]. Like the chaotic inflation picture, new expanding universe bubbles are forming all the time; but one of them is the universe region where the bubble was formed, this being possible because closed timelike lines are allowed, so 'the universe is its own mother'. This region of closed timelike lines is separated from the later causally regular regions by a Cauchy horizon.

There are thus a variety of ingenious and intriguing options which, in a sense, allow avoidance of setting initial conditions. But this is really a technicality: the issue still arises as to why in each case one particular initial state 'existed' or came into being rather than any of the other options. Some particular solutions of the equations have been implemented rather than the other possibilities; boundary conditions choosing one set of solutions over others have still been set, even if they are not technically initial conditions set at a finite time in the past.

3.9.1.3   *Initial conditions are irrelevant because they all happened*

The idea of an ensemble of universes, mentioned earlier, is one approach that sidesteps the problem of choice of specific initial data, because by hypothesis *all that can occur has then occurred*. Anthropic arguments select the particular universe in which we live from all those in this vast family (see e.g. [57,70]). This is again an intriguing and ingenious idea, extending to a vast scale the Feynman approach to quantum theory. However, there are several problems.

First, it is not clear that the selection of universes from this vast family by anthropic arguments will necessarily result in as large and as isotropic a universe as we see today; here one runs up against the unsolved problem of justifying a choice of probabilities in this family of universes. Second, this proposal suffers from complete lack of verifiability. In my view, this means this is a metaphysical rather than scientific proposal, because it is completely untestable. And in the end, this suggestion does not solve the basic issue in any case, because then one can ask: *Why does this particular ensemble exist, rather than a different ensemble with different properties?*; and the whole series of fundamental questions arises all over again, in an even more unverifiable form than before.

3.9.2   The explanation of initial conditions

The explanation of initial conditions has been the aim of the family of theories one can label collectively as *quantum cosmology* and the more recent studies of *string cosmology*.

3.9.2.1   *Explanation of initial conditions from a previous state of a different nature*

One option has been explaining the universe as we see it as arising from some completely different initial state, for example:

- proposals for *creation of the universe as a bubble formed in a flat spacetime or de Sitter spacetime*, for example Tryon's vacuum fluctuations and Gott's open bubble universes; or
- Vilenkin's *tunnelling universe* which arises from a state with no classical analogue (described as 'creation of the universe from nothing', but this is inaccurate).

These proposals (like the proposals by Hartle and Hawking, Hawking and Turok, and Gott and Liu previously mentioned; for a comparative discussion and references, see Gott and Liu [63]) are based on the quantum cosmology idea of *the wavefunction of the universe*, taken to obey the Wheeler–de Witt equation (a generalization to the cosmological context of the Schrödinger equation) (see e.g. [67]). This approach faces considerable technical problems, related to

- the meaning of time, because vanishing of the Hamiltonian of general relativity means that the wavefunction appears to be explicitly independent of time;
- divergences in the path-integrals often used to formulate the solutions to the Wheeler–de-Witt equation;
- the meaning of the wavefunction of the universe, in a context where probabilities are ill defined [56];
- the fundamentally important issue of the meaning of measurement in quantum theory (when does 'collapse of the wavefunction' take place, in a context where a classical 'observer' does not exist);
- the conditions which will lead to these quantum equations having classical-like behaviour at some stage in the history of the universe [65]; and
- the way in which this reduced set of equations, taken to be valid irrespective of the nature of the full quantum theory of gravity, relates to that as yet unknown theory.

The alternative is to work with the best current proposal for such a theory, taken by many to be *M-theory*, which aims to unite the previously disparate superstring theories into a single theory, with the previously separate theories related to each other by a series of symmetries called dualities. There is a rapidly growing literature on *superstring cosmology*, relating this theory to cosmology [89]. In particular, much work is taking place on two approaches:

- The *pre big-bang proposal*, where a 'pre big-bang' branch of the universe is related to a 'post big-bang' era by a duality: $a(t) \rightarrow 1/a(t), t \rightarrow -t$, and dimensional reduction results in a scalar field (a 'dilaton') occurring in the field equations (see Gasperini [58] for updated references).

This approach has major technical difficulties to solve, particularly related to the transition from the 'pre big-bang' phase to the 'post big-bang' phase, and to the transition from that phase to a standard cosmological expansion. In additionally it faces fine-tuning problems related to its initial conditions. So this too is very much a theory in the course of development, rather than a fully viable proposal.

- The *brane cosmology proposal*, where the physical universe is confined to a four-dimensional 'brane' in a five-dimensional universe. The physics of this proposal are very speculative, and issues arise as to why the initial conditions in the 5D space had the precise nature so as to confine matter to this lower-dimensional subspace; and then the confinement problem is why they remain there.

Supposing these technical difficulties can be overcome in each case, it is still unclear that these proposals avoid the real problem of origins. It can be claimed they simply postpone facing it, for one now has to ask all the same questions of origins and uniqueness about the supposed prior state to the present hot big bang expansion phase: Why did this previous state have the properties it had? (whether or not it had a classical analogue)? This 'pre-state' should be added to one's cosmology, and then the same basic questions as before now arise regarding this completed model.

3.9.2.2   *Explanation of initial conditions from 'nothing'*

Attempts at an 'explanation' of a true origin, i.e. not arising from some pre-existing state (whether it has a classical analogue or not), are difficult even to formulate.

They may depend on *assuming a pre-existing set of physical laws* that are similar to those that exist once spacetime exists, for they rely on an array of properties of quantum field theory and of fields (existence of Hilbert spaces and operators, validity of variational principles and symmetry principles, and so on) that seem to hold sway independently of the existence of the universe and of space and time (for the universe itself, and so space and time, is to arise out of their validity). This issue arises, for example, in the case of Vilenkin's tunnelling universes: not only do they come from a pre-existent state, as remarked previously, but they also take the whole apparatus of quantum theory for granted. This is far from 'nothing'—it is a very complex structure; but there is no clear locus for those laws to exist in or material for them to act on. The manner of their existence or other grounds for their validity in this context are unclear—and we run into the problems noted before: there are problems with the concepts of 'occurred', 'circumstances' and even 'when'—for we are talking *inter alia* about the existence of spacetime. Our language can hardly deal with this. Given the feature that no spacetime exists before such a beginning, brave attempts to define a 'physics of creation' stretch the meaning of 'physics'. There cannot be a prior physical explanation, precisely because physics and the causality associated with physics does not exist there/then.

Perhaps the most radical proposal is that

> order arises out of nothing: all order, including the laws of physics, somehow arises out of chaos,

in the true sense of that word—namely a total lack of order and structure of any kind (e.g. [1]). However, this does not seem fully coherent as a proposal. If the pre-ordered state is truly chaotic and without form, I do not see how order can arise therefrom when physical action is as yet unable to take place, or even how we can meaningfully contemplate that situation. We cannot assume any statistical properties would hold in that regime, for example; even formulating a description of states seems well nigh impossible, for that can only be done in terms of concepts that have a meaning only in a situation of some stability and underlying order such as is characterized by physical laws.

3.9.3   The irremovable problem

Thus a great variety of possibilities is being investigated. However, the same problem arises in every approach: even if a literal creation does not take place, as is the case in various of the present proposals, this does not resolve the underlying issue. Apart from all the technical difficulties, and the lack of experimental support for these proposals, none of these can get around the basic problem: given any specific proposal,

> *How was it decided that this particular kind of universe would be the one that was actually instantiated and what fixed its parameters?*

A choice between different contingent possibilities has somehow occurred; the fundamental issue is what underlies this choice. Why does the universe have one specific form rather than another, when other forms seem perfectly possible? Why should any one of these approaches have occurred if all the others are possibilities? This issue arises even if we assume an ensemble of universes exists: for then we can ask why this particular ensemble, and not another one?

All approaches face major problems of verifiability, for the underlying dynamics relevant to these times can never be tested. Here we inevitably reach the limits to what the scientific study of the cosmos can ever say—if we assume that such studies must of necessity involve an ability to observationally or experimentally check the relevant physical theories. However we can attain some checks on these theories by examining their predictions for the present state of the universe—its large-scale structure, smaller scale structure and observable features such as gravitational waves emitted at very early times. These are important restrictions, and are very much under investigation at the present time; we need to push our observations as far as we can, and this is indeed happening at present (particularly through deep galactic observations; much improved CBR observations; and the prospect of new generation gravitational wave detectors coming on line).

If it could be shown that only one of all these options was compatible with observations of the present day universe, this would be a major step forward: it would select one dynamical evolution from all the possibilities. However, this does not seem likely, particularly because of the proliferation of auxiliary functions that can be used to fit the data to the models, as noted before. In addition, even if this was achieved, it would not show why that one had occurred rather than any of the others. This would be achieved if it could be eventually shown that only one of these possibilities is self-consistent: that, in fact, fatal flaws in all the others reduce the field of possibilities to one. We are nowhere near this situation at present, indeed possibilities are proliferating rather than reducing.

Given these problems, any progress is of necessity based on specific philosophical positions, which decide which of the many possible physical and metaphysical approaches is to be preferred. These philosophical positions should be identified as such and made explicit [37, 88]. As explained earlier, no experimental test can determine the nature of any mechanisms that may be in operation in circumstances where even the concepts of cause and effect are suspect. Initial conditions cannot be determined by the laws of physics alone—for if they were so determined they would no longer be contingent conditions, the essential feature of initial data, but rather would be necessary. A purely scientific approach cannot succeed in explaining this specific nature of the universe.

Consequent on this situation, it follows that unavoidably, whatever approach one may take to issues of cosmological origins, metaphysical issues inevitably arise in cosmology: philosophical choices are needed in order to shape the theory. That feature should be explicitly recognized, and then sensibly developed in the optimal way by carefully examining the best way to make such choices.

3.10   Conclusion

There is a tension between theory and observation in cosmology. The issue we have considered here is, Which family of models is consistent with observations? To answer this demands an equal sophistication of geometry and physics, whereas in the usual approaches there is a major imbalance: very sophisticated physics and very simple geometry. We have looked here at tools to deal with the geometry in a resaonably sophisticated way, and summarized some of the results that are obtained by using them. This remains an interesting area of study, particularly in terms of relating realistic inhomogeneous models to the smoothed out standard FL models of cosmology.

Further problems arise in considering the physics of the extremely early universe, and any pre-physics determining initial conditions for the universe. We will need to develop approaches to these topics that explicitly recognizes the limitations of the scientific method—assuming that this method implies the possibility of verification of our theories.

References

[1] Anandan J 1998 *Preprint* quant-phy/9808045
[2] Bardeen J 1980 *Phys. Rev.* D **22** 1882
[3] Barrow J and Tipler F J 1986 *The Anthropic Cosmological Principle* (Oxford: Oxford University Press)
[4] Boerner G and Gottlober S (ed) 1997 *The Evolution of the Universe* (New York: Wiley)
[5] Bondi H 1947 *Mon. Not. R. Astron. Soc.* **107** 410
[6] Bondi H 1960 *Cosmology* 1960 (Cambridge: Cambridge University Press)
[7] Bonnor W B and Ellis G F R 1986 *Mon. Not. R. Astron. Soc.* **218** 605

[8] Bridle S L, Zehavi I, Dekel A, Lahav O, Hobson M P and Lasenby A N 2000 *Mon. Not. R. Astron. Soc.* **321** 333

[9] Challinor A 2000 *Class. Quantum Grav.* **17** 871 (astro-ph/9906474)

[10] Challinor A and Lasenby A 1998 *Phys. Rev.* D **58** 023001

[11] Cohn P M 1961 *Lie Algebras* (Cambridge: Cambridge University Press).

[12] Collins C B 1971 *Commun. Math. Phys.* **23** 137

[13] Collins C B 1977 *J. Math. Phys.* **18** 2116

[14] Collins C B 1985 *J. Math. Phys.* **26** 2009

[15] Collins C B and Ellis G F R 1979 *Phys. Rep.* **56** 63

[16] Collins C B and Hawking S W 1973 *Astrophys. J.* **180** 317

[17] Cornish N J and Spergel D N 1999 *Phys. Rev.* D **92** 087304

[18] Cornish N J, Spergel D N and Starkman G 1966 *Phys. Rev. Lett.* **77** 215

[19] d'Inverno R 1992 *Introducing Einstein's Relativity* (Oxford: Oxford Univerity Press)

[20] Dunsby P K S, Bassett B A C and Ellis G F R 1996 *Class. Quantum Grav.* **14** 1215

[21] Ehlers J 1961 *Akad. Wiss. Lit. Mainz, Abhandl. Math.-Nat. Kl.* **11** 793 (Engl. transl. 1993 *Gen. Rel. Grav.* **25** 1225)

[22] Ehlers J, Geren P and Sachs R K 1968 *J. Math. Phys.* **9** 1344

[23] Einstein A and Straus E G 1945 *Rev. Mod. Phys.* **17** 120

[24] Eisenhart L P 1933 *Continuous Groups of Transformations* (Princeton, NJ: Princeton University Press) reprinted: 1961 (New York: Dover)

[25] Ellis G F R 1967 *J. Math. Phys.* **8** 1171

[26] Ellis G F R 1971 General relativity and cosmology *Proc. XLVII Enrico Fermi Summer School* ed R K Sachs (New York: Academic Press)

[27] Ellis G F R 1971 *Gen. Rel. Grav.* **2** 7

[28] Ellis G F R 1973 *Cargèse Lectures in Physics* vol 6, ed E Schatzman (New York: Gordon and Breach)

[29] Ellis G F R 1975 *Q. J. R. Astron. Soc.* **16** 245

[30] Ellis G F R 1979 *Gen. Rel. Grav.* **11** 281

[31] Ellis G F R 1980 *Ann. New York Acad. Sci.* **336** 130

[32] Ellis G F R 1984 *General Relativity and Gravitation* ed B Bertotti *et al* (Dordrecht: Reidel) p 215

[33] Ellis G F R 1987 *Vth Brazilian School on Cosmology and Gravitation* ed M Novello (Singapore: World Scientific)

[34] Ellis G F R 1987 *Theory and Observational Limits in Cosmology* ed W Stoeger (Vatican Observatory) pp 43–72
  Ellis G F R 1995 *Galaxies and the Young Universe* ed H von Hippelein, K Meisenheimer and J H Roser (Berlin: Springer) p 51

[35] Ellis G F R 1990 *Modern Cosmology in Retrospect* ed B Bertotti *et al* (Cambridge: Cambridge University Press) p 97

[36] Ellis G F R 1991 *Mem. Ital. Ast. Soc.* **62** 553–605

[37] Ellis G F R 1999 *Astron. Geophys.* **40** 4.20
  Ellis G F R 2000 *Toward a New Millenium in Galaxy Morphology* ed D Block *et al* (Dordrecht: Kluwer)
  Ellis G F R 1999 *Astrophysics and Space Science* **269–279** 693

[38] Ellis G F R and Baldwin J 1984 *Mon. Not. R. Astron. Soc.* **206** 377–81

[39] Ellis G F R and Bruni M 1989 *Phys. Rev.* D **40** 1804

[40] Ellis G F R, Bruni M and Hwang J C 1990 *Phys. Rev.* D **42** 1035

[41] Ellis G F R and van Elst H 1999 *Theoretical and Observational Cosmology (Nato Science Series C, 541)* ed M Lachieze-Rey (Dordrecht: Kluwer) p 1
[42] Ellis G F R and King A R 1974 *Commun. Math. Phys.* **38** 119
[43] Ellis G F R, Maartens R and Nel S D 1978 *Mon. Not. R. Astron. Soc.* **184** 439–65
[44] Ellis G F R, Nel S D, Stoeger W, Maartens R and Whitman A P 1985 *Phys. Rep.* **124** 315
[45] Ellis G F R and MacCallum M A H 1969 *Commun. Math. Phys.* **12** 108
[46] Ellis G F R and Madsen M S 1991 *Class. Quantum Grav.* **8** 667
[47] Ellis G F R, Perry J J and Sievers A 1984 *Astron. J.* **89** 1124
[48] Ellis G F R and Schreiber G 1986 *Phys. Lett.* A **115** 97–107
[49] Ellis G F R and Sciama D W 1972 *General Relativity (Synge Festschrift)* ed L O'Raifeartaigh (Oxford: Oxford University Press)
[50] Ellis G F R and Stoeger W R 1988 *Class. Quantum Grav.* **5** 207
[51] van Elst H and Ellis G F R 1996 *Class. Quantum Grav.* **13** 1099
[52] van Elst H and Ellis G F R 1998 *Class. Quantum Grav.* **15** 3545
[53] van Elst H and Ellis G F R 1999 *Phys. Rev.* D **59** 024013
[54] Ellis G F R and van Elst H 1999 *On Einstein's Path: Essays in Honour of Englebert Schucking* ed A Harvey (Berlin: Springer) pp 203–26
[55] van Elst H and Uggla C 1997 *Class. Quantum Grav.* **14** 2673
[56] Fink H and Lesche H 2000 *Found. Phys. Lett.* **13** 345
[57] Garriga J and Vilenkin A 2000 *Phys. Rev.* D **61** 083502
[58] Gasperini M 1999 *String Cosmology* http://www.to.infin.it/~gasperini
[59] Gebbie T and Ellis G F R 2000 *Ann. Phys.* **282** 285
     Gebbie T, Dunsby P K S and Ellis G F R 2000 *Ann. Phys.* **282** 321
[60] Gödel K 1949 *Rev. Mod. Phys.* **21** 447
[61] Gödel K 1952 *Proc. Int. Cong. Math. (Am. Math. Soc.)* **175**
[62] Goliath M and Ellis G F R 1999 *Phys. Rev.* D **60** 023502 (gr-qc/9811068)
[63] Gott J R and Liu L 1999 *Phys. Rev.* D **58** 023501 (astro-ph/9712344)
[64] Harrison E R 1981 *Cosmology: The Science of the Universe* (Cambridge: Cambridge University Press)
[65] Hartle J B 1996 Quantum mechanics at the Planck scale *Physics at the Planck Scale* ed J Maharana, A Khare and M Kumar (Singapore: World Scientific)
[66] Hawking S W 1966 *Astrophys. J.* **145** 544
[67] Hawking S W 1993 *Hawking on the Big Bang and Black Holes* (Singapore: World Scientific)
[68] Hawking S W and Ellis G F R 1973 *The Large Scale Structure of Spacetime* (Cambridge: Cambridge University Press)
[69] Hawking S W and Penrose R 1970 *Proc. R. Soc.* A **314** 529
[70] Hogan C J 1999 *Preprint* astro-ph/9909295
[71] Hoyle F, Burbidge G and Narlikar J 1993 *Astrophys. J.* **410** 437
[72] Hoyle F, Burbidge G and Narlikar J 1995 *Proc. R. Soc.* A **448** 191
[73] Hsu L and Wainwright J 1986 *Class. Quantum Grav.* **3** 1105
[74] Isham C J 1997 *Lectures on Quantum Theory: Mathematical and Structural Foundations* (London: Imperial College Press, Singapore: World Scientific)
[75] Jaffe A H *et al* 2001 *Phys. Rev. Lett.* **86** 3475
[76] Kantowski R and Sachs R K 1966 *J. Math. Phys.* **7** 443
[77] King A R and Ellis G F R 1973 *Commun. Math. Phys.* **31** 209
[78] Kinney W H, Melchiorri A and Riotto A 2001 *Phys. Rev.* D **63** 023505

[79] Kolb E W and Turner M S 1990 *The Early Universe* (New York: Wiley)

[80] Kompaneets A S and Chernov A S 1965 *Sov. Phys.–JETP* **20** 1303

[81] Lachieze R M and Luminet J P 1995 *Phys. Rep.* **254** 136

[82] Lewis A, Challinor A and Lasenby A 2000 *Astrophys. J.* **538** 3273 (astro-ph/9911177)

[83] Kramer D, Stephani H, MacCallum M A H and Herlt E 1980 *Exact Solutions of Einstein's Field Equations* (Cambridge: Cambridge University Press)

[84] Krasiński A 1983 *Gen. Rel. Grav.* **15** 673

[85] Krasiński A 1996 *Physics in an Inhomogeneous Universe* (Cambridge: Cambridge University Press)

[86] Kristian J and Sachs R K 1966 *Astrophys. J.* **143** 379

[87] Lemaître G 1933 *Ann. Soc. Sci. Bruxelles I* A **53** 51 (Engl. transl. 1997 *Gen. Rel. Grav.* **29** 641)

[88] Leslie J (ed) 1998 *Modern Cosmology and Philosophy* (Amherst, NY: Prometheus Books)

[89] Lidsey J E, Wands D and Copeland E J 2000 Superstring cosmology *Phys. Rep.* **337** 343–492

[90] Linde A D 1990 *Particle Physics and Inflationary Cosmology* (Chur, Switzerland: Harwood Academic)

[91] Maartens R 1997 *Phys. Rev.* D **55** 463

[92] MacCallum M A H 1973 *Cargèse Lectures in Physics* vol 6, ed E Schatzman (New York: Gordon and Breach)

[93] MacCallum M A H 1979 *General Relativity, An Einstein Centenary Survey* ed S W Hawking and W Israel (Cambridge: Cambridge University Press)

[94] Madsen M and Ellis G F R 1988 *Mon. Not. R. Astron. Soc.* **234** 67

[95] Maor I, Brustein R and Steinhardt P J 2001 *Phys. Rev. Lett.* **86** 6

[96] Martin J and Brandenberger R H 2001 *Phys. Rev.* D **63** 123501

[97] Misner C W 1968 *Astrophys. J.* **151** 431

[98] Mustapha N, Hellaby C and Ellis G F R 1999 *Mon. Not. R. Astron. Soc.* **292** 817

[99] Nilsson U S, Uggla C and Wainwright J 1999 *Astrophys. J. Lett.* **522** L1 (gr-qc/9904252)

[100] Nilsson U S, Uggla C and Wainwright J 2000 *Gen. Rel. Grav.* **32** 1319 (gr-qc/9908062)

[101] Oszvath I and Schücking E 1962 *Nature* **193** 1168

[102] Pan J and Coles P 2000 *Mon. Not. R. Astron. Soc.* **318** L51

[103] Peacocke J R 1999 *Cosmological Physics* (Cambridge: Cambridge University Press)

[104] Peebles P J E, Schramm D N, Turner E L and Kron R G 1991 *Nature* **352** 769

[105] Penrose R 1989 *Proc. 14th Texas Symposium on Relativistic Astrophysics (Ann. New York Acad. Sci.)* ed E Fenves

[106] Penrose R 1989 *The Emperor's New Mind* (Oxford: Oxford University Press) ch 7

[107] Pirani F A E 1956 *Acta Phys. Polon.* **15** 389

[108] Pirani F A E 1957 *Phys. Rev.* **105** 1089

[109] Raychaudhuri A and Modak B 1988 *Class. Quantum Grav.* **5** 225

[110] Rindler W 1956 *Mon. Not. R. Astron. Soc.* **116** 662

[111] Rothman A and Ellis G F R 1986 *Phys. Lett.* B **180** 19

[112] Schucking E 1954 *Z. Phys.* **137** 595

[113] Senovilla J M, Sopuerta C and Szekeres P 1998 *Gen. Rel. Grav.* **30** 389

[114] Smolin L 1992 *Class. Quantum Grav.* **9** 173
[115] Stabell R and Refsdal S 1966 *Mon. Not. R. Astron. Soc.* **132** 379
[116] Stephani H 1987 *Class. Quantum Grav.* **4** 125
[117] Stephani H 1990 *General Relativity* (Cambridge: Cambridge University Press)
[118] Stewart J M and Ellis G F R 1968 *J. Math. Phys.* **9** 1072
[119] Stoeger W, Maartens R and Ellis G F R 1995 *Astrophys. J.* **443** 1
[120] Szekeres P 1965 *J. Math. Phys.* **6** 1387
[121] Szekeres P 1975 *Commun. Math. Phys.* **41** 55
[122] Szekeres P 1975 *Phys. Rev.* D **12** 2941
[123] Tegmark M 1998 *Ann. Phys., NY* **270** 1
[124] Tipler F J, Clarke C J S and Ellis G F R 1980 *General Relativity and Gravitation: One Hundred Years after the Birth of Albert Einstein* vol 2, ed A Held (New York: Plenum)
[125] Tolman R C 1934 *Proc. Natl Acad. Sci., USA* **20** 69
[126] Wainwright J 1988 *Relativity Today* ed Z Perjes (Singapore: World Scientific)
[127] Wainright J, Coley A A, Ellis G F R and Hancock M 1998 *Class. Quantum Grav.* **15** 331
[128] Wainwright J and Ellis G F R (ed) 1997 *Dynamical Systems in Cosmology* (Cambridge: Cambridge University Press)
[129] Wald R M 1984 *General Relativity* (Chicago, IL: University of Chicago Press)
[130] Wald R M 1983 *Phys. Rev.* D **28** 2118
[131] Weinberg S W 1972 *Gravitation and Cosmology* (New York: Wiley)
[132] Wheeler J A 1968 *Einstein's Vision* (Berlin: Springer)

Chapter 4

Inflationary cosmology and creation of matter in the universe

Andrei D Linde
Department of Physics, Stanford University, Stanford, USA

4.1 Introduction

The typical lifetime of a new trend in high-energy physics and cosmology is nowadays about 5–10 years. If it has survived for a longer time, the chances are that it will be with us for quite a while. Inflationary theory by now is 20 years old, and it is still very much alive. It is the only theory which explains why our universe is so homogeneous, flat and isotropic, and why its different parts began their expansion simultaneously. It provides a mechanism explaining galaxy formation and solves numerous different problems at the intersection between cosmology and particle physics. It seems to be in a good agreement with observational data and it does not have any competitors. Thus we have some reasons for optimism.

According to the standard textbook description, inflation is a stage of exponential expansion in a supercooled false vacuum state formed as a result of high-temperature phase transitions in Grand Unified Theories (GUTs). However, during the last 20 years inflationary theory has changed quite substantially. New versions of inflationary theory typically do not require any assumptions about initial thermal equilibrium in the early universe, supercooling and exponential expansion in the false vacuum state. Instead of this, we are thinking about chaotic initial conditions, quantum cosmology and the theory of a self-reproducing universe.

Inflationary theory was proposed as an attempt to resolve problems of the big bang theory. In particular, inflation provides a simple explanation of the extraordinary homogeneity of the observable part of the universe. But it can make the universe extremely inhomogeneous on a much greater scale. Now we believe that instead of being a single, expanding ball of fire produced in the big bang, the universe looks like a huge growing fractal. It consists of many inflating balls that produce new balls, which in turn produce more new balls, ad infinitum. Even now we continue learning new things about inflationary cosmology, especially about the stage of reheating of the universe after inflation.

In this chapter we will briefly describe the history of inflationary cosmology and then we will give a review of some recent developments.

4.2   Brief history of inflation

The first inflationary model was proposed by Alexei Starobinsky in 1979 [1]. It was based on investigation of conformal anomaly in quantum gravity. This model was rather complicated, it did not aim on solving homogeneity, horizon and monopole problems, and it was not easy to understand the beginning of inflation in this model. However, it did not suffer from the graceful exit problem and, in this sense, it can be considered the first working model of inflation. The theory of density perturbations in this model was developed in 1981 by Mukhanov and Chibisov [2]. This theory does not differ much from the theory of density perturbations in new inflation, which was proposed later by Hawking, Starobinsky, Guth, Pi, Bardeen, Steinhardt, Turner and Mukhanov [3,4].

A much simpler model with a very clear physical motivation was proposed by Alan Guth in 1981 [5]. His model, which is now called 'old inflation', was based on the theory of supercooling during the cosmological phase transitions [6]. It was so attractive that even now all textbooks on astronomy and most of the popular books on cosmology describe inflation as exponential expansion of the universe in a supercooled false vacuum state. It is seductively easy to explain the nature of inflation in this scenario. False vacuum is a metastable state without any fields or particles but with a large energy density. Imagine a universe filled with such 'heavy nothing'. When the universe expands, empty space remains empty, so its energy density does not change. The universe with a constant energy density expands exponentially, thus we have inflation in the false vacuum.

Unfortunately this explanation is somewhat misleading. Expansion in the false vacuum in a certain sense is false: de Sitter space with a constant vacuum energy density can be considered either expanding, or contracting, or static, depending on the choice of a coordinate system [7]. The absence of a preferable hypersurface of decay of the false vacuum is the main reason why the universe after inflation in this scenario becomes very inhomogeneous [5]. After many attempts to overcome this problem, it was concluded that the old inflation scenario cannot be improved [8].

Fortunately, this problem was resolved with the invention of the new inflationary theory [9]. In this theory, just as in the Starobinsky model, inflation may begin in the false vacuum. This stage of inflation is not very useful, but it prepares a stage for the next stage, which occurs when the inflaton field ϕ driving inflation moves away from the false vacuum and slowly rolls down to the minimum of its effective potential. The motion of the field away from the false vacuum is of crucial importance: density perturbations produced during inflation are inversely proportional to $\dot{\phi}$ [2, 3]. Thus the key difference between the new inflationary scenario and the old one is that the useful part of inflation in the new scenario, which is responsible for homogeneity of our universe, does *not* occur in the false vacuum state.

The new inflation scenario was plagued by its own problems. This scenario works only if the effective potential of the field ϕ has a very flat plateau near $\phi = 0$, which is somewhat artificial. In most versions of this scenario the inflaton field originally could not be in a thermal equilibrium with other matter fields. The theory of cosmological phase transitions, which was the basis for old and new inflation, simply did not work in such a situation. Moreover, thermal equilibrium requires many particles interacting with each other. This means that new inflation could explain why our universe was so large only if it was very large and contained many particles from the very beginning. Finally, inflation in this theory begins very late, and during the preceding epoch the universe could easily collapse or become so inhomogeneous that inflation may never happen [7]. Because of all these difficulties no realistic versions of the new inflationary universe scenario have been proposed so far.

From a more general perspective, old and new inflation represented a substantial but incomplete modification of the big bang theory. It was still assumed that the universe was in a state of thermal equilibrium from the very beginning, that it was relatively homogeneous and large enough to survive until the beginning of inflation, and that the stage of inflation was just an intermediate stage of the evolution of the universe. At the beginning of the 1980s these assumptions seemed most natural and practically unavoidable. That is why it was so difficult to overcome a certain psychological barrier and abandon all of these assumptions. This was done with the invention of the chaotic inflation scenario [10]. This scenario resolved all the problems of old and new inflation. According to this scenario, inflation may occur even in the theories with simplest potentials such as $V(\phi) \sim \phi^n$. Inflation may begin even if there was no thermal equilibrium in the early universe, and it may start even at the Planckian density, in which case the problem of initial conditions for inflation can be easily resolved [7].

4.2.1 Chaotic inflation

To explain the basic idea of chaotic inflation, let us consider the simplest model of a scalar field ϕ with a mass m and with the potential energy density $V(\phi) = (m^2/2)\phi^2$, see figure 4.1. Since this function has a minimum at $\phi = 0$, one may expect that the scalar field ϕ should oscillate near this minimum. This is indeed the case if the universe does not expand. However, one can show that in a rapidly expanding universe the scalar field moves down very slowly, as a ball in a viscous liquid, viscosity being proportional to the speed of expansion.

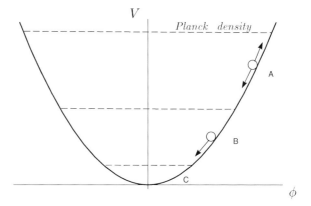

Figure 4.1. Motion of the scalar field in the theory with $V(\phi) = \frac{1}{2}m^2\phi^2$. Several different regimes are possible, depending on the value of the field ϕ. If the potential energy density of the field is greater than the Planck density $M_P^4 \sim 10^{94}$ g cm^{-3}, quantum fluctuations of spacetime are so strong that one cannot describe it in usual terms. Such a state is called spacetime foam. At a somewhat smaller energy density (region A: $mM_P^3 < V(\phi) < M_P^4$) quantum fluctuations of spacetime are small, but quantum fluctuations of the scalar field ϕ may be large. Jumps of the scalar field due to quantum fluctuations lead to a process of eternal self-reproduction of inflationary universe which we are going to discuss later. At even smaller values of $V(\phi)$ (region B: $m^2M_P^2 < V(\phi) < mM_P^3$) fluctuations of the field ϕ are small; it slowly moves down as a ball in a viscous liquid. Inflation occurs both in the region A and region B. Finally, near the minimum of $V(\phi)$ (region C) the scalar field rapidly oscillates, creates pairs of elementary particles, and the universe becomes hot.

There are two equations which describe evolution of a homogeneous scalar field in our model, the field equation

$$\ddot{\phi} + 3H\dot{\phi} = -V'(\phi), \tag{4.1}$$

and the Einstein equation

$$H^2 + \frac{k}{a^2} = \frac{8\pi}{3M_P^2}\left(\frac{1}{2}\dot{\phi}^2 + V(\phi)\right). \tag{4.2}$$

Here $H = \dot{a}/a$ is the Hubble parameter in the universe with a scale factor $a(t)$, $k = -1, 0, 1$ for an open, flat or closed universe respectively, M_P is the Planck mass. In the case $V = m^2\phi^2/2$, the first equation becomes similar to the equation of motion for a harmonic oscillator, where instead of $x(t)$ we have $\phi(t)$, with a friction term $3H\dot{\phi}$:

$$\ddot{\phi} + 3H\dot{\phi} = -m^2\phi. \tag{4.3}$$

If the scalar field ϕ initially was large, the Hubble parameter H was large too, according to the second equation. This means that the friction term in the first equation was very large, and therefore the scalar field was moving very slowly, as a ball in a viscous liquid. Therefore at this stage the energy density of the scalar field, unlike the density of ordinary matter, remained almost constant, and expansion of the universe continued with a much greater speed than in the old cosmological theory. Due to the rapid growth of the scale of the universe and a slow motion of the field ϕ, soon after the beginning of this regime one has $\ddot{\phi} \ll 3H\dot{\phi}$, $H^2 \gg (k/a^2)$, $\dot{\phi}^2 \ll m^2\phi^2$, so the system of equations can be simplified:

$$3\frac{\dot{a}}{a}\dot{\phi} = -m^2\phi, \qquad \frac{\dot{a}}{a} = \frac{2m\phi}{M_P}\sqrt{\frac{\pi}{3}}. \tag{4.4}$$

The last equation shows that the size of the universe in this regime grows approximately as e^{Ht}, where

$$H = \frac{2m\phi}{M_P}\sqrt{\frac{\pi}{3}}.$$

More exactly, these equations lead to following solutions for ϕ and a:

$$\phi(t) = \phi_0 - \frac{m M_P t}{\sqrt{12\pi}}, \tag{4.5}$$

$$a(t) = a_0 \exp\frac{2\pi}{M_P^2}(\phi_0^2 - \phi^2(t)). \tag{4.6}$$

This stage of exponentially rapid expansion of the universe is called inflation. In realistic versions of inflationary theory its duration could be as short as 10^{-35} s. When the field ϕ becomes sufficiently small, viscosity becomes small, inflation ends, and the scalar field ϕ begins to oscillate near the minimum of $V(\phi)$. As any rapidly oscillating classical field, it loses its energy by creating pairs of elementary particles. These particles interact with each other and come to a state of thermal equilibrium with some temperature T. From this time on, the corresponding part of the universe can be described by the standard hot universe theory.

The main difference between inflationary theory and the old cosmology becomes clear when one calculates the size of a typical inflationary domain at the end of inflation. Investigation of this question shows that even if the initial size of inflationary universe was as small as the Plank size $l_P \sim 10^{-33}$ cm, after 10^{-35} s of inflation the universe acquires a huge size of $l \sim 10^{10^{12}}$ cm!

This number is model-dependent, but in all realistic models the size of the universe after inflation appears to be many orders of magnitude greater than the size of the part of the universe which we can see now, $l \sim 10^{28}$ cm. This immediately solves most of the problems of the old cosmological theory.

Our universe is almost exactly homogeneous on large scale because all inhomogeneities were stretched by a factor of $10^{10^{12}}$. The density of primordial monopoles and other undesirable 'defects' becomes exponentially diluted by inflation. The universe becomes enormously large. Even if it was a closed universe of a size $\sim 10^{-33}$ cm, after inflation the distance between its 'South' and 'North' poles becomes many orders of magnitude greater than 10^{28} cm. We see only a tiny part of the huge cosmic balloon. That is why nobody has ever seen how parallel lines cross. That is why the universe looks so flat.

If one considers a universe which initially consisted of many domains with chaotically distributed scalar field ϕ (or if one considers different universes with different values of the field), then domains in which the scalar field was too small never inflated. The main contribution to the total volume of the universe will be given by those domains which originally contained large scalar field ϕ. Inflation of such domains creates huge homogeneous islands out of initial chaos. Each homogeneous domain in this scenario is much greater than the size of the observable part of the universe.

The first models of chaotic inflation were based on the theories with polynomial potentials, such as

$$V(\phi) = \pm \frac{m^2}{2}\phi^2 + \frac{\lambda}{4}\phi^4.$$

But the main idea of this scenario is quite generic. One should consider any particular potential $V(\phi)$, polynomial or not, with or without spontaneous symmetry breaking, and study all possible initial conditions without assuming that the universe was in a state of thermal equilibrium, and that the field ϕ was in the minimum of its effective potential from the very beginning [10]. This scenario strongly deviated from the standard lore of the hot big bang theory and was psychologically difficult to accept. Therefore during the first few years after invention of chaotic inflation many authors claimed that the idea of chaotic initial conditions is unnatural, and made attempts to realize the new inflation scenario based on the theory of high-temperature phase transitions, despite numerous problems associated with it. Gradually, however, it became clear that the idea of chaotic initial conditions is most general, and it is much easier to construct a consistent cosmological theory without making unnecessary assumptions about thermal equilibrium and high temperature phase transitions in the early universe.

Many other versions of inflationary cosmology have been proposed since 1983. Most of them are based not on the theory of high-temperature phase transitions, as in old and new inflation, but on the idea of chaotic initial conditions, which is the definitive feature of the chaotic inflation scenario.

4.3 Quantum fluctuations in the inflationary universe

The vacuum structure in the exponentially expanding universe is much more complicated than in ordinary Minkowski space. The wavelengths of all vacuum fluctuations of the scalar field ϕ grow exponentially during inflation. When the wavelength of any particular fluctuation becomes greater than H^{-1}, this fluctuation stops oscillating, and its amplitude freezes at some non-zero value

$\delta\phi(x)$ because of the large friction term $3H\dot\phi$ in the equation of motion of the field ϕ. The amplitude of this fluctuation then remains almost unchanged for a very long time, whereas its wavelength grows exponentially. Therefore, the appearance of such a frozen fluctuation is equivalent to the appearance of a classical field $\delta\phi(x)$ that does not vanish after averaging over macroscopic intervals of space and time.

Because the vacuum contains fluctuations of all wavelengths, inflation leads to the continuous creation of new perturbations of the classical field with wavelengths greater than H^{-1}, i.e. with momentum k smaller than H. One can easily understand on dimensional grounds that the average amplitude of perturbations with momentum $k \sim H$ is $O(H)$. A more accurate investigation shows that the average amplitude of perturbations generated during a time interval H^{-1} (in which the universe expands by a factor of e) is given by [7]

$$|\delta\phi(x)| \approx \frac{H}{2\pi}. \qquad (4.7)$$

Some of the most important features of inflationary cosmology can be understood only with an account taken of these quantum fluctuations. That is why in this section we will discuss this issue. We will begin this discussion on a rather formal level, and then we will suggest a simple interpretation of our results.

First of all, we will describe inflationary universe with the help of the metric of a flat de Sitter space,

$$ds^2 = dt^2 - e^{2Ht}\,dx^2. \qquad (4.8)$$

We will assume that the Hubble constant H practically does not change during the process, and for simplicity we will begin with investigation of a massless field ϕ.

To quantize the massless scalar field ϕ in de Sitter space in the coordinates (4.8) in much the same way as in Minkowski space [11]. The scalar field operator $\phi(x)$ can be represented in the form

$$\phi(x, t) = (2\pi)^{-3/2} \int d^3 p\, [a_p^+ \psi_p(t)e^{ipx} + a_p^- \psi_p^*(t)e^{-ipx}], \qquad (4.9)$$

where $\psi_p(t)$ satisfies the equation

$$\ddot\psi_p(t) + 3H\dot\psi_p(t) + p^2 e^{-2Ht}\psi_p(t) = 0. \qquad (4.10)$$

The term $3H\dot\psi_p(t)$ originates from the term $3H\dot\phi$ in equation (4.1), the last term appears because of the gradient term in the Klein–Gordon equation for the field ϕ. Note, that p is a comoving momentum, which, just like the coordinates x, does not change when the universe expands.

In Minkowski space, $\psi_p(t)\frac{1}{\sqrt{2p}}e^{-ipt}$, where $p = \sqrt{p^2}$. In de Sitter space (4.8), the general solution of (4.10) takes the form

$$\psi_p(t) = \frac{\sqrt{\pi}}{2} H\eta^{3/2}[C_1(p)H_{3/2}^{(1)}(p\eta) + C_2(p)H_{3/2}^{(2)}(p\eta)], \qquad (4.11)$$

where $\eta = -H^{-1}e^{-Ht}$ is the conformal time, and the $H^{(i)}_{3/2}$ are Hankel functions:

$$H^{(2)}_{3/2}(x) = [H^{(1)}_{3/2}(x)]^* = -\sqrt{\frac{2}{\pi x}}e^{-ix}\left(1 + \frac{1}{ix}\right). \qquad (4.12)$$

Quantization in de Sitter space and Minkowski space should be identical in the high-frequency limit, i.e. $C_1(p) \to 0$, $C_2(p) \to -1$ as $p \to \infty$. In particular, this condition is satisfied† for $C_1 \equiv 0$, $C_2 \equiv -1$. In that case,

$$\psi_p(t) = \frac{iH}{p\sqrt{2p}}\left(1 + \frac{p}{iH}e^{-Ht}\right)\exp\left(\frac{ip}{H}e^{-Ht}\right). \qquad (4.13)$$

Note that at sufficiently large t (when $pe^{-Ht} < H$), $\psi_p(t)$ ceases to oscillate, and becomes equal to $iH/p\sqrt{2p}$.

The quantity $\langle\phi^2\rangle$ may be simply expressed in terms of ψ_p:

$$\langle\phi^2\rangle = \frac{1}{(2\pi)^3}\int|\psi_p|^2\,d^3p = \frac{1}{(2\pi)^3}\int\left(\frac{e^{-2Ht}}{2p} + \frac{H^2}{2p^3}\right)d^3p. \qquad (4.14)$$

The physical meaning of this result becomes clear when one transforms from the conformal momentum p, which is time-independent, to the conventional physical momentum $k = pe^{-Ht}$, which decreases as the universe expands:

$$\langle\phi^2\rangle = \frac{1}{(2\pi)^3}\int\frac{d^3k}{k}\left(\frac{1}{2} + \frac{H^2}{2k^2}\right). \qquad (4.15)$$

The first term is the usual contribution of vacuum fluctuations in Minkowski space with $H = 0$. This contribution can be eliminated by renormalization. The second term, however, is directly related to inflation. Looked at from the standpoint of quantization in Minkowski space, this term arises because of the fact that de Sitter space, apart from the usual quantum fluctuations that are present when $H = 0$, also contains ϕ-particles with occupation numbers

$$n_k = \frac{H^2}{2k^2}. \qquad (4.16)$$

It can be seen from (4.15) that the contribution to $\langle\phi^2\rangle$ from long-wave fluctuations of the ϕ field diverges.

However, the value of $\langle\phi^2\rangle$ for a massless field ϕ is infinite only in eternally existing de Sitter space with $H = $ constant, and not in the inflationary universe, which expands (quasi)exponentially starting at some time $t = 0$ (for example, when the density of the universe becomes smaller than the Planck density).

---

† It is important that if the inflationary stage is long enough, all physical results are independent of the specific choice of functions $C_1(p)$ and $C_2(p)$ if $C_1(p) \to 0$, $C_2(p) \to -1$ as $p \to \infty$.

Indeed, the spectrum of vacuum fluctuations (4.15) strongly differs from the spectrum in Minkowski space when $k \ll H$. If the fluctuation spectrum before inflation has a cut-off at $k \leq k_0 \sim T$ resulting from high-temperature effects, or at $k \leq k_0 \sim H$ due to a small initial size $\sim H^{-1}$ of an inflationary region, then the spectrum will change at the time of inflation, due to exponential growth in the wavelength of vacuum fluctuations. The spectrum (4.15) will gradually be established, but only at momenta $k \geq k_0 e^{-Ht}$. There will then be a cut-off in the integral (4.14). Restricting our attention to contributions made by long-wave fluctuations with $k \leq H$, which are the only ones that will subsequently be important for us, and assuming that $k_0 = O(H)$, we obtain

$$
\begin{aligned}
\langle \phi^2 \rangle &\approx \frac{H^2}{2(2\pi)^3} \int_{He^{-Ht}}^{H} \frac{d^3 k}{k} = \frac{H^2}{4\pi^2} \int_{-Ht}^{0} d \ln \frac{k}{H} \\
&\equiv \frac{H^2}{4\pi^2} \int_{0}^{Ht} d \ln \frac{p}{H} = \frac{H^3}{4\pi^2} t.
\end{aligned}
\tag{4.17}
$$

A similar result is obtained for a massive scalar field ϕ. In that case, long-wave fluctuations with $m^2 \ll H^2$ behave as

$$
\langle \phi^2 \rangle = \frac{3H^4}{8\pi^2 m^2} \left[1 - \exp\left(-\frac{2m^2}{3H} t \right) \right].
\tag{4.18}
$$

When $t \leq 3H/m^2$, the term $\langle \phi^2 \rangle$ grows linearly, just as in the case of the massless field (4.17), and it then tends to its asymptotic value

$$
\langle \phi^2 \rangle = \frac{3H^4}{8\pi^2 m^2}.
\tag{4.19}
$$

Let us now try to provide an intuitive physical interpretation of these results. First, note that the main contribution to $\langle \phi^2 \rangle$ (4.17) comes from integrating over exponentially small k (with $k \sim H \exp(-Ht)$). The corresponding occupation numbers n_k (4.16) are then exponentially large. One can show that for large $l = |\mathbf{x} - \mathbf{y}| e^{Ht}$, the correlation function $\langle \phi(x)\phi(y) \rangle$ for the massless field ϕ is

$$
\langle \phi(\mathbf{x}, t)\phi(\mathbf{y}, t) \rangle \approx \langle \phi^2(\mathbf{x}, t) \rangle \left(1 - \frac{1}{Ht} \ln Hl \right).
\tag{4.20}
$$

This means that the magnitudes of the fields $\phi(x)$ and $\phi(y)$ will be highly correlated out to exponentially large separations $l \sim H^{-1} \exp(Ht)$, and the corresponding occupation numbers will be exponentially large. By all these criteria, long-wave quantum fluctuations of the field ϕ with $k \ll H^{-1}$ behave like a weakly inhomogeneous (quasi)classical field ϕ generated during the inflationary stage.

Analogous results also hold for a massive field with $m^2 \ll H^2$. There, the principal contribution to $\langle \phi^2 \rangle$ comes from modes with exponentially small momenta $k \sim H \exp(-3H^2/2\,m^2)$, and the correlation length is of order $H^{-1} \exp(3H^2/2m^2)$.

Later on we will develop a stochastic formalism which will allow us to describe various properties of the motion of the scalar field.

4.4   Quantum fluctuations and density perturbations

Fluctuations of the field ϕ lead to adiabatic density perturbations $\delta\rho \sim V'(\phi)\delta\phi$, which grow after inflation. The theory of inflationary density perturbations is rather complicated, but one can make an estimate of their post-inflationary magnitude in the following intuitively simple way: Fluctuations of the scalar field lead to a local delay of the end of inflation by the time $\delta t \sim \delta\phi/\dot\phi$. Density of the universe after inflation decreases as t^{-2}, so the local time delay δt leads to density contrast $|\delta\rho/\rho| \sim |2\delta t/t|$. If one takes into account that $\delta\phi \sim H/2\pi$ and that at the end of inflation $t^{-1} \sim H$, one obtains an estimate

$$\frac{\delta\rho}{\rho} \sim \frac{H^2}{2\pi\dot\phi}. \tag{4.21}$$

Needless to say, this is a very rough estimate. Fortunately, however, it gives a very good approximation to the correct result which can be obtained by much more complicated methods [2–4, 7]:

$$\frac{\delta\rho}{\rho} = C\frac{H^2}{2\pi\dot\phi}, \tag{4.22}$$

where the parameter C depends on equation of state of the universe. For example, $C = 6/5$ for the universe dominated by cold dark matter [4]. Then equations $3H\dot\phi = V'$ and $H^2 = 8\pi V/3M_{\mathrm{P}}^2$ imply that

$$\frac{\delta\rho}{\rho} = \frac{16\sqrt{6\pi}}{5}\frac{V^{3/2}}{V'}. \tag{4.23}$$

Here ϕ is the value of the classical field $\phi(t)$ (4), at which the fluctuation we consider has the wavelength $l \sim k^{-1} \sim H^{-1}(\phi)$ and becomes frozen in amplitude. In the simplest theory of the massive scalar field with $V(\phi) = \frac{1}{2}m^2\phi^2$ one has

$$\frac{\delta\rho}{\rho} = \frac{8\sqrt{3\pi}}{5}m\phi^2. \tag{4.24}$$

Taking into account (4.4) and also the expansion of the universe by about 10^{30} times after the end of inflation, one can obtain the following result for the density perturbations with the wavelength l (cm) at the moment when these perturbations begin growing and the process of the galaxy formation starts:

$$\frac{\delta\rho}{\rho} \sim m \ln l \; (\mathrm{cm}). \tag{4.25}$$

The definition of $\delta\rho/\rho$ used in [7] corresponds to COBE data for $\delta\rho/\rho \sim 5 \times 10^{-5}$. This gives $m \sim 10^{-6}$, in Planck units, which is equivalent to 10^{13} GeV.

An important feature of the spectrum of density perturbations is its flatness: $\delta\rho/\rho$ in our model depends on the scale l only logarithmically. For the theories with exponential potentials, the spectrum can be represented as

$$\frac{\delta\rho}{\rho} \sim l^{(1-n)/2}. \tag{4.26}$$

This representation is often used for the phenomenological description of various inflationary models. Exact flatness of the spectrum implies $n = 1$. Usually $n < 1$, but the models with $n > 1$ are also possible. In most of the realistic models of inflation one has $n = 1 \pm 0.2$.

Flatness of the spectrum of $\delta\rho/\rho$ together with flatness of the universe ($\Omega = 1$) constitute the two most robust predictions of inflationary cosmology. It is possible to construct models where $\delta\rho/\rho$ changes in a very peculiar way, and it is also possible to construct theories where $\Omega \neq 1$, but it is extremely difficult to do so.

4.5 From the big bang theory to the theory of eternal inflation

A significant step in the development of inflationary theory which I would like to discuss here is the discovery of the process of self-reproduction of inflationary universe. This process was known to exist in old inflationary theory [5] and in the new one [12], but it is especially surprising and leads to most profound consequences in the context of the chaotic inflation scenario [13]. It appears that in many models large scalar field during inflation produces large quantum fluctuations which may locally increase the value of the scalar field in some parts of the universe. These regions expand at a greater rate than their parent domains, and quantum fluctuations inside them lead to the production of new inflationary domains which expand even faster. This surprising behaviour leads to an eternal process of self-reproduction of the universe.

To understand the mechanism of self-reproduction one should remember that the processes separated by distances l greater than H^{-1} proceed independently of one another. This is so because during exponential expansion the distance between any two objects separated by more than H^{-1} is growing with a speed exceeding the speed of light. As a result, an observer in the inflationary universe can see only the processes occurring inside the horizon of the radius H^{-1}.

An important consequence of this general result is that the process of inflation in any spatial domain of radius H^{-1} occurs independently of any events outside it. In this sense any inflationary domain of initial radius exceeding H^{-1} can be considered as a separate mini-universe.

To investigate the behaviour of such a mini-universe, with an account taken of quantum fluctuations, let us consider an inflationary domain of initial radius

H^{-1} containing sufficiently homogeneous field with initial value $\phi \gg M_{\rm P}$. Equation (4.4) implies that during a typical time interval $\Delta t = H^{-1}$ the field inside this domain will be reduced by $\Delta\phi = M_{\rm P}^2/4\pi\phi$. By comparison this expression with

$$|\delta\phi(x)| \approx \frac{H}{2\pi} = \sqrt{\frac{2V(\phi)}{3\pi M_{\rm P}^2}} \sim \frac{m\phi}{3M_{\rm P}},$$

one can easily see that if ϕ is much less than

$$\phi^* \sim \frac{M_{\rm P}}{3}\sqrt{\frac{M_{\rm P}}{m}},$$

then the decrease of the field ϕ due to its classical motion is much greater than the average amplitude of the quantum fluctuations $\delta\phi$ generated during the same time. But for $\phi \gg \phi^*$ one has $\delta\phi(x) \gg \Delta\phi$. Because the typical wavelength of the fluctuations $\delta\phi(x)$ generated during the time is H^{-1}, the whole domain after $\Delta t = H^{-1}$ effectively becomes divided into $e^3 \sim 20$ separate domains (mini-universes) of radius H^{-1}, each containing almost homogeneous field $\phi - \Delta\phi + \delta\phi$. In almost a half of these domains the field ϕ grows by $|\delta\phi(x)| - \Delta\phi \approx |\delta\phi(x)| = H/2\pi$, rather than decreases. This means that the total volume of the universe containing *growing* field ϕ increases 10 times. During the next time interval $\Delta t = H^{-1}$ the situates repeats. Thus, after the two time intervals H^{-1} the total volume of the universe containing the growing scalar field increases 100 times, etc. The universe enters eternal process of self-reproduction.

This effect is very unusual. Its investigation still brings us new unexpected results. For example, for a long time it was believed that self-reproduction in the chaotic inflation scenario can occur only if the scalar field ϕ is greater than ϕ^* [13]. However, it was shown in [14] that if the size of the initial inflationary domain is large enough, then the process of self-reproduction of the universe begins for all values of the field ϕ for which inflation is possible (for $\phi > M_{\rm P}$ in the theory $2m^2\phi^2$). This result is based on the investigation of quantum jumps with amplitude $\delta\phi \gg H/2\pi$.

Until now we have considered the simplest inflationary model with only one scalar field, which had only one minimum of its potential energy. Meanwhile, realistic models of elementary particles propound many kinds of scalar fields. For example, in the unified theories of weak, strong and electromagnetic interactions, at least two other scalar fields exist. The potential energy of these scalar fields may have several different minima. This means that the same theory may have different 'vacuum states', corresponding to different types of symmetry breaking between fundamental interactions, and, as a result, to different laws of low-energy physics.

As a result of quantum jumps of the scalar fields during inflation, the universe may become divided into infinitely many exponentially large domains that have different laws of low-energy physics. Note that this division occurs even if the whole universe originally began in the same state, corresponding to one particular minimum of potential energy.

To illustrate this scenario, we present here the results of computer simulations of the evolution of a system of two scalar fields during inflation. The field ϕ is the inflaton field driving inflation; it is shown by the height of the distribution of the field $\phi(x, y)$ in a two-dimensional slice of the universe. The field χ determines the type of spontaneous symmetry breaking which may occur in the theory. We paint the surface black if this field is in a state corresponding to one of the two minima of its effective potential; we paint it white if it is in the second minimum corresponding to a different type of symmetry breaking, and therefore to a different set of laws of low-energy physics.

In the beginning of the process the whole inflationary domain was black, and the distribution of both fields was very homogeneous. Then the domain became exponentially large (but it has the same size in comoving coordinates, as shown in figure 4.1). Each peak of the mountains corresponds to nearly Planckian density and can be interpreted as a beginning of a new 'big bang'. The laws of physics are rapidly changing there, but they become fixed in the parts of the universe where the field ϕ becomes small. These parts correspond to valleys in figure 4.2. Thus quantum fluctuations of the scalar fields divide the universe into exponentially large domains with different laws of low-energy physics, and with different values of energy density.

If this scenario is correct, then physics alone cannot provide a complete explanation for all the properties of our part of the universe. The same physical theory may yield large parts of the universe that have diverse properties. According to this scenario, we find ourselves inside a four-dimensional domain with our kind of physical laws not because domains with different dimensionality and with alternate properties are impossible or improbable, but simply because our kind of life cannot exist in other domains.

This consideration is based on the anthropic principle, which was not very popular among physicists for two main reasons. First of all, it was based on the assumption that the universe was created many times until the final success. Second, it would be much easier (and quite sufficient) to achieve this success in a small vicinity of the solar system rather than in the whole observable part of our universe.

Both objections can be answered in the context of the theory of eternal inflation. First of all, the universe indeed reproduces itself in all its possible versions. Second, if the conditions suitable for the existence of life appear in a small vicinity of the solar system, then because of inflation the same conditions will exist in a domain much greater than the observable part of the universe. This means that inflationary theory for the first time provides real physical justification of the anthropic principle.

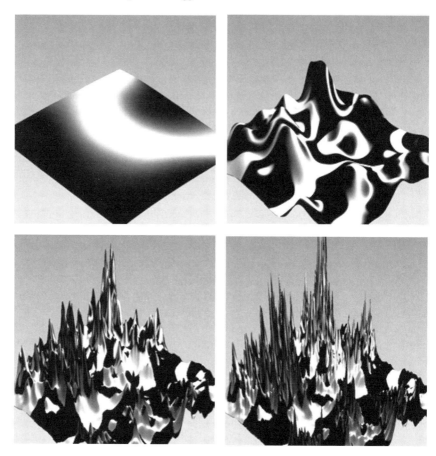

Figure 4.2. Evolution of scalar fields ϕ and χ during the process of self-reproduction of the universe. The height of the distribution shows the value of the field ϕ which drives inflation. The surface is painted black in those parts of the universe where the scalar field χ is in the first minimum of its effective potential, and white where it is in the second minimum. The laws of low-energy physics are different in the regions of different colour. The peaks of the 'mountains' correspond to places where quantum fluctuations bring the scalar fields back to the Planck density. Each such place in a certain sense can be considered as the beginning of a new big bang.

4.6   (P)reheating after inflation

The theory of the universe reheating after inflation is the most important application of the quantum theory of particle creation, since almost all matter constituting the universe was created during this process.

At the stage of inflation all energy is concentrated in a classical slowly moving inflaton field ϕ. Soon after the end of inflation this field begins to oscillate near the minimum of its effective potential. Eventually it produces many elementary particles, they interact with each other and come to a state of thermal equilibrium with some temperature T_r.

Elementary theory of this process was developed many years ago [15]. It was based on the assumption that the oscillating inflaton field can be considered as a collection of non-interacting scalar particles, each of which decays separately in accordance with perturbation theory of particle decay. However, it was recently understood that in many inflationary models the first stages of reheating occur in a regime of a broad parametric resonance. To distinguish this stage from the subsequent stages of slow reheating and thermalization, it was called *pre-heating* [16]. The energy transfer from the inflaton field to other bose fields and particles during pre-heating is extremely efficient.

To explain the main idea of the new scenario we will consider first the simplest model of chaotic inflation with the effective potential $V(\phi) = \frac{1}{2}m^2\phi^2$, and with the interaction Lagrangian $-\frac{1}{2}g^2\phi^2\chi^2$. We will take $m = 10^{-6}M_P$, as required by microwave background anisotropy [7] and, in the beginning, we will assume for simplicity that χ particles do not have a bare mass, i.e. $m_\chi(\phi) = g|\phi|$.

In this model inflation occurs at $|\phi| > 0.3M_P$ [7]. Suppose for definiteness that initially ϕ is large and negative, and inflation ends at $\phi \sim -0.3M_P$. After that the field ϕ rolls to $\phi = 0$, and then it oscillates about $\phi = 0$ with a gradually decreasing amplitude.

For the quadratic potential $V(\phi) = \frac{1}{2}m\phi^2$ the amplitude after the first oscillation becomes only $0.04M_P$, i.e. it drops by a factor of ten during the first oscillation. Later on, the solution for the scalar field ϕ asymptotically approaches the regime

$$\phi(t) = \Phi(t)\sin mt$$
$$\Phi(t) = \frac{M_P}{\sqrt{3\pi}mt} \sim \frac{M_P}{2\pi\sqrt{3\pi}N}. \tag{4.27}$$

Here $\Phi(t)$ is the amplitude of oscillations, N is the number of oscillations since the end of inflation. For simple estimates which we will make later one may use

$$\Phi(t) \approx \frac{M_P}{3mt} \approx \frac{M_P}{20N}. \tag{4.28}$$

The scale factor averaged over several oscillations grows as $a(t) \approx a_0(t/t_0)^{2/3}$. Oscillations of ϕ in this theory are sinusoidal, with the decreasing amplitude

$$\Phi(t) = \frac{M_P}{3}\left(\frac{a_0}{a(t)}\right)^{3/2}.$$

The energy density of the field ϕ decreases in the same way as the density of non-relativistic particles of mass m:

$$\rho_\phi = \frac{1}{2}\dot{\phi}^2 + \frac{1}{2}m^2\phi^2 \sim a^{-3}.$$

Hence the coherent oscillations of the homogeneous scalar field correspond to the matter-dominated effective equation of state with vanishing pressure.

We will assume that $g > 10^{-5}$ [16], which implies $gM_P > 10^2 m$ for the realistic value of the mass $m \sim 10^{-6} M_P$. Thus, immediately after the end of inflation, when $\phi \sim M_P/3$, the effective mass $g|\phi|$ of the field χ is much greater than m. It decreases when the field ϕ moves down, but initially this process remains adiabatic, $|\dot{m}_\chi| \ll m_\chi^2$.

Particle production occurs at the time when the adiabaticity condition becomes violated, i.e. when $|\dot{m}_\chi| \sim g|\dot{\phi}|$ becomes greater than $m_\chi^2 = g^2\phi^2$. This happens only when the field ϕ rolls close to $\phi = 0$. The velocity of the field at that time was $|\dot{\phi}_0| \approx mM_P/10 \approx 10^{-7} M_P$. The process becomes non-adiabatic for $g^2\phi^2 < g|\dot{\phi}_0|$, i.e. for $-\phi_* < \phi < \phi_*$, where $\phi_* \sim \sqrt{|\dot{\phi}_0|/g}$ [16]. Note that for $g \gg 10^{-5}$ the interval $-\phi_* < \phi < \phi_*$ is very narrow: $\phi_* \ll M_P/10$. As a result, the process of particle production occurs nearly instantaneously, within the time

$$\Delta t_* \sim \frac{\phi_*}{|\dot{\phi}_0|} \sim (g|\dot{\phi}_0|)^{-1/2}. \tag{4.29}$$

This time interval is much smaller than the age of the universe, so all effects related to the expansion of the universe can be neglected during the process of particle production. The uncertainty principle implies in this case that the created particles will have typical momenta $k \sim (\Delta t_*)^{-1} \sim (g|\dot{\phi}_0|)^{1/2}$. The occupation number n_k of χ particles with momentum k is equal to zero all the time when it moves toward $\phi = 0$. When it reaches $\phi = 0$ (or, more exactly, after it moves through the small region $-\phi_* < \phi < \phi_*$) the occupation number suddenly (within the time Δt_*) acquires the value [16]

$$n_k = \exp\left(-\frac{\pi k^2}{g|\dot{\phi}_0|}\right), \tag{4.30}$$

and this value does not change until the field ϕ rolls to the point $\phi = 0$ again.

To derive this equation one should first represent quantum fluctuations of the scalar field $\hat{\chi}$ minimally interacting with gravity in the following way:

$$\hat{\chi}(t, x) = \frac{1}{(2\pi)^{3/2}} \int d^3k\, (\hat{a}_k\, \chi_k(t) e^{-ikx} + \hat{a}_k^+ \chi_k^*(t) e^{ikx}), \tag{4.31}$$

where $\hat{a}_k$ and $\hat{a}_k^+$ are annihilation and creation operators. In general, one should write equations for these fluctuations taking into account expansion of the universe. However, in the beginning we will neglect expansion. Then the functions χ_k obey the following equation:

$$\ddot{\chi}_k + (k^2 + g^2\phi^2(t))\chi_k = 0. \tag{4.32}$$

Equation (4.32) describes an oscillator with a variable frequency $\omega_k^2 = k^2 + g^2\phi^2(t)$. If ϕ does not change in time, then one has the usual solution $\chi_k =$

$e^{-i\omega_k t}/\sqrt{2\omega_k}$. However, when the field ϕ changes, the solution becomes different, and this difference can be interpreted in terms of creation of particles χ.

The number of created particles is equal to the energy of particles $\frac{1}{2}|\dot{\chi}_k|^2 + \frac{1}{2}\omega_k^2|\chi_k|^2$ divided by the energy ω_k of each particle:

$$n_k = \frac{\omega_k}{2}\left(\frac{|\dot{\chi}_k|^2}{\omega_k^2} + |\chi_k|^2\right) - \frac{1}{2}. \tag{4.33}$$

The subtraction $\frac{1}{2}$ is needed to eliminate vacuum fluctuations from the counting. To calculate this number, one should solve equation (4.32) and substitute the solutions to equation (4.33). One can easily check that for the usual quantum fluctuations $\chi_k = e^{-i\omega_k t}/\sqrt{2\omega_k}$ one finds $n_k = 0$. In the case described earlier, when the particles are created by the rapidly changed field ϕ in the regime of strong violation of adiabaticity condition, one can solve equation (4.32) analytically and find the number of produced particles given by equation (4.30).

One can also solve equations for quantum fluctuations and calculate n_k numerically. In figure 4.3 we show the growth of fluctuations of the field χ and the number of particles χ produced by the oscillating field ϕ in the case when the mass of the field ϕ (i.e. the frequency of its oscillations) is much smaller than the average mass of the field χ given by $g\phi$.

The time evolution in figure 4.3 is shown in units $m/2\pi$, which corresponds to the number of oscillations N of the inflaton field ϕ. The oscillating field $\phi(t) \sim \Phi \sin mt$ is zero at integer and half-integer values of the variable $mt/2\pi$. This allows us to identify particle production with time intervals when $\phi(t)$ is very small.

During each oscillation of the inflaton field ϕ, the field χ oscillates many times. Indeed, the effective mass $m_\chi(t) = g\phi(t)$ is much greater than the inflaton mass m for the main part of the period of oscillation of the field ϕ in the broad resonance regime with $q^{1/2} = g\Phi/2m \gg 1$. As a result, the typical frequency of oscillation $\omega(t) = \sqrt{k^2 + g^2\phi^2(t)}$ of the field χ is much higher than that of the field ϕ. That is why during the most of this interval it is possible to talk about an adiabatically changing effective mass $m_\chi(t)$. But this condition breaks at small ϕ, and particles χ are produced there.

Each time the field ϕ approaches the point $\phi = 0$, new χ particles are being produced. Bose statistics implies that the number of particles produced each time will be proportional to the number of particles produced before. This leads to explosive process of particle production out of the state of thermal equilibrium. We called this process *pre-heating* [16].

This process does not occur for all momenta. It is most efficient if the field ϕ comes to the point $\phi = 0$ in phase with the field χ_k, which depends on k; see phases of the field χ_k for some particular values of k for which the process is most efficient on the upper panel of figure 4.3. Thus we deal with the effect of the exponential growth of the number of particles χ due to parametric resonance.

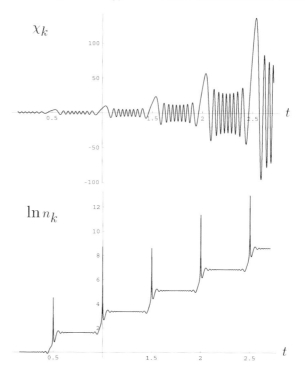

Figure 4.3. Broad parametric resonance for the field χ in Minkowski space in the theory $\frac{1}{2}m^2\phi^2$. For each oscillation of the field $\phi(t)$ the field χ_k oscillates many times. Each peak in the amplitude of the oscillations of the field χ corresponds to a place where $\phi(t) = 0$. At this time the occupation number n_k is not well defined, but soon after that time it stabilizes at a new, higher level, and remains constant until the next jump. A comparison of the two parts of this figure demonstrates the importance of using proper variables for the description of pre-heating. Both χ_k and the integrated dispersion $\langle \chi^2 \rangle$ behave erratically in the process of parametric resonance. Meanwhile n_k is an adiabatic invariant. Therefore, the behaviour of n_k is relatively simple and predictable everywhere except at the short intervals of time when $\phi(t)$ is very small and the particle production occurs.

Expansion of the universe modifies this picture for many reasons. First of all, expansion of the universe's redshifts produced particles, making their momenta smaller. More importantly, the amplitude of oscillations of the field ϕ decreases because of the expansion. Therefore the frequency of oscillations of the field χ also decreases. This may destroy the parametric resonance because it changes, in an unpredictable way, the phase of the oscillations of the field χ each moment that ϕ becomes close to zero.

That is why the number of created particles χ may either increase or decrease each time when the field ϕ becomes zero. However, a more detailed investigation

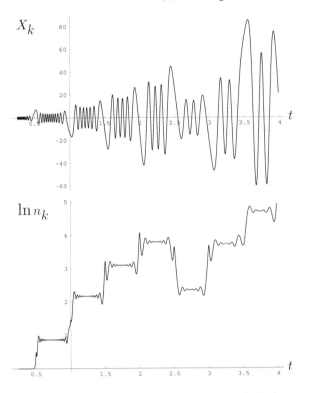

Figure 4.4. Early stages of parametric resonance in the theory $\frac{1}{2}m^2\phi^2$ in an expanding universe with scale factor $a \sim t^{2/3}$ for $g = 5 \times 10^{-4}$, $m = 10^{-6}M_{\rm P}$. Note that the number of particles n_k in this process typically increases, but it may occasionally decrease as well. This is a distinctive feature of stochastic resonance in an expanding universe. A decrease in the number of particles is a purely quantum mechanical effect which would be impossible if these particles were in a state of thermal equilibrium.

shows that it increases three times more often than it decreases, so the total number of produced particles grows exponentially, though in a rather specific way, see figure 4.4. We called this regime *stochastic resonance*.

In the course of time the amplitude of the oscillations of the field ϕ decreases, and when $g\phi$ becomes smaller than m, particle production becomes inefficient and their number stops growing.

In reality the situation is even more complicated. First of all, created particles change the frequency of oscillations of the field ϕ because they give a contribution $\sim g^2\langle\chi^2\rangle$ to the effective mass squared of the inflaton field [16]. Also, these particles scatter on each other and on the oscillating scalar field ϕ, which leads to additional particle production. As a result, it becomes extremely difficult to describe analytically the last stages of the process of the parametric resonance,

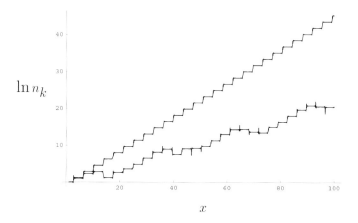

Figure 4.5. Development of the resonance in the theory $\frac{1}{2}m^2\phi^2 + \frac{1}{4}\lambda\phi^4 + \frac{1}{2}g^2\phi^2\chi^2$ for $g^2/\lambda = 5200$. The upper curve corresponds to the massless theory, the lower curve describes stochastic resonance with a theory with a mass m which is chosen to be much smaller than $\sqrt{\lambda}\phi$ during the whole period of calculations. Nevertheless, the presence of a small mass term completely changes the development of the resonance.

even though in many cases it is possible to estimate the final results. In particular, one can show that the whole process of parametric resonance typically takes only few dozen of oscillations, and the final occupation numbers of particles grow up to $n_k \sim 10^2 g^{-2}$ [16]. But a detailed description of the last stages of pre-heating requires lattice simulations, as proposed by Khlebnikov and Tkachev [18].

The theory of pre-heating is very sensitive to the choice of the model. For example, in the theory $\frac{1}{4}\lambda\phi^4 + \frac{1}{2}g^2\phi^2\chi^2$ the resonance does not become stochastic despite expansion of the universe. However, if one adds to this theory even a very small term $m^2\phi^2$, the resonance becomes stochastic [17].

This conclusion is illustrated by figure 4.5, where we show the development of the resonance both for the massless theory with $g^2/\lambda \sim 5200$, and for the theory with a small mass m. As we see, in the purely massless theory the logarithm of the number density n_k for the leading growing mode increases linearly in time x, whereas in the presence of a mass m, which we took to be much smaller than $\sqrt{\lambda}\phi$ during the whole process, the resonance becomes stochastic.

In fact, the development of the resonance is rather complicated even for smaller g^2/λ. The resonance for a massive field with $m \ll \sqrt{\lambda}\phi$ in this case is not stochastic, but it may consist of stages of regular resonance separated by the stages without any resonance, see figure 4.6.

Thus we see that the presence of the mass term $\frac{1}{2}m^2\phi^2$ can modify the nature of the resonance even if this term is much smaller than $\frac{1}{4}\lambda\phi^4$. This is a rather unexpected conclusion, which is an additional manifestation of the non-perturbative nature of pre-heating.

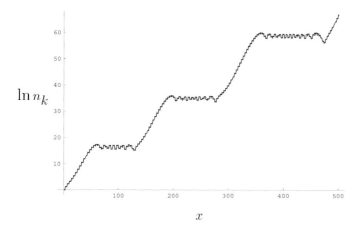

x

Figure 4.6. Development of the resonance in the theory $\frac{1}{2}m^2\phi^2 + \frac{1}{4}\lambda\phi^4 + \frac{1}{2}g^2\phi^2\chi^2$ with $m^2 \ll \lambda\phi^2$ for $g^2/\lambda = 240$. In this particular case the resonance is not stochastic. As time x grows, the relative contribution of the mass term to the equation describing the resonance also grows. This shifts the mode from one instability band to another.

Different regimes of parametric resonance in the theory

$$\tfrac{1}{2}m^2\phi^2 + \tfrac{1}{4}\lambda\phi^4 + \tfrac{1}{2}g^2\phi^2\chi^2$$

are shown in figure 4.7. We suppose that immediately after inflation the amplitude Φ of the oscillating inflaton field is greater than $m/sqrt\lambda$. If $g/\sqrt{\lambda} < \sqrt{\lambda}M_P/m$, the χ-particles are produced in the regular stable resonance regime until the amplitude $\Phi(t)$ decreases to $m/\sqrt{\lambda}$, after which the resonance occurs as in the theory $\frac{1}{2}m^2\phi^2 + \frac{1}{2}g^2\phi^2\chi^2$ [16]. The resonance never becomes stochastic.

If $g\sqrt{/\lambda} > \sqrt{\lambda}M_P/m$, the resonance originally develops as in the conformally invariant theory $\frac{1}{4}\lambda\phi^4 + \frac{1}{2}g^2\phi^2\chi^2$, but with a decrease of $\Phi(t)$ the resonance becomes stochastic. Again, for $\Phi(t) < m/\sqrt{\lambda}$ the resonance occurs as in the theory $\frac{1}{2}m^2\phi^2 + \frac{1}{2}g^2\phi^2\chi^2$. In all cases the resonance eventually disappears when the field $\Phi(t)$ becomes sufficiently small. Reheating in this class of models can be complete only if there is a symmetry breaking in the theory, i.e. $m^2 < 0$, or if one adds interaction of the field ϕ with fermions. In both cases the last stages of reheating are described by perturbation theory [17].

Adding fermions does not alter substantially the description of the stage of parametric resonance. Meanwhile the change of sign of m^2 does lead to substantial changes in the theory of pre-heating, see figure 4.8. Here we will briefly describe the structure of the resonance in the theory $-\frac{1}{2}m^2\phi^2 + \frac{1}{4}\lambda\phi^4 + \frac{1}{2}g^2\phi^2\chi^2$ for various g^2 and λ neglecting effects of backreaction.

First of all, at $\Phi \gg m/\sqrt{\lambda}$ the field ϕ oscillates in the same way as in the massless theory $\frac{1}{4}\lambda\phi^4 + \frac{1}{2}g^2\phi^2\chi^2$. The condition for the resonance to be

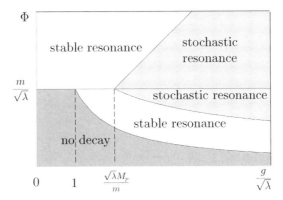

Figure 4.7. Schematic representation of different regimes which are possible in the theory $\frac{1}{2}m^2\phi^2 + \frac{1}{4}\lambda\phi^4 + \frac{1}{2}g^2\phi^2\chi^2$ for $m/\sqrt{\lambda} \ll 10^{-1} M_P$ and for various relations between g^2 and λ in an expanding universe. The theory developed in this chapter describes the resonance in the white area above the line $\Phi = m/\sqrt{\lambda}$. The theory of pre-heating for $\Phi < m/\sqrt{\lambda}$ is given in [16]. A complete decay of the inflaton is possible only if additional interactions are present in the theory which allow one inflaton particle to decay to several other particles, for example, an interaction with fermions $\bar{\psi}\psi\phi$.

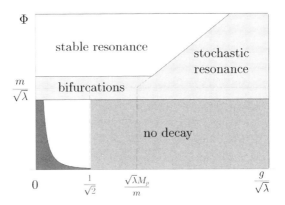

Figure 4.8. Schematic representation of different regimes which are possible in the theory $-\frac{1}{2}m^2\phi^2 + \frac{1}{4}\lambda\phi^4 + \frac{1}{2}g^2\phi^2\chi^2$. White regions correspond to the regime of a regular stable resonance, a small dark region in the left-hand corner near the origin corresponds to the perturbative decay $\phi \to \chi\chi$. Unless additional interactions are included (see figure 4.7), a complete decay of the inflaton field is possible only in this small area.

stochastic is

$$\Phi < \frac{g}{\sqrt{\lambda}} \frac{\pi^2 m^2}{3\lambda M_P}.$$

However, as soon as the amplitude Φ drops down to $m/\sqrt{\lambda}$, the situation changes dramatically. First of all, depending on the values of parameters the field rolls to one of the minima of its effective potential at $\phi = \pm m/\sqrt{\lambda}$. The description of this process is rather complicated. Depending on the values of parameters and on the relation between $\sqrt{\langle\phi^2\rangle}$, $\sqrt{\langle\chi^2\rangle}$ and $\sigma \equiv m/\sqrt{\lambda}$, the universe may become divided into domains with $\phi = \pm\sigma$, or it may end up in a single state with a definite sign of ϕ. After this transitional period the field ϕ oscillates near the minimum of the effective potential at $\phi = \pm m/\sqrt{\lambda}$ with an amplitude $\Phi \ll \sigma = m/\sqrt{\lambda}$. These oscillations lead to parametric resonance with χ-particle production. For definiteness we will consider here the regime $\lambda^{3/2} M_P < m \ll \lambda^{1/2} M_P$. The resonance in this case is possible only if $g^2/\lambda < \frac{1}{2}$. Using the results of [16] one can show that the resonance is possible only for

$$\frac{g}{\sqrt{\lambda}} > \left(\frac{m}{\sqrt{\lambda} M_P}\right)^{1/4}.$$

(The resonance may terminate somewhat earlier if the particles produced by the parametric resonance give a considerable contribution to the energy density of the universe.) However, this is not the end of reheating, because the perturbative decay of the inflaton field remains possible. It occurs with the decay rate $\Gamma(\phi \rightarrow \chi\chi) = g^4 m/8\pi\lambda$. This is the process which is responsible for the last stages of the decay of the inflaton field. It occurs only if one ϕ-particle can decay into two χ-particles, which implies that $g^2/\lambda < \frac{1}{2}$.

Thus we see that pre-heating is an incredibly rich phenomenon. Interestingly, complete decay of the inflaton field is not by any means guaranteed. In most of the models not involving fermions the decay never completes. Efficiency of pre-heating and, consequently, efficiency of baryogenesis, depends in a very non-monotonic way on the parameters of the theory. This may lead to a certain 'unnatural selection' of the theories where all necessary conditions for creation of matter and the subsequent emergence of life are satisfied.

Bosons produced at that stage are far away from thermal equilibrium and have enormously large occupation numbers. Explosive reheating leads to many interesting effects. For example, specific non-thermal phase transitions may occur soon after pre-heating, which are capable of restoring symmetry even in the theories with symmetry breaking on the scale $\sim 10^{16}$ GeV [19]. These phase transitions are capable of producing topological defects such as strings, domain walls and monopoles [20]. Strong deviation from thermal equilibrium and the possibility of production of superheavy particles by oscillations of a relatively light inflaton field may resurrect the theory of GUT baryogenesis [21] and may considerably change the way baryons are produced in the Affleck–Dine scenario [22], and in the electroweak theory [23].

Usually only a small fraction of the energy of the inflaton field $\sim 10^{-2}g^2$ is transferred to the particles χ when the field ϕ approaches the point $\phi = 0$ for the first time [24]. The role of the parametric resonance is to increase this energy exponentially within several oscillations of the inflaton field. But suppose that the particles χ interact with fermions ψ with the coupling $h\bar{\psi}\psi\chi$. If this coupling is strong enough, then χ particles may decay to fermions before the oscillating field ϕ returns back to the minimum of the effective potential. If this happens, parametric resonance does not occur. However, something equally interesting may occur instead of it: the energy density of the χ particles at the moment of their decay may become much greater than their energy density at the moment of their creation. This may be sufficient for a complete reheating.

Indeed, prior to their decay the number density of χ particles produced at the point $\phi = 0$ remains practically constant [16], whereas the effective mass of each χ particle grows as $m_\chi = g\phi$ when the field ϕ rolls up from the minimum of the effective potential. Therefore their total energy density grows. One may say that χ particles are 'fattened', being fed by the energy of the rolling field ϕ. The fattened χ particles tend to decay to fermions at the moment when they have the greatest mass, i.e. when ϕ reaches its maximal value $\sim 10^{-1} M_{\rm P}$, just before it begins rolling back to $\phi = 0$.

At that moment χ particles can decay to two fermions with mass up to $m_\psi \sim \frac{1}{2}g 10^{-1} M_{\rm P}$, which can be as large as 5×10^{17} GeV for $g \sim 1$. This is five orders of magnitude greater than the masses of the particles which can be produced by the usual decay of ϕ particles. As a result, the chain reaction $\phi \rightarrow \chi \rightarrow \psi$ considerably enhances the efficiency of transfer of energy of the inflaton field to matter.

More importantly, superheavy particles ψ (or the products of their decay) may eventually dominate the total energy density of matter even if in the beginning their energy density was relatively small. For example, the energy density of the oscillating inflaton field in the theory with the effective potential $\frac{1}{4}\lambda\phi^4$ decreases as a^{-4} in an expanding universe with a scale factor $a(t)$. Meanwhile the energy density stored in the non-relativistic particles ψ (prior to their decay) decreases only as a^{-3}. Therefore their energy density rapidly becomes dominant even if originally it was small. A subsequent decay of such particles leads to a complete reheating of the universe.

Thus in this scenario the process of particle production occurs within less than one oscillation of the inflaton field. We called it *instant pre-heating* [24]. This mechanism is very efficient even in the situation when all other mechanisms fail. Consider, for example, models where the post-inflationary motion of the inflaton field occurs along a flat direction of the effective potential. In such theories the standard scenario of reheating does not work because the field ϕ does not oscillate. Until the invention of the instant pre-heating scenario the only mechanism of reheating discussed in the context of such models was based on the gravitational production of particles [25]. The mechanism of instant pre-heating in such models is typically much more efficient. After the moment when χ particles are produced their energy density grows due to the growth of the field ϕ. Meanwhile the energy density of the field ϕ moving along a flat direction of $V(\phi)$ decreases extremely rapidly, as $a^{-6}(t)$. Therefore very soon all energy becomes concentrated in the particles produced at the end of inflation, and reheating completes.

As we see, the theory of creation of matter in the universe is much more interesting and complicated than we expected few years ago.

4.7 Conclusions

During the last 20 years inflationary theory gradually became the standard paradigm of modern cosmology. In addition to resolving many problems of the standard big bang theory, inflation made several important predictions. In particular:

(1) The universe must be flat. In most models $\Omega_{total} = 1 \pm 10^{-4}$.
(2) Perturbations of the metric produced during inflation are adiabatic. (Topological defects produce isocurvature perturbations.)
(3) These perturbations should have flat spectrum. In most of the models one has $n = 1 \pm 0.2$.
(4) These perturbations should be Gaussian. (Topological defects produce non-Gaussian perturbations.)
(5) There should be no (or almost no) vector perturbations after inflation. (They may appear in the theory of topological defects.)

At the moment all of these predictions seem to be in a good agreement with observational data [26], and no other theory is available that makes all of these predictions.

This does not mean that all difficulties are over and we can relax. First of all, inflation is still a scenario which changes with every new idea in particle theory. Do we really know that inflation began at Planck density 10^{94} g cm^{-3}? What if our space has large internal dimensions, and energy density could never rise above the electroweak density 10^{25} g cm^{-3}? Was there any stage before inflation? Is it possible to implement inflation in string theory/M-theory?

We do not know which version of inflationary theory will survive ten years from now. It is absolutely clear than new observational data are going to rule out 99% of all inflationary models. But it does not seem likely that they will rule out the basic idea of inflation. Inflationary scenario is very versatile, and now, after 20 years of persistent attempts of many physicists to propose an alternative to inflation, we still do not know any other way to construct a consistent cosmological theory. For the time being, we are taking the position suggested long ago by Sherlock Holmes: 'When you have eliminated the impossible, whatever remains, however improbable, must be the truth' [27]. Did we really eliminate the impossible? Do we really know the truth? It is for you to find the answer.

References

[1] Starobinsky A A 1979 *JETP Lett.* **30** 682

Starobinsky A A 1980 *Phys. Lett.* B **91** 99

[2] Mukhanov V F and Chibisov G V 1981 *JETP Lett.* **33** 532

[3] Hawking S W 1982 *Phys. Lett.* B **115** 295

Starobinsky A A 1982 *Phys. Lett.* B **117** 175

Guth A H and Pi S-Y 1982 *Phys. Rev. Lett.* **49** 1110

Bardeen J, Steinhardt P J and Turner M S 1983 *Phys. Rev.* D **28** 679

[4] Mukhanov V F 1985 *JETP Lett.* **41** 493

Mukhanov V F, Feldman H A and Brandenberger R H 1992 *Phys. Rep.* **215** 203

[5] Guth A H 1981 *Phys. Rev.* D **23** 347

[6] Kirzhnits D A 1972 *JETP Lett.* **15** 529

Kirzhnits D A and Linde A D 1972 *Phys. Lett.* B **42** 471

Kirzhnits D A and Linde A D 1974 *Sov. Phys.–JETP* **40** 628

Kirzhnits D A and Linde A D 1976 *Ann. Phys.* **101** 195

Weinberg S 1974 *Phys. Rev.* D **9** 3320

Dolan L and Jackiw R 1974 *Phys. Rev.* D **9** 3357

[7] Linde A D 1990 *Particle Physics and Inflationary Cosmology* (Chur, Switzerland: Harwood)

[8] Guth A H and Weinberg E J 1983 *Nucl. Phys.* B **212** 321

[9] Linde A D 1982 *Phys. Lett.* B **108** 389

Linde A D 1982 *Phys. Lett.* B **114** 431

Linde A D 1982 *Phys. Lett.* B **116** 335, 340

Albrecht A and Steinhardt P J 1982 *Phys. Rev. Lett.* **48** 1220

[10] Linde A D 1983 *Phys. Lett.* B **129** 177

[11] Vilenkin A and Ford L H 1982 *Phys. Rev.* D **26** 1231

Linde A D 1982 *Phys. Lett.* B **116** 335

Starobinsky A A 1982 *Phys. Lett.* B **117** 175

[12] Steinhardt P J 1982 *The Very Early Universe* ed G W Gibbons, S W Hawking and S Siklos (Cambridge: Cambridge University Press) p 251

Linde A D 1982 *Nonsingular Regenerating Inflationary Universe* (Cambridge: Cambridge University Press)

Vilenkin A 1983 *Phys. Rev.* D **27** 2848

[13] Linde A D 1986 *Phys. Lett.* B **175** 395

Goncharov A S, Linde A and Mukhanov V F 1987 *Int. J. Mod. Phys.* A **2** 561

[14] Linde A D, Linde D A and Mezhlumian A 1994 *Phys. Rev.* D **49** 1783

[15] Dolgov A D and Linde A D 1982 *Phys. Lett.* B **116** 329

Abbott L F, Fahri E and Wise M 1982 *Phys. Lett.* B **117** 29

[16] Kofman L A, Linde A D and Starobinsky A A 1994 *Phys. Rev. Lett.* **73** 3195

Kofman L, Linde A and Starobinsky A A 1997 *Phys. Rev.* D **56** 3258–95

[17] Greene P B, Kofman L, Linde A D and Starobinsky A A 1997 *Phys. Rev.* D **56** 6175–92 (hep-ph/9705347)

[18] Khlebnikov S and Tkachev I 1996 *Phys. Rev. Lett.* **77** 219

Khlebnikov S and Tkachev I 1997 *Phys. Lett.* B **390** 80

Khlebnikov S and Tkachev I 1997 *Phys. Rev. Lett.* **79** 1607

Khlebnikov S and Tkachev I 1997 *Phys. Rev.* D **56** 653

Prokopec T and Roos T G 1997 *Phys. Rev.* D **55** 3768

Greene B R, Prokopec T and Roos T G 1997 *Phys. Rev.* D **56** 6484

[19] Kofman L A, Linde A D and Starobinsky A A 1996 *Phys. Rev. Lett.* **76** 1011

Tkachev I 1996 *Phys. Lett.* B **376** 35

[20] Tkachev I, Kofman L A, Linde A D, Khlebnikov S and Starobinsky A A in
      preparation
[21] Kolb E W, Linde A and Riotto A 1996 *Phys. Rev. Lett.* **77** 4960
[22] Anderson G W, Linde A D and Riotto A 1996 *Phys. Rev. Lett.* **77** 3716
[23] García-Bellido J, Grigorev D, Kusenko A and Shaposhnikov M 1999 *Phys. Rev.* D
      60 123504
      García-Bellido J and Linde A 1998 *Phys. Rev.* D **57** 6075
[24] Felder G, Kofman L A and LindeA D 1999 *Phys. Rev.* D **59** 123523
[25] Ford L H 1987 *Phys. Rev.* D **35** 2955
      Spokoiny B 1993 *Phys. Lett.* B **315** 40
      Joyce M 1997 *Phys. Rev.* D **55** 1875
      Joyce M and Prokopec T 1998 *Phys. Rev.* D **57** 6022
      Peebles P J E and Vilenkin A 1999 *Phys. Rev.* D **59** 063505
[26] Jaffe A H *et al* 2001 Cosmology from Maxima-1, Boomerang and COBE/DMR CMB
      observations *Phys. Rev. Lett.* **86** 3475
[27] Conan Doyle A *The Sign of Four* ch 6 (http://www.litrix.com/sign4/signo006.htm)

Chapter 5

Dark matter and particle physics

Antonio Masiero and Silvia Pascoli
SISSA, Trieste, Italy

Dark matter constitutes a key problem at the interface between particle physics, astrophysics and cosmology. Indeed, the observational facts which have been accumulated in the last years on dark matter point to the existence of an amount of non-baryonic dark matter. Since the Standard Model (SM) of particle physics does not possess any candidate for such non-baryonic dark matter, this problem constitutes a major indication for new physics beyond the SM.

We analyse the most important candidates for non-baryonic dark matter in the context of extensions of the SM (in particular supersymmetric models). Recent hints of the presence of a large amount of unclustered 'vacuum' energy (cosmological constant?) are discussed from the astrophysical and particle physics perspective.

5.1   Introduction

The electroweak SM is now approximately 30 years old and it enjoys a full maturity with an extraordinary success in reproducing the many electroweak tests which have been going on since its birth. Not only have its characteristic gauge bosons, W and Z, been discovered but also the top quark has been found in the mass range expected by the electroweak radiative corrections, but the SM has been able to account for an impressively long and very accurate series of measurements. Indeed, in particular at LEP, some of the electroweak observables have been tested with precisions reaching the per mille level without finding any discrepancy with the SM predictions. At the same time, the SM has successfully passed another very challenging class of exams, namely it has so far accounted for all the very suppressed or forbidden processes where flavour-changing neutral currents (FCNC) are present.

By now we can firmly state that no matter what physics should lies beyond the SM, necessarily such new physics will necessarily have to reproduce the SM with great accuracy at energies of the order of 100 GeV.

And, yet, in spite of all this glamorous success of the SM in reproducing an impressive set of experimental electroweak results, we are deeply convinced of the existence of new physics beyond this model. We see two main motivations pushing us beyond the SM.

First, we have theoretical 'particle physics' reasons to believe that the SM is not the whole story. The SM does not truly unify the elementary interactions (if nothing else, gravity is left out of the game), it leaves the problem of fermion masses and mixings completely unsolved and it exhibits the gauge hierarchy problem in the scalar sector (namely, the scalar Higgs mass is not protected by any symmetry and, hence, it would tend to acquire large values of the order of the energy scale at which the new physics sets in). This first class of motivation for new physics is well known to particle physicists. Less familiar is a second class of reasons which finds its origin in some relevant issues of astroparticle physics. We refer to the problems of the solar and atmospheric neutrino deficits, baryogenesis, inflation and dark matter (DM). In a sense these aspects (or at least some of them, in particular the solar and atmospheric neutrino problems and DM) may be considered as the only 'observational' evidence that we have at the moment for physics beyond the SM.

As for baryogenesis, if it is true that in the SM it is not possible to give rise to a sufficient amount of baryon–antibaryon asymmetry, still one may debate whether baryogenesis should have a dynamical origin and, indeed, whether primordial antimatter is absent. Coming to inflation, again one has to admit that in the SM there seems to be no room for an inflationary epoch in its scalar sector, but, as nice as inflation is in coping with several crucial cosmological problems, its presence in the history of the universe is still debatable. Finally, let me come to the main topic of this chapter, namely the relation between the DM issue and physics beyond the SM.

There exists little doubt that a conspicuous amount of the DM has to be in non-baryonic nature. This is supported both by the upper bound on the amount of baryonic matter from nucleosynthesis and by studies of galaxy formation. The SM does not have any viable candidate for such non-baryonic DM. Hence the DM issue constitutes a powerful probe in our search for new physics beyond the SM.

In this chapter we will briefly review the following aspects.

- The main features of the SM such as its spectrum, the Lagrangian and its symmetries, the Higgs mechanism, the successes and shortcomings of the SM.
- The experimental evidence for the existence of DM.
- Two major particle physics candidates for DM: massive (light) neutrinos and the lightest supersymmetric (SUSY) particle in SUSY extensions of the SM with R parity (to be defined later on). Light neutrinos and the lightest

sparticle are 'canonical' examples of the hot and cold DM, respectively. This choice does not mean that these are the only interesting particle physics candidates for DM. For instance axions are still of great interest as CDM candidates and the experimental search for them is proceeding at full steam.

- The possibility of warm dark matter which has recently attracted much attention in relation to the possibility of light gravitinos (as WDM candidates) in a particular class of SUSY models known as gauge-mediated SUSY breaking schemes.
- Finally the problem of the cosmological constant Λ in relation to the structure formation in the universe as in the ΛCDM or QCDM models.

5.2   The SM of particle physics

In particle physics the fundamental interactions are described by the Glashow–Weinberg–Salam standard theory (GSW) for the electroweak interactions [1–3] (for a recent review see [4]) and QCD for the strong one. GWS and QCD are gauge theories based, respectively, on the gauge groups $SU(2)_L \times U(1)_Y$ and $SU(3)_c$ where L refers to left, Y to hypercharge and c to colour. We recall that a gauge theory is invariant under a local symmetry and requires the existence of vector gauge fields living in the adjoint representation of the group. Therefore in our case we have:

(i)   three gauge fields W^1_μ, W^2_μ, W^3_μ for $SU(2)_L$;
(ii)   one gauge field B_μ for $U(1)_Y$; and
(iii)   eight gauge bosons λ^a_μ for $SU(3)_c$.

The SM fermions live in the irreducible representations of the gauge group and are reported in table 5.1: the indices L and R indicate, respectively, the left and right fields, $b = 1, 2, 3$ the generation, the colour is not shown.

The Lagrangian of the SM is dictated by the invariance under the Lorentz group and gauge group and the request of renormalizability. It is given by the sum of the kinetic fermionic part $\mathcal{L}_{K\,mat}$ and the gauge one $\mathcal{L}_{K\,gauge}$: $\mathcal{L} = \mathcal{L}_{K\,mat} + \mathcal{L}_{K\,gauge}$. The fermionic part reads for one generation:

$$\mathcal{L}_{K\,mat} = i\overline{Q}_L\gamma^\mu\left(\partial_\mu + ig W^a_\mu T_a + i\frac{g'}{6}B_\mu\right)Q_L + i\overline{d}_R\gamma^\mu\left(\partial_\mu - i\frac{g'}{3}B_\mu\right)d_R$$

$$+ i\overline{u}_R\gamma^\mu\left(\partial_\mu + i\frac{2g'}{3}B_\mu\right)u_R + i\overline{E}_L\gamma^\mu\left(\partial_\mu + ig W^a_\mu T_a - i\frac{g'}{2}B_\mu\right)E_L$$

$$+ i\overline{e}_R\gamma^\mu\left(\partial_\mu - ig'B_\mu\right)e_R \tag{5.1}$$

where the matrices $T_a = \sigma_a/2$, σ_a are the Pauli matrices, g and g' are the coupling constants of the groups $SU(2)_L$ and $U(1)_Y$ respectively. The Dirac matrices γ^μ are defined as usual. The colour and generation indeces are not specified. This Lagrangian $\mathcal{L}_{K\,mat}$ is invariant under two global accidental symmetries, the

Table 5.1. The fermionic spectrum of the SM.

Fermions	Generations			$SU(2)_L \otimes U(1)_Y$
	I	II	III	
$E_{bL} \equiv \begin{pmatrix} \nu_b \\ e_b^- \end{pmatrix}_L$	$\begin{pmatrix} \nu_e \\ e^- \end{pmatrix}_L$	$\begin{pmatrix} \nu_\mu \\ \mu^- \end{pmatrix}_L$	$\begin{pmatrix} \nu_\tau \\ \tau^- \end{pmatrix}_L$	$(2, -1)$
e_{bR}	e_R^-	μ_R^-	τ_R^-	$(1, -2)$
$Q_{bL} \equiv \begin{pmatrix} u_b \\ d_b \end{pmatrix}_L$	$\begin{pmatrix} u \\ d \end{pmatrix}_L$	$\begin{pmatrix} c \\ s \end{pmatrix}_L$	$\begin{pmatrix} t \\ b \end{pmatrix}_L$	$(2, 1/3)$
u_{bR}	u_R	c_R	t_R	$(1, 4/3)$
d_{bR}	d_R	s_R	b_R	$(1, -2/3)$

leptonic number and the baryonic one: the fermions belonging to the fields E_{bL} and e_{bR} are called leptons and trasform under the leptonic symmetry $U(1)_L$ while the ones belonging to Q_{bL}, u_{bR} and d_{bR} baryons and trasform under $U(1)_B$.

The Lagrangian for the gauge fields reads:

$$
\begin{aligned}
\mathcal{L}_{\text{K gauge}} = & -\tfrac{1}{4}(\partial_\mu W_\nu^a - \partial_\nu W_\mu^a + \epsilon^{abc} W_\mu^b W_\nu^c) \\
& \times (\partial^\mu W^{\nu a} - \partial^\nu W^{\mu a} + \epsilon^{ab'c'} W_\nu^{b'} W_\mu^{c'}) \\
& -\tfrac{1}{4}(\partial_\mu B_\nu - \partial_\nu B_\mu)(\partial^\mu B^\nu - \partial^\nu B^\mu).
\end{aligned}
\tag{5.2}
$$

5.2.1 The Higgs mechanism and vector boson masses

The gauge symmetry protects the gauge bosons from having mass. Unfortunately the weak interactions require massive gauge bosons in order to explain the experimental behaviour. However, adding a direct mass term for gauge bosons breaks explicitly the gauge symmetry and spoils renormalizability. To preserve such nice feature of gauge theories, it is necessary to break spontaneously the symmetry. This is achieved through the Higgs mechanism. We introduce in the spectrum a scalar field H, which transforms as a doublet under $SU(2)_L$, carries hypercharge while it is colourless. The Higgs doublet has the following potential V_{Higgs}, kinetic terms $\mathcal{L}_{\text{KH}}$ and Yukawa couplings with the fermions $\mathcal{L}_{\text{Hf}}$:

$$
V_{\text{Higgs}} = -\mu^2 H^\dagger H + \lambda (H^\dagger H)^2
$$

$$
\mathcal{L}_{\text{KH}} = -\left(\partial_\mu H + ig W_\mu^a T_a H + i\frac{g'}{2} B_\mu H\right)^\dagger \left(\partial_\mu H + ig W_\mu^a T_a H + i\frac{g'}{2} B_\mu H\right)
$$

$$\mathcal{L}_{\text{Hf}} = -\sum_{b,c}^{\text{gener.}} (\lambda_{bc}^d \, \overline{Q}_{Lb} H D_{Rc} + \lambda_{bc}^u \, \overline{Q}_{Lb} \widetilde{H} U_{Rc} + \lambda_{bc}^e \, \overline{E}_{Lb} H E_{Rc}) + \text{h.c.} \quad (5.3)$$

where the parameters μ and λ are real constants, λ_{bc}^d, λ_{bc}^u and λ_{bc}^e are 3×3 matrices on the generation space. $\widetilde{H}$ indicates the charge conjugate of H: $\widetilde{H}^a = \epsilon^{ab} H_b^{\dagger}$.

While the Lagrangian is invariant under gauge symmetry the vacuum is not and the neutral component of the doublet H develops a vacuum expectation value (vev):

$$\langle H^0 \rangle = \begin{pmatrix} 0 \\ v \end{pmatrix}. \quad (5.4)$$

This breaks the symmetry $SU(2)_L \otimes U(1)_Y$ down to $U(1)_{EM}$. We recall that when a global symmetry is spontaneously broken, a massless Goldstone boson appears in the theory; if the symmetry is local (gauge) these Goldstone bosons become the longitudinal components of the vector bosons (it is said that they are eaten up by the gauge bosons). The gauge bosons relative to the broken symmetry acquire a mass as shown in $\mathcal{L}_{\text{M gauge}}$:

$$\mathcal{L}_{\text{M gauge}} = -\frac{1}{2}\frac{v^2}{4}[g^2 (W_\mu^1)^2 + g^2 (W_\mu^2)^2 + (-g W_\mu^3 + g' B_\mu)^2]. \quad (5.5)$$

Therefore there are three massive vectors $W_\mu^{\pm}$ and Z_μ^0:

$$W_\mu^{\pm} = \frac{1}{\sqrt{2}} (W_\mu^1 \mp i W_\mu^2), \quad (5.6)$$

$$Z_\mu^0 = \frac{1}{\sqrt{g^2 + g'^2}} (g W_\mu^3 - g' B_\mu), \quad (5.7)$$

whose masses are given by

$$m_W = g\frac{v}{2}, \quad (5.8)$$

$$m_Z = \sqrt{(g^2 + g'^2)}\frac{v}{2}, \quad (5.9)$$

while the gauge boson

$$A_\mu \equiv \frac{1}{\sqrt{g^2 + g'^2}} (g W_\mu^3 + g' B_\mu),$$

relative to $U(1)_{EM}$, remains massless as imposed by the gauge symmetry. Such a mechanism is called the Higgs mechanism and preserves renormalizability.

5.2.2 Fermion masses

Fermions are spinors with respect to the Lorentz group $SU(2) \otimes SU(2)$. Weyl spinors are two-component spinors which transform under the Lorentz group:

$$\chi_L \qquad \text{as } (\tfrac{1}{2}, 0) \tag{5.10}$$

$$\eta_R \qquad \text{as } (0, \tfrac{1}{2}) \tag{5.11}$$

and therefore are said to be left-handed and right-handed respectively.

A fermion mass term must be invariant under the Lorentz group. We have two possibilities:

(i) A Majorana mass term couples just one spinor with itself:

$$\chi^\alpha \chi^\beta \epsilon_{\alpha\beta} \qquad \text{or} \qquad \eta^{\dot\alpha} \eta^{\dot\beta} \epsilon_{\dot\alpha\dot\beta}. \tag{5.12}$$

It is not invariant under any local or global symmetry under which the field transforms not trivially;

(ii) A Dirac mass term involves two different spinors χ_L and η_R:

$$\chi^\alpha \bar\eta^\beta \epsilon_{\alpha\beta} \qquad \text{or} \qquad \bar\chi^{\dot\alpha} \eta^{\dot\beta} \epsilon_{\dot\alpha\dot\beta}. \tag{5.13}$$

This can be present even if the fields carry quantum numbers.

In the SM Majorana masses are forbidden by the gauge symmetry; in fact we have that, for example,

$$e_L e_L \Rightarrow Q \neq 0$$
$$\nu_L \nu_L \Rightarrow SU(2)_L \neq$$

and $SU(2)_L$ forbids Dirac mass terms:

$$\overline{e_L} e_R \Rightarrow SU(2)_L \neq . \tag{5.14}$$

Therefore no direct mass term can be present for fermions in the SM.

However, when the gauge symmetry breaks spontaneously the Yukawa couplings provide Dirac mass terms to fermions which read:

$$\mathcal{L}_{\text{M mat}} = +\frac{1}{\sqrt{2}} \lambda^e v \bar{e}_L e_R + \frac{1}{\sqrt{2}} \lambda^u v \bar{u}_L u_R + \frac{1}{\sqrt{2}} \lambda^d v \bar{d}_L d_R + \text{h.c.} \tag{5.15}$$

with masses:

$$m_e = \frac{1}{\sqrt{2}} \lambda_e v \qquad m_u = \frac{1}{\sqrt{2}} \lambda_u v \qquad m_d = \frac{1}{\sqrt{2}} \lambda_d v. \tag{5.16}$$

We note that neutrinos are massless and so remain at any order in perturbation theory:

(i)   lacking of the right component they cannot have a Dirac mass term; and
(ii) belonging to a $SU(2)_L$ doublet, they cannot have a Majorana mass term.

However, from experimental data we can infer that neutrinos are massive and that their mass is very small compared to the other mass scales in the SM. The SM cannot provide such a mass to neutrinos and hence this consitutes a proof of the existence of a physics beyond the SM. The problem of ν masses will be addressed in more detail in section 5.4.2.

5.2.3   Successes and difficulties of the SM

The SM has been tested widely at accelerators receiving strong confirmations of its validity. Up to now there are no appreciable deviations from its expectations even if some observables have been tested at the per mille level. In the future LHC will reach higher energies and will have the possibility to discover new physics beyond the SM if this one lies at the TeV scale. Another promising class of experiments to detect deviations from the SM predictions are rare processes which are very suppressed or forbidden in the SM such as flavour-changing neutral currents phenomena or CP-violation ones. All these tests up to now are compatible with SM expectations.

However, we see good reasons to expect the existence of physics beyond the SM. From a theoretical point of view the SM cannot give an explanation of the existence of three families, of the hierarchy present among their masses, of the fine tuning of some of its parameters, of the lack of unification of the three fundamental interactions (considering the behaviour of the coupling constants, we see that they unify at a scale $M_X \sim 10^{15}$ GeV where a unified simple group might arise), of the hierarchy problem of the scalar masses which tend to become as large as the highest mass scale in the theory. From an experimental point of view, the measured neutrino masses are a proof of a physics beyond SM even if what the type of physics is still an open question to be addressed. Cosmology also gives strong hints in favour of a physics beyond the SM: in particular baryogenesis cannot find a satisfactory explanation in the SM, inflation is not predicted by SM and finally we have the dark matter problem.

5.3   The dark matter problem: experimental evidence

Let us define Ω (for a review see [5] and [6]) as the ratio between the density ρ and the critical density

$$\rho_{cr} = \frac{3H_0^2}{8\pi G} = 1.88h_0^2 \times 10^{-29} \text{ g cm}^{-3}$$

where H_0 is the Hubble constant, G the gravitational constant:

$$\Omega = \frac{\rho}{\rho_{cr}}. \tag{5.17}$$

The Ω_{lum} due to the contribution of the luminous matter (stars, emitting clouds of gases) is given by

$$\Omega_{lum} \leq 0.01. \tag{5.18}$$

The first evidence for dark matter (DM) comes from observations of galactic rotation curves (circular orbital velocity versus radial distance from the galactic centre) using stars and clouds of neutral hydrogen. These curves show an increasing profile for small values of the radial distance r while for larger ones it becomes flat, finally decreasing again. According to Newtonian mechanics this behaviour can be explained if the enclosed mass rises linearly with galactocentric distance. However, the light falls off more rapidly and therefore we are forced to assume that the main part of the matter in the galaxies is made of non-shining matter or DM which extends for a much bigger region than the luminous one. The limit on $\Omega_{galactic}$ which can be inferred from the study of these curves is

$$\Omega_{galactic} \geq 0.1. \tag{5.19}$$

The simplest idea is to suppose that the DM is due to baryonic objects which do not shine. However big-bang nucleosynthesis (BBN) and, in particular, a precise determination of the primeval aboundance of deuterium provide strong limits on the value of the baryon density [7] $\Omega_B = \rho_B/\rho_{cr}$:

$$\Omega_B = (0.019 \pm 0.001)h_0^{-2} \simeq 0.045 \pm 0.005. \tag{5.20}$$

One-third of the BBN baryon density is given by stars and the cold and warm gas present in galaxies. The other two-thirds are probably in hot intergalactic gas, warm gas in galaxies and dark stars such as low-mass objects which do not shine (brown dwarfs and planets) or the result of stellar evolution (neutron stars, black holes, white dwarfs). The latter ones are called MACHOS (MAssive Compact Halo Objects) and can be detected in our galaxy through microlensing.

Anyway from cluster observations the ratio of baryons to total mass is $f = (0.075 \pm 0.002)h_0^{-3/2}$; assuming that clusters provide a good sample of the universe, from f and Ω_B in (5.20) we can infer that:

$$\Omega_M \sim 0.35 \pm 0.07. \tag{5.21}$$

Such a value for Ω_M is supported by evidence, from the evolution, of the abundance of clusters and measurements of the power spectrum of large-scale structures.

Hence the major part of DM is non-baryonic [8]. The crucial point is that the SM does not possess any candidate for such non-baryonic relics of the early universe. Hence the demand for non-baryonic DM implies the existence of a new physics beyond the SM. Non-baryonic DM divides into two classes [23, 26]: cold DM (CDM), made of neutral heavy particles called WIMPS (Weakly Interacting Massive Particles) or very light ones such as axions, hot DM (HDM) made of relativistic particles as neutrinos or even warm dark matter (WDM) with intermediate characteristics such as gravitinos.

5.4 Lepton number violation and neutrinos as HDM candidates

Neutrinos are the first candidate for DM we review which can account for HDM [30]: particles that were relativistic at their decoupling from the thermal bath when their rate of interaction became smaller than the expansion rate and they froze out (or, to be more precise, at the time galaxy formation starts at $T \sim 300$ eV). The SM has no candidate for HDM; however, it is now well established from experimental data that neutrinos are massive and very light. Therefore they can account for HDM. We briefly discuss their characteristics.

5.4.1 Experimental limits on neutrino masses

The recent atmospheric neutrino data from Super-Kamiokande provide strong evidence of neutrino oscillations which can take place only if neutrinos are massive. The parameters relevant in ν-oscillations are the mixing angle θ and the mass-squared differences which can be measured in atmospheric neutrinos, solar neutrinos, short-baseline and long-baseline experiments (for a review see [9, 10]):

(i) In atmospheric neutrino experiments, to account for the deficit of the ν_μ flux expected towards the ν_e one from cosmic rays and its zenith dependence, it is necessary to call for $\nu_\mu \to \nu_\tau$ oscillations with

$$\sin^2 2\theta_{\text{atm}} \geq 0.82 \tag{5.22}$$
$$\Delta m^2_{\text{atm}} \simeq (1.5\text{--}8.0) \times 10^{-3} \text{ eV}^2 \tag{5.23}$$

from Super-Kamiokande data at 99% C.L. [11].

(ii) The solar ν anomaly arises from the fact that the ν_e flux coming from the Sun is sensibly less than the one predicted by the solar SM: this problem can also be explained in terms of neutrino oscillations. The recent Super-Kamiokande data [12] favour the LMA (large mixing angle) solution with

$$\tan^2 \theta_\odot \simeq 0.15\text{--}4 \tag{5.24}$$
$$\Delta m^2_\odot \simeq (1.5\text{--}10) \times 10^{-5} \text{ eV}^2 \tag{5.25}$$

at 99% C.L., even if the small mixing angle ($\tan^2 \theta_\odot \sim 10^{-4}$) and the LOW ($\tan^2 \theta_\odot \sim 0.4\text{--}4$) solutions cannot be excluded and the oscillations into sterile neutrinos are strongly disfavoured.

(iii) Reactor [13] and short- and long-baseline experiments constrain further the parameters and, in particular, the mixing angles.

(iv) Finally the LSND experiment has evidence of $\bar{\nu}_\mu \to \bar{\nu}_e$ oscillations with $\Delta m^2_{\text{LSND}} \simeq 0.1\text{--}2$ eV^2, the Karmen experiment has no positive results for the same oscillation and then restricts the LSND allowed region [14].

In the near future several long-baseline experiments will be held to test ν-oscillations directly and measure the relevant parameters: K2K in Japan is already looking for missing ν_μ in $\nu_\mu \rightarrow \nu_\tau$ oscillations, MINOS (in US) and OPERA (with neutrino beam from CERN to Gran Sasso) are long-baseline experiments devoted to this aim, which are under construction.

The tritium beta-decay experiments are searching directly for the effective electron neutrino mass m_β and the present Troitzk [15] and Mainz [16] limits give $m_\beta \leq 2.5–2.9$ eV; there are perspectives to increase the sensitivity down to 1 eV.

The $\beta\beta_{0\nu}$ decay predicted if neutrinos are Majorana particles will indicate the value of the effective mass $|\langle m \rangle|$, the present Heidelberg–Moscow bound is (see, for example, [17]):

$$|\langle m \rangle| \equiv \left| \sum_i U_{ei}^2 m_i \right| \leq 0.2–1 \text{ eV} \qquad (5.26)$$

but in the near future there are perpectives to reach $|\langle m \rangle| \sim 0.01$ eV.

Finally the direct search for m_ν at accelerators has so far given negative results leading to upper bounds [18]:

$$m_{\nu_\mu} < 0.19 \text{ MeV}, \qquad m_{\nu_\tau} < 18.2 \text{ MeV} \qquad (5.27)$$

from LEP at 90% C.L. and 95% C.L. respectively.

From all these experiments we can conclude that neutrinos have masses and that their values must be much lower than the other mass scales in the SM.

5.4.2 Neutrino masses in the SM and beyond

The SM cannot account for neutrino masses: we cannot construct either a Dirac mass term as there is only a left-handed neutrino and no right-handed component, or a Majorana mass term because such a mass would violate the lepton number and the gauge symmetry.

To overcome this problem, many possibilities have been suggested:

(1) Within the SM spectrum we can form an $SU(2)_L$ singlet with ν_L using a triplet formed by two Higgs field H as $\nu_L \nu_L H H$. When the Higgs field H develops a vev, this term gives rise to a Majorana mass term. However, this term is not renormalizable, breaks the leptonic symmetry and does not give an explanation of the smallness of the neutrino masses.

(2) We can introduce a new Higgs triplet Δ and produce a Majorana mass term as in the previous case when Δ acquires a vev.

(3) However, the most economical way to extend the SM is to introduce a right-handed component N_R, a singlet under the gauge group, which couples with the left-handed neutrinos. The lepton number L can be either conserved or violated. In the former option neutrinos acquire a 'regular' Dirac mass like all other charged fermions of the SM. The left- and right-handed components of the neutrino combine together to give rise to a massive four-component Dirac fermion. The problem is that the extreme lightness of the neutrinos (in particular

of the electron-neutrino) requires an exceedingly small neutrino Yukawa coupling of $O(10^{-11})$ or so. Although quite economical, we do not consider this option particularly satisfactory.

(4) The other possibility is to link the presence of neutrino masses to the violation of L. In this case one introduces a new mass scale, in addition to the electroweak Fermi scale, into the problem. Indeed, lepton number can be violated at a very high or a very low mass scale. The former choice represents, in our view, the most satisfactory way to have massive neutrinos with a very small mass. The idea (see-saw mechanism [19,20]) is to introduce a right-handed neutrino into the fermion mass spectrum with a Majorana mass M much larger than M_W. Indeed, being the right-handed neutrino, a singlet under the electroweak symmetry group, its mass is not chirally protected. The simultaneous presence of a very large chirally unprotected Majorana mass for the right-handed component together with a 'regular' Dirac mass term (which can be at most of $O(100 \text{ GeV})$ gives rise to two Majorana eigenstates with masses very far apart.

The Lagrangian for neutrino masses is given by

$$\mathcal{L}_{\text{mass}} = -\frac{1}{2}(\bar{\nu}_L \overline{N}_L^c) \begin{pmatrix} 0 & m_D \\ m_D & M \end{pmatrix} \begin{pmatrix} \nu_R^c \\ N_R \end{pmatrix} + \text{h.c.} \qquad (5.28)$$

where ν_R^c is the CP-conjugated of ν_L and N_L^c of N_R. It holds that $m_D \ll M$. Diagonalizing the mass matrix we find two Majorana eigenstates n_1 and n_2 with masses very far apart:

$$m_1 \simeq \frac{m_D^2}{M}, \qquad m_2 \simeq M.$$

The light eigenstate n_1 is mainly in the ν_L direction and is the neutrino that we 'observe' experimentally while the heavy one n_2 is in the N_R one. The key point is that the smallness of its mass (in comparison with all the other fermion masses in the SM) finds a 'natural' explanation in the appearance of a new, large mass scale where L is violated explicitly (by two units) in the right-handed neutrino mass term.

5.4.3 Thermal history of neutrinos

Let us consider a stable massive neutrino (of mass less than 1 MeV) (see for example [5]). If its mass is less than 10^{-4} eV it is still relativistic today and its contribution to Ω_M is negligible. In the opposite case it is non-relativistic and its contribution to the energy density of the universe is simply given by its number density multiplied by its mass. The number density is determined by the temperature at which the neutrino decouples and, hence, by the strength of the weak interactions. Neutrinos decouple when their mean free path exceeds the horizon size or equivalently $\Gamma < H$. Using natural units ($c = \hbar = 1$), we have that

$$\Gamma \sim \sigma_\nu n_{e^\pm} \sim G_F^2 T^5 \qquad (5.29)$$

and

$$H \sim \frac{T^2}{M_P} \tag{5.30}$$

so that

$$T_{\nu d} \sim M_P^{-1/3} G_F^{-2/3} \sim 1 \text{ MeV}, \tag{5.31}$$

where G_F is the Fermi constant, T denotes the temperature, M_P is the Planck mass. Since this decoupling temperature $T_{\nu d}$ is higher than the electron mass, then the relic neutrinos are slightly colder than the relic photons which are 'heated' by the energy released in the electron–positron annihilation. The neutrino number density turns out to be linked to the number density of relic photons n_γ by the relation:

$$n_\nu = \tfrac{3}{22} g_\nu n_\gamma, \tag{5.32}$$

where $g_\nu = 2$ or 4 according to the Majorana or Dirac nature of the neutrino, respectively.

Then one readily obtains the ν contribution to Ω_M:

$$\Omega_\nu = 0.01 \times m_\nu(\text{eV}) h_0^{-2} \frac{g_\nu}{2} \left(\frac{T_0}{2.7} \right)^3. \tag{5.33}$$

Imposing $\Omega_\nu h_0^2$ to be less than one (which comes from the lower bound on the lifetime of the universe), one obtains the famous upper bound of $200(g_\nu)^{-1}$ eV on the sum of the masses of the light and stable neutrinos:

$$\sum_i m_{\nu_i} \leq 200(g_\nu)^{-1} \text{ eV}. \tag{5.34}$$

Clearly from equation (5.33) one easily sees that it is enough to have one neutrino with a mass in the 1–20 eV range to obtain Ω_ν in the 0.1–1 range of interest for the DM problem.

5.4.4 HDM and structure formation

Hence massive neutrinos with mass in the eV range are very natural candidates to contribute to an Ω_M larger than 0.1. The actual problem for neutrinos as viable DM candidates concerns their role in the process of large-scale structure formation. The crucial feature of HDM is the erasure of small fluctuations by free-streaming: neutrinos stream relativistically for quite a long time until their temperature drops to $T \sim m_\nu$. Therefore a neutrino fluctuation in order to be preserved must be larger than the distance d_ν travelled by neutrinos during such an interval. The mass contained in that space volume is of the order of the supercluster masses:

$$M_{J,\nu} \sim d_\nu^3 m_\nu n_\nu (T = m_\nu) \sim 10^{15} M_\odot, \tag{5.35}$$

where n_ν is the number density of the relic neutrinos, $M_\odot$ is the solar mass. Therefore the first structures to form are superclusters and smaller structures such as galaxies arise from fragmentation in a typical top-down scenario. Unfortunately in these schemes one obtains too many structures at superlarge scales. The possibility of improving the situation by adding the seeds for small-scale structure formation using topological defects (cosmic strings) are essentially ruled out at present [21,22]. Hence schemes of pure HDM are strongly disfavoured by the demand of a viable mechanism for large-structure formation.

5.5 Low-energy SUSY and DM

Another kind of DM, widely studied, called cold DM (CDM) is made of particles which were non-relativistic at their decoupling. Natural candidates for such DM are Weakly Interacting Massive Particles (WIMPs), which are very heavy if compared to neutrinos. The SM does not have non-baryonic neutral particles which can account for CDM and therefore we need to consider extensions of the SM as supersymmetric SM in which there are heavy neutral particles remnants of annichilations such as neutralinos (for a review see [36]).

5.5.1 Neutralinos as the LSP in SUSY models

One of the major shortcomings of the SM concerns the protection of the scalar masses once the SM is embedded into some underlying theory (and at least at the Planck scale such new physics should set in to incorporate gravity into the game). Since there is no typical symmetry protecting the scalar masses (while for fermions there is the chiral symmetry and for gauge bosons there are gauge symmetries), the clever idea which was introduced in the early 1980s to prevent scalar masses from becoming too large was to have a supersymmetry (SUSY) unbroken down to the weak scale. Since fermion masses are chirally protected and as long as SUSY is unbroken there must be a degeneracy between the fermion and scalar components of a SUSY multiplet; then, having a low-energy SUSY, it is possible to have an 'induced protection' on scalar masses (for a review see [34, 35]).

However, the mere supersymmetrization of the SM faces an immediate problem. The most general Lagrangian contains terms which violate baryon and lepton numbers producing a proton decay which is too rapid. To prevent this catastrophic result we have to add some symmetry which forbids all or part of these dangerous terms with L or B violations. The most familiar solution is the imposition of a discrete symmetry, called R matter parity, which forbids all these dangerous terms. It reads over the fields contained in the theory:

$$R = (-1)^{3(B-L)+2s}. \tag{5.36}$$

R is a multiplicative quantum number reading -1 over the SUSY particles and $+1$ over the ordinary particles. Clearly in models with R parity the lightest

SUSY particle can never decay. This is the famous LSP (lightest SUSY particle) candidate for CDM.

Note that proton decay does not call directly for R parity. Indeed this decay entails the violation of both B and L. Hence, to prevent a fast proton decay one may impose a discrete symmetry which forbids all the B violating terms in the SUSY Lagrangian, while allowing for terms with L violation (the reverse is also viable). Models with such alternative discrete symmetries are called SUSY models with broken R parity. In such models the stability of the LSP is no longer present and the LSP cannot be a candidate for stable CDM. We will comment later on these alternative models in relation to the DM problem, but we turn now to the more 'orthodox' situation with R parity. The favourite LSP is the lightest neutralino.

5.5.2 Neutralinos in the minimal supersymmetric SM

If we extend the SM in the minimal way, adding for each SM particle a supersymmetric partner with the same quantum numbers, we obtain the so called Minimal Supersymmetric Standard Model (MSSM). In this context the neutralinos are the eigenvectors of the mass matrix of the four neutral fermions partners of the W_3, B, H_1^0 and H_2^0 called, respectively, wino $\tilde{W}_3$, bino $\tilde{B}$, higgsinos $\tilde{H}_1^0$ and $\tilde{H}_2^0$. There are four parameters entering the mass matrix, M_1, M_2, μ and $\tan \beta$:

$$M = \begin{pmatrix} M_2 & 0 & m_Z \cos\theta_w \cos\beta & -m_Z \cos\theta_w \sin\beta \\ 0 & M_1 & -m_Z \sin\theta_w \cos\beta & m_Z \sin\theta_w \sin\beta \\ m_Z \cos\theta_w \cos\beta & -m_Z \sin\theta_w \cos\beta & 0 & -\mu \\ -m_Z \cos\theta_w \sin\beta & m_Z \sin\theta_w \sin\beta & -\mu & 0 \end{pmatrix}$$
(5.37)

where $m_Z = 91.19 \pm 0.002$ GeV is the mass of the Z boson, θ_w is the weak mixing angle, $\tan\beta \equiv v_2/v_1$ with v_1, v_2 vevs of the scalar fields H_1^0 and H_2^0 respectively.

In general M_1 and M_2 are two independent parameters, but if one assumes that a grand unification scale takes place, then at grand unification $M_1 = M_2 = M_3$, where M_3 is the gluino mass at that scale. Then at the M_W scale one obtains:

$$M_1 = \tfrac{5}{3} \tan^2 \theta_w M_2 \simeq \tfrac{1}{2} M_2,$$
(5.38)

$$M_2 = \frac{g_2^2}{g_3^2} m_{\tilde{g}} \simeq m_{\tilde{g}}/3,$$
(5.39)

where g_2 and g_3 are the $SU(2)$ and $SU(3)$ gauge coupling constants, respectively, and $m_{\tilde{g}}$ is the gluino mass.

The relation (5.38) between M_1 and M_2 reduces to three the number of independent parameters which determine the lightest neutralino composition and mass: $\tan\beta$, μ and M_2. The neutralino eigenstates are usually denoted by $\tilde{\chi}_i^0$, $\tilde{\chi}_1^0$ being the lightest one.

If $|\mu| > M_1, M_2$ then $\tilde{\chi}_1^0$ is mainly a gaugino and, in particular, a bino if $M_1 > m_Z$, if $M_1, M_2 > |\mu|$ then $\tilde{\chi}_1^0$ is mainly a higgsino. The corresponding phenomenology is drastically different leading to different predictions for CDM.

For fixed values of $\tan \beta$ one can study the neutralino spectrum in the (μ, M_2) plane. The major experimental inputs to exclude regions in this plane are the request that the lightest chargino be heavier than $m_Z/2$ and the limits on the invisible width of the Z hence limiting the possible decays $Z \rightarrow \tilde{\chi}_1^0 \tilde{\chi}_1^0, \tilde{\chi}_1^0 \tilde{\chi}_2^0$. Moreover, if the GUT assumption is made, then the relation (5.38) between M_2 and $m_{\tilde{g}}$ implies a severe bound on M_2 from the experimental lower bound on $m_{\tilde{g}}$ of CDF (roughly $m_{\tilde{g}} > 220$ GeV, hence implying $M_2 > 70$ GeV); the theoretical demand that the electroweak symmetry be broken radiatively, i.e. due to the renormalization effects on the Higgs masses when going from the superlarge scale of supergravity breaking down to M_W, further constrains the available (μ, M_2) region. The first important outcome of this analysis is that the lightest neutralino mass exhibits a lower bound of roughly 30 GeV. The actual bound on the mass of the lightest neutralino $\tilde{\chi}_1^0$ from LEP2 is:

$$m_{\tilde{\chi}_1^0} \geq 40 \text{ GeV} \tag{5.40}$$

for any value of $\tan \beta$. This bound becomes stronger if we put further constraints on the MSSM. The strongest limit is obtained in the so-called Constrained MSSM (CMSSM) where we have only four independent SUSY parameters plus the sign of the μ parameter (see equation (5.37)): $m_{\tilde{\chi}_1^0} \geq 95$ GeV [29].

There are many experiments already running or approved to detect WIMPS; however, they rely on different techniques:

(i) DAMA and CDMS use the scattering of WIMPS on nuclei measuring the recoil energy; in particular DAMA [31] exploits an annual modulation of the signal which could be explained in terms of WIMPS;

(ii) ν-telescopes (AMANDA) are held to detect ν fluxes coming from the annihilation of WIMPS which accumulate in celestial bodies such as the Earth or the Sun;

(iii) experiments (AMS, PAMELA) which detect low-energy antiprotons and γ-rays from $\tilde{\chi}_1^0$ annihilation in the galactic halo.

5.5.3   Thermal history of neutralinos and Ω_{CDM}

Let us focus now on the role played by $\tilde{\chi}_1^0$ as a source of CDM. The lightest neutralino $\tilde{\chi}_1^0$ is kept in thermal equilibrium through its electroweak interactions not only for $T > m_{\tilde{\chi}_1^0}$, but even when T is below $m_{\tilde{\chi}_1^0}$. However for $T < m_{\tilde{\chi}_1^0}$ the number of $\tilde{\chi}_1^0$s rapidly decreases because of the appearance of the typical Boltzmann suppression factor $\exp(-m_{\tilde{\chi}_1^0}/T)$. When T is roughly $m_{\tilde{\chi}_1^0}/20$ the number of $\tilde{\chi}_1^0$ diminishes so much that they no longer interact, i.e. they decouple. Hence their contribution to Ω_{CDM} of $\tilde{\chi}_1^0$ is determined by two parameters: $m_{\tilde{\chi}_1^0}$ and the temperature at which $\tilde{\chi}_1^0$ decouples ($T_{\chi d}$) which fixes the number of surviving $\tilde{\chi}_1^0$s. As for the determination of $T_{\chi d}$ itself, one has to compute the $\tilde{\chi}_1^0$ annihilation rate and compare it with the cosmic expansion rate.

Several annihilation channels are possible with the exchange of different SUSY or ordinary particles, $\tilde{f}$, H, Z, etc. Obviously the relative importance of the channels depends on the composition of $\tilde{\chi}_1^0$:

(i) If $\tilde{\chi}_1^0$ is mainly a gaugino (say at least at the 90% level) then the annihilation goes through $\tilde{f}$ or $\tilde{l}_R$ exchange and the sfermion mass $m_{\tilde{f}}$ plays a crucial role. The actual limits from LEP2 are roughly:

$$m_{\tilde{\nu}} \geq 43\,\text{GeV} \qquad \text{and} \qquad m_{\tilde{e}}, m_{\tilde{q}} \geq 90\,\text{GeV}. \tag{5.41}$$

The contribution to Ω due to neutralinos $\Omega_{\tilde{\chi}_1^0}$ is given by

$$\Omega_{\tilde{\chi}_1^0} h_0^2 \simeq \frac{m_{\tilde{\chi}_1^0}^2 + m_{\tilde{l}_R}^2}{(1\,\text{TeV})^2 m_{\tilde{\chi}_1^0}^2} \; \frac{1}{\left(1 - \dfrac{m_{\tilde{\chi}_1^0}^2}{m_{\tilde{\chi}_1^0}^2 + m_{\tilde{l}_R}^2}\right)^2 + \dfrac{m_{\tilde{\chi}_1^0}^4}{(m_{\tilde{\chi}_1^0}^2 + m_{\tilde{l}_R}^2)^2}}. \tag{5.42}$$

If sfermions are light the $\tilde{\chi}_1^0$ annihilation rate is fast and $\Omega_{\tilde{\chi}_1^0}$ is negligible. However, if $\tilde{f}$ (and hence $\tilde{l}$, in particular) is heavier than 150 GeV, the annihilation rate of $\tilde{\chi}_1^0$ is sufficiently suppressed so that $\Omega_{\tilde{\chi}_1^0}$ can be in the right ball park for Ω_{CDM}. In fact if all the $\tilde{f}$s are heavy, say above 500 GeV and for $m_{\tilde{\chi}_1^0} \ll m_{\tilde{f}}$, then the suppression of the annihilation rate can become too efficient yielding $\Omega_{\tilde{\chi}_1^0}$ unacceptably large.

(ii) If $\tilde{\chi}_1^0{}'$ is mainly a combination of $\tilde{H}_1^0$ and $\tilde{H}_2^0$ it means that M_1 and M_2 have to be much larger than μ. Invoking the relation (5.38) one concludes that, in this case, we expect heavy gluinos, typically in the TeV range. As for the number of surviving $\tilde{\chi}_1^0$s in this case, what is crucial is whether $m_{\tilde{\chi}_1^0}$ is larger or smaller than M_W. Indeed, for $m_{\tilde{\chi}_1^0} > M_W$ the annihilation channels $\tilde{\chi}_1^0 \tilde{\chi}_1^0 \to$ WW, ZZ, $t\bar{t}$ reduce $\Omega_{\tilde{\chi}_1^0}$ too much. If $m_{\tilde{\chi}_1^0} < M_W$ then acceptable contributions of $\tilde{\chi}_1^0$ to Ω_{CDM} are obtainable in rather wide areas of the $(\mu - m_Z)$ parameter space;

(iii) Finally it turns out that if $\tilde{\chi}_1^0$ results from a large mixing of the gaugino ($\tilde{W}_3$ and $\tilde{B}$) and higgsino ($\tilde{H}_1^0$ and $\tilde{H}_2^0$) components, then the annihilation is too efficient to allow the surviving $\tilde{\chi}_1^0$ to provide a large enough $\Omega_{\tilde{\chi}_1^0}$. Typically in this case $\Omega_{\tilde{\chi}_1^0} < 10^{-2}$ and hence $\tilde{\chi}_1^0$ is not a good CDM candidate.

In the minimal SUSY standard model there are five new parameters in addition to those already present in the non-SUSY case. Imposing the electroweak radiative breaking further reduces this number to four. Finally, in simple supergravity realizations the soft parameters A and B are related. Hence we end up with only three new, independent parameters. One can use the constraint that the relic $\tilde{\chi}_1^0$ abundance provides a correct Ω_{CDM} to restrict the allowed area in this three-dimensional space. Or, at least, one can eliminate points of this space which would lead to $\Omega_{\tilde{\chi}_1^0} > 1$, hence overclosing the universe. For $\tilde{\chi}_1^0$ masses up to 150 GeV it is possible to find sizable regions in the SUSY parameter space where $\Omega_{\tilde{\chi}_1^0}$ acquires interesting values for the DM problem. The interested reader can find a thorough analysis in the review [36] and the original papers therein quoted.

Finally a comment on models without R parity. From the point of view of DM, the major implication is that in this context the LSP is no longer a viable CDM candidate since it decays. There are very special circumstances under which this decay may be so slow that the LSP can still constitute a CDM candidate. The very slow decay of $\tilde{\chi}_1^0$ may have testable consequences. For instance in some schemes the LSP could decay emitting a neutrino and a photon. The negative result of the search for such neutrino line at Kamiokande resulted in an improved lower bound on the $\tilde{\chi}_1^0$ lifetime.

5.5.4   CDM models and structure formation

In the pure CDM model, almost all of the energy density needed to reach the critical one (the remaining few percent being given by the baryons) was provided by CDM alone. However, some observational facts (in particular the results of COBE) put this model into trouble, showing that it cannot correctly reproduce the power spectrum of density perturbations at all scales. At the same time it became clear that some CDM was needed anyway in order to obtain a successful scheme for large-scale structure formation.

A popular option is that of a flat universe realized with the total energy density mostly provided by two different matter components, CDM and HDM in a convenient fraction. These models, which have been called mixed DM (MDM) [33], succeeded in fitting the entire power spectrum quite well. A little amount of HDM has a dramatic effect on CDM models because the free-streaming of relativistic neutrinos washes out any inhomogeneities in their spatial distribution which will become galaxies. Therefore their presence slows the growth rates of the density inhomogeneities which will lead to galaxies.

Another interesting possibility for improving CDM models consists in the introduction of some late-time decaying particle [50]. The injection of non-thermal radiation due to such decays and the consequent increase in the horizon length at the equivalence time could lead to a convenient suppression of the excessive power at small scales (hence curing the major disease of the pure CDM SM). As appealing as this proposal may be from the cosmological point of view, its concrete realization in particle physics models meets several difficulties. Indeed, after considering cosmological and astrophysical bounds on such late decays, it turns out that only a few candidates survive as viable solutions.

These schemes beyond pure CDM which presently enjoy most scientific favour accompany CDM with a conspicous amount of 'vacuum' energy density, a form of unclustered energy which could be due to the presence of a cosmological constant Λ. We will deal with this interesting class of DM models, called ΛCDM models, in the final part of this report.

5.6 Warm dark matter

Another route which has been followed in the attempt to go beyond the pure CDM proposal is the possibility of having some form of warm DM (WDM). The implementation of this idea is quite attractive in SUSY models where the breaking of SUSY is conveyed by gauge interactions instead of gravity (these are the so-called gauge-mediated SUSY breaking (GMSB) models, for a review see [32]). This scenario had already been critically considered in the old days of the early constructions of SUSY models and was subject to renewed interest with the proposal in [37–39], where some guidelines for the realization of low-energy SUSY breaking are provided. In these schemes, the gravitino mass ($m_{3/2}$) loses its role of fixing the typical size of soft breaking terms and we expect it to be much smaller than that in models with a hidden sector. Indeed, given the well-known relation [34] between $m_{3/2}$ and the scale of SUSY breaking $\sqrt{F}$, i.e. $m_{3/2} = O(F/M)$, where M is the reduced Planck scale, we expect $m_{3/2}$ in the keV range for a scale $\sqrt{F}$ of $O(10^6 \text{ GeV})$ that has been proposed in models with low-energy SUSY breaking in a visible sector.

In the following we briefly report some implications of SUSY models with a light gravitino (in the keV range) in relation with the dark matter (DM) problem. We anticipate that a gravitino of that mass behaves as a warm dark matter (WDM) particle [24, 25, 27], that is, a particle whose free-streaming scale involves a mass comparable to that of a galaxy, $\sim 10^{11-12} M_\odot$.

5.6.1 Thermal history of light gravitinos and WDM models

Suppose that the gravitinos were once in thermal equilibrium and were frozen out at the temperature T_{ψ_μ}d during the cosmic expansion. It can be shown that the density parameter Ω_{ψ_μ} contributed by relic thermal gravitinos is:

$$\Omega_{\psi_\mu} h_0^2 = 1.17 \left(\frac{m_{3/2}}{1 \text{ keV}} \right) \left(\frac{g_*(T_{\psi_\mu}\text{d})}{100} \right)^{-1}, \tag{5.43}$$

where $g_*(T_{\psi_\mu}\text{d})$ represents the effective massless degrees of freedom at the temperature T_{ψ_μ}d. Therefore, a gravitino in the previously mentioned keV range provides a significant portion of the mass density of the present universe.

As for the redshift Z_{NR} at which gravitinos become non-relativistic, it corresponds to the epoch at which their temperature becomes $m_{3/2}/3$, that is:

$$Z_{NR} \simeq \left(\frac{g_*(T_{\psi_\mu d})}{g_{*S}(T_0)} \right)^{1/3} \frac{m_{3/2}/3}{T_0}$$

$$= 4.14 \times 10^6 \times \left(\frac{g_*(T_{\psi_\mu d})}{100} \right)^{1/3} \left(\frac{m_{3/2}}{1 \text{ keV}} \right), \qquad (5.44)$$

where $T_0 = 2.726$ K is the temperature of the CMB at present time. Once Z_{NR} is known, one can estimate the free-streaming length until the epoch of the matter–radiation equality, λ_{FS}, which represents a quantity of crucial relevance for the formation of large-scale cosmic structures.

The free-streaming length for the thermal gravitinos is about 1 Mpc (for $Z_{NR} \sim 4 \times 10^6$) which, in turn, corresponds to $\sim 10^{12} M_\odot$, if it is required to provide a density parameter close to unity. This explicitly shows that light gravitinos are actually WDM candidates. We also note that, taking $h = 0.5$, the requirement of not overclosing the universe turns into $m_{3/2} \leq 200$ eV.

However, critical density models with pure WDM are known to suffer from serious troubles [41]. Indeed, a WDM scenario behaves much like CDM on scales above λ_{FS}. Therefore, we expect in the light gravitino scenario that the level of cosmological density fluctuations on the scale of galaxy clusters ($\sim 10 h_0^{-1}$ Mpc) to be almost the same as in CDM. As a consequence, the resulting number density of galaxy clusters is predicted to be much larger than what is observed [42].

This is potentially a critical test for any WDM-dominated scheme, the abundance of high-redshift galaxies having been already recognized as a non-trivial constraint for several DM models. It is, however, clear that quantitative conclusions on this point would at least require the explicit computation of the fluctuation power spectrum for the whole class of WDM scenarios.

5.7 Dark energy, ΛCDM and xCDM or QCDM

The expansion of the universe is described by two parameters, the Hubble constant H_0 and the deceleration parameter q_0:

(i) $H_0 \equiv \dot{R}(t_0)/R(t_0)$, where $R(t_0)$ is the scale factor, t_0 the age of the universe at present epoch, and we have

$$H_0 = 65 \pm 5 \text{ km s}^{-1} \text{ Mpc}^{-1} \qquad (h = 0.65 \pm 0.05); \qquad (5.45)$$

(ii) $q_0 \equiv -(\ddot{R}(t_0)/H_0^2)R(t_0)$ states whether the universe is accelerating or decelerating. q_0 is related to Ω_0 as follows

$$q_0 = \frac{\Omega_0}{2} + \frac{3}{2} \sum_i \Omega_i w_i \qquad (5.46)$$

where $\Omega_0 \equiv \sum_i \rho_i/\rho_{cr}$, Ω_i is the fraction of critical density due to the component i, $p_i = w_i \rho_i$ is the pressure of the component i, $\rho_{cr} = \frac{3H_0^2}{8\pi G} = 1.88h^2 \times 10^{-29}$ g cm^{-3}.

Measurements of q_0 from high-Z Type Ia SuperNovae (SNeIa) [44, 45] give strong indications in favour of an accelerating universe. CMB data [46] and cluster mass distribution [47] seem to favour models in which the energy density contributed by the negative pressure component should be roughly twice as much as the energy of the matter, thus leading to a flat universe ($\Omega_{tot} = 1$) with $\Omega_M \sim 0.4$ and $\Omega_\Lambda \sim 0.6$. Therefore the universe should be presently dominated by a smooth component with effective negative pressure; this is, in fact, the most general requirement in order to explain the observed accelerated expansion. The most straightforward candidate for that is, of course, a 'true' cosmological constant [48]. A plausible alternative that has recently received much attention is a dynamical vacuum energy given by a scalar field rolling down its potential: a cosmological scalar field, depending on its dynamics, can easily fulfil the condition of an equation of state $w_Q = p_Q/\rho_Q$ between -1 (which corresponds to the cosmological constant case) and 0 (that is the equation of state of matter). Since it is useful to have a short name for the rather long definition of this dynamical vacuum energy, we follow the literature in calling it briefly 'quintessence' [49].

5.7.1 ΛCDM models

At the moment models with $\Omega_\Lambda \sim 0.6$ seem to be favoured (see for example [28]). Ω_Λ is given by

$$\Omega_\Lambda \equiv \frac{8\pi G \Lambda}{3H_0^2} \tag{5.47}$$

where Λ is the cosmological constant, which appears in the most general form of the Einstein equation. The equation of state for Λ is $p = -\rho$ or, equivalently, $w = -1$. In order to have $\Omega_\Lambda \sim O(1)$, Λ has to be:

$$\Lambda \sim (2 \times 10^{-3} \text{ eV})^4. \tag{5.48}$$

Being a constant there is no reason in particle physics why this constant should be so small and not receive corrections at the highest mass scale present in the theory. This constitutes the most severe hierarchy problem in particle physics and there are no hints as to how to solve it.

If $\Lambda \neq 0$, in the early universe the density of energy and matter is dominant over the vacuum energy contribution, while the universe expands the average matter density decreases and at low redshifts the Λ term becomes important. At the end the universe starts inflating under the influence of the Λ term.

At present there are models based on the presence of Λ called ΛCDM models or ΛCHDM if we allow the presence of a small amount of HDM. Such models provide a good fit of the observed universe even if they need further study and more data confirmations.

5.7.2 Scalar field cosmology and quintessence

The role of the cosmological constant in accelerating the universe expansion could be played by any smooth component with a negative equation of state $p_Q/\rho_Q = w_Q - 0.6$ [49,52], as in the so-called 'quintessence' models (QCDM) [49], otherwise known as xCDM models [51].

A natural candidate for quintessence is given by a rolling scalar field Q with potential $V(Q)$ and equation of state:

$$w_Q = \frac{\dot{Q}^2/2 - V(Q)}{\dot{Q}^2/2 + V(Q)}, \qquad (5.49)$$

which—depending on the amount of kinetic energy—could, in principle, take any value from -1 to $+1$. Study of scalar field cosmologies has shown [53,54] that, for certain potentials, there exist attractor solutions that can be of the 'scaling' [55–57] or 'tracker' [58,59] type; this means that for a wide range of initial conditions the scalar field will rapidly join a well-defined late-time behaviour.

In the case of an exponential potential, $V(Q) \sim \exp(-Q)$, the solution $Q \sim \ln t$ is, under very general conditions, a 'scaling' attractor in a phase space characterized by $\rho_Q/\rho_B \sim$ constant [55–57]. This could potentially solve the so called 'cosmic coincidence' problem, providing a dynamical explanation for the order of magnitude equality between matter and scalar field energy today. Unfortunately, the equation of state for this attractor is $w_Q = w_B$, which cannot explain the acceleration of the universe neither during radiation domination ($w_{rad} = 1/3$) nor during matter domination ($w_m = 0$). Moreover, BBNS constrains the field energy density to values much smaller than the required $\sim 2/3$ [54,56,57].

If, instead, an inverse power-law potential is considered, $V(Q) = M^{4+\alpha} Q^{-\alpha}$, with $\alpha > 0$, the attractor solution is $Q \sim t^{1-n/m}$, where $n = 3(w_Q + 1)$, $m = 3(w_B + 1)$; and the equation of state turns out to be $w_Q = (w_B \alpha - 2)/(\alpha + 2)$, which is always negative during matter domination. The ratio of the energies is no longer constant but scales as $\rho_Q/\rho_B \sim a^{m-n}$ thus growing during the cosmological evolution, since $n < m$. ρ_Q could then have been safely small during nucleosynthesis and grown later into the phenomenologically interesting values. These solutions are then good candidates for quintessence and have been called 'tracker' solutions in the literature [54,58,59].

The inverse power-law potential does not improve the cosmic coincidence problem with respect to the cosmological constant case. Indeed, the scale M has to be fixed from the requirement that the scalar energy density today is exactly what is needed. This corresponds to choosing the desired tracker path. An important difference exists in this case though. The initial conditions for the physical variable ρ_Q can vary between the present critical energy density ρ_{cr}

and the background energy density ρ_B at the time of beginning (this range can span many tens of orders of magnitude, depending on the initial time), and will anyway end on the tracker path before the present epoch, due to the presence of an attractor in the phase space. In contrast, in the cosmological constant case, the physical variable ρ_Λ is fixed once and for all at the beginning. This allows us to state that in the quintessence case the fine-tuning issue, even if still far from being solved, is at least weakened.

Much effort has recently been devoted to finding ways to constrain such models with present and future cosmological data in order to distinguish quintessence from Λ models [60, 61]. An even more ambitious goal is the partial reconstruction of the scalar field potential from measuring the variation of the equation of state with increasing redshift [62].

Natural candidates for these scalar fields are pseudo-Goldstone bosons, axions, e.g. scalar fields with a scalar potential decreasing to zero for an infinite value of the fields. Such behaviour occurs naturally in models of dynamical SUSY breaking: in SUSY models scalar potentials have many flat directions, that is directions in the field's space where the potential vanishes. After dynamical SUSY breaking the degeneracy of the flat potential is lifted but it is restored for infinite values of the scalar fields.

However, the investigation of quintessence models from the particle physics point of view is just in a preliminary stage and a realistic model is not yet available (see, for example, [63–66]). There are two classes of problems: the construction of a field theory model with the required scalar potential and the interaction of the quintessence field with SM fields [67]. The former problem has already been considered by Binétruy [63], who pointed out that scalar inverse power law potentials appear in supersymmetric QCD theories (SQCD) [68] with N_c colours and $N_f < N_c$ flavours. The latter seems the toughest. Indeed the quintessence field today has typically a mass of order $Q_0 \sim 10^{-33}$ eV. Then, in general, it would mediate long range interactions of gravitational strength, which are phenomenologically unacceptable.

References

[1] Salam A 1967 *Elementary Particle Theory* ed N Svartholm (Stockholm: Almquist and Wiksells)
[2] Weinberg S 1967 *Phys. Rev. Lett.* **19** 1264
[3] Glashow S L 1961 *Nucl. Phys.* **22** 579
[4] Peskin M E and Schroeder D V 1995 *An Introduction to Quantum Field Theory* (Reading, MA: Addison-Wesley)
[5] For an introduction to the DM problem, see, for instance: Kolb R and Turner S 1990 *The Early Universe* (New York: Addison-Wesley)
    Srednicki M (ed) 1989 *Dark Matter* (Amsterdam: North-Holland)
    Primack J, Seckel D and Sadoulet B 1988 *Annu. Rev. Nucl. Part. Sci.* **38** 751
[6] For a recent review see: Primack J 2000 *Preprint* astro-ph/0007187

[7] Burles S *et al* 1999 *Phys. Rev. Lett.* **82** 4176

[8] Freese K, Fields B D and Graff D S 2000 *Proc. MPA/ESO Workshop on the First Stars (Garching, Germany, August 4–6, 1999)* (astro-ph/0002058)

[9] For a review see: Ellis J 2000 *Nucl. Phys. Proc. Suppl.* **91** 903

[10] See http://www.hep.anl.gov/NDK/Hypertext/nuindustry.html

[11] Kajita T 2000 *Now2000: Europhysics Neutrino Oscillation Workshop (Otranto, Italy, September 9–16)*

[12] Suzuki Y 2000 *Neutrino2000: XIX Int. Conf. on Neutrino Physics and Astrophysics (Sudbury, Canada, June 16–21, 2000)*

[13] Apollonio M *et al* (CHOOZ Collaboration) 1999 *Phys. Lett.* B **466** 415

[14] Mills G 2000 *Neutrino2000: XIX Int. Conf. on Neutrino Physics and Astrophysics (Sudbury, Canada, June 16–21, 2000)*

[15] Lobashev V M *et al* 1999 *Phys. Lett.* B **460** 227

[16] Weinheimer C *et al* 1999 *Phys. Lett.* B **460** 219

[17] See, for example, Bilenky S M *et al* 1999 *Phys. Lett.* B **465** 193

[18] Particle Data Book, Caso C *et al* 2000 *Eur. Phys. J.* C **15** 1

[19] Tanagida T 1979 *The Unified Theories and the Baryon Number in the Universe* ed O Sawada and A Sugamoto (Tsukuba: KEK)

[20] Gell-Mann M, Ramond P and Slansky R 1979 *Supergravity* ed P Van Nieuwenhuizen and D Z Freedman

[21] Pen U L, Seljak U and Turok N 1997 *Phys. Rev. Lett.* **79** 1611

[22] Albrecht A, Battye R A and Robinson J 1999 *Phys. Rev.* D **59** 023508

[23] Bond J R, Centrella J, Szalay A S and Wilson J R 1984 *Formation and Evolution of Galaxies and Large Structures in the Universe* ed J Audouze, J Tran Thanh Van (Dordrecht: Reidel) pp 87–99

[24] Pagels H and Primack J R 1982 *Phys. Rev. Lett.* **48** 223

[25] Blumenthal G R, Pagels H and Primack J R 1982 *Nature* **299** 37

[26] Blumenthal G R and Primack J R 1984 *Formation and Evolution of Galaxies and Large Structures in the Universe* ed J Audouze and J Tran Thanh Van (Dordrecht: Reidel) pp 163–83

[27] Bond J R, Szalay A S and Turner M S 1982 *Phys. Rev. Lett.* **48** 1636

[28] Blumenthal G R, Faber S M, Primack J R and Rees M J 1984 *Nature* **311** 517

[29] Ellis J *et al* 2001 *Phys. Lett.* B **502** 171

[30] Primack J R and Gross M A K 2000 *Current Aspect of Neutrino Physics* ed D O Caldwell (Berlin: Springer)

[31] Bernabei R *et al* (DAMA Collaboration) 2000 *Phys. Lett.* B **480** 23
Bottino A, Donato F, Fornengo N and Scopel S 2000 *Phys. Rev.* D **62** 056006

[32] Giudice G and Rattazzi R 1999 *Phys. Rep.* **322** 419

[33] Shafi Q and Stecker F W 1984 *Phys. Lett.* B **53** 1292
Bonometto S A and Valdarnini R 1985 *Astrophys. J.* **299** L71
Achilli S, Occhionero F and Scaramella R 1985 *Astrophys. J.* **299** 577
Holtzman J A 1981 *Astrophys. J. Suppl.* **71** 1
Taylor A N and Rowan-Robinson M 1992 *Nature* **359** 396
Holtzman J A and Primach J 1992 *Astrophys. J.* **396** 113
Pogosyan D Yu and Starobinoski A A 1993 *Preprint*
Klypin A, Holtzman J, Primach J and Regös E 1993 *Astrophys. J.* **415** 1

[34] For a review, see: Nilles H P 1984 *Phys. Rep.* C **110** 1 (1984);
Haber H and Kane G 1985 *Phys. Rep.* C **117** 1

[35] Cremmer E, Ferrara S, Girardello L and van Proeyen A 1982 *Phys. Lett.* B **116** 231
Cremmer E, Ferrara S, Girardello L and van Proeyen A 1983 *Nucl. Phys.* B **212** 413

[36] Jungman G, Kamionkowski M and Griest K 1996 *Phys. Rep.* **267** 195 and references therein
Bottino A and Fornengo N 1999 *Preprint* hep-ph/9904469

[37] Dine M, Nelson A, Nir Y and Shirman Y 1996 *Phys. Rev.* D **53** 2658

[38] Dine M and Nelson A E 1993 *Phys. Rev.* D **48** 1277
Dine M, Nelson A E and Shirman Y 1995 *Phys. Rev.* D **51** 1362

[39] Dvali G, Giudice G F and Pomarol A 1996 *Nucl. Phys.* B **478** 31

[40] Ambrosanio S, Kane G L, Kribs G D, Martin S P and Mrenna S 1996 *Phys. Rev. Lett.* **76** 3498

[41] Colombi S, Dodelson S and Widrow L M 1996 *Astrophys. J.* **458** 1
Pierpaoli E, Borgani S, Masiero A and Yamaguchi M 1998 *Phys. Rev.* D **57** 2089

[42] White S D M, Efstathiou G and Frenk C S 1993 *Mon. Not. R. Astron. Soc.* **262** 1023
Biviano A, Girardi M, Giuricin G, Mardirossian F and Mezzetti M 1993 *Astrophys. J.* **411** L13
Viana T P V and Liddle A R 1996 *Mon. Not. R. Astron. Soc.* **281** 323

[43] White M, Viana P T P, Liddle A R and Scott D 1996 *Mon. Not. R. Astron. Soc.* **283** 107

[44] Perlmutter S *et al* 1999 *Astrophys. J.* **517** 565
Perlmutter S *et al* 1997 *Bull. Am. Astron. Soc.* **29** 1351
Perlmutter S *et al* 1998 *Nature* **391** 51
See also http://www-super-nova.LBL.gov/

[45] Riess A G *et al* 1998 *Astron. J.* **116** 1009
Filippenko A V and Riess A G 1998 *Phys. Rep.* **307** 31
Filippenko A V and Riess A G 1999 *Preprint* astro-ph/9905049
Garnavich P M *et al* 1998 *Astrophys. J.* **501** 74
Leibundgut B, Contardo G, Woudt P and Spyromilio J 1998 *Dark '98* ed H Klapoloz-Kleingzothaus and L Baudis (Singapore: World Scientific)
See also http://cfa-www.harvard.edu/cfa/oir/Research/supernova/HighZ.html

[46] Bartlett J G, Blanchard A, Le Dour M, Douspis M and Barbosa D 1998 *Preprint* astro-ph/9804158
Efstathiou G 1999 *Preprint* astro-ph/9904356
Efstathiou G, Bridle S L, Lasenby A N, Hobson M P and Ellis R S 1999 *Mon. Not. R. Astron. Soc.* **303** L47
Lineweaver C 1998 *Astrophys. J.* **505** L69

[47] Carlberg R G, Yee H K C and Ellingson E 1997 *Astrophys. J.* **478** 462
Carlstrom J 1999 *Phys. Scr.* in press

[48] See, for example: Carroll S M, Press W H and Turner E L 1992 *Annu. Rev. Astron. Astrophys.* **30** 499

[49] Caldwell R R, Dave R and Steinhardt P J 1998 *Phys. Rev. Lett.* **80** 1582

[50] Kim H B and Kim J E 1995 *Nucl. Phys.* B **433** 421
Masiero A, Montanino D and Peloso M 2000 *Astropart. Phys.* **12** 351

[51] Turner M S and White M 1997 *Phys. Rev.* D **56** 4439
Chiba T, Sugiyama N and Nakamura T 1997 *Mon. Not. R. Astron. Soc.* **289** L5

[52] Frieman J A and Waga I 1998 *Phys. Rev.* D **57** 4642

[53] Peebles P J E and Ratra B 1988 *Astrophys. J.* **325** L17
Ratra B and Peebles P J E 1988 *Phys. Rev.* D **37** 3406

[54] Liddle A R and Scherrer R J 1999 *Phys. Rev.* D **59** 023509

[55] Wetterich C 1988 *Nucl. Phys.* B **302** 668

[56] Copeland E J, Liddle A R and Wands D 1998 *Phys. Rev.* D **57** 4686

[57] Ferreira P G and Joyce M 1997 *Phys. Rev. Lett.* **79** 4740
Ferreira P G and Joyce M 1998 *Phys. Rev.* D **58** 023503

[58] Zlatev I, Wang L and Steinhardt P J 1999 *Phys. Rev. Lett.* **82** 896

[59] Steinhardt P J, Wang L and Zlatev I 1999 *Phys. Rev.* D **59** 123504

[60] Baccigalupi C and Perrotta F 1999 *Phys. Rev.* D **59** 123508
Hu W, Eisenstein D J, Tegmark M and White M 1999 *Phys. Rev.* D **59** 023512
Cooray A R and Huterer D 1999 *Astrophys. J.* **513** L95
Wang L and Steinhardt P J 1998 *Astrophys. J.* **508** 483
Hui L 1999 *Astrophys. J.* **519** L9
Ratra B, Stompor R, Ganga K, Rocha G, Sugiyama N and Górski K M 1999
Astrophys. J. **517** 549
van de Bruck C and Priester W 1998 *Preprint* astro-ph/9810340
Alcaniz J S and Lima J A S 1999 *Astrophys. J.* **521** L87

[61] Wang L, Caldwell R R, Ostriker J P and Steinhardt P J 2000 *Astrophys. J.* **530** 17
Huey G, Wang L, Dave R, Caldwell R R and Steinhardt P J 1999 *Phys. Rev.* D **59**
063005
Perlmutter S, Turner M S and White M 1999 *Phys. Rev. Lett.* **83** 670
Chiba T, Sugiyama N and Nakamura T 1998 *Mon. Not. R. Astron. Soc.* **301** 72

[62] Huterer D and Turner M S 2000 *Phys. Rev.* D **60** 081301
Nakamura T and Chiba T 1999 *Mon. Not. R. Astron. Soc.* **306** 696
Chiba T and Nakamura T 1998 *Prog. Theor. Phys.* **100** 1077

[63] Binétruy P 1999 *Phys. Rev.* D **60** 063502
Masiero A, Pietroni M and Rosati F 2000 *Phys. Rev.* D **61** 023504

[64] Frieman J A, Hill C T, Stebbins A and Waga I 1995 *Phys. Rev. Lett.* **75** 2077
Choi K 2000 *Phys. Rev.* D **62** 043509
Kim J E 1999 *JHEP* **9905** 022

[65] Kolda C and Lyth D H 1999 *Phys. Lett.* B **458** 197

[66] Brax P and Martin J 1999 *Phys. Lett.* B **468** 40

[67] Carroll S M 1998 *Phys. Rev. Lett.* **81** 3067

[68] Taylor T R, Veneziano G and Yankielowicz S 1983 *Nucl. Phys.* B **218** 493
Affleck I, Dine M and Seiberg N 1983 *Phys. Rev. Lett.* **51** 1026
Affleck I, Dine M and Seiberg N 1984 *Nucl. Phys.* B **241** 493
For a pedagogical introduction, see also: Peskin M E 1997 *Preprint* hep-th/9702094,
TASI 96 lectures

Chapter 6

Supergravity and cosmology

Renata Kallosh
Department of Physics, Stanford University, Stanford, USA

6.1 M/string theory and supergravity

Supergravity is a low-energy limit of a fundamental M/string theory. At present there is no well-established M/string theory cosmology. However, there are some urgent issues in cosmology which require a knowledge of the fundamental theory. Those issues are related to expanding universe, dark matter, inflation, creation of particles after inflation, etc. The basic problem is that general relativity which is required for explanation of the cosmology and an expanding universe is not yet combined with any relativistic quantum theory and particle physics to the extent in which a full description of the early universe would be possible. Superstring theory offers a consistent theory of quantum gravity at least at the level of the string theory perturbation theory in ten-dimensional target space. The non-perturbative string theory which includes the D-branes is much less understood, since these objects are charged under so-called Ramond–Ramond charges which can be incorporated only at the non-perturbative level. The main attempts during the last few years have been focused on understanding the M-theory, which represents a string theory at strong coupling, when an additional dimension is decompactified. M-theory has as a low-energy limit the 11-dimensional supergravity and has two types of extended objects: two-branes and five-branes.

The radical aspect of major attempts to construct quantum gravity is the concept that the spacetime $x^\mu = \{t, \boldsymbol{x}\}$ is not fundamental. The coordinates x^μ are not labels but fields which are defined by the dynamics of the the world-volume of a p-brane so that they depend on world-volume coordinates, $x^\mu(\sigma^0, \sigma^1, \ldots, \sigma^p)$. A two-dimensional object, a string is an one-brane with $x^\mu(\sigma^0, \sigma^1)$, a two-brane is a three-dimensional object with $x^\mu(\sigma^0, \sigma^1, \sigma^2)$, a four-dimensional object called a three-brane and has $x^\mu(\sigma^0, \sigma^1, \sigma^2, \sigma^3)$, etc. M-theory/string theory includes a theory of branes of various dimensions. The fields $x^\mu(\sigma)$ have their own dynamics. The zero modes of the excitations of such extended objects are coordinates of spacetime, $x^\mu(\sigma) = x^\mu_{\text{constant}} + \cdots$. Thus the concept of spacetime is an approximation to a full quantum theory of gravity!

Supergravity (gravity + supersymmetry) may be viewed as an approximate effective description of a fundamental theory when the dependence on coordinates of the world-volume is ignored. The smallest theory of supergravity includes two types of fields, the graviton and the gravitino. Supergravity interacting with matter multiplets includes also scalars, spinors and vectors. All these fields are functions of the usual spacetime coordinates t, x in a four-dimensional spacetime. The fundamental M-theory, which should encompass both supergravity and string theory, at present experiences rapid changes. Over the last few years M-theory and string theory focused its main attention on the superconformal theories and adS/CFT (anti-de Sitter/conformal field theory) correspondence [1]. It has been discovered that IIB string theory on $adS_5 \times S^5$ is related to $SU(2, 2|4)$ superconformal symmetry. In particular, one finds the $SU(2, 2|1)$ superconformal algebra from the anti-de Sitter compactification of the string theory with one-quarter of the unbroken supersymmetry. These recent developments in M-theory and non-perturbative string theory suggest that we should take a *fresh look at the superconformal formulation underlying the supergravity*.

The 'phenomenological supergravity' based on the most general $N = 1$ supergravity [2] has an underlying superconformal structure. This has been known for a long time but only recently the complete most general $N = 1$ gauge theory superconformally coupled to supergravity was introduced [4]. The theory has local $SU(2, 2|1)$ symmetry and no dimensional parameters. The phase of this theory with spontaneously broken conformal symmetry gives various formulations of $N = 1$ supergravity interacting with matter, depending on the choice of the R-symmetry fixing.

The relevance of supergravity to cosmology is that it gives a framework of an effective field theory in the background of the expanding universe and time-dependent scalar fields. Let us remind here that the early universe is described by an FRW metric which can be written in a form which is conformal to a flat metric:

$$\mathrm{d}s^2 = a^2(\eta)[-\mathrm{d}\eta^2 + \gamma_{ij}\,\mathrm{d}x^i\,\mathrm{d}x^j]. \tag{6.1}$$

This fact leads to an interest in the superconformal properties of supergravity.

6.2   Superconformal symmetry, supergravity and cosmology

The most general four-dimensional $N = 1$ supergravity [2] describes a supersymmetric theory of gravity interacting with scalars, spinors and vectors of a supersymmetric gauge theory. It is completely defined by the choice of the three functions: the superpotential $W[\phi]$ and the vector coupling $f_{ab}[\phi]$ which are holomorphic functions of the scalar fields (depend on ϕ^i and do not depend on ϕ_i^*) and the Kähler potential $K[\phi, \phi*]$. These functions from the perspective of supergravity are arbitrary. One may hope that they will be defined eventually from the fundamental M/string theory.

The potential V of the scalar fields is given by

$$M_P^{-2} e^K [-3WW^* + (\mathcal{D}^i W) g^{-1}{}_i{}^j (\mathcal{D}_j W^*)] + \tfrac{1}{2} (\text{Re}(f)_{\alpha\beta}) D^\alpha D^\beta, \qquad (6.2)$$

here D^α are the D-components of the vector superfields, which may take some non-vanishing values. The metric of the Kähler space, $g_i{}_j$ which depends on $\phi, \phi*$, is the metric of the moduli space which defines the kinetic term for the scalar fields:

$$g_i{}^j \partial_\mu \phi^i \partial^\mu \phi_j^*. \qquad (6.3)$$

The properties of the Kähler space in M/string theory are related to the Calabi–Yau spaces on which the theory is compactified to four dimensions.

One of the problems related to the gravitino is the issue of the conformal invariance of the gravitino and the possibility of non-thermal gravitino production in the early universe.

Many observable properties of the universe are, to a large extent, determined by the underlying conformal properties of the fields. One may consider inflaton scalar field(s) ϕ which drive inflation, inflaton fluctuations which generate cosmological metric fluctuations, gravitational waves generated during inflation, photons in the cosmic microwave background (CMB) radiation which (almost) freely propagate from the last scattering surface, etc. If the conformal properties of any of these fields were different, the universe would also look quite different. For example, the theory of the usual massless electromagnetic field is conformally invariant. This implies, in particular, that the strength of the magnetic field in the universe decreases as $a^{-2}(\eta)$. As a result, all vector fields become exponentially small after inflation. Meanwhile the theory of the inflaton field(s) should not be conformally invariant, because otherwise these fields would rapidly disappear and inflation would never happen.

Superconformal supergravity is particularly suitable to study the conformal properties of various fields, because in this framework all fields initially are conformally covariant; this invariance becomes spontaneously broken only when one uses a particular gauge which requires that some combination of scalar fields becomes equal to M_P^2.

The issue of conformal invariance of the gravitino remained rather obscure for a long time. One could argue that a massless gravitino should be conformally invariant. Once we introduce a scalar field driving inflation, the gravitino acquires a mass $m_{3/2} = e^{K/2} |W|/M_P^2$. Thus, one could expect that the conformal invariance of gravitino equations should be broken only by the small gravitino mass $m_{3/2}$, which is suppressed by the small gravitational coupling constant M_P^{-2}. This is indeed the case for the gravitino component with helicity $\pm 3/2$. However, breaking of conformal invariance for the gravitino component with helicity $\pm 1/2$, which appears due to the super-Higgs effect, is much stronger.

In the first approximation in the weak gravitational coupling, it is related to the chiral fermion mass scale [3].

This locally superconformal theory is useful for describing the physics of the early universe with a conformally flat FRW metric.

Superconformal theory underlying supergravity has no dimensional parameters and one extra chiral superfield, the conformon. This superfield can be gauged away using local conformal symmetry and S-supersymmetry. The mechanism can be explained using a simple example: an arbitrary gauge theory with Yang–Mills fields W_μ coupled to fermions λ and gravity:

$$S^{\text{conf}} = \int d^4x \sqrt{g}(\tfrac{1}{2}(\partial_\mu\phi)(\partial_\nu\phi)g^{\mu\nu} - \tfrac{1}{12}\phi^2 R$$
$$- \tfrac{1}{4}\operatorname{Tr} F_{\mu\nu}g^{\mu\rho}g^{\nu\sigma}F_{\rho\sigma} - \tfrac{1}{2}\bar\lambda\gamma^\mu D_\mu\lambda). \tag{6.4}$$

The field ϕ is a conformon. The last two terms in the action represent super-Yang–Mills theory coupled to gravity. The action is conformal invariant under the following local transformations:

$$g'_{\mu\nu} = e^{-2\sigma(x)}g_{\mu\nu}, \qquad \phi' = e^{\sigma(x)}\phi, \qquad W'_\mu = W_\mu, \qquad \lambda' = e^{\frac{3}{2}\sigma(x)}\lambda. \tag{6.5}$$

The gauge symmetry (6.5) with one local gauge parameter can be gauge fixed. If we choose the $\phi = \sqrt{6}M_P$ gauge, the ϕ-terms in (6.4) reduce to the Einstein action, which is no longer conformally invariant:

$$S^{\text{conf}}_{\text{g.f.}} \sim \int d^4x \sqrt{g}(-\tfrac{1}{2}M_P^2 R - \tfrac{1}{4}F_{\mu\nu}g^{\mu\rho}g^{\nu\sigma}F_{\rho\sigma} + \tfrac{1}{2}\bar\lambda\gamma^\mu D_\mu\lambda). \tag{6.6}$$

Here $M_P \equiv M_{\text{Planck}}/\sqrt{8\pi} \sim 2 \times 10^{18}$ GeV. In this action, the transformation (6.5) no longer leaves the Einstein action invariant. The R-term transforms with derivatives of $\sigma(x)$, which in the action (6.4) were compensated by the kinetic term of the compensator field. However, the actions of the Yang–Mills sector of the theory, i.e. spin-$\tfrac{1}{2}$ and spin-1 fields interacting with gravity, remain conformally invariant. Only the conformal properties of the gravitons are affected by the removal of the compensator field. A supersymmetric version of this mechanism requires adding a few more symmetries, so that the $SU(2, 2|1)$ symmetric theory is constructed. The non-conformal properties of the gravitino can be followed from this starting point, as shown in [4].

Few applications of superconformal theory to cosmology include the study of (i) particle production after inflation, in particular the study of the non-conformal helicity $\pm 1/2$ states of gravitino; (ii) the super-Higgs effect in cosmology and the derivation of the equations for the gravitino interacting with any number of chiral and vector multiplets in the gravitational background with varying scalar fields; and (iii) the weak coupling limit of supergravity $M_P \to \infty$ and gravitino–goldstino equivalence. This explains why gravitino production in the early universe in general is not suppressed in the limit of weak gravitational coupling.

6.3 Gravitino production after inflation

During the last couple of years there has been a growing interest in understanding gravitino production in the early universe [3, 14]. The general consensus is that gravitinos can be produced during pre-heating after inflation due to a combined effect of interactions with an oscillating inflaton field and because the helicity $\pm 1/2$ gravitino have equations of motion which break conformal invariance. In general the probability of gravitino production is *not* suppressed by the small gravitational coupling. This may lead to a copious production of gravitinos after inflation. The efficiency of the new non-thermal mechanism of gravitino production is very sensitive to the choice of the underlying theory. This may put strong constraints on certain classes of inflationary models.

A formal reason why the effect may be strong even at $M_P \rightarrow \infty$ is the following: in Minkowski space the constraint which the massive gravitino satisfies has the form

$$\gamma^\mu \psi_\mu = 0. \tag{6.7}$$

In an expanding universe, the analogue of equation (6.7) looks as follows:

$$\gamma^0 \psi_0 - \hat{A} \gamma^i \psi_i = 0 \tag{6.8}$$

where, in the limit $M_P \rightarrow \infty$,

$$\hat{A} = \frac{p}{\rho} + \gamma_0 \frac{2\dot{W}}{\rho}, \qquad |\hat{A}|^2 = 1. \tag{6.9}$$

Matrix $\hat{A}$ rotates twice during each oscillation of the field ϕ. The non-adiabaticity of the gravitino field ψ_0 (related to helicity $\pm 1/2$ is determined not by the mass of the gravitino but by the mass of the chiral fermion $\mu = W_{\phi\phi}$. This equation was obtained in the framework of a simple model of the supergravity theory interacting with one chiral multiplet. The gauge-fixing of the spontaneously broken supersymmetry was relatively easy, the only one available in the model chiral fermion, a goldstino field, was chosen to vanish and the massive gravitino was described by helicity $\pm 3/2$ as well as helicity $\pm 1/2$ states.

A physical reason for gravitino production is a gravitino–goldstino equivalence theorem which, however, had to be properly understood in the cosmological context.

One of the major problems with studies of gravitino production after inflation was to consider the theories with few chiral multiplets. It become clear that one cannot simply apply the well-known super-Higgs mechanism of supergravity in the flat background to the situation in which we have a curved metric of the early universe.

6.4   Super-Higgs effect in cosmology

We would like to choose a gauge in which a goldstino equals zero. The question is which field is this goldstino: we start with the gravitino ψ_μ and some number of left- and right-handed chiral fermions χ^i, χ_i. In the past, this has been sought for constant backgrounds [2], but in cosmological applications the scalar fields are time-dependent in the background. Therefore we need a modification.

In the action there are a few terms where gravitinos mix with the other fermions, and these as well as the supersymmetry transformations should give us the possibility of finding the correct goldstino in the cosmological time-dependent background. We want to obtain a combination whose variation is always non-zero for spontaneously broken supersymmetry. This leads to the following definition of a goldstino:

$$\upsilon = \xi^{\dagger i}\chi_i + \xi_i^\dagger\chi^i + \tfrac{1}{2}\mathrm{i}\gamma_5 D_\alpha\lambda^\alpha, \tag{6.10}$$

where the λ^α are gauginos, the D_α are auxiliary fields from the vector multiplets and

$$\xi^{\dagger i} \equiv \mathrm{e}^{K/2}D^i W - \gamma_0 g_j{}^i\dot\phi^j, \qquad \xi_i^\dagger \equiv \mathrm{e}^{K/2}D_i W - \gamma_0 g_i{}^j\dot\phi_j. \tag{6.11}$$

The goldstino defined here differs from the one in the flat background by the presence of the time-dependent derivatives of the scalar fields.

Goldstino is non-vanishing in the vacuum supersymmetry transformation:

$$\delta\upsilon = -\tfrac{3}{2}(H^2 + m_{3/2}^2)\epsilon. \tag{6.12}$$

Here H is the Hubble 'constant':

$$\left(\frac{\dot a}{a}\right)^2 = H^2 = \frac{\rho}{3M_\mathrm{P}^2}. \tag{6.13}$$

This has important implications.   First of all, it shows that, in a conformally flat universe (6.1), the parameter α is strictly positive. To avoid misunderstandings, we should note that, in general, one may consider situations in which the energy density ρ is negative. The famous example is anti-de Sitter space with a negative cosmological constant. However, in the context of inflationary cosmology, the *energy density never can turn negative*, so anti-de Sitter space cannot appear. The reason is that inflation makes the universe almost exactly flat. As a result, the term k/a^2 drops out from the Einstein equation for the scale factor independently of whether the universe is closed, open or flat. Then gradually the energy density decreases, but it can never become negative even if a negative cosmological constant is present, as in anti-de Sitter space. Indeed, the equation

$$\left(\frac{\dot a}{a}\right)^2 = \frac{\rho}{3M_\mathrm{P}^2}$$

implies that as soon as the energy density becomes zero, expansion stops. Then the universe recollapses, and the energy density becomes positive again. This implies that supersymmetry is *always broken*. The symmetry breaking is associated, to an equal extent, with the expansion of the universe and with the non-vanishing gravitino mass (the term $(H^2 + m_{3/2}^2)$). This is an interesting result because usually supersymmetry breaking is associated with the existence of the gravitino mass. Here we see that, in an expanding universe, the Hubble parameter H plays an equally important role.

The progress achieved in understanding the super-Higgs effect in an expanding universe has allowed us to find the equations for the gravitino in the most general theory of supergravity interacting with chiral and vector multiplets [4]. Analysis of these equations in various inflationary models and the estimates of the scale of gravitino production remains to be done.

Consider, for example, the hybrid inflation model. In this model all coupling constants are of order 10^{-1}, so there should be no suppression of the production of chiral fermions as compared to the other particles. One can expect, therefore, that

$$\frac{n_{3/2}}{s} \sim 10^{-1}-10^{-2}. \tag{6.14}$$

This would violate the cosmological bound by 13 orders of magnitude! However, one should check whether these gravitinos will survive until the end or turn into the usual fermions.

Thus supergravity theory and its underlying superconformal structures provide the framework for studies of the production of particles in supersymmetric theories in the early universe.

6.5 $M_P \to \infty$ limit

The complete equations of motion for the gravitino in a cosmological background were derived in [4] with an account of the gravitational effects. However, in [11] some part of these equations, corresponding to the vanishing Hubble constant and vanishing gravitino mass, was derived in the framework of a gauge theory, i.e. from rigid supersymmetric theory without gravity. To find the relation between these two equations one has to understand how to take the limit $M_P \to \infty$ in supergravity. This is a very subtle issue, if one starts with the fields of phenomenological supergravity. One has to do various rescaling of the fields with different powers of the M_P to be able to compare these two sets of equations. Surprisingly, the full set of rescalings reproduces exactly the fields of the underlying superconformal theory. These are the fields which survive in the weak coupling limit of supergravity.

Thus at present there are indications that a description of the cosmology of the early universe may be achieved in the framework of superconformal theory only after the gauge-fixing of conformal symmetry is equivalent to supergravity. The super-Higgs mechanism in cosmology and the goldstino–gravitino equivalence theorem have a clear origin in this $SU(2, 2|1)$ symmetric theory of gravity.

References

[1] Maldacena J 1998 The large N limit of superconformal field theories and supergravity *Adv. Theor. Math. Phys.* **2** 231 (hep-th/9711200)
[2] Cremmer E, Ferrara S, Girardello L and Van Proeyen A 1983 *Nucl. Phys.* B **212** 413
[3] Kallosh R, Kofman L, Linde A and Van Proeyen A 2000 Gravitino production after inflation *Phys. Rev.* D **61** 103503 (hep-th/9907124)
[4] Kallosh R, Kofman L, Linde A and Van Proeyen A 2000 Superconformal symmetry, supergravity and cosmology *Class. Quantum Grav.* **17** 4269
[5] Moroi T 1995 Effects of the gravitino on the inflationary universe *PhD Thesis* Tohoku, Japan (hep-ph/9503210)
[6] Maroto A L and Mazumdar A 2000 Production of spin 3/2 particles from vacuum fluctuations *Phys. Rev. Lett.* **84** 1655 (hep-ph/9904206)
[7] Lemoine M 1999 Gravitational production of gravitinos *Phys. Rev.* D **60** 103522 (hep-ph/9908333)
[8] Giudice G F, Tkachev I and Riotto A 1999 Non-thermal production of dangerous relics in the early universe *JHEP* **9908** 009 (hep-ph/9907510)
[9] Lyth D H 1999 Abundance of moduli, modulini and gravitinos produced by the vacuum fluctuation *Phys. Lett.* B **469** 69 (hep-ph/9909387)
[10] Lyth D H 2000 The gravitino abundance in supersymmetric 'new' inflation models *Phys. Lett.* B **488** 417
[11] Giudice G F, Riotto A and Tkachev I 1999 Thermal and non-thermal production of gravitinos in the early universe *JHEP* **9911** 036 (hep-ph/9911302)
[12] Maroto A L and Pelaez J R 2000 The equivalence theorem and the production of gravitinos after inflation *Phys. Rev.* D **62** 023518
[13] Lyth D H 2000 Late-time creation of gravitinos from the vacuum *Phys. Lett.* B **476** 356 (hep-ph/9912313)
[14] Bastero-Gil M and Mazumdar A 2000 Gravitino production in hybrid inflationary models *Phys. Rev.* D **62** 083510

Chapter 7

The cosmic microwave background

Arthur Kosowsky
Rutgers University, Piscataway, New Jersey, USA

It is widely accepted that the field of cosmology is entering an era dubbed 'precision cosmology'. Data directly relevant to the properties and evolution of the universe are flooding in by the terabyte (or soon will be). Such vast quantities of data were the purview only of high-energy physics just a few years ago; now expertise from this area is being coopted by some astronomers to help deal with our wealth of information. In the past decade, cosmology has gone from a data-starved science in which often highly speculative theories went unconstrained to a data-driven pursuit where many models have been ruled out and the remaining 'standard cosmology' will be tested with stringent precision.

The cosmic microwave background (CMB) radiation is at the centre of this revolution. The radiation present today as a 2.7 K thermal background originated when the universe was denser by a factor of 10^9 and younger by a factor of around 5×10^4. The radiation provides the most distant direct image of the universe we can hope to see, at least until gravitational radiation becomes a useful astronomical data source. The microwave background radiation is extremely uniform, varying in temperature by only a few parts in 10^5 over the sky (apart from an overall dipole variation arising from our peculiar motion through the microwave background's rest frame); its departure from a perfect blackbody spectrum has yet to be detected.

The very existence of the microwave background provides crucial support for the hot big bang cosmological model: the universe began in a very hot, dense state from which it expanded and cooled. The microwave background visible today was once in thermal equilibrium with the primordial plasma of the universe, and the universe at that time was highly uniform. Crucially, the universe could not have been perfectly uniform at that time or no structures would have formed subsequently. The study of small temperature and polarization fluctuations in the microwave background, reflecting small variations in density and velocity

219

in the early universe, have the potential to provide the most precise constraints on the overall properties of the universe of any data source. The reasons are that (1) the universe was very simple at the time imaged by the microwave background and is extremely well described by linear perturbation theory around a completely homogeneous and isotropic cosmological spacetime; and (2) the physical processes relevant at that time are all simple and very well understood. The microwave background is essentially unique among astrophysical systems in these regards.

The goal behind this chapter is to provide a qualitative description of the physics of the microwave background, an appreciation for the microwave background's cosmological importance, and an understanding of what kinds of constraints may be placed on cosmological models. It is not intended to be a definitive technical reference to the microwave background. Unfortunately, such a reference does not really exist at this time, but I have attempted to provide pedagogically useful references to other literature. I have also not attempted to give a complete bibliography; please do not consider this article to give definitive references to any topics mentioned. A recent review of the microwave background with a focus on potential particle physics constraints is Kamionkowski and Kosowsky (1999). A more general review of the microwave background and large-scale structure with references to many early microwave background articles is White *et al* (1994).

7.1   A brief historical perspective

The story of the serendipidous discovery of the microwave background in 1965 is widely known, so I will only briefly summarize it here. A recent book by the historian of science Helge Kragh (1996) is a careful and authoritative reference on the history of cosmology, from which much of the information in this section was obtained. Arno Penzias and Robert Wilson, two radio astronomers at Bell Laboratories in Crawford, New Jersey, were using a sensitive microwave horn radiometer originally intended for talking to the early Telstar telecommunications satellites. When Bell Laboratories decided to get out of the communications satellite business in 1963, Penzias and Wilson began to use the radiometer to measure radio emission from the Cassiopeia A supernova remnant. They detected a uniform noise source, which was assumed to come from the apparatus. But after many months of checking the antenna and the electronics (including removal of a bird's nest from the horn), they gradually concluded that the signal might actually be coming from the sky. When they heard about a talk given by P J E Peebles of Princeton predicting a 10 K blackbody cosmological background, they got in touch with the group at Princeton and realized that they had detected the cosmological radiation. At the time, Peebles was collaborating with Dicke, Roll and Wilkinson in a concerted effort to detect the microwave background. The Princeton group wound up confirming the Bell Laboratories discovery a few months later. Penzias and Wilson published their result in a brief paper with the unassuming title of 'A Measurement of Excess Antenna Temperature at $\lambda = 7.3$ cm' (Penzias and Wilson 1965); a companion paper by the Princeton group explained the cosmological significance of the measurement (Dicke *et al* 1965). The microwave background detection was a stunning success of the hot big bang model, which to that point had been well outside the mainstream of theoretical physics. The following years saw an explosion of work related to the big bang model of the expanding universe. To the best of my knowledge, the Penzias and Wilson paper was the second-shortest ever to garner a Nobel Prize, awarded in 1978. (Watson and Crick's renowned double helix paper wins by a few lines.)

Less well known is the history of earlier probable detections of the microwave background which were not recognized as such. Tolman's classic monograph on thermodynamics in an expanding universe was written in 1934, but a blackbody relic of the early universe was not predicted theoretically until 1948 by Alpher and Herman, a by-product of their pioneering work on nucleosynthesis in the early universe. Prior to this, Andrew McKellar (1940) had observed the population of excited rotational states of CN molecules in interstellar absorption lines, concluding that it was consistent with being in thermal equilibrium with a temperature of around 2.3 K. Walter Adams also made similar measurements (1941). Its significance was unappreciated and the result essentially forgotten, possibly because the Second World War had begun to divert much of the world's physics talent towards military problems.

Alpher and Herman's prediction of a 5 K background contained no suggestion of its detectability with available technology and had little impact. Over the next decade, George Gamow and collaborators, including Alpher and Herman, made a variety of estimates of the background temperature which fluctuated between 3 and 50 K (e.g. Gamow 1956). This lack of a definitive temperature might have contributed to an impression that the prediction was less certain than it actually was, because it aroused little interest among experimenters even though microwave technology had been highly developed through radar work during the war. At the same time, the incipient field of radio astronomy was getting started. In 1955, Emile Le Roux undertook an all-sky survey at a wavelength of $\lambda = 33$ cm, finding an isotropic emission corresponding to a blackbody temperature of $T = 3 \pm 2$ K (Denisse *et al* 1957). This was almost certainly a detection of the microwave background, but its significance was unrealized. Two years later, T A Shmaonov observed a signal at $\lambda = 3.2$ cm corresponding to a blackbody temperature of 4 ± 3 K independent of direction (see Sharov and Novikov 1993, p 148). The significance of this measurement was not realized, amazingly, until 1983! (Kragh 1996). Finally in the early 1960s the pieces began to fall into place: Doroshkevich and Novikov (1964) emphasized the detectability of a microwave blackbody as a basic test of Gamow's hot big bang model. Simultaneously, Dicke and collaborators began searching for the radiation, prompted by Dicke's investigations of the physical consequences of the Brans–Dicke theory of gravitation. They were soon scooped by Penzias and Wilson's discovery.

As soon as the microwave background was discovered, theorists quickly realized that fluctuations in its temperature would have fundamental significance as a reflection of the initial perturbations which grew into galaxies and clusters. Initial estimates of the amplitude of temperature fluctuations were a part in a hundred; this level of sensitivity was attained by experimenters after a few years with no observed fluctuations. Thus began a quarter-century chase after temperature anisotropies in which the theorists continually revised their estimates of the fluctuation amplitude downwards, staying one step ahead of the experimenters' increasingly stringent upper limits. Once the temperature fluctuations were shown to be less than a part in a thousand, baryonic density fluctuations did not have time to evolve freely into the nonlinear structures visible today, so theorists invoked a gravitationally dominant DM component (structure formation remains one of the strongest arguments in favour of non-baryonic DM). By the end of the 1980s, limits on temperature fluctuations were well below a part in 10^4 and theorists scrambled to reconcile standard cosmology with this small level of primordial fluctuations. Ideas like late-time phase transitions at redshifts less than $z = 1000$ were taken seriously as a possible way to evade the microwave background limits (see, e.g., Jaffe *et al* 1990). Finally, the COBE satellite detected fluctuations at the level of a few parts in 10^5 (Smoot *et al* 1990), just consistent with structure formation in inflation-motivated Cold Dark Matter cosmological models. The COBE results were soon confirmed by numerous ground-based and balloon measurements, sparking the intense theoretical and experimental interest in the microwave background over the past decade.

7.2   Physics of temperature fluctuations

The minute temperature fluctuations present in the microwave background contain a wealth of information about the fundamental properties of the universe. In order to understand the reasons for this and the kinds of information available, an appreciation of the underlying physical processes generating temperature and polarization fluctuations is required. This section and the following one give a general description of all basic physics processes involved in producing microwave background fluctuations.

First, one practical matter. Throughout this chapter, common cosmological units will be employed in which $\hbar = c = k_b = 1$. All dimensionful quantities can then be expressed as powers of an energy scale, commonly taken as GeV. In particular, length and time both have units of $[\text{GeV}]^{-1}$, while Newton's constant G has units of $[\text{GeV}]^{-2}$ since it is defined as equal to the square of the inverse Planck mass. These units are very convenient for cosmology, because many problems deal with widely varying scales simultaneously. For example, any computation of relic particle abundances (e.g. primordial nucleosynthesis) involves both a quantum mechanical scale (the interaction cross section) and a cosmological scale (the time scale for the expansion of the universe). Conversion between these cosmological units and physical (cgs) units can be achieved by inserting needed factors of $\hbar$, c, and k_b. The standard textbook by Kolb and Turner (1990) contains an extremely useful appendix on units.

7.2.1   Causes of temperature fluctuations

Blackbody radiation in a perfectly homogeneous and isotropic universe, which is always adopted as a zeroth-order approximation, must be at a uniform temperature, by assumption. When perturbations are introduced, three elementary physical processes can produce a shift in the apparent blackbody temperature of the radiation emitted from a particular point in space. All temperature fluctuations in the microwave background are due to one of the following three effects.

The first is simply a change in the intrinsic temperature of the radiation at a given point in space. This will occur if the radiation density increases via adiabatic compression, just as with the behaviour of an ideal gas. The fractional temperature perturbation in the radiation just equals the fractional density perturbation.

The second is equally simple: a Doppler shift if the radiation at a particular point is moving with respect to the observer. Any density perturbations within the horizon scale will necessarily be accompanied by velocity perturbations. The induced temperature perturbation in the radiation equals the peculiar velocity (in units of c, of course), with motion towards the observer corresponding to a positive temperature perturbation.

The third is a bit more subtle: a difference in gravitational potential between a particular point in space and an observer will result in a temperature shift of the radiation propagating between the point and the observer due to gravitational redshifting. This is known as the Sachs–Wolfe effect, after the original paper describing it (Sachs and Wolfe, 1967). This paper contains a completely straightforward general relativistic calculation of the effect, but the details are lengthy and complicated. A far simpler and more intuitive derivation has been given by Hu and White (1997) making use of gauge transformations. The Sachs–Wolfe effect is often broken into two parts, the usual effect and the so-called Integrated Sachs–Wolfe effect. The latter arises when gravitational potentials are evolving with time: radiation propagates into a potential well, gaining energy and blueshifting in the process. As it climbs out, it loses energy and redshifts, but if the depth of the potential well has increased during the time the radiation propagates through it, the redshift on exiting will be larger than the blueshift on entering, and the radiation will gain a net redshift, appearing cooler than it started out. Gravitational potentials remain constant in time in a matter–dominated universe, so to the extent the universe is matter dominated during the time the microwave background radiation freely propagates, the Integrated Sachs–Wolfe effect is zero. In models with significantly less than critical density in matter (i.e.

the currently popular ΛCDM models), the redshift of matter–radiation equality occurs late enough that the gravitational potentials are still evolving significantly when the microwave background radiation decouples, leading to a non-negligible Integrated Sachs–Wolfe effect. The same situation also occurs at late times in these models; gravitational potentials begin to evolve again as the universe makes a transition from matter domination to either vacuum energy domination or a significantly curved background spatial metric, giving an additional Integrated Sachs–Wolfe contribution.

7.2.2   A formal description

The early universe at the epoch when the microwave background radiation begins propagating freely, around a redshift of $z = 1100$, is a conceptually simple place. Its constituents are 'baryons' (including protons, helium nuclei and electrons, even though electrons are not baryons), neutrinos, photons and DM particles. The neutrinos and DM can be treated as interacting only gravitationally since their weak interaction cross sections are too small at this energy scale to be dynamically or thermodynamically relevant. The photons and baryons interact electromagnetically, primarily via Compton scattering of the radiation from the electrons. The typical interaction energies are low enough for the scattering to be well approximated by the simple Thomson cross section. All other scattering processes (e.g. Thomson scattering from protons, Rayleigh scattering of radiation from neutral hydrogen) have small enough cross-sections to be insignificant, so we have four species of matter with only one relevant (and simple) interaction process among them. The universe is also very close to being homogeneous and isotropic, with small perturbations in density and velocity on the order of a part in 10^5. The tiny size of the perturbations guarantees that linear perturbation theory around a homogeneous and isotropic background universe will be an excellent approximation.

Conceptually, the formal description of the universe at this epoch is quite simple. The unperturbed background cosmology is described by the Friedmann–Robertson–Walker (FRW) metric, and the evolution of the cosmological scale factor $a(t)$ in this metric is given by the Friedmann equation (see the lectures by Peacock in this volume). The evolution of the free electron density n_e is determined by the detailed atomic physics describing the recombination of neutral hydrogen and helium; see Seager *et al* (2000) for a detailed discussion. At a temperature of around 0.5 eV, the electrons combine with the protons and helium nuclei to make neutral atoms. As a result, the photons cease Thomson scattering and propagate freely to us. The microwave background is essentially an image of the 'surface of last scattering'. Recombination must be calculated quite precisely because the temperature and thickness of this surface depend sensitively on the ionization history through the recombination process.

The evolution of first-order perturbations in the various energy density components and the metric are described with the following sets of equations:

- The photons and neutrinos are described by distribution functions $f(x, p, t)$. A fundamental simplifying assumption is that the energy dependence of both is given by the blackbody distribution. The space dependence is generally Fourier transformed, so the distribution functions can be written as $\Theta(k, \hat{n}, t)$, where the function has been normalized to the temperature of the blackbody distribution and $\hat{n}$ represents the direction in which the radiation propagates. The time evolution of each is given by the Boltzmann equation. For neutrinos, collisions are unimportant so the Boltzmann collision term on the right hand side is zero; for photons, Thomson scattering off electrons must be included.

- The DM and baryons are, in principle, described by Boltzmann equations as well, but a fluid description incorporating only the lowest two velocity moments of the distribution functions is adequate. Thus each is described by the Euler and continuity equations for their densities and velocities. The baryon Euler equation must include the coupling to photons via Thomson scattering.

- Metric perturbation evolution and the connection of the metric perturbations to the matter perturbations are both contained in the Einstein equations. This is where the subtleties arise. A general metric perturbation has 10 degrees of freedom, but four of these are unphysical gauge modes. The physical perturbations include two degrees of freedom constructed from scalar functions, two from a vector, and two remaining tensor perturbations (Mukhanov *et al* 1992). Physically, the scalar perturbations correspond to gravitational potential and anisotropic stress perturbations; the vector perturbations correspond to vorticity and shear perturbations; and the tensor perturbations are two polarizations of gravitational radiation. Tensor and vector perturbations do not couple to matter evolving only under gravitation; in the absence of a 'stiff source' of stress energy, like cosmic defects or magnetic fields, the tensor and vector perturbations decouple from the linear perturbations in the matter.

A variety of different variable choices and methods for eliminating the gauge freedom have been developed. The subject can be fairly complicated. A detailed discussion and comparison between the Newtonian and synchronous gauges, along with a complete set of equations, can be found in Ma and Bertschinger (1995); also see Hu *et al* (1998). An elegant and physically appealing formalism based on an entirely covariant and gauge-invariant description of all physical quantities has been developed for the microwave background by Challinor and Lasenby (1999) and Gebbie *et al* (2000), based on earlier work by Ehlers (1993) and Ellis and Bruni (1989). A more conventional gauge-invariant approach was originated by Bardeen (1980) and developed by Kodama and Sasaki (1984).

The Boltzmann equations are partial differential equations, which can be converted to hierarchies of ordinary differential equations by expanding their directional dependence in Legendre polynomials. The result is a large set of coupled, first-order linear ordinary differential equations which form a well-posed initial value problem. Initial conditions must be specified. Generally they are taken to be so-called adiabatic perturbations: initial curvature perturbations with equal fractional perturbations in each matter species. Such perturbations arise naturally from the simplest inflationary scenarios. Alternatively, isocurvature perturbations can also be considered: these initial conditions have fractional density perturbations in two or more matter species whose total spatial curvature perturbation cancels. The issue of numerically determining initial conditions is discussed later in section 7.4.2.

The set of equations are numerically stiff before last scattering, since they contain the two widely discrepant time scales: the Thomson scattering time for electrons and photons and the (much longer) Hubble time. Initial conditions must be set with high accuracy and an appropriate stiff integrator must be employed. A variety of numerical techniques have been developed for evolving the equations. Particularly important is the line-of-sight algorithm first developed by Seljak and Zaldarriaga (1996) and then implemented by them in the publicly available CMBFAST code (see http://www.sns.ias.edu/~matiasz/CMBFAST/cmbfast.html).

This discussion is intentionally heuristic and somewhat vague because many of the issues involved are technical and not particularly illuminating. My main point is an appreciation for the detailed and precise physics which goes into computing microwave background fluctuations. However, all of this formalism should not obscure several basic physical processes which determine the ultimate form of the fluctuations. A widespread understanding of most of the physical processes detailed have followed from a seminal paper by Hu and Sugiyama (1996), a classic of the microwave background literature.

7.2.3 Tight coupling

Two basic time scales enter into the evolution of the microwave background. The first is the photon scattering time scale t_s, the mean time between Thomson scatterings. The other is the expansion time scale of the universe, H^{-1}, where $H = \dot{a}/a$ is the Hubble parameter. At temperatures significantly greater than 0.5 eV, hydrogen and helium are completely ionized and $t_s \ll H^{-1}$. The Thomson scatterings which couple the electrons and photons occur much more rapidly than the expansion of the universe; as a result, the baryons and photons behave as a single 'tightly coupled' fluid. During this period, the fluctuations in the photons mirror the fluctuations in the baryons. (Note that recombination occurs at around 0.5 eV rather than 13.6 eV because of the huge photon–baryon ratio; the universe contains somewhere around 10^9 photons for each baryon, as we know from primordial nucleosynthesis. It is a useful exercise to work out the approximate recombination temperature.)

The photon distribution function for scalar perturbations can be written as $\Theta(\boldsymbol{k}, \mu, t)$ where $\mu = \hat{\boldsymbol{k}} \cdot \hat{\boldsymbol{n}}$ and the scalar character of the fluctuations guarantees the distribution cannot have any azimuthal directional dependence. (The azimuthal dependence for vector and tensor perturbations can also be included in a similar decomposition). The moments of the distribution are defined as

$$\Theta(\boldsymbol{k}, \mu, t) = \sum_{l=0}^{\infty} (-i)^l \Theta_l(\boldsymbol{k}, t) P_l(\mu); \qquad (7.1)$$

sometimes other normalizations are used. Tight coupling implies that $\Theta_l = 0$ for $l > 1$. Physically, the $l = 0$ moment corresponds to the photon energy density perturbation, while $l = 1$ corresponds to the bulk velocity. During tight coupling, these two moments must match the baryon density and velocity perturbations. Any higher moments rapidly decay due to the isotropizing effect of Thomson scattering; this follows immediately from the photon Boltzmann equation.

7.2.4 Free-streaming

In the other regime, for temperatures significantly lower than 0.5 eV, $t_s \gg H^{-1}$ and photons on average never scatter again until the present time. This is known as the 'free-streaming' epoch. Since the radiation is no longer tightly coupled to the electrons, all higher moments in the radiation field develop as the photons propagate. In a flat background spacetime, the exact solution is simple to derive. After scattering ceases, the photons evolve according to the Liouville equation

$$\Theta' + ik\mu\Theta = 0 \qquad (7.2)$$

with the trivial solution

$$\Theta(\boldsymbol{k}, \mu, \eta) = e^{-ik\mu(\eta-\eta_*)}\Theta(\boldsymbol{k}, \mu, \eta_*), \qquad (7.3)$$

where we have converted to conformal time defined by $d\eta = dt/a(t)$ and η_* corresponds to the time at which free-streaming begins. Taking moments of both sides results in

$$\Theta_l(\boldsymbol{k}, \eta) = (2l + 1)[\Theta_0(\boldsymbol{k}, \eta_*) j_l(k\eta - k\eta_*) + \Theta_1(\boldsymbol{k}, \eta_*) j_l'(k\eta - k\eta_*)] \quad (7.4)$$

with j_l a spherical Bessel function. The process of free-streaming essentially maps spatial variations in the photon distribution at the last-scattering surface (wavenumber k) into angular variations on the sky today (moment l).

7.2.5 Diffusion damping

In the intermediate regime during recombination, $t_s \simeq H^{-1}$. Photons propagate a characteristic distance L_D during this time. Since some scattering is still occurring, baryons experience a drag from the photons as long as the ionization fraction is appreciable. A second-order perturbation analysis shows that the result is damping of baryon fluctuations on scales below L_D, known as Silk damping or diffusion damping. This effect can be modelled by the replacement

$$\Theta_0(\boldsymbol{k}, \eta_*) \rightarrow \Theta_0(\boldsymbol{k}, \eta_*)e^{-(kL_D)^2} \tag{7.5}$$

although detailed calculations are needed to define L_D precisely. As a result of this damping, microwave background fluctuations are exponentially suppressed on angular scales significantly smaller than a degree.

7.2.6 The resulting power spectrum

The fluctuations in the universe are assumed to arise from some random statistical process. We are not interested in the exact pattern of fluctuations we see from our vantage point, since this is only a single realization of the process. Rather, a theory of cosmology predicts an underlying distribution, of which our visible sky is a single statistical realization. The most basic statistic describing fluctuations is their power spectrum. A temperature map on the sky $T(\hat{\boldsymbol{n}})$ is conventionally expanded in spherical harmonics,

$$\frac{T(\hat{\boldsymbol{n}})}{T_0} = 1 + \sum_{l=1}^{\infty} \sum_{m=-l}^{l} a_{(lm)}^{T} Y_{(lm)}(\hat{\boldsymbol{n}}) \tag{7.6}$$

where

$$a_{(lm)}^{T} = \frac{1}{T_0} \int d\hat{\boldsymbol{n}}\, T(\hat{\boldsymbol{n}}) Y_{(lm)}^{*}(\hat{\boldsymbol{n}}) \tag{7.7}$$

are the temperature multipole coefficients and T_0 is the mean CMB temperature. The $l = 1$ term in equation (7.6) is indistinguishable from the kinematic dipole and is normally ignored. The temperature angular power spectrum C_l is then given by

$$\langle a_{(lm)}^{T*} a_{(l'm')}^{T} \rangle = C_l^{T} \delta_{ll'} \delta_{mm'}, \tag{7.8}$$

where the angled brackets represent an average over statistical realizations of the underlying distribution. Since we have only a single sky to observe, an unbiased estimator of C_l is constructed as

$$\hat{C}_l^{T} = \frac{1}{2l + 1} \sum_{m=-l}^{l} a_{lm}^{T*} a_{lm}^{T}. \tag{7.9}$$

The statistical uncertainty in estimating C_l^{T} by a sum of $2l + 1$ terms is known as 'cosmic variance'. The constraints $l = l'$ and $m = m'$ follow from the assumption of statistical isotropy: C_l^{T} must be independent of the orientation of the coordinate system used for the harmonic expansion. These conditions can be verified via an explicit rotation of the coordinate system.

A given cosmological theory will predict C_l^{T} as a function of l, which can be obtained from evolving the temperature distribution function as described earlier.

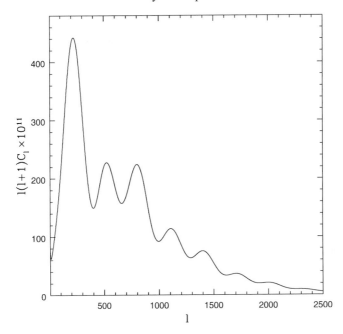

Figure 7.1. The temperature angular power spectrum for a cosmological model with mass density $\Omega_0 = 0.3$, vacuum energy density $\Omega_\Lambda = 0.7$, Hubble parameter $h = 0.7$, and a scale-invariant spectrum of primordial adiabatic perturbations.

This prediction can then be compared with data from measured temperature differences on the sky. Figure 7.1 shows a typical temperature power spectrum from the inflationary class of models, described in more detail later. The distinctive sequence of peaks arise from coherent acoustic oscillations in the fluid during the tight coupling epoch and are of great importance in precision tests of cosmological models; these peaks will be discussed in section 7.4. The effect of diffusion damping is clearly visible in the decreasing power above $l = 1000$. When viewing angular power spectrum plots in multipole space, keep in mind that $l = 200$ corresponds approximately to fluctuations on angular scales of a degree, and the angular scale is inversely proportional to l. The vertical axis is conventionally plotted as $l(l + 1)C_l^{\mathrm{T}}$ because the Sachs–Wolfe temperature fluctuations from a scale-invariant spectrum of density perturbations appears as a horizontal line on such a plot.

7.3   Physics of polarization fluctuations

In addition to temperature fluctuations, the simple physics of decoupling inevitably leads to non-zero polarization of the microwave background radiation as well, although quite generically the polarization fluctuations are expected to be significantly smaller than the temperature fluctuations. This section reviews the physics of polarization generation and its description. For a more detailed pedagogical discussion of microwave background polarization, see Kosowsky (1999), from which this section is excerpted.

7.3.1   Stokes parameters

Polarized light is conventionally described in terms of the Stokes parameters, which are presented in any optics text. If a monochromatic electromagnetic wave propagating in the z-direction has an electric field vector at a given point in space given by

$$E_x = a_x(t) \cos[\omega_0 t - \theta_x(t)], \qquad E_y = a_y(t) \cos[\omega_0 t - \theta_y(t)], \qquad (7.10)$$

then the Stokes parameters are defined as the following time averages:

$$I \equiv \langle a_x^2 \rangle + \langle a_y^2 \rangle; \qquad (7.11)$$

$$Q \equiv \langle a_x^2 \rangle - \langle a_y^2 \rangle; \qquad (7.12)$$

$$U \equiv \langle 2a_x a_y \cos(\theta_x - \theta_y) \rangle; \qquad (7.13)$$

$$V \equiv \langle 2a_x a_y \sin(\theta_x - \theta_y) \rangle. \qquad (7.14)$$

The averages are over times long compared to the inverse frequency of the wave. The parameter I gives the intensity of the radiation which is always positive and is equivalent to the temperature for blackbody radiation. The other three parameters define the polarization state of the wave and can have either sign. Unpolarized radiation, or 'natural light', is described by $Q = U = V = 0$.

The parameters I and V are physical observables independent of the coordinate system, but Q and U depend on the orientation of the x and y axes. If a given wave is described by the parameters Q and U for a certain orientation of the coordinate system, then after a rotation of the x–y plane through an angle ϕ, it is straightforward to verify that the same wave is now described by the parameters

$$Q' = Q \cos(2\phi) + U \sin(2\phi),$$
$$U' = -Q \sin(2\phi) + U \cos(2\phi). \qquad (7.15)$$

From this transformation it is easy to see that the quantity $P^2 \equiv Q^2 + U^2$ is invariant under rotation of the axes, and the angle

$$\alpha \equiv \frac{1}{2} \tan^{-1} \frac{U}{Q} \qquad (7.16)$$

defines a constant orientation parallel to the electric field of the wave. The Stokes parameters are a useful description of polarization because they are *additive* for incoherent superposition of radiation; note this is not true for the magnitude or orientation of polarization. Note that the transformation law in equation (7.15) is characteristic not of a vector but of the second-rank *tensor*

$$\rho = \frac{1}{2} \begin{pmatrix} I+Q & U-iV \\ U+iV & I-Q \end{pmatrix}, \tag{7.17}$$

which also corresponds to the quantum mechanical density matrix for an ensemble of photons (Kosowsky 1996). In kinetic theory, the photon distribution function $f(x, p, t)$ discussed in section 7.2.2 must be generalized to $\rho_{ij}(x, p, t)$, corresponding to this density matrix.

7.3.2   Thomson scattering and the quadrupolar source

Non-zero linear polarization in the microwave background is generated around decoupling because the Thomson scattering which couples the radiation and the electrons is not isotropic but varies with the scattering angle. The total scattering cross-section, defined as the radiated intensity per unit solid angle divided by the incoming intensity per unit area, is given by

$$\frac{d\sigma}{d\Omega} = \frac{3\sigma_T}{8\pi} |\hat{\varepsilon}' \cdot \hat{\varepsilon}|^2 \tag{7.18}$$

where σ_T is the total Thomson cross section and the vectors $\hat{\varepsilon}$ and $\hat{\varepsilon}'$ are unit vectors in the planes perpendicular to the propogation directions which are aligned with the outgoing and incoming polarization, respectively. This scattering cross section can give no net circular polarization, so $V = 0$ for cosmological perturbations and will not be discussed further. Measurements of V polarization can be used as a diagnostic of systematic errors or microwave foreground emission.

It is a straightforward but slightly involved exercise to show that these relations imply that an incoming unpolarized radiation field with the multipole expansion equation (7.6) will be Thomson scattered into an outgoing radiation field with Stokes parameters

$$Q(\hat{n}) - iU(\hat{n}) = \frac{3\sigma_T}{8\pi\sigma_B} \sqrt{\frac{\pi}{5}} a_{20} \sin^2 \beta \tag{7.19}$$

if the incoming radiation field has rotational symmetry around its direction of propagation, as will hold for individual Fourier modes of scalar perturbations. Explicit expressions for the general case of no symmetry can be derived in terms of Wigner D-symbols (Kosowsky 1999).

In simple and general terms, unpolarized incoming radiation will be Thomson scattered into linearly polarized radiation if and only if the incoming radiation has a non-zero quadrupolar directional dependence. This single fact is sufficient to understand the fundamental physics behind polarization of the microwave background. During the tight-coupling epoch, the radiation field has only monopole and dipole directional dependences as explained earlier; therefore, scattering can produce no net polarization and the radiation remains unpolarized. As tight coupling begins to break down as recombination begins, a quadrupole moment of the radiation field will begin to grow due to free-streaming of the photons. Polarization is generated during the brief interval when a significant quadrupole moment of the radiation has built up, but the scattering electrons have not yet all recombined. Note that if the universe recombined instantaneously, the net polarization of the microwave background would be zero. Due to this competition between the quadrupole source building up and the density of scatterers declining, the amplitude of polarization in the microwave background is generically suppressed by an order of magnitude compared to the temperature fluctuations.

Before polarization generation commences, the temperature fluctuations have either a monopole dependence, corresponding to density perturbations, or a dipole dependence, corresponding to velocity perturbations. A straightforward solution to the photon free-streaming equation (in terms of spherical Bessel functions) shows that for Fourier modes with wavelengths large compared to a characteristic thickness of the last-scattering surface, the quadrupole contribution through the last scattering surface is dominated by the velocity fluctuations in the temperature, not the density fluctuations. This makes intuitive sense: the dipole fluctuations can free stream directly into the quadrupole, but the monopole fluctuations must stream through the dipole first. This conclusion breaks down on small scales where either monopole or dipole can be the dominant quadrupole source, but numerical computations show that on scales of interest for microwave background fluctuations, the dipole temperature fluctuations are always the dominant source of quadrupole fluctuations at the last scattering-surface. Therefore, polarization fluctuations reflect mainly velocity perturbations at last scattering, in contrast to temperature fluctuations which predominantly reflect density perturbations.

7.3.3   Harmonic expansions and power spectra

Just as the temperature on the sky can be expanded into spherical harmonics, facilitating the computation of the angular power spectrum, so can the polarization. The situation is formally parallel, although in practice it is more complicated: while the temperature is a scalar quantity, the polarization is a second-rank tensor. We can define a polarization tensor with the correct transformation properties, equation (7.15), as

$$P_{ab}(\hat{\boldsymbol{n}}) = \frac{1}{2} \begin{pmatrix} Q(\hat{\boldsymbol{n}}) & -U(\hat{\boldsymbol{n}}) \sin \theta \\ -U(\hat{\boldsymbol{n}}) \sin \theta & -Q(\hat{\boldsymbol{n}}) \sin^2 \theta \end{pmatrix}. \tag{7.20}$$

The dependence on the Stokes parameters is the same as for the density matrix, equation (7.17); the extra factors are convenient because the usual spherical coordinate basis is orthogonal but not orthonormal. This tensor quantity must be expanded in terms of tensor spherical harmonics which preserve the correct transformation properties. We assume a complete set of orthonormal basis functions for symmetric trace-free 2×2 tensors on the sky,

$$\frac{P_{ab}(\hat{n})}{T_0} = \sum_{l=2}^{\infty} \sum_{m=-l}^{l} [a_{(lm)}^{G} Y_{(lm)ab}^{G}(\hat{n}) + a_{(lm)}^{C} Y_{(lm)ab}^{C}(\hat{n})], \tag{7.21}$$

where the expansion coefficients are given by

$$a_{(lm)}^{G} = \frac{1}{T_0} \int d\hat{n} \, P_{ab}(\hat{n}) Y_{(lm)}^{Gab*}(\hat{n}), \tag{7.22}$$

$$a_{(lm)}^{C} = \frac{1}{T_0} \int d\hat{n} \, P_{ab}(\hat{n}) Y_{(lm)}^{Cab*}(\hat{n}), \tag{7.23}$$

which follow from the orthonormality properties

$$\int d\hat{n} \, Y_{(lm)ab}^{G*}(\hat{n}) Y_{(l'm')}^{Gab}(\hat{n}) = \int d\hat{n} \, Y_{(lm)ab}^{C*}(\hat{n}) Y_{(l'm')}^{Cab}(\hat{n}) = \delta_{ll'} \delta_{mm'}, \tag{7.24}$$

$$\int d\hat{n} \, Y_{(lm)ab}^{G*}(\hat{n}) Y_{(l'm')}^{Cab}(\hat{n}) = 0. \tag{7.25}$$

These tensor spherical harmonics are not as exotic as they might sound; they are used extensively in the theory of gravitational radiation, where they naturally describe the radiation multipole expansion. Tensor spherical harmonics are similar to vector spherical harmonics used to represent electromagnetic radiation fields, familiar from chapter 16 of Jackson (1975). Explicit formulas for tensor spherical harmonics can be derived via various algebraic and group theoretic methods; see Thorne (1980) for a complete discussion. A particularly elegant and useful derivation of the tensor spherical harmonics (along with the vector spherical harmonics as well) is provided by differential geometry: the harmonics can be expressed as covariant derivatives of the usual spherical harmonics with respect to an underlying manifold of a two-sphere (i.e. the sky). This construction has been carried out explicitly and applied to the microwave background polarization (Kamionkowski *et al* 1996).

The existence of two sets of basis functions, labelled here by 'G' and 'C', is due to the fact that the symmetric traceless 2×2 tensor describing linear polarization is specified by two independent parameters. In two dimensions, any symmetric traceless tensor can be uniquely decomposed into a part of the form $A_{;ab} - (1/2)g_{ab}A_{;c}^{c}$ and another part of the form $B_{;ac}\epsilon_{b}^{c} + B_{;bc}\epsilon_{a}^{c}$ where A and B are two scalar functions and semicolons indicate covariant derivatives. This decomposition is quite similar to the decomposition of a vector field into a part which is the gradient of a scalar field and a part which is the curl of a vector field; hence we use the notation G for 'gradient' and C for 'curl'. In fact, this correspondence is more than just cosmetic: if a linear polarization field is visualized in the usual way with headless 'vectors' representing the amplitude and orientation of the polarization, then the G harmonics describe the portion of the polarization field which has no handedness associated with it, while the C harmonics describe the other portion of the field which does have a handedness (just as with the gradient and curl of a vector field). Note that Zaldarriaga and Seljak (1997) label these harmonics E and B, with a slightly different normalization than defined here (see Kamionkowski *et al* 1996).

We now have three sets of multipole moments, $a^{\mathrm{T}}_{(lm)}$, $a^{\mathrm{G}}_{(lm)}$, and $a^{\mathrm{C}}_{(lm)}$, which fully describe the temperature/polarization map of the sky. These moments can be combined quadratically into various power spectra analogous to the temperature C^{T}_l. Statistical isotropy implies that

$$\langle a^{\mathrm{T}*}_{(lm)} a^{\mathrm{T}}_{(l'm')} \rangle = C^{\mathrm{T}}_l \delta_{ll'} \delta_{mm'}, \qquad \langle a^{\mathrm{G}*}_{(lm)} a^{\mathrm{G}}_{(l'm')} \rangle = C^{\mathrm{G}}_l \delta_{ll'} \delta_{mm'},$$

$$\langle a^{\mathrm{C}*}_{(lm)} a^{\mathrm{C}}_{(l'm')} \rangle = C^{\mathrm{C}}_l \delta_{ll'} \delta_{mm'}, \qquad \langle a^{\mathrm{T}*}_{(lm)} a^{\mathrm{G}}_{(l'm')} \rangle = C^{\mathrm{TG}}_l \delta_{ll'} \delta_{mm'},$$

$$\langle a^{\mathrm{T}*}_{(lm)} a^{\mathrm{C}}_{(l'm')} \rangle = C^{\mathrm{TC}}_l \delta_{ll'} \delta_{mm'}, \qquad \langle a^{\mathrm{G}*}_{(lm)} a^{\mathrm{C}}_{(l'm')} \rangle = C^{\mathrm{GC}}_l \delta_{ll'} \delta_{mm'}, \qquad (7.26)$$

where the angle brackets are an average over all realizations of the probability distribution for the cosmological initial conditions. Simple statistical estimators of the various C_ls can be constructed from maps of the microwave background temperature and polarization.

For fluctuations with Gaussian random distributions (as predicted by the simplest inflation models), the statistical properties of a temperature/polarization map are specified fully by these six sets of multipole moments. In addition, the scalar spherical harmonics $Y_{(lm)}$ and the G tensor harmonics $Y^{\mathrm{G}}_{(lm)ab}$ have parity $(-1)^l$, but the C harmonics $Y^{\mathrm{C}}_{(lm)ab}$ have parity $(-1)^{l+1}$. If the large-scale perturbations in the early universe were invariant under parity inversion, then $C^{\mathrm{TC}}_l = C^{\mathrm{GC}}_l = 0$. So generally, microwave background fluctuations are characterized by the four power spectra C^{T}_l, C^{G}_l, C^{C}_l, and C^{TG}_l. The end result of the numerical computations described in section 7.2.2 are these power spectra. Polarization power spectra C^{G}_l and C^{TG}_l for scalar perturbations in a typical inflation-like cosmological model, generated with the CMBFAST code (Seljak and Zaldarriaga 1996), are displayed in figure 7.2. The temperature power spectrum in figure 7.1 and the polarization power spectra in figure 7.2 come from the same cosmological model. The physical source of the features in the power spectra is discussed in the next section, followed by a discussion of how cosmological parameters can be determined to high precision via detailed measurements of the microwave background power spectra.

7.4   Acoustic oscillations

Before decoupling, the matter in the universe has significant pressure because it is tightly coupled to radiation. This pressure counteracts any tendency for matter to collapse gravitationally. Formally, the Jeans mass is greater than the mass within a horizon volume for times earlier than decoupling. During this epoch,

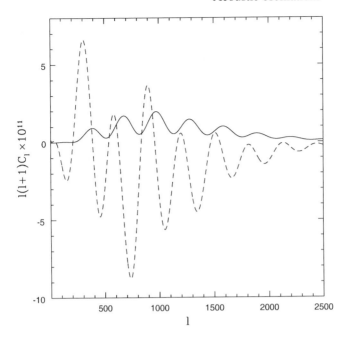

Figure 7.2. The G polarization power spectrum (full curve) and the cross-power TG between temperature and polarization (dashed curve), for the same model as in figure 7.1.

density perturbations will set up standing acoustic waves in the plasma. Under certain conditions, these waves leave a distinctive imprint on the power spectrum of the microwave background, which in turn provides the basis for precision constraints on cosmological parameters. This section reviews the basics of the acoustic oscillations.

7.4.1 An oscillator equation

In their classic 1996 paper, Hu and Sugiyama transformed the basic equations describing the evolution of perturbations into an oscillator equation. Combining the zeroth moment of the photon Boltzmann equation with the baryon Euler equation for a given k-mode in the tight-coupling approximation (mean baryon velocity equals mean radiation velocity) gives

$$\ddot{\Theta}_0 + H\frac{R}{1+R}\dot{\Theta}_0 + k^2 c_s^2 \Theta_0 = -\ddot{\Phi} - H\frac{R}{1+R}\dot{\Phi} - \frac{1}{3}k^2\Psi, \qquad (7.27)$$

where Θ_0 is the zeroth moment of the temperature distribution function (proportional to the photon density perturbation), $R = 3\rho_b/4\rho_\gamma$ is proportional to the scale factor a, $H = \dot{a}/a$ is the conformal Hubble parameter, and the sound speed is given by $c_s^2 = 1/(3 + 3R)$. (All overdots are derivatives with respect to conformal time.) Φ and Ψ are the scalar metric perturbations in the Newtonian gauge; if we neglect the anisotropic stress, which is generally small in conventional cosmological scenarios, then $\Psi = -\Phi$. But the details are not very important. The equation represents damped, driven oscillations of the radiation density, and the various physical effects are easily identified. The second term on the left-hand side is the damping of oscillations due to the expansion of the universe. The third term on the left-hand side is the restoring force due to the pressure, since $c_s^2 = dP/d\rho$. On the right-hand side, the first two terms depend on the time variation of the gravitational potentials, so these two are the source of the Integrated Sachs–Wolfe effect. The final term on the right-hand side is the driving term due to the gravitational potential perturbations. As Hu and Sugiyama emphasized, these damped, driven acoustic oscillations account for all of the structure in the microwave background power spectrum.

A WKB approximation to the homogeneous equation with no driving source terms gives the two oscillation modes (Hu and Sugiyama 1996)

$$\Theta_0(k, \eta) \propto \begin{cases} (1 + R)^{-1/4} \cos kr_s(\eta) \\ (1 + R)^{-1/4} \sin kr_s(\eta) \end{cases} \tag{7.28}$$

where the sound horizon r_s is given by

$$r_s(\eta) \equiv \int_0^\eta c_s(\eta') \, d\eta'. \tag{7.29}$$

Note that at times well before matter–radiation equality, the sound speed is essentially constant, $c_s = 1/\sqrt{3}$, and the sound horizon is simply proportional to the causal horizon. In general, any perturbation with wavenumber k will set up an oscillatory behaviour in the primordial plasma described by a linear combination of the two modes in equation (7.28). The relative contribution of the modes will be determined by the initial conditions describing the perturbation.

Equation (7.27) appears to be simpler than it actually is, because Φ and Ψ are the total gravitational potentials due to all matter and radiation, including the photons which the left-hand side is describing. In other words, the right-hand side of the equation contains an implicit dependence on Θ_0. At the expense of pedagogical transparency, this situation can be remedied by considering separately the potential from the photon–baryon fluid and the potential from the truly external sources, the DM and neutrinos. This split has been performed by Hu and White (1996). The resulting equation, while still an oscillator equation, is much more complicated, but must be used for a careful physical analysis of acoustic oscillations.

7.4.2   Initial conditions

The initial conditions for radiation perturbations for a given wavenumber k can be broken into two categories, according to whether the gravitational potential perturbation from the baryon–photon fluid, $\Phi_{b\gamma}$, is non-zero or zero as $\eta \to 0$. The former case is known as 'adiabatic' (which is somewhat of a misnomer since adiabatic technically refers to a property of a time-dependent process) and implies that n_b/n_γ, the ratio of baryon to photon number densities, is a constant in space. This case must couple to the cosine oscillation mode since it requires $\Theta_0 \neq 0$ as $\eta \to 0$. The simplest (i.e. single-field) models of inflation produce perturbations with adiabatic initial conditions.

The other case is termed 'isocurvature' since the fluid gravitational potential perturbation $\Phi_{b\gamma}$, and hence the perturbations to the spatial curvature, are zero. In order to arrange such a perturbation, the baryon and photon densities must vary in such a way that they compensate each other: n_b/n_γ varies, and thus these perturbations are in entropy, not curvature. At an early enough time, the temperature perturbation in a given k mode must arise entirely from the Sachs–Wolfe effect, and thus isocurvature perturbations couple to the sine oscillation mode. These perturbations arise from causal processes like phase transitions: a phase transition cannot change the energy density of the universe from point to point, but it can alter the relative entropy between various types of matter depending on the values of the fields involved. The potentially most interesting cause of isocurvature perturbations is multiple dynamical fields in inflation. The fields will exchange energy during inflation, and the field values will vary stochastically between different points in space at the end of the phase transition, generically giving isocurvature along with adiabatic perturbations (Polarski and Starobinsky 1994).

The numerical problem of setting initial conditions is somewhat tricky. The general problem of evolving perturbations involves linear evolution equations for around a dozen variables, outlined in section 7.2.2. Setting the correct initial conditions involves specifying the value of each variable in the limit as $\eta \to 0$. This is difficult for two reasons: the equations are singular in this limit, and the equations become increasingly numerically stiff in this limit. Simply using the leading-order asymptotic behaviour for all of the variables is only valid in the high-temperature limit. Since the equations are stiff, small departures from this limiting behaviour in any of the variables can lead to numerical instability until the equations evolve to a stiff solution, and this numerical solution does not necessarily correspond to the desired initial conditions. Numerical techniques for setting the initial conditions to high accuracy at temperaturesare currently being developed.

7.4.3 Coherent oscillations

The characteristic 'acoustic peaks' which appear in figure 7.1 arise from acoustic oscillations which are phase coherent: at some point in time, the phases of all of the acoustic oscillations were the same. This requires the same initial condition for *all* k-modes, including those with wavelengths longer than the horizon. Such a condition arises naturally for inflationary models, but is very hard to reproduce in models producing perturbations causally on scales smaller than the horizon. Defect models, for example, produce acoustic oscillations, but the oscillations generically have incoherent phases and thus display no peak structure in their power spectrum (Seljak *et al* 1997). Simple models of inflation which produce only adiabatic perturbations insure that all perturbations have the same phase at $\eta = 0$ because all of the perturbations are in the cosine mode of equation (7.28).

A glance at the k dependence of the adiabatic perturbation mode reveals how the coherent peaks are produced. The microwave background images the radiation density at a fixed time; as a function of k, the density varies like $\cos(kr_s)$, where r_s is fixed. Physically, on scales much larger than the horizon at decoupling, a perturbation mode has not had enough time to evolve. At a particular smaller scale, the perturbation mode evolves to its maximum density in potential wells, at which point decoupling occurs. This is the scale reflected in the first acoustic peak in the power spectrum. Likewise, at a particular still smaller scale, the perturbation mode evolves to its maximum density in potential wells and then turns around, evolving to its minimum density in potential wells; at that point, decoupling occurs. This scale corresponds to that of the second acoustic peak. (Since the power spectrum is the square of the temperature fluctuation, both compressions and rarefactions in potential wells correspond to peaks in the power spectrum.) Each successive peak represents successive oscillations, with the scales of odd-numbered peaks corresponding to those perturbation scales which have ended up compressed in potential wells at the time of decoupling, while the even-numbered peaks correspond to the perturbation scales which are rarefied in potential wells at decoupling. If the perturbations are not phase coherent, then the phase of a given k-mode at decoupling is not well defined, and the power spectrum just reflects some mean fluctuation power at that scale.

In practice, two additional effects must be considered: a given scale in k-space is mapped to a range of l-values; and radiation velocities as well as densities contribute to the power spectrum. The first effect broadens out the peaks, while the second fills in the valleys between the peaks since the velocity extrema will be exactly out of phase with the density extrema. The amplitudes of the peaks in the power spectrum are also suppressed by Silk damping, as mentioned in section 7.2.5.

7.4.4   The effect of baryons

The mass of the baryons creates a distinctive signature in the acoustic oscillations (Hu and Sugiyama 1996). The zero-point of the oscillations is obtained by setting Θ_0 constant in equation (7.27): the result is

$$\Theta_0 \simeq \frac{1}{3c_s^2}\Phi = (1 + a)\Phi. \tag{7.30}$$

The photon temperature Θ_0 is not itself observable, but must be combined with the gravitational redshift to form the 'apparent temperature' $\Theta_0 - \Phi$, which oscillates around $a\Phi$. If the oscillation amplitude is much larger than $a\Phi = 3\rho_b\Phi/4\rho_\gamma$, then the oscillations are effectively about the mean temperature. The positive and negative oscillations are of the same amplitude, so when the apparent temperature is squared to form the power spectrum, all of the peaks have the same height. However, if the baryons contribute a significant mass so that $a\Phi$ is a significant fraction of the oscillation amplitude, then the zero point of the oscillations are displaced, and when the apparent temperature is squared to form the power spectrum, the peaks arising from the positive oscillations are higher than the peaks from the negative oscillations. If $a\Phi$ is larger than the amplitude of the oscillations, then the power spectrum peaks corresponding to the negative oscillations disappear entirely. The physical interpretation of this effect is that the baryon mass deepens the potential well in which the baryons are oscillating, increasing the compression of the plasma compared to the case with less baryon mass. In short, as the baryon density increases, the power spectrum peaks corresponding to compressions in potential wells get higher, while the alternating peaks corresponding to rarefactions get lower. This alternating peak height signature is a distinctive signature of baryon mass, and allows the precise determination of the cosmological baryon density with the measurement of the first several acoustic peak heights.

7.5 Cosmological models and constraints

The cosmological interpretation of a measured microwave background power spectrum requires, to some extent, the introduction of a particular space of models. A very simple, broad and well-motivated set of models are motivated by inflation: a universe described by a homogeneous and isotropic background with phase-coherent, power-law initial perturbations which evolve freely. This model space excludes, for example, perturbations caused by topological defects or other 'stiff' sources, arbitrary initial power spectra, or any departures from the standard background cosmology. This set of models has the twin virtues of being relatively simple to calculate and best conforming to current power spectrum measurements. (In fact, most competing cosmological models, like those employing cosmic defects to make structure, are essentially ruled out by current microwave background and large-scale structure measurements.) This section will describe the parameters defining the model space and discuss the extent to which the parameters can be constrained through the microwave background.

7.5.1 A space of models

The parameters defining the model space can be broken into three types: cosmological parameters describing the background spacetime; parameters describing the initial conditions; and other parameters describing miscellaneous additional physical effects. Background cosmological parameters are as follows.

- Ω, the ratio of the total energy density to the critical density $\rho_{cr} = 8\pi/3H^2$. This parameter determines the spatial curvature of the universe: $\Omega = 1$ is a flat universe with critical density. Smaller values of Ω correspond to a negative spatial curvature, while larger values correspond to positive curvature. Current microwave background measurements constrain Ω to be roughly within the range 0.8–1.2, consistent with a critical-density universe.
- Ω_b, the ratio of the baryon density to the critical density. Observations of the abundance of deuterium in high redshift gas clouds and comparison with predictions from primordial nucleosynthesis place strong constraints on this parameter (Tytler *et al* 2000).
- Ω_m, the ratio of the DM density to the critical density. Dynamical constraints, gravitational lensing, cluster abundances and numerous other lines of evidence all point to a total matter density in the neighbourhood of $\Omega_0 = \Omega_m + \Omega_b = 0.3$.
- Ω_Λ, the ratio of vacuum energy density Λ to the critical density. This is the notorious cosmological constant. Several years ago, almost no cosmologist advocated a cosmological constant; now almost every cosmologist accepts its existence. The shift was precipitated by the Type Ia supernova Hubble diagram (Perlmutter *et al* 1999, Riess *et al* 1998) which shows an apparent acceleration in the expansion of the universe. Combined with strong constraints on Ω, a cosmological constant now seems unavoidable, although high-energy theorists have a difficult time accepting it. Strong gravitational lensing of quasars places upper limits on Ω_Λ (Falco *et al* 1998).
- The present Hubble parameter h, in units of 100 km s^{-1}/Mpc^{-1}. Distance ladder measurements (Mould *et al* 2000) and supernova Ia measurements (Riess *et al* 1998) give consistent estimates for h of around 0.70, with systematic errors on the order of 10%.
- Optionally, further parameters describing additional contributions to the energy density of the universe; for example, the 'quintessence' models (Caldwell *et al* 1998) which add one or more scalar fields to the universe.

Parameters describing the initial conditions are:

- The amplitude of fluctuations Q, often defined at the quadrupole scale. COBE fixed this amplitude to high accuracy (Bennett *et al* 1996).
- The power law index n of initial adiabatic density fluctuations. The scale-invariant Harrison–Zeldovich spectrum is $n = 1$. Comparison of microwave background and large-scale structure measurements shows that n is close to unity.
- The relative contribution of tensor and scalar perturbations r, usually defined as the ratio of the power at $l = 2$ from each type of perturbation. The fact that prominent features are seen in the power spectrum (presumably arising from scalar density perturbations) limits the power spectrum contribution of tensor perturbations to roughly 20% of the scalar amplitude.
- The power law index n_T of tensor perturbations. Unfortunately, tensor power

spectra are generally defined so that $n_T = 0$ corresponds to scale invariant, in contrast to the scalar case.

- Optionally, more parameters describing either departures of the scalar perturbations from a power law (e.g. Kosowsky and Turner 1995) or a small admixture of isocurvature perturbations.

Other miscellaneous parameters include:

- A significant neutrino mass m_ν. None of the current neutrino oscillation results favour a cosmologically interesting neutrino mass.
- The effective number of neutrino species N_ν. This quantity includes any particle species which is relativistic when it decouples or can model entropy production prior to last scattering.
- The redshift of reionization, z_r. Spectra of quasars at redshift $z = 5$ show that the universe has been reionized at least since then.

A realistic parameter analysis might include at least eight free parameters. Given a particular microwave background measurement, deciding on a particular set of parameters and various priors on those parameters is as much art as science. For the correct model, parameter values should be insensitive to the size of the parameter space or the particular priors invoked. Several particular parameter space analyses are mentioned in section 7.5.5.

7.5.2 Physical quantities

While these parameters are useful and conventional for characterizing cosmological models, the features in the microwave background power spectrum depend on various physical quantities which can be expressed in terms of the parameters. Here the physical quantities are summarized, and their dependence on parameters given. This kind of analysis is important for understanding the model space of parameters as more than just a black box producing output power spectra. All of the physical dependences discussed here can be extracted from Hu and Sugiyama (1996). By comparing numerical solutions with the evolution equations, Hu and Sugiyama demonstrated that they had accounted for all relevant physical processes.

Power-law initial conditions are determined in a straightforward way by the appropriate parameters Q, n, r and n_T, if the perturbations are purely adiabatic. Additional parameters must be used to specify any departure from power-law spectra or to specify an additional admixture of isocurvature initial conditions (e.g. Bucher *et al* 1999). These parameters directly express physical quantities.

However, the physical parameters determining the evolution of the initial perturbations until decoupling involve a few specific combinations of cosmological parameters. First, note that the density of radiation is fixed by the current microwave background temperature which is known from COBE, as well as the density of the neutrino backgrounds. The gravitational potentials describing scalar perturbations determine the size of the Sachs–Wolfe effect and also magnitude of the forces driving the acoustic oscillations. The potentials are determined by $\Omega_0 h^2$, the matter density as a fraction of critical density. The baryon density, $\Omega_b h^2$, determines the degree to which the acoustic peak amplitudes are modulated as previousy described in section 7.4.4.

The time of matter–radiation equality is obviously determined solely by the total matter density $\Omega_0 h^2$. This quantity affects the size of the DM fluctuations, since DM starts to collapse gravitationally only after matter–radiation equality. Also, the gravitational potentials evolve in time during radiation domination and not during matter domination: the later matter–radiation equality occurs, the greater the time evolution of the potentials at decoupling, increasing the Integrated Sachs–Wolfe effect. The power spectrum also has a weak dependence on Ω_0 in models with Ω_0 significantly less than unity, because at late times the evolution of the background cosmology will be dominated not by matter, but rather by vacuum energy (for a flat universe with Λ) or by curvature (for an open universe). In either case, the gravitational potentials once again begin to evolve with time, giving an additional late-time integrated Sachs–Wolfe contribution, but this tends to affect only the largest scales for which the constraints from measurements are least restrictive due to cosmic variance (see the discussion in section 7.5.4).

The sound speed, which sets the sound horizon and thus affects the wavelength of the acoustic modes (cf equation (7.28)), is completely determined by the baryon density $\Omega_b h^2$. The horizon size at recombination, which sets the overall scale of the acoustic oscillations, depends only on the total mass density $\Omega_0 h^2$. The damping scale for diffusion damping depends almost solely on the baryon density $\Omega_b h^2$, although numerical fits give a slight dependence on Ω_b alone (Hu and Sugiyama 1996). Finally, the angular diameter distance to the last-scattering surface is determined by $\Omega_0 h$ and Λh; the angular diameter sets the angular scale on the sky of the acoustic oscillations.

In summary, the physical dependence of the temperature perturbations at last scattering depends on $\Omega_0 h^2$, $\Omega_b h^2$, $\Omega_0 h$, and Λh instead of the individual cosmological parameters Ω_0, Ω_b, h and Λ. When analysing constraints on cosmological models from microwave background power spectra, it may be more meaningful and powerful to constrain these physical parameters rather than the cosmological ones.

7.5.3   Power spectrum degeneracies

As might be expected from the previous discussion, not all of the parameters considered here are independent. In fact, one nearly exact degeneracy exists if Ω_0, Ω_b, h and Λ are taken as independent parameters. To see this, consider a shift in Ω_0. In isolation, such a shift will produce a corresponding stretching of the power spectrum in l-space. But this effect can be compensated by first shifting h to keep $\Omega_0 h^2$ constant, then shifting Ω_b to keep $\Omega_b h^2$ constant, and finally shifting Λ to keep the angular diameter distance constant. This set of shifted parameters will, in linear perturbation theory, produce almost exactly the same microwave background power spectra as the original set of parameters. The universe with shifted parameters will generally not be flat, but the resulting late-time Integrated Sachs–Wolfe effect only weakly break the degeneracy. Likewise, gravitational lensing has only a very weak effect on the degeneracy.

But all is not lost. The required shift in Λ is generally something like eight times larger than the original shift in Ω_0, so although the degeneracy is nearly exact, most of the degenerate models represent rather extreme cosmologies. Good taste requires either that $\Lambda = 0$ or that $\Omega = 1$, in other words that we disfavour models which have both a cosmological constant and are not flat. If such models are disallowed, the degeneracy disappears. Finally, other observables not associated with the microwave background break the degeneracy: the acceleration parameter $q_0 = \Omega_0/2 - \Lambda$, for example, is measured directly by the high-redshift supernova experiments. So in practice, this fundamental degeneracy in the microwave background power spectrum between Ω and Λ is not likely to have a great impact on our ability to constrain cosmological parameters.

Other approximate degeneracies in the temperature power spectrum exist between Q and r, and between z_r and n. The first is illusory: the amplitudes of the scalar and tensor power spectra can be used in place of their sum and ratio, which eliminates the degeneracy. The power spectrum of large-scale structure will lift the latter degeneracy if bias is understood well enough, as will polarization measurements and small-scale second-order temperature fluctuations (the Ostriker–Vishniac effect, see Gnedin and Jaffe 2000) which are both sensitive to z_r.

Finally, many claims have been made about the ability of the microwave background to constrain the effective number of neutrino species or neutrino masses. The effective number of massless degrees of freedom at decoupling can be expressed in terms of the effective number of neutrino species N_ν (which does not need to be an integer). This is a convenient way of parameterizing ignorance about fundamental particle constituents of nature. Contributors to N_ν could include, for example, an extra sterile neutrino sometimes invoked in neutrino oscillation models, or the thermal background of gravitons which would exist if inflation did not occur. This parameter can also include the effects of entropy increases due to decaying or annihilating particles; see chapter 3 of Kolb and Turner (1990) for a detailed discussion. As far as the microwave background is concerned, N_ν determines the radiation energy density of the universe and thus modifies the time of matter–radiation equality. It can, in principle, be distinguished from a change in $\Omega_0 h^2$ because it affects other physical parameters like the baryon density or the angular diameter distance differently than a shift in either Ω_0 or h.

Neutrino masses cannot provide the bulk of the DM, because their free-streaming greatly suppresses fluctuation power on galaxy scales, leading to a drastic mismatch with observed large-scale structure. But models with some small fraction of dark matter as neutrinos have been advocated to improve the agreement between the predicted and observed large-scale structure power spectrum. Massive neutrinos have several small effects on the microwave background, which have been studied systematically by Dodelson *et al* (1996). They can slightly increase the sound horizon at decoupling due to their transition from relativistic to non-relativistic behaviour as the universe expands. More importantly, free-streaming of massive neutrinos around the time of last scattering leads to a faster decay of the gravitational potentials, which in turn means more forcing of the acoustic oscillations and a resulting increase in the monopole perturbations. Finally, since matter–radiation equality is slightly delayed for neutrinos with cosmologically interesting masses of a few eV, the gravitational potentials are less constant and a larger Integrated Sachs–Wolfe effect is induced. The change in sound horizon and shift in matter–radiation equality due to massive neutrinos cannot be distinguished from changes in $\Omega_b h^2$ and $\Omega_0 h^2$, but the alteration of the gravitational potential's time dependence due to neutrino free-streaming cannot be mimicked by some other change in parameters. In principle the effect of neutrino masses can be extracted from the microwave background, although the effects are very small.

7.5.4   Idealized experiments

Remarkably, the microwave background power spectrum contains enough information to constrain numerous parameters simultaneously (Jungman *et al* 1996). We would like to estimate quantitatively just how well the space of parameters described earlier can be constrained by ideal measurements of the microwave background. The question has been studied in some detail; this section outlines the basic methods and results, and discusses how good various approximations are. For simplicity, only temperature fluctuations are considered in this section; the corresponding formalism for the polarization power spectra is developed in Kamionkowski *et al* (1997a, b).

Given a pixelized map of the microwave sky, we need to determine the contribution of pixelization noise, detector noise, and beam width to the multipole moments and power spectrum. Consider a temperature map of the sky $T^{\mathrm{map}}(\hat{n})$ which is divided into N_{pix} equal-area pixels. The observed temperature in pixel j is due to a cosmological signal plus noise, $T_j^{\mathrm{map}} = T_j + T_j^{\mathrm{noise}}$. The multipole coefficients of the map can be constructed as

$$
\begin{aligned}
d_{lm}^{\mathrm{T}} &= \frac{1}{T_0} \int \mathrm{d}\hat{n}\, T^{\mathrm{map}}(\hat{n}) Y_{lm}(\hat{n}) \\
&\simeq \frac{1}{T_0} \sum_{j=1}^{N_{\mathrm{pix}}} \frac{4\pi}{N_{\mathrm{pix}}} T_j^{\mathrm{map}} Y_{lm}(\hat{n}_j),
\end{aligned}
\tag{7.31}
$$

where $\hat{n}_j$ is the direction vector to pixel j. The map moments are written as d_{lm} to distinguish them from the moments of the cosmological signal a_{lm}; the former include the effects of noise. The extent to which the second line in equations (7.31) is only an approximate equality is the pixelization noise. Most current experiments oversample the sky with respect to their beam, so the pixelization noise is negligible. Now assume that the noise is uncorrelated between pixels and is well represented by a normal distribution. Also, assume that the map is created with a Gaussian beam with width θ_b. Then it is straightforward to show that the variance of the temperature moments is given by (Knox 1995)

$$\langle d_{lm}^{\mathrm{T}} d_{l'm'}^{\mathrm{T}*} \rangle = (C_l e^{-l^2 \sigma_b^2} + w^{-1}) \delta_{ll'} \delta_{mm'}, \tag{7.32}$$

where $\sigma_b = 0.007\,42(\theta_b/1°)$ and

$$w^{-1} = \frac{4\pi}{N_{\mathrm{pix}}} \frac{\langle (T_i^{\mathrm{noise}})^2 \rangle}{T_0^2} \tag{7.33}$$

is the inverse statistical weight per unit solid angle, a measure of experimental sensitivity independent of the pixel size.

Now the power spectrum can be estimated via equation (7.32) as

$$C_l^{\mathrm{T}} = (D_l^{\mathrm{T}} - w^{-1}) e^{l^2 \sigma_b^2} \tag{7.34}$$

where

$$D_l^{\mathrm{T}} = \frac{1}{2l+1} \sum_{m=-l}^{l} d_{lm}^{\mathrm{T}} d_{lm}^{\mathrm{T}*}. \tag{7.35}$$

The individual coefficients d_{lm}^{T} are Gaussian random variables. This means that C_l^{T} is a random variable with a χ_{2l+1}^2 distribution, and its variance is (Knox 1995)

$$(\Delta C_l^{\mathrm{T}})^2 = \frac{2}{2l+1} (C_l + w^{-1} e^{l^2 \sigma_b^2}). \tag{7.36}$$

Note that even for $w^{-1} = 0$, corresponding to zero noise, the variance is non-zero. This is the cosmic variance, arising from the fact that we have only one sky to observe: the estimator in equation (7.35) is the sum of $2l + 1$ random variables, so it has a fundamental fractional variance of $(2l + 1)^{-1/2}$ simply due to Poisson statistics. This variance provides a benchmark for experiments: if the goal is to determine a power spectrum, it makes no sense to improve resolution or sensitivity beyond the level at which cosmic variance is the dominant source of error.

Equation (7.36) is extremely useful: it gives an estimate of how well the power spectrum can be determined by an experiment with a given beam size and detector noise. If only a portion of the sky is covered, the variance estimate should be divided by the fraction of the total sky covered. With these variances in hand, standard statistical techniques can be employed to estimate how well a given measurement can recover a given set *s* of cosmological parameters. Approximate the dependence of C_l^{T} on a given parameter as linear in the parameter; this will always be true for some sufficiently small range of parameter values. Then the parameter space curvature matrix (also known as the Fisher information matrix) is specified by

$$\alpha_{ij} = \sum_l \frac{\partial C_l^{\mathrm{T}}}{\partial s_i} \frac{\partial C_l^{\mathrm{T}}}{\partial s_j} \frac{1}{(\Delta C_l^{\mathrm{T}})^2}. \tag{7.37}$$

The variance in the determination of the parameter s_i from a set of C_l^{T} with variances ΔC_l^{T} after marginalizing over all other parameters is given by the diagonal element i of the matrix α^{-1}.

Estimates of this kind were first made by Jungman *et al* (1996) and subsequently refined by Zaldarriaga *et al* (1997) and Bond *et al* (1997), among others. The basic result is that a map with pixels of a few arcminutes in size and a signal-to-noise ratio of around one per pixel can determine Ω, $\Omega_b h^2$, $\Omega_m h^2$, Λh^2, Q, n, and z_r at the few percent level *simultaneously*, up to the one degeneracy mentioned earlier (see the table in Bond *et al* 1997). Significant constraints will also be placed on r and N_ν. This prospect has been the primary reason that the microwave background has generated such excitement. Note that Ω, h, Ω_b, and Λ are the classical cosmological parameters. Decades of painstaking astronomical observations have been devoted to determining the values of these parameters. The microwave background offers a completely independent method of determining them with comparable or significantly greater accuracy, and with fewer astrophysical systematic effects to worry about. The microwave background is also the only source of precise information about the spectrum and character of the primordial perturbations from which we arose. Of course, these exciting possibilities hold only if the universe is accurately represented by a model in the assumed model space. The model space is, however, quite broad. Model-independent constraints which the microwave background provides are discussed in section 7.6.

The estimates of parameter variances based on the curvature matrix would be exact if the power spectrum always varied linearly with each parameter. This, of course, is not true in general. Given a set of power spectrum data, we want to know two pieces of information about the cosmological parameters: (1) What parameter values provide the best-fit model? (2) What are the error bars on these parameters, or more precisely, what is the region of parameter space which defines a given confidence level? The first question can be answered easily using standard methods of searching parameter space; generally such a search requires evaluating the power spectrum for fewer than 100 different models. This shows that the parameter space is generally without complicated structure or many false minima. The second question is more difficult. Anything beyond the curvature matrix analysis requires looking around in parameter space near the best-fit model. A specific Monte Carlo technique employing a Metropolis algorithm has recently been advocated (Christensen and Meyer 2000); such techniques will certainly prove more flexible and efficient than recent brute-force grid searches (Tegmark and Zaldarriaga 2000). As upcoming data-sets contain more information and consequently have greater power to constrain parameters, efficient techniques of parameter space exploration will become increasingly important.

To this point, the discussion has assumed that the microwave background power spectrum is perfectly described by linear perturbation theory. Since the temperature fluctuations are so small, parts in a hundred thousand, linear theory is a very good approximation. However, on small scales, nonlinear effects become important and can dominate over the linear contributions. The most important nonlinear effects are the Ostriker–Vishniac effect coupling velocity and density perturbations (Jaffe and Kamionkowski 1998, Hu 2000), gravitational lensing by large-scale structure (Seljak 1996), the Sunyaev–Zeldovich effect which gives spectral distortions when the microwave background radiation passes through hot ionized regions (Birkinshaw 1999) and the kinetic Sunyaev–Zeldovich effect which Doppler shifts radiation passing through plasma with bulk velocity (Gnedin and Jaffe 2000). All three effects are measurable and give important additional constraints on cosmology, but more detailed descriptions are outside the scope of this chapter.

Finally, no discussion of parameter determination would be complete without mention of galactic foreground sources of microwave emission. Dust radiates significantly at microwave frequencies, as do free–free and synchrotron emission; point source microwave emission is also a potential problem. Dust emission generally has a spectrum which rises with frequency, while free–free and synchrotron emission have falling frequency spectra. The emission is not uniform on the sky, but rather concentrated in the galactic plane, with fainter but pervasive diffuse emission in other parts of the sky. The dust and synchrotron/free–free emission spectra cross each other at a frequency of around 90 GHz. Fortunately for cosmologists, the amplitude of the foreground emission at this frequency is low enough to create a frequency window in which the cosmological temperature fluctuations dominate the foreground temperature fluctuations. At other frequencies, the foreground contribution can be effectively separated from the cosmological blackbody signal by measuring in several different frequencies and projecting out the portion of the signal with a flat frequency spectrum. The foreground situation for polarization is less clear, both in amplitude and spectral index, and could potentially be a serious systematic limit to the quality of cosmological polarization data. However,, it may be no greater problem for polarization fluctuations than for temperature fluctuations. For an overview of issues surrounding foreground emission, see Bouchet and Gispert 1999 or the WOMBAT web site, http://astro.berkeley.edu/wombat.

7.5.5 Current constraints and upcoming experiments

As the Como School began, results from the high-resolution balloon-born experiment MAXIMA (Hanany *et al* 2000) were released, complementing the week-old data from BOOMERanG (de Bernardis *et al* 2000) and creating a considerable buzz at coffee breaks. The derived power spectrum estimates are

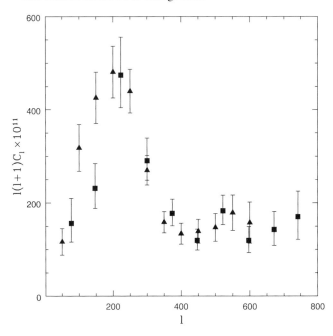

Figure 7.3.   Two current measurements of the microwave background radiation temperature power spectrum. Triangles are BOOMERanG measurements multiplied by 1.21; squares are MAXIMA measurements multiplied by 0.92. The normalization factors are within the calibration uncertainties of the experiments, and were chosen by Hanany *et al* (2000) to give the most consistent results between the two experiments.

shown in figure 7.3. The data from the two measurements appear consistent up to calibration uncertainties, and for simplicity will be referred to here as 'balloon data' and discussed as a single result. While a few experimenters and data analysers were members of both experimental teams, the measurements and data reductions were done essentially independently. Earlier data from the previous year (Miller *et al* 1999) had clearly demonstrated the existence and angular scale of the first peak in the power spectrum and produced the first maps of the microwave background at angular scales below a degree. But the new results from balloon experiments utilizing extremely sensitive bolometric detectors represent a qualitative step forward. These experiments begin to exploit the potential of the microwave background for 'precision cosmology'; their power spectra put strong constraints on several cosmological parameters simultaneously and rule out many variants of cosmological models. In fact, what is most interesting is that, at face value, these measurements put significant pressure on all of the standard models outlined earlier.

The balloon data show two major features: first, a large peak in the power spectrum centred around $l = 200$ with an amplitude of approximately $l^2 C_l =$

36 000 μK^2, and second, a broad plateau between $l = 400$ and $l = 700$ with an amplitude of approximately $l^2 C_l = 10\,000\ \mu K^2$. The first peak is clearly delineated and provides good evidence that the universe is spatially flat, i.e. $\Omega = 1$. The issue of a second acoustic peak is much less clear. In most flat universe models with acoustic oscillations, the second peak is expected to appear at an angular scale of around $l = 400$. The angular resolution of the balloon experiments is certainly good enough to see such a peak, but the power spectrum data show no evidence for one. I argue that a flat line is an excellent fit to the data past $l = 300$, and that any model which shows a peak in this region will be a worse fit than a flat line. This does not necessarily mean that no peak is present; the error bars are too large to rule out a peak, but the amplitude of such a peak is fairly strongly constrained to be lower than expected given the first peak.

What does this mean for cosmological models? Within the model space outlined in the previous section, there are three ways to suppress the second peak. The first would be to have a power spectrum index n substantially less than one. This solution would force abandonment of the assumption of power-law initial fluctuations, in order to match the observed amplitude of large-scale structure at smaller scales. While this is certainly possible, it represents a drastic weakening in the predictive power of the microwave background: essentially, a certain feature is reproduced by arbitrarily changing the primordial power spectrum. While no physical principle requires power-law primordial perturbations, we should wait for microwave background measurements on a much wider range of scales combined with overlapping large-scale structure measurements before resorting to departures from power-law initial conditions. If the universe really did possess an initial power spectrum with a variety of features in it, most of the promise of precision cosmology is lost. Recent power spectra extracted from the IRAS Point Source Survey Redshift Catalogue (Hamilton and Tegmark 2000), which show a remarkable power law behaviour spanning three orders of magnitude in wavenumber, seem to argue against this possibility.

The second possibility is a drastic amount of reionization. It is not clear the extent to which this might be compatible with the height of the first peak and still suppress the second peak sufficiently. This possibility seems unlikely as well, but would show clear signatures in the microwave background polarization.

The most commonly discussed possibility is that the very low second peak amplitude reflects an unexpectedly large fraction of baryons relative to DM in the universe. The baryon signature discussed in section 7.4.4 gives a suppression of the second peak in this case. However, primordial nucleosynthesis also constrains the baryon–photon ratio. Recent high-precision measurements of deuterium absorption in high-redshift neutral hydrogen clouds (Tytler *et al* 2000) give a baryon–photon number ratio of $\eta = 5.1 \pm 0.5 \times 10^{10}$, which translates to $\Omega_b h^2 = 0.019 \pm 0.002$ assuming that the entropy (i.e. photon number) per comoving volume remains constant between nucleosynthesis and the present. Requiring Ω_b to satisfy this nucleosynthesis constraint leads to microwave background power spectra which are not particularly good fits to the data. An alternative is that the entropy per comoving volume has *not* remained fixed between nucleosynthesis and recombination (see, e.g., Kaplinghat and Turner 2000). This could be arranged by having a DM particle which decays to photons, although such a process must evade limits from the lack of microwave background spectral distortions (Hu and Silk 1993). Alternately, a large chemical potential for the neutrino background could lead to larger inferred values for the baryon–photon ratio from nucleosynthesis (Esposito *et al* 2000). Either way, if both the microwave background measurements and the high-redshift deuterium abundances hold up, the discrepancy points to new physics. Of course, a final explanation for the discrepancies is simply that the balloon data have significant systematic errors.

I digress for a brief editorial comment about data analysis. Straightforward searches of the conventional cosmological model space described earlier for good fits to the balloon data give models with very low DM densities, high baryon fractions and very large cosmological constants (see model P1 in table 1 of Lange *et al* 2000). Such models violate other observational constraints on age, which must be at least 12 billion years (see, e.g., Peacock *et al* 1998), and quasar and radio source strong lensing number counts, which limit a cosmological constant to $\Lambda \leq 0.7$ (Falco *et al* 1998). The response to this situation so far has been to invoke Bayesian prior probability distributions on various quantities like Ω_b and the age. This leads to a best-fit model with a nominally acceptable χ^2 (Lange *et al* 2000, Tegmark *et al* 2000 and others). But be wary of this procedure when the priors have a large effect on the best-fit model! The microwave background will soon provide tighter constraints on most parameters than any other source of prior information. Priors probabilities on a given parameter are useful and justified when the microwave background data have little power to constrain that parameter; in this case, the statistical quality of the model fit to the microwave background data will not be greatly affected by imposing the prior. However, something fishy is probably going on when a prior pulls a parameter multiple sigma away from its best-fit value without the prior. This is what happens presently with Ω_b when the nucleosynthesis prior is enforced. If your priors make a big difference, it is likely either that some of the data are incorrect or that the model space does not include the correct model. Both the microwave background measurements and the high-redshift deuterium detections are taxing observations dominated by systematic effects, so it is certainly possible that one or both are wrong. However, MAXIMA and BOOMERanG are consistent with each other while using different instruments, different parts of the sky, and different analysis pipelines, and the deuterium measurements are consistent for several different clouds. This suggests possible missing physics ingredients, like extreme reionization or an entropy increase mentioned earlier, or perhaps significant contributions from cosmic defects. It has even been suggested by otherwise sober and reasonable people that the microwave background results, combined with various difficulties related to dynamics of spiral galaxies, may point towards a radical revision of the standard cosmology (Sellwood and Kosowsky 2000).

We should not rest lightly until the cosmological model preferred by microwave background measurements is comfortably consistent with all relevant priors derived from other data sources of comparable precision.

The picture will come into sharper relief over the next two years. The MAP satellite (http://map.gsfc.nasa.gov), launched by NASA on 30 June 2001, will map the full microwave sky in five frequency channels with an angular resolution of around 15 arc minutes and a temperature sensitivity per pixel of a part in a million. Space missions offer unequalled sky coverage and control of systematics and, if it works as advertized, MAP will be a benchmark experiment. Prior to its launch, expect to see the first interferometric microwave data at angular scales smaller than a half degree from the CBI interferometer experiment (http://www.astro.caltech.edu/~tjp/CBI/). In this same time frame, we also may have the first detection of polarization. The most interesting power spectrum feature to focus on will be the existence and amplitude of a third acoustic peak. If a third peak appears with amplitude significantly higher than the putative second peak, this almost certainly indicates conventional acoustic oscillations with a high baryon fraction and possibly new physics to reconcile the result with the deuterium measurements. If, however, the power spectrum remains flat or falls further past the second peak region, then all bets are off. In a time frame of the next 5 to 10 years, we can reasonably expect to have a cosmic-variance limited temperature power spectrum down to scales of a few arcminutes (say, $l = 4000$), along with significant polarization information (though probably not cosmic-variance limited power spectra). In particular, ESA's Planck satellite mission (http://astro.estec.esa.nl/SA-general/Projects/Planck/) will map the microwave sky in nine frequency bands at significantly better resolution and sensitivity than the MAP mission. For a comprehensive listing of past and planned microwave background measurements, see Max Tegmark's experiments web page, http://www.hep.upenn.edu/~max/cmb/experiments.html.

7.6   Model-independent cosmological constraints

Most analysis of microwave background data and predictions about its ability to constrain cosmology have been based on the cosmological parameter space described in section 7.5.1. This space is motivated by inflationary cosmological scenarios, which generically predict power-law adiabatic perturbations evolving only via gravitational instability. Considering that this space of models is broad and appears to fit all current data far better than any other proposed models, such an assumed model space is not very restrictive. In particular, proposed extensions tend to be rather *ad hoc*, adding extra elements to the model without providing any compelling underlying motivation for them. Examples which have been discussed in the literature include multiple types of DM with various properties, non-standard recombination, small admixtures of topological defects, production of excess entropy, or arbitrary initial power spectra. None of these possibilities are attractive from an aesthetic point of view: all add significant complexity and freedom to the models without any corresponding restrictions on the original parameter space. The principle of Occam's Razor should cause us to be sceptical about any such additions to the space of models.

However, it is possible that some element *is* missing from the model space, or that the actual cosmological model is radically different in some respect. The microwave background is the probe of cosmology most tightly connected to the fundamental properties of the universe and least influenced by astrophysical complications, and thus the most capable data source for deciding whether the universe actually is well described by some model in the usual model space. An interesting question is the extent to which the microwave background can determine various properties of the universe independent from particular models. While any cosmological interpretation of temperature fluctuations in the microwave sky requires some kind of minimal assumptions, all of the conclusions outlined later can be drawn without invoking a detailed model of initial conditions or structure formation. These conclusions are in contrast to precision determination of cosmological parameters, which does require the assumption of a particular space of models and which can vary significantly depending on the space.

7.6.1 Flatness

The Friedmann–Robertson–Walker spacetime describing homogeneous and isotropic cosmology comes in three flavours of spatial curvature: positive, negative and flat, corresponding to $\Omega > 1$, $\Omega < 1$ and $\Omega = 1$ respectively. One of the most fundamental questions of cosmology, dating to the original relativistic cosmological models, is the curvature of the background spacetime. The fate of the universe quite literally depends on the answer: in a cosmology with only matter and radiation, a positively curved universe will eventually recollapse in a fiery 'Big Crunch' while flat and negatively curved universes will expand forever, meeting a frigid demise. Note these fates are at least 40 billion years in the future. (A cosmological constant or other energy density component with an unusual equation of state can alter these outcomes, causing a closed universe eventually to enter an inflationary stage.)

The microwave background provides the cleanest and most powerful probe of the geometry of the universe (Kamionkowski *et al* 1994). The surface of last scattering is at a high enough redshift that photon geodesics between the last scattering surface and the Earth are significantly curved if the geometry of the universe is appreciably different than flat. In a positively curved space, two geodesics will bend towards each other, subtending a larger angle at the observer than in the flat case; likewise, in a negatively curved space two geodesics bend away from each other, resulting in a smaller observed angle between the two. The operative quantity is the angular diameter distance; Weinberg (2000) gives a pedagogical discussion of its dependence on Ω. In a flat universe, the horizon length at the time of last scattering subtends an angle on the sky of around two degrees. For a low-density universe with $\Omega = 0.3$, this angle becomes smaller by half, roughly.

A change in angular scale of this magnitude will change the apparent scale of all physical scales in the microwave background. A model-independent determination of Ω thus requires a physical scale of known size to be imprinted on the primordial plasma at last scattering; this physical scale can then be compared with its apparent observed scale to obtain a measurement of Ω. The microwave background fluctuations actually depend on two basic physical scales. The first is the sound horizon at last scattering, r_s (cf equation (7.29)). If coherent acoustic oscillations are visible, this scale sets their characteristic wavelengths. Even if coherent acoustic oscillations are not present, the sound horizon represents the largest scale on which any causal physical process can influence the primordial plasma. Roughly, if primordial perturbations appear on all scales, the resulting microwave background fluctuations appear as a featureless power law at large scales, while the scale at which they begin to depart from this assumed primordial behaviour corresponds to the sound horizon. This is precisely the behaviour observed by current measurements, which show a prominent power spectrum peak at an angular scale of a degree ($l = 200$), arguing strongly for a flat universe. Of course, it is logically possible that the primordial power spectrum has power on scales only significantly smaller than the horizon at last scattering. In this case, the largest scale perturbations would appear at smaller angular scales for a given geometry. But then the observed power-law perturbations at large angular scales must be reproduced by the Integrated Sachs–Wolfe effect, and resulting models are contrived. If the microwave background power spectrum exhibits acoustic oscillations, then the spacing of the acoustic peaks depends only on the sound horizon independent of the phase of the oscillations; this provides a more general and precise probe of flatness than the first peak position.

The second physical scale provides another test: the Silk damping scale is determined solely by the thickness of the surface of last scattering, which in turn depends only on the baryon density $\Omega_b h^2$, the expansion rate of the universe and standard thermodynamics. Observation of an exponential suppression of power at small scales gives an estimate of the angular scale corresponding to the damping scale. Note that the effects of reionization and gravitational lensing must both be accounted for in the small-scale dependence of the fluctuations. If the reionization redshift can be accurately estimated from microwave background polarization (see later) and the baryon density is known from primordial nucleosynthesis or from the alternating peak heights signature (section 7.4.4), only a radical modification of the standard cosmology altering the time dependence of the scale factor or modifying thermodynamic recombination can change the physical damping scale. If the estimates of Ω based on the sound horizon and damping scales are consistent, this is a strong indication that the inferred geometry of the universe is correct.

7.6.2   Coherent acoustic oscillations

If a series of peaks equally spaced in l is observed in the microwave background temperature power spectrum, it strongly suggests we are seeing the effects of coherent acoustic oscillations at the time of last scattering. Microwave background polarization provides a method for confirming this hypothesis. As explained in section 7.3.2, polarization anisotropies couple primarily to velocity perturbations, while temperature anisotropies couple primarily to density perturbations. Now coherent acoustic oscillations produce temperature power spectrum peaks at scales where a mode of that wavelength has either maximum or minimum compression in potential wells at the time of last scattering. The fluid velocity for the mode at these times will be zero, as the oscillation is turning around from expansion to contraction (envision a mass on a spring.) At scales intermediate between the peaks, the oscillating mode has zero density contrast but a maximum velocity perturbation. Since the polarization power spectrum is dominated by the velocity perturbations, its peaks will be at scales interleaved with the temperature power spectrum peaks. This alternation of temperature and polarization peaks as the angular scale changes is characteristic of acoustic oscillations (see Kosowsky (1999) for a more detailed discussion). Indeed, it is almost like seeing the oscillations directly: it is difficult to imagine any other explanation for density and velocity extrema on alternating scales. The temperature-polarization cross-correlation must also have peaks with corresponding phases. This test will be very useful if a series of peaks is detected in a temperature power spectrum which is not a good fit to the standard space of cosmological models. If the peaks turn out to reflect coherent oscillations, we must then modify some aspect of the underlying cosmology, while if the peaks are not coherent oscillations, we must modify the process by which perturbations evolve.

If coherent oscillations are detected, any cosmological model must include a mechanism for enforcing coherence. Perturbations on all scales, in particular on scales outside the horizon, provide the only natural mechanism: the phase of the oscillations is determined by the time when the wavelength of the perturbation becomes smaller than the horizon, and this will clearly be the same for all perturbations of a given wavelength. For any source of perturbations inside the horizon, the source itself must be coherent over a given scale to produce phase-coherent perturbations on that scale. This cannot occur without artificial fine-tuning.

7.6.3   Adiabatic primordial perturbations

If the microwave background temperature and polarization power spectra reveal coherent acoustic oscillations and the geometry of the universe can also be determined with some precision, then the phases of the acoustic oscillations can be used to determine whether the primordial perturbations are adiabatic or isocurvature. Quite generally, equation (7.28) shows that adiabatic and isocurvature power spectra must have peaks which are out of phase. While current measurements of the microwave background and large-scale structure rule out models based entirely on isocurvature perturbations, some relatively small admixture of isocurvature modes with dominant adiabatic modes is possible. Such mixtures arise naturally in inflationary models with more than one dynamical field during inflation (see, e.g., Mukhanov and Steinhardt 1998).

7.6.4 Gaussian primordial perturbations

If the temperature perturbations are well approximated as a Gaussian random field, as microwave background maps so far suggest, then the power spectrum C_l contains all statistical information about the temperature distribution. Departures from Gaussianity take myriad different forms; the business of providing general but useful statistical descriptions is a complicated one (see, e.g., Ferreira *et al* 1997). Tiny amounts of non-Gaussianity will arise inevitably from the nonlinear evolution of fluctuations, and larger non-Gaussian contributions can be a feature of the primordial perturbations or can be induced by 'stiff' stress–energy perturbations such as topological defects. As explained later, defect theories of structure formation seem to be ruled out by current microwave background and large-scale structure measurements, so interest in non-gaussianity has waned. But the extent to which the temperature fluctuations are actually Gaussian is experimentally answerable and, as observations improve, this will become an important test of inflationary cosmological models.

7.6.5 Tensor or vector perturbations

As described in section 7.3.3, the tensor field describing microwave background polarization can be decomposed into two components corresponding to the gradient-curl decomposition of a vector field. This decomposition has the same physical meaning as that for a vector field. In particular, any gradient-type tensor field, composed of the G-harmonics, has no curl, and thus may not have any handedness associated with it (meaning the field is even under parity reversal), while the curl-type tensor field, composed of the C-harmonics, does have a handedness (odd under parity reversal).

This geometric interpretation leads to an important physical conclusion. Consider a universe containing only scalar perturbations, and imagine a single Fourier mode of the perturbations. The mode has only one direction associated with it, defined by the Fourier vector k; since the perturbation is scalar, it must be rotationally symmetric around this axis. (If it were not, the gradient of the perturbation would define an independent physical direction, which would violate the assumption of a scalar perturbation.) Such a mode can have no physical handedness associated with it and, as a result, the polarization pattern it induces in the microwave background couples only to the G harmonics. Another way of

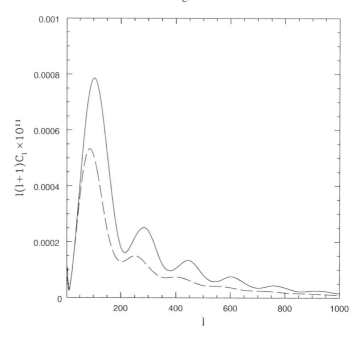

Figure 7.4. Polarization power spectra from tensor perturbations: the full curve is C_l^G and the broken curve is C_l^C. The amplitude gives a 10% contribution to the COBE temperature power spectrum measurement at low l. Note that scalar perturbations give no contribution to C_l^C.

stating this conclusion is that primordial density perturbations produce *no* C-type polarization as long as the perturbations evolve linearly. However, primordial tensor or vector perturbations produce both G-type and C-type polarization of the microwave background (provided that the tensor or vector perturbations themselves have no intrinsic net polarization associated with them).

Measurements of cosmological C-polarization in the microwave background are free of contributions from the dominant scalar density perturbations and thus can reveal the contribution of tensor modes in detail. For roughly scale-invariant tensor perturbations, most of the contribution comes at angular scales larger than $2°$ ($2 < l < 100$). Figure 7.4 displays the C and G power spectra for scale-invariant tensor perturbations contributing 10% of the COBE signal on large scales. A microwave background map with forseeable sensitivity could measure gravitational wave perturbations with amplitudes smaller than 10^{-3} times the amplitude of density perturbations (Kamionkowski and Kosowsky 1998). The C-polarization signal also appears to be the best hope for measuring the spectral index n_T of the tensor perturbations.

7.6.6 Reionization redshift

Reionization produces a distinctive microwave background signature. It suppresses temperature fluctuations by increasing the effective damping scale, while it also increases large-angle polarization due to additional Thomson scattering at low redshifts when the radiation quadrupole fluctuations are much larger. This enhanced polarization peak at large angles will be significant for reionization prior to $z = 10$ (Zaldarriaga 1997). Reionization will also greatly enhance the Ostriker–Vishniac effect, a second-order coupling between density and velocity perturbations (Jaffe and Kamionkowski 1998). The non-uniform reionization inevitable if the ionizing photons come from point sources, as seems likely, may also create an additional feature at small angular scales (Hu and Gruzinov 1998, Knox *et al* 1998). Taken together, these features are clear indicators of the reionization redshift z_r independent of any cosmological model.

7.6.7 Magnetic fields

Primordial magnetic fields would be clearly indicated if cosmological Faraday rotation were detected in the microwave background polarization. A field with comoving field strength of 10^{-9} Gauss would produce a signal with a few degrees of rotation at 30 GHz, which is likely just detectable with future polarization experiments (Kosowsky and Loeb 1996). Faraday rotation has the effect of mixing G-type and C-type polarization, and would be another contributor to the C-polarization signal, along with tensor perturbations. Depolarization will also result from Faraday rotation in the case of significant rotation through the last-scattering surface (Harari *et al* 1996). Additionally, the tensor and vector metric perturbations produced by magnetic fields result in further microwave background fluctuations. A distinctive signature of such fields is that for a range of power spectra, the polarization fluctuations from the metric perturbations is comparable to, or larger than, the corresponding temperature fluctuations (Kahniashvili *et al* 2000). Since the microwave background power spectra vary as the fourth power of the magnetic field amplitude, it is unlikely that we can detect magnetic fields with comoving amplitudes significantly below 10^{-9} Gauss. However, if such fields do exist, the microwave background provides several correlated signatures which will clearly reveal them.

7.6.8 The topology of the universe

Finally, one other microwave background signature of a very different character deserves mention. Most cosmological analyses make the implicit assumption that the spatial extent of the universe is infinite or, in practical terms, at least much larger than our current Hubble volume so that we have no way of detecting the bounds of the universe. However, this need not be the case. The requirement that the unperturbed universe be homogeneous and isotropic determines the spacetime metric to be of the standard Friedmann–Robertson–Walker form, but this is only a *local* condition on the spacetime. Its global structure is still unspecified. It is possible to construct spacetimes which at every point have the usual homogeneous and isotropic metric, but which are spatially compact (have finite volumes). The most familiar example is the construction of a three-torus from a cubical piece of the flat spacetime by identifying opposite sides. Classifying the possible topological spaces which locally have the metric structure of the usual cosmological spacetimes (i.e. have the Friedmann–Robertson–Walker spacetimes as a topological covering space) has been studied extensively. The zero-curvature and positive-curvature cases have only a handful of possible topological spaces associated with them, while the negative curvature case has an infinite number with a very rich classification. See Weeks (1998) for a review.

If the topology of the universe is non-trivial and the volume of the universe is smaller than the volume contained by a sphere with radius equal to the distance to the surface of last scattering, then it is possible to detect the topology. Cornish *et al* (1998) pointed out that because the last scattering surface is always a sphere in the covering space, any small topology will result in matched circles of temperature on the microwave sky. The two circles represent photons originating from the same physical location in the universe but propagating to us in two different directions. Of course, the temperatures around the circles will not match exactly, but only the contributions coming from the Sachs–Wolfe effect and the intrinsic temperature fluctuations will be the same; the velocity and Integrated Sachs–Wolfe contributions will differ and constitute a noise source. Estimates show the circles can be found efficiently via a direct search of full-sky microwave background maps. Once all matching pairs of circles have been discovered, their number and relative locations on the sky strongly overdetermine the topology of the universe in most cases. Remarkably, the microwave background essentially allows us to determine the size of the universe if it is smaller than the current horizon volume in any dimension.

7.7   Finale: testing inflationary cosmology

In summary, the CMB radiation is a remarkably interesting and powerful source of information about cosmology. It provides an image of the universe at an early time when the relevant physical processes were all very simple, so the dependence of anisotropies on the cosmological model can be calculated with high precision. At the same time, the universe at decoupling was an interesting enough place that small differences in cosmology will produce measurable differences in the anisotropies.

The microwave background has the ultimate potential to determine fundamental cosmological parameters describing the universe with percent-level precision. If this promise is realized, the standard model of cosmology would compare with the standard model of particle physics in terms of physical scope, explanatory power and detail of confirmation. But in order for such a situation to come about, we must first choose a model space which includes the correct model for the universe. The accuracy with which cosmological parameters can be determined is of course limited by the accuracy with which some model in the model space represents the actual universe.

The space of models discussed in section 7.5.1 represents universes which we would expect to arise from the mechanism of inflation. These models have become the standard testing ground for comparisons with data because they are simple, general and well motivated. So far, these types of models fit the data well, much better than any competing theories. Future measurements may remain perfectly consistent with inflationary models, may reveal inconsistencies which can be remedied via minor extensions or modifications of the parameter space or may require more serious departures from these types of models.

For the sake of a concluding discussion about the power of the microwave background, assume that the universe actually is well described by inflationary cosmology, and that it can be modelled by the parameters in section 7.5.1. For an overview of inflation and the problems it solves, see Kolb and Turner (1990, ch 8) or the chapter by A Linde in this volume. To what extent can we hope to verify inflation, a process which likely would have occurred at an energy scale of 10^{16} GeV when the universe was 10^{-38} s old? Direct tests of physics at these energy scales are unimaginable, leaving cosmology as the only likely way to probe this physics.

Inflation is not a precise theory, but rather a mechanism for exponential expansion of the universe which can be realized in a variety of specific physical models. Cosmology in general and the cosmic microwave background, in particular, can hope to test the following predictions of inflation (see Kamionkowski and Kosowsky 1999 for a more complete discussion of inflation and its observable microwave background properties):

- The most basic prediction of inflation is a spatially flat universe. The flatness problem was one of the fundamental motivations for considering inflation in the first place. While it is possible to construct models of inflation which result in a non-flat universe, they all must be finely tuned for inflation to end at just the right time for a tiny but non-negligible amount of curvature to remain. The geometry of the universe is one of the fundamental pieces of physics which can be extracted from the microwave background power spectra. Recent measurements make a strong case that the universe is indeed flat.

- Inflation generically predicts primordial perturbations which have a Gaussian statistical distribution. The microwave background is the only precision test of this prediction. Primordial Gaussian perturbations will still be almost precisely Gaussian at recombination, whereas they will have evolved significant non-Gaussianity by the time the local large-scale structure forms, due to gravitational collapse. Other methods of probing Gaussianity, like number densities of galaxies or other objects, inevitably depend significantly on astrophysical modelling.

- The simplest models of inflation, with a single dynamical scalar field, give adiabatic primordial perturbations. The only real test of this prediction comes from the microwave background power spectrum. More complex models of inflation with multiple dynamical fields generically result in dominant adiabatic fluctuations with some admixture of isocurvature fluctuations. Limits on isocurvature fluctuations obtained from microwave background measurements could be used to place constraints on the size of couplings between different fields at inflationary energy scales.

- Inflation generically predicts primordial perturbations on all scales, including scales outside the horizon. Of course we can never test directly whether perturbations on scales larger than the horizon exist, but the microwave background can reveal perturbations at recombination on scales comparable to the horizon scale. Zaldarriaga and Spergel (1997) have argued that inflation generically gives a peak in the polarization power spectrum at angular scales larger than $2°$, and that no causal perturbations at the epoch of last scattering can produce a feature at such large scales. Inflation further predicts that the primordial power spectrum should be close to a scale-invariant power law (e.g. Huterer and Turner 2000), although complicated models can lead to power spectra with features or significant departures from scale invariance. The microwave background can probe the primordial power spectrum over three orders of magnitude.

- Inflationary perturbations result in phase-coherent acoustic oscillations. The coherence arises because on any given scale, the perturbations start in the same state determined only by their character outside the horizon. For a discussion in the language of squeezed quantum states, see Albrecht (2000). It is extremely difficult to produce coherent oscillations by any mechanism other than perturbations outside the horizon. The microwave background temperature and polarization power spectra will together clearly reveal coherent oscillations.

- Inflation finally predicts potentially measurable relationships between the amplitudes and power law indices of the primordial density and gravitational wave perturbations (see Lidsey *et al* 1997 for a comprehensive overview), and measuring a C_l^C power spectrum appears to be the only way to obtain precise enough measurements of the tensor perturbations to test these predictions, thanks to the fact that the density perturbations do not contribute to C_l^C. Detection of inflationary tensor perturbations would reveal the energy scale at which inflation occurred, while confirming the inflationary relationships between scalar and tensor perturbations would provide a strong consistency check on inflation.

The potential power of the microwave background is demonstrated by the fact that inflation, a theoretical mechanism which likely would occur at energy scales not too different from the Planck scale, would result in several distinctive signatures in the microwave background radiation. Current measurements beautifully confirm a flat universe and are fully consistent with Gaussian perturbations; the rest of the tests will come into clearer view over the coming years. If inflation actually occurred, we can expect to have very strong circumstantial supporting evidence from the above signatures, along with precision measurements of the cosmological parameters describing our universe. However, if inflation did not occur, the universe will likely look different in some respects from the space of models in section 7.5.1. In this case, we may not be able to recover cosmological parameters as precisely, but the microwave background will be equally important in discovering the correct model of our universe.

Acknowledgments

I thank the organizers for a stimulating and enjoyable Summer School. The preparation of this chapter has been supported by a grant from NASA and through the Cotrell Scholars Program of the Research Corporation.

References

Adams W S 1941 *Astrophys. J.* **93** 11

Albrecht A 2000 *Structure Formation in the Universe* ed R Crittenden and N Turok (Dordrecht: Kluwer) to appear (astro-ph/0007247)

Alpher R A and Herman R C 1949 *Phys. Rev.* **75** 1089

Bardeen J M 1980. *Phys. Rev.* D **22** 1882

Bennett C L *et al* 1996 *Astrophys. J.* **464** L1

Birkinshaw M 1999 *Phys. Rep.* **310** 97

Bond J R, Efstathiou G and Tegmark M 1997 *Mon. Not. R. Astron. Soc.* **291** L33

Bouchet F R and Gispert R 1999 *New Astron.* **4** 443

Bucher M, Moodley K, and Turok N 1999 *Phys. Rev.* D **62** 083508

Caldwell R R, Dave R, and Steinhardt P J 1998 *Phys. Rev. Lett.* **80** 1582

Challinor A and Lasenby A 1999 *Astrophys. J.* **513** 1

Christensen N and Meyer R 2000 *Preprint* astro-ph/0006401

Cornish N J, Spergel D N and Starkman G D 1998 *Phys. Rev.* D **57** 5982

de Bernardis P *et al* 2000 *Nature* **404** 955

Denisse J F, Le Roux E and Steinberg J C 1957 *C. R. Acad. Sci., Paris* **244** 3030 (in French)

Dicke R H, Peebles P J E, Roll P G and Wilkinson D T 1965 *Astrophys. J.* **142** 414

Dodelson S, Gates E and Stebbins A 1996 *Astrophys. J.* **467** 10

Doroshkevich A G and Novikov I D 1964 *Sov. Phys. Dokl.* **9** 111

Ehlers J 1993 *Gen. Rel. Grav.* **25** 1225

Ellis G F R and Bruni 1989 *Phys. Rev.* D **40** 1804

Esposito S, Mangano G, Miele G, and Pisanti O 2000 *J. High Energy Phys.* **9** 038

Falco E E, Kochanek C S, and Munoz J A 1998 *Astrophys. J.* **494** 47

Ferreira P G, Magueijo J and Silk J 1997 *Phys. Rev.* D **56** 4592

Gamow G 1956 *Vistas Astron.* **2** 1726

Gebbie T, Dunsby P and Ellis G F R 2000 *Ann. Phys.* **282** 321

Gnedin N Y and Jaffe A H 2001 *Astrophys. J.* **551** 3

Hamilton A J and Tegmark M 2000 *Preprint* astro-ph/0008392

Hanany S *et al* 2000 *Astrophys. J.* **545** L5

Harari D D, Hayward J and Zaldarriaga M 1996 *Phys. Rev.* D **55** 1841

Hu W 2000 *Astrophys. J.* **529** 12

Hu W and Gruzinov A 1998 *Astrophys. J.* **508** 435

Hu W and Silk J 1993 *Phys. Rev. Lett.* **70** 2661

Hu W and Sugiyama N 1996 *Astrophys. J.* **471** 542

Hu W and White M 1996 *Astrophys. J.* **471** 30

——1997 *Astron. Astrophys.* **321** 8

Hu W, Seljak U, White M and Zaldarriaga M 1998 **57** 3290

Huterer D and Turner M S 2000 *Phys. Rev.* D **62** 063503

Jackson J D 1975 *Classical Electrodynamics* 2nd edn (New York: Wiley)

Jaffe A H and Kamionkowski M 1998 *Phys. Rev.* D **58** 043001

Jaffe A H, Stebbins A and Frieman J A 1994 *Astrophys. J.* **420** 9

Jungman G, Kamionkowski M, Kosowsky A and Spergel D N 1996 *Phys. Rev.* D **54** 1332

Kahniashvili T, Mack A, Kosowsky A and Durrer R 2000 *Cosmology and Particle Physics 2000* ed J Garcia-Bellido, R Durrer and M Shaposhnikov to appear

Kamionkowski M and Kosowsky A 1998 *Phys. Rev.* D **67** 685

——1999 *Annu. Rev. Nucl. Part. Sci.* **49** 77

Kamionkowski M, Kosowsky A and Stebbins A 1997a *Phys. Rev. Lett.* **78** 2058

——1997b *Phys. Rev.* D **55** 7368

Kamionkowski M, Spergel D N and Sugiyama N 1994 *Astrophys. J. Lett.* **426** L57

Kaplinghat M and Turner M S 2001 *Phys. Rev. Lett.* **86** 385

Knox L 1995 *Phys. Rev.* D **52** 4307

Knox L, Scoccimaro R and Dodelson S 1998 *Phys. Rev. Lett.* **81** 2004

Kodama H and Sasaki M 1984 *Prog. Theor. Phys. Suppl.* **78** 1

Kolb E W and Turner M S 1990 *The Early Universe* (Redwood City, CA: Addison-Wesley)

Kosowsky A 1996 *Ann. Phys.* **246** 49

——1999 *New Astron. Rev.* **43** 157

Kosowsky A and Loeb A 1996 *Astrophys. J.* **469** 1

Kosowsky A and Turner M S 1995 *Phys. Rev.* D **52** 1739

Kragh, H 1996 *Cosmology and Controversy* (Princeton, NJ: Princeton University Press)

Lange A *et al* 2001 *Phys. Rev.* D **63** 042001

Lidsey J E *et al* 1997 *Rev. Mod. Phys.* **69** 373

Ma C P and Bertschinger E 1995 *Astrophys. J.* **455** 7

McKellar A 1940 *Proc. Astron. Soc. Pac.* **52** 187

Miller A *et al* 1999 *Astrophys. J.* **524** L1

Mould J, Kennicut R C and Freedman W 2000 *Rep. Prog. Phys.* **63** 763

Mukhanov V F, Feldman H A and Brandenberger R H 1992 *Phys. Rep.* **215** 203

Mukhanov V F and Steinhardt P J 1998 *Phys. Lett.* B **422** 52

Peacock J A *et al* 1998 *Mon. Not. R. Astron. Soc.* **296** 1089

Penzias A A and Wilson R W 1965 *Astrophys. J.* **142** 419

Perlmutter S *et al* 1999 *Astrophys. J.* **517** 565

Polarski D and Starobinsky A A 1994 *Phys. Rev.* D **50** 6123

Riess A G *et al* 1998 *Astron. J.* **116** 1009

Sachs R K and Wolfe A M 1967 *Astrophys. J.* **147** 73

Seager S, Sasselov D and Scott D 2000 *Astrophys. J. Suppl.* **128** 407

Seljak U 1996 *Astrophys. J.* **463** 1

Seljak U, Pen U and Turok N 1997 *Phys. Rev. Lett.* **79** 1615

Seljak U and Zaldarriaga M 1996 *Astrophys. J.* **469** 437

Sellwood J and Kosowsky A 2000 *Gas and Galaxy Evolution* ed J E Hibbard, M P Rupen and J van Gorkom in press

Sharov A S and Novikov I D 1993 *Edwin Hubble, the Discoverer of the Big Bang Universe* (Cambridge: Cambridge University Press)

Smoot G F *et al* 1990 *Astrophys. J.* **360** 685

Tegmark M and Zaldarriaga M 2000 *Astrophys. J.* **544** 30

Tegmark M, Zaldarriaga M and Hamilton A J S 2001 *Phys. Rev.* D **63** 043007

Thorne K S 1980 *Rev. Mod. Phys.* **52** 299

Tolman R C 1934 *Relativity, Thermodynamics, and Cosmology* (Oxford: Oxford University Press)

Tytler D *et al* 2000 *Phys. Scr.* in press (astro-ph/0001318)

Weeks J R 1998 *Class. Quantum Grav.* **15** 2599

Weinberg S 2000 *Preprint* astro-ph/0005265

White M, Scott D and Silk J 1994 *Annu. Rev. Astron. Astrophys.* **32** 319

Zaldarriaga 1997 *Phys. Rev.* D **55** 1822

Zaldarriaga M and Seljak U 1997 *Phys. Rev.* D **55** 1830

Zaldarriaga M, Seljak U and Spergel D N 1997 *Astrophys. J.* **488** 1

Zaldarriaga M and Spergel D N 1997 *Phys. Rev. Lett.* **79** 2180

Chapter 8

Dark matter search with innovative techniques

Andrea Giuliani
University of Insubria at Como, Italy

The evidence that most of the matter in the universe does not shine has firmly established the concept of dark matter (DM). It is by now clear that there is room in our galactic halo for DM in the form of exotic particles (WIMPs—Weakly Interacting Massive Particles—or axions) [1,2], whose supposed properties make their experimental observation within the reach of frontier detection methods. This stimulates the creativity of experimental physicists, who are induced to push the existing techniques to their extreme limits or to elaborate new ones in order to attempt DM detection.

The scope of this chapter is to give a survey of the most innovative detection techniques (sections 8.3 and 8.4), comparing their potential with existing results, after a brief elementary introduction on the general concepts of CDM direct detection (section 8.1). Since I consider the approach based on phonon-mediated particle detection one of the most promising, an entire section (8.2) is devoted to this subject.

8.1 CDM direct detection

8.1.1 Status of the DM problem

The abundance of the luminous matter in the universe, inferred by direct observations, is in the range $0.002 < \Omega_{lum} < 0.005$, if a reduced Hubble constant $h = 0.65$ is taken as a reference value. In contrast, primordial nucleosynthesis suggests $0.015 < \Omega_{baryon} < 0.025$, while gravitational effects lead to $\Omega_{matter} > 0.3$. This scenario [3] shows that there are two separate DM problems: the gap between Ω_{lum} and Ω_{baryon} requires baryonic matter in some exotic form (like MACHOs or hot intergalactic gas), while the gap between Ω_{baryon} and Ω_{matter}

can admit particle physics solutions. In particular, axions and neutralinos look like plausible candidates and their detection is within the reach of the present technologies.

Recent observational achievements, suggesting an accelerating universe expansion and a flat universe, lead to a scenario which accommodates an important contribution from the vacuum energy ($\Omega_\Lambda \simeq 2/3$), leaving some room for baryonic and non-baryonic DM, since it is expected that $\Omega_\Lambda \simeq 1/3$. Which features do we require for the particles which are supposed to form, at least in part, the non-baryonic fraction of the matter that escapes our observation? They should be

- neutral,
- massive,
- weakly interacting,
- steady, or at least long living with respect to the universe age, and
- with a relic abundance $\Omega \simeq 0.1$–1.

DM is usually classified as cold dark matter (CDM) and hot dark matter (HDM), consisting, respectively, of fast and slow moving particles (for a review see for example [4]). Neutrinos with masses below 30 eV are an example of HDM, since they were relativistic at the decoupling time. The mechanism of galaxy formation requires, however, a substantial amount of CDM; therefore neutrinos cannot represent a complete solution for the DM problem. Axions and neutralinos are examples of CDM. Axions, although their mass is expected to lie in the range 10^{-6}–10^{-3} eV, are slow moving since they were never in thermal equilibrium and were non-relativistic since their first appearance at 1 GeV temperature [5]. Techniques for axion detection [6] are beyond the scope of this chapter and will not discussed here. Neutralinos will be briefly introduced in the next subsection.

8.1.2 Neutralinos

Neutralinos (χ) [2, 7–9] are supersymmetric Majorana fermions consisting of four mass eigenstates, defined as the linear superposition of the two neutral gauginos and higgsinos. The lowest mass eigenstate may play the role of the lightest supersymmetric particle (LSP) and constitute a viable CDM candidate. Supersymmetric models involve several free parameters, whose choice fixes the neutralino properties, such as the χ–χ annihilation rates and interaction rates with ordinary matter. It is therefore possible, once an assumption has been made about the free parameters, to calculate the neutralino relic density Ω_χ and the cross section with atomic nuclei. There are wide regions in the parameter space which correspond to Ω_χ values relevant for the DM problem ($\Omega_\chi \simeq 0.1$–1) and to measurable interaction rates with reasonable mass detectors. Typical neutralino masses are in the range 30–300 GeV, where the lower limit is due to accelerator constraints.

Neutralinos are supposed to interact with quarks within the nucleons [10,11]. This interaction can be described by a total χ–nucleon cross section σ_p. The parameter experimentally accessible is of course the χ–nucleus cross section σ_0, that can, in a very general way, be expressed as

$$\sigma_0 \propto \frac{g_\chi^2 g_N^2}{M_E^4} \mu^2 k,$$

where M_E is the mass of a virtual particle exchanged between the neutralino and the nucleus in a t-channel interaction, g_χ and g_N the coupling constants of this particle with neutralino and nucleus respectively, μ the reduced mass of the neutralino–nucleus system and k a dimensionless constant. Since g_χ and g_N are weak interaction couplings and M_E is in the Fermi scale (it is, for example, one of the Higgs masses in the case of Higgs boson exchange), the total cross section has a typical weak size: for this reason, neutralinos are sometimes referred to by the more generic term 'WIMPs'. Two types of couplings are usually discussed:

- scalar spin-independent (SI) coupling, for which

$$k = A^2 F_N,$$

  where A is the nucleon number and F_N [12] a nuclear form factor; the term A^2 describes an enhancement of the cross section determined by the coherent interaction with the nucleons;
- axial spin-dependent (SD) coupling, which requires odd A (non-zero nuclear spin); in this case

$$k = (\lambda C_W)^2 J(J+1),$$

  where λ and C_W [12] are nuclear form factors and J the nuclear spin.

Due to the coherence effect, SI coupling is expected to lead to much higher cross sections. Knowledge of the nuclear form factors allows us to express σ_0 in terms of the χ–nucleon cross section σ_p. This makes comparisons among experiments with different nuclear targets possible.

8.1.3   The galactic halo

There is kinematic evidence that there is a halo of DM around spiral galaxies. The evidence comes from the observation of the galactic rotation curves, in which the velocity of the galactic objects is expressed as a function of the object distance from the galactic centre. Since this function is flat sufficiently far way from the centre, instead of the Keplerian decline expected from the distribution of the luminous matter, it is inferred that an invisible mass $M(R)$ is contained in a radius R, with $M(R) \propto R$.

Many uncertainties, however, affect the shape profile and the mass distribution in the halo. Moreover, a substantial component could be of baryonic origin (MACHOs). Standard assumptions [12] are the following:

- $\rho_l = 0.3 \, \mathrm{GeV \, cm^{-3}}$, where ρ_l is the local halo density (at the sun position); and
- $\rho_\chi = \xi \rho_l$, with $\xi < 1$, where ξ is the neutralino fraction of the halo density.

The neutralino velocity distribution is unknown; it is usually taken is Maxwellian:

$$\mathrm{d}n \propto (\pi v_0^2)^{-3/2} \exp\left[-\left(\frac{v}{v_0}\right)^2\right] \mathrm{d}^3 v.$$

To be more exact, v^2 should be replaced by $|v + v_\mathrm{E}|^2$, where v_E is the Earth velocity with respect to the DM distribution. In addition, the Maxwellian should be truncated at $|v + v_\mathrm{E}| = v_\mathrm{esc}$, v_esc being the galactic escape velocity. The usual assumptions for the Maxwellian parameters are $v_0 = 230 \, \mathrm{km \, s^{-1}}$ and $v_\mathrm{esc} = 600 \, \mathrm{km \, s^{-1}}$. A complete discussion about the halo structure and the possible choices for the Maxwellian parameters can be found in [12].

An important point for DM direct detection concerns the motion of the Earth inside the DM distribution [12]. This motion is the composition of the Sun's motion in the galaxy and of the orbital terrestrial motion. The velocity of the sun in the halo affects the WIMP flux as seen by a terrestrial detector (one speaks about a 'WIMP wind'); in addition, the terrestrial orbital velocity adds to the Sun's velocity in summer and subtracts from it in winter. This determines an expected seasonal modulation (typically up to 7%) in the WIMP interaction rate in terrestrial detectors, with a maximum on 2 June. As we shall see in section 8.1.4, this modulation may be a signature for DM identification. The rotational motion of the Earth can also be responsible for a diurnal modulation in the average impact direction of the WIMPs. This effect, much more difficult to detect but also much more pronounced (the modulation would be of the order of some 10%), can also constitute a precious tool for DM detection [13, 31].

8.1.4 Strategies for WIMP direct detection

The interaction of the WIMPs supposed to compose part of the galactic halo determines a nuclear recoil rate in a terrestrial detector. In the case of elastic scattering, isotropic in the centre of mass, the differential energy spectrum of the nuclear recoil $\mathrm{d}R/\mathrm{d}E_\mathrm{R}$ can be easily evaluated [12]. It is exactly exponential in case of stationary Earth:

$$\frac{\mathrm{d}R}{\mathrm{d}E_\mathrm{R}} = \frac{R_0}{E_0 r} \exp\left[-\left(\frac{E_\mathrm{R}}{E_0 r}\right)\right], \tag{8.1}$$

where E_R is the recoil energy, R_0 the total rate, r a kinematic factor given by

$$r = \frac{4 M_\chi M_\mathrm{N}}{(M_\chi + M_\mathrm{N})^2}$$

(with M_χ is the neutralino mass and M_N the target nucleus mass) and E_0 a characteristic WIMP velocity expressed by

$$E_0 = \tfrac{1}{2} M_\chi v_0^2.$$

When the finite velocity of the Earth in the Galaxy is accounted for, equation (8.1) no longer holds and must be replaced by a more complicate expression [12], which preserves anyway an almost exponential shape. Therefore, the expected energy spectrum is featureless and dangerously similar to any sort of radioactive background, which can often be well represented by an exponential tail at low energies. The typical energies over which the spectrum extends can be estimated from the expected M_χ and from the nuclear target mass. It is easy to check with equation (8.1) that most of the counts are expected below 20 keV in typical situations, for example with $M_\chi = 40\,\text{GeV}$ and $A = 127$ (iodine-based detector). This means that the spectrum must be searched for in a region very close to the physical threshold of most conventional nuclear detectors.

In the simplified assumptions that $v_E = 0$ and $v_{esc} = \infty$, the total recoil rate is given by [12]

$$R_0 = \left(\frac{2}{\pi^{1/2}}\right) \left(\frac{N_{av}\,1000}{A}\right) \left(\frac{\rho_\chi v_0}{M_\chi}\right) \sigma_0, \qquad (8.2)$$

where, after a numerical factor, we can identify the number of targets in one kilogram (second factor), the neutralino flow (third factor) and the cross section for each target (last factor). Equation (8.2) predicts rates so low as to represent a formidable challenge for experimentalists. Since neutralinos relevant for the solution of the DM problem are expected to have a nucleon cross section lower than 10^{-41} cm^2, total rates lower than 1 event/(day kilogram) and 10^{-3} event/(day kilogram) are predicted for SI and SD couplings, respectively.

Now that we know the features of what we are looking for, it is possible to conceive an ideal device for WIMP detection. We need a *low-energy nuclear detector* with the following characteristics:

- A very low-*energy threshold* for nuclear recoils (given the nearly exponential shape of the spectrum, a gain in threshold corresponds to a relevant increase in sensitivity). Thresholds of $\sim$10 keV are reachable with conventional devices, while with phonon-mediated detectors (see section 8.2) thresholds down to 300 eV have already been demonstrated.
- Very low *raw radioactive background* at low energies. In general, it requires hard work in terms of material selection and cleaning to reduce raw background below 1 event/(day kilogram keV). Backgrounds lower than 10^{-1} event/(day kilogram keV) have already been demonstrated. Furthermore, an underground site is necessary to host high sensitivity experiments, since cosmic rays produce a huge number of counts at low energies.

- Sensitivity to a *recoil-specific observable*. This allows the ordinary γ and β background for which the energy deposition comes from a primary fast electron to be rejected. When such an observable is available, the only relevant background source left consists in fast neutrons.
- Sensitivity to a *WIMP-specific observable*; it is necessary for an undisputable signature and consists typically in the seasonal modulation of the rate.

A simple measurement of a background level performed with a low-energy nuclear detector produces information on the neutralinos in the galactic halo. Usually, this information is expressed in the form of an *exclusion plot* in a $(\xi \sigma_p, M_\chi)$ plane. The challenge is to test those regions in this plane which are populated by points corresponding to neutralinos viable for DM composition, in the sense explained in section 8.1.2. A simple background measurement cannot prove the existence of neutralinos; it can only exclude neutralinos with given features.

The parameters which affect the shape of the exclusion plot are the threshold, the background spectrum and the target mass. The exclusion plot is constructed by first fixing a neutralino mass: given the nuclear target mass, this allows the recoil spectrum shape apart from a normalization factor to be determined using the exact version of equation (8.1); the value of $\xi \sigma_p$ which leads the recoil spectrum to 'touch' the background spectrum at one point constitutes the upper limit to $\xi \sigma_p$ for that neutralino mass. (Higher values of $\xi \sigma_p$ would produce a recoil spectrum with more counts in one energy bin than those experimentally observed.) The repetition of this procedure over the whole mass range provides the exclusion plot.

The effect on the exclusion plot of the relevant detector parameters can be so summarized: reducing the background improves the exclusion plot for any WIMP mass; reducing the nuclear target mass, the exclusion plot improves at low WIMP masses, but worsens at high WIMP masses; reducing the threshold improves the exclusion plot mainly at low WIMP masses. It is useless nowadays to operate detectors with low target masses (say $A < 50$), since in this case the region with higher sensitivity is already excluded by accelerator constraints. It is important to point out that the exclusion plot does not improve with longer exposition times or with higher detector masses. Relevant results can therefore be achieved even with small detectors and short measurements, provided the background level is low.

In order to get a DM signature, it is important to realize detectors sensitive to a WIMP-specific observable, like the seasonal modulation. For a detailed discussion of this subject, see [12, 14]. Here, we shall follow the simplified discussion reported in [15]. In the presence of halo WIMP interactions, a component of the background must present a seasonal modulation with very specific features, hard to mimic with fake effects:

- the modulation must be present only in a *definite energy region*;
- the modulation must be ruled by a *cosine function*;
- the *proper period* is $T = 1$ year;

- the *proper phase* is 152.5th day in the year (2 June); and
- the *proper modulation amplitude* is <7% in the maximum sensitivity region.

In order to have a signal at the 1σ level, we require:

$$S_{\text{sum}} + B_{\text{sum}} - (S_{\text{win}} + B_{\text{win}}) > (S_{\text{sum}} + B_{\text{sum}} + S_{\text{win}} + B_{\text{win}})^{1/2}, \qquad (8.3)$$

where S_{sum} and B_{sum} are the signal and background counts in summer, while S_{win} and B_{win} represent the corresponding observables in winter. Equation (8.3) ensures that the difference between the summer and winter number of counts is statistically significant. If one assumes that

$$B_{\text{sum}} = B_{\text{win}}$$
$$S_{\text{sum}} - S_{\text{win}} = a(\text{d}R/\text{d}E)M_{\text{det}}T\Delta E$$
$$S_{\text{sum}} + S_{\text{win}} = 2(\text{d}R/\text{d}E)M_{\text{det}}T\Delta E$$
$$B_{\text{sum}} + B_{\text{win}} = 2BM_{\text{det}}T\Delta E,$$

where a is the relative modulation amplitude, B a background coefficient that is expressed in event/(day kilogram keV), $(\text{d}R/\text{d}E)$ an average signal rate per unit mass and energy, also expressed in event/(day kilogram keV), M_{det} the detector mass, T the experiment duration and ΔE the energy range relevant for the signal expressed in keV. Inserting these observables in (8.3), one has as a condition on a:

$$a > \left[\frac{2}{(\text{d}R/\text{d}E)\Delta E} \right]^{1/2} \left[1 + \frac{B}{(\text{d}R/\text{d}E)} \right]^{1/2} \frac{1}{(M_{\text{det}}T)^{1/2}}. \qquad (8.4)$$

The second term in the inequality (8.4) represents the lower limit for the modulation amplitude. The sensitivity of the experiment scales therefore as $(M_{\text{det}}T)^{1/2}$, since the signal, growing as $(M_{\text{det}}T)$, is in competition with background fluctuations growing as $(M_{\text{det}}T)^{1/2}$.

Unlike experiments aiming at exclusion plot production, searches for a real signal imply large detectors and long exposure time. Of course, the same set-up can produce an exclusion plot both from a background measurement and from the non-observation of a modulation amplitude. Increasing the detector mass and the exposure time, the second method becomes more stringent than the first, since in the first case the sensitivity is constant, while in the second one it grows with $(M_{\text{det}}T)^{1/2}$. If we take, for example, $A = 127$, an energy threshold $\simeq 20$ keV, $B \simeq 1.5$ event/(day kilogram keV), a modulation analysis requires a detector mass around 100 kg to get the same sensitivity as a background analysis, assuming $M_\chi \simeq 40$ GeV.

In sections 8.2 and 8.3, we shall focus attention on how detectors which are sensitive to a recoil-specific observable can be realized, with total masses high enough to ensure a significant sensitivity to a seasonal modulation.

Table 8.1. Nuclear quenching factors.

Q_n	Detector	Recoiling nucleus
0.25	Ge diode	Ge
0.30	Si diode	Si
0.30	NaI(Tl) scint.	Na
0.09	NaI(Tl) scint.	I
0.80	Liquid Xe scint.	Xe

8.2 Phonon-mediated particle detection

Conventional nuclear detectors [16] (like scintillators and semiconductor diodes) are sensitive to the amount of ionization that an energetic particle produce in them. Since a slow nuclear recoil (like those produced by WIMP interactions) is a scarcely ionizing particle, the response of a conventional device to such an event is much lower than the response to an electron depositing the same energy. An important quantity characterizing a WIMP detector is, therefore, the nuclear quenching factor Q_n, defined by

$$Q_n(E) = \frac{R_n(E)}{R_e(E)}$$

where $R_n(E)$ and $R_e(E)$ are the responses of the detector (measured for example in volts, since detectors have typically voltage outputs) to a nuclear recoil and to an electron respectively, for a deposited energy E. In principle Q_n depends on energy, but it can be considered constant with an excellent approximation over the energy range of interest for WIMPs. Q_n can also depend on the type of recoiling nucleus. Some experimentally important values are reported in table 8.1.

Since a detector is usually calibrated by means of β and γ sources, the obtained energy scale must be divided by Q_n in order to get the nuclear recoil energy scale. The real threshold is therefore higher than that determined by the calibration; as a trade-off, the background, if not due to fast neutrons, is reduced by a factor Q_n, since to an energy interval ΔE in the electron scale there corresponds an energy interval $\Delta E / Q_n$ in the nuclear recoil energy scale.

Phonon-mediated detectors have the unique feature [17] that their Q_n is very close to one [18]. Joined with the extraordinary energy sensitivity of these devices, this property allows these detectors to reach impressively low energy thresholds. On the other side, the raw β and γ background is a serious problem. One possible solution consists of developing a detector which combines a phonon-mediated with a conventional read-out. The remarkable advantages of this approach are reported in section 8.3. In this section, as an introduction, we shall present briefly the basic principle of a phonon-mediated detector (PMD).

Over the last few years, PMDs have provided better energy resolution, lower energy thresholds and wider material choice than conventional detectors for many applications.

8.2.1 Basic principles

PMDs were proposed initially as perfect calorimeters, i.e. as devices able to thermalize thoroughly the energy released by the impinging particle [19, 20]. In this approach, the energy deposited by a single quantum into an energy absorber (weakly connected to a heat sink) determines an increase of its temperature T. This temperature variation corresponds simply to the ratio between the energy released by the impinging particle and the heat capacity C of the absorber. The only requirements are therefore to work at low temperatures (usually <0.1 K and sometimes <0.015 K) in order to make the heat capacity of the device low enough, and to have a sensitive enough thermometer coupled to the energy absorber. The thermometer is usually a high sensitivity thermistor consisting either in a properly doped semiconductor thermistor (ST) or in a superconductive film kept at the transition edge, usually called the transition edge sensor (TES).

8.2.2 The energy absorber

The energy-absorbing part of the detector is usually a diamagnetic dielectric material in order to avoid dangerous contributions to the specific heat in addition to the Debye term, proportional to T^3 at low temperatures. In such devices, the energy resolution can be fantastically high and close to the so (but not properly) called 'thermodynamic limit' $\sqrt{kT^2C}$ [20]. However, the constraint set by the heat capacity limits the maximum mass for the energy absorber to about 1 kg.

In fact, the real situation is far more complicated. The interaction of an elementary particle with a solid-detecting medium produces excitations of its elastic field; in other terms, the energy spectrum of the target phonon system is modified. Only when the time elapsed after the interaction is long enough to allow the phonon system to relax on a new equilibrium energy distribution, does the detector really work as a calorimeter. In contrast, if the sensor response is very fast, excess non-equilibrium phonons are detected long before they thermalize. (In this case, the sensing element should be defined a 'phonon sensor' rather than a 'thermometer'). In many experimental situations, it is difficult to distinguish between these two extreme cases, and the nature of the detection mechanism is still poorly known. Nevertheless, even when PMDs are not pure calorimeters, their intrinsic energy resolution is better than for conventional detectors, since the typical energy of the excitations produced (high-frequency phonons) is the order of the Debye energy ($\sim$10 meV), instead of 1 eV or more as in ordinary devices (in conventional Ge diodes, for instance, the energy required to produce an electron–hole pair is around 3 eV). Since the energy resolution is limited intrinsically by the fluctuations of the excitation number, its value scales as the square root of the energy required on the average to produce a single excitation.

Detection of non-equilibrium phonons is very attractive because it can, in principle, provide information about interaction position (space resolution has already been proved with this method), discrimination about different types of interacting radiation and the direction of the primary recoil in the target material. The last two points remain to be proved.

8.2.3 Phonon sensors

As anticipated, the commonly used phonon sensors are STs and TESs. STs consist usually of Ge or Si small crystals with a dopant concentration slightly below the metal–insulator transition [21, 22]. This implies a steep dependence of the sensor resistivity on temperature at low temperatures, where the variable range hopping conduction mechanism dominates.

TESs are much more sensitive devices, since their resistivity changes rapidly from a finite value to zero in a very narrow temperature interval. Normally, the superconductive film is deposited on the absorber crystal, with a typical thickness of few hundred nanometres, and the shape is defined after deposition by photolithography and wet etching. With a rectangular shape the normal resistance near the critical temperature is typically between several mΩ and several Ω, and SQUID technology is required for the readout, but with meander shape resistances of ~10 kΩ can be obtained, and a standard voltage-sensitive preamplifier can be used. Films are usually made of a single superconductor. (The most interesting results have been obtained with tungsten [23, 24].) In another approach, the film consists of two layers (a normal metal in contact with a superconductor): this structure allows the critical temperature to be tuned.

8.3 Innovative techniques based on phonon-mediated devices

8.3.1 Basic principles of double readout detectors

An important feature of PMDs is that a high response is expected for energies deposited by slow (<100 keV) nuclear recoils, which are difficult to detect with conventional devices because of their scarce ionizing power. In a perfect calorimeter, a nuclear recoil produces the same signal of a fast electron of the same energy, since it deposits the same amount of heat. In spite of the naiveness of this approach, it has been proven with *ad hoc* measurements that the detecting efficiency for recoiling nuclei and electrons is indeed the same within 2% in dielectric ST-PMDs [18]. In other terms, as already anticipated, $Q_n \simeq 1$ for PMDs. As a consequence, impressively low thresholds can be achieved in large amounts of low specific heat material (typically sapphire). If properly operated in a low radioactive background environments, these low threshold PMDs can be very sensitive DM detectors. The CRESST experiment has installed in the Gran

Sasso laboratory 4×250 g sapphire-TES detectors with a threshold of about 300 eV, which is well beyond the reach of any conventional scheme [23].

Perhaps, the best strategy for PMDs around WIMP search is the achievement of an active rejection of background through the recognition of nuclear recoils, expected from WIMP interactions. The basic idea consists of realizing a detector with both a phonon-mediated and a conventional readout, which could be a charge signal (in the case of semiconductor diodes) or a light signal (in the case of scintillators). The charge signal is proportional to the number of electron–hole pairs, while the light signal is proportional to the amount of scintillation produced by the interacting particle. I will define the *non-phonon signal* S_{np} as the output provided by the conventional (charge or light) readout, and the *phonon signal* S_{ph} the output given by the phonon sensor.

The basic point is that the same event produces, in general, both a phonon and a non-phonon signal. If we consider the observable:

$$R = \frac{S_{np}}{S_{ph}} \tag{8.5}$$

the value of R depends on the *type* of primary interaction. In the case of slow nuclear recoil R is significantly higher than for a fast electron of the same energy, since the non-phonon component, connected to the amount of ionization, is much less important. The parameter R defined in (8.5) therefore represents a powerful recoil-specific observable in the sense exposed in section 8.1.4.

In practice, a nuclear detector which follows all the specifications of section 8.1.4 could consist in one of the two following possibilities:

- An array of large Ge or Si diodes operated as conventional semiconductor devices with an additional phonon sensor. The total mass must be large enough to make the detector competitive in terms of seasonal modulation sensitivity (WIMP-specific observable). Therefore, the array must consist of tens of individual elements. The double readout provides the recoil-specific observable R. The raw background and the energy threshold must be conveniently low.
- An array of large scintillators with an additional phonon sensor and with the same features as in the previous point in terms of total mass, threshold and background. A remarkable technical difficulty consists of the necessity to operate a light detector at very low temperatures.

Three collaborations in the world are successfully developing detectors fulfilling these two requirements. That is the topic of the next section.

8.3.2   CDMS, EDELWEISS and CRESST experiments

The American collaboration CDMS ('Cold Dark Matter Search') [24] is realizing silicon and germanium detectors cooled to 20 mK and capable of measuring both the charge and the phonon component of any single energy release. The charge is measured by means of conventional charge amplifier technology [16], whereas the phonon measurement is performed with two different technologies. One is based on eutecticly bonded Ge ST, and the other on W TES elements sensitive to non-equilibrium phonons. In the second approach [25], non-equilibrium phonons created by particle interactions break Cooper pairs in superconductive Al films which cover a large fraction of the crystal surface. The created quasiparticles are then trapped in a W film (with a critical temperature around 70 mK) which is grown above the Al films. The W film is operated as a TES and, heated by the trapped quasiparticles, provides the signal, proportional to the initially deposited energy. The system of Al and W films presents a pattern which allows reasonable space resolution (of the order of 1 mm) in the plane where the films lie (the crystal surface) to be achieved. In the dimension orthogonal to this plane, space resolution is also possible exploiting the risetime of the phonon signal. This allows the events which occur close to the crystal surface to be recognized. This detector capability helps substantially in background identification. The point is that background events generated by β contamination in the surface can mimic nuclear recoil events, since events at the surface suffer from incomplete charge collection, while the phonon signal is, of course, unchanged. The space resolution permits us to identify these close-to-surface events and to reject them. Therefore, at the price of an acceptable loss of sensitive volume, the background identification is much safer. In preliminary tests, a rejection capability better than 99% was achieved down to 20 keV. In figure 8.1, the points in the upper band correspond to γ interactions, while the ones in the lower band to nuclear recoils. In these tests, the nuclear recoils are induced by means of external sources of fast neutrons.

The French collaboration EDELWEISS [26] adopts a scheme similar to the first type of CDMS detector. The best results were obtained with a 70 g high-purity Ge detector with a disk shape. The charge signal is provided by a conventional readout, based on charge amplifier technology, while the phonon signal comes from a Ge ST glued on the disk. In the range 15–70 keV a raw background of about 40 event/(day kilogram keV) is reduced down to 0.6 event/(day kilogram keV). This collaboration aims at operating a large mass experiment, realized by means of many independent detectors, in the Frejus underground laboratory (France).

The German–English collaboration CRESST [27] is developing a detector sensitive to phonons and scintillation light. A test device was realized, consisting of a 1 cm^3 CaWO$_4$ crystal scintillator. A W film (with a critical temperature around 11 mK) is deposited on the crystal and operated as a TES. The scintillation photons which escape from the crystal are collected by auxiliary Al$_2$O$_3$ PMDs which surround the scintillator. Due to the very low threshold of the auxiliary detectors, a few photons can be detected by them, allowing a safe threshold to be set down to 15 keV (for nuclear recoils). The rejection capability at this energy is, impressively enough, 99.7%. This result can be appreciated in figure 8.2, where

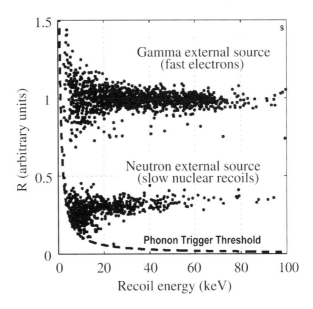

Figure 8.1. Discrimination capability of the CDMS experiment.

the parameter R in (8.5) is given by the slope of the bands. Even in this case, the total detector mass can be increased only through the realization of a large array of independent devices, which should be operated underground, for example in the Gran Sasso Laboratory (Italy).

8.3.3   Discussion of the CDMS results

The CDMS collaboration was able to perform up to now the most sensitive experiment in terms of exclusion plot [24] (see figure 8.4). This shows clearly the potential of the double readout technique based on phonon-mediated detection. The CDSM experiment, even if largely preliminary, is particularly important since it allows us to probe, at least partially, the region in the $(\xi\sigma_p, M_\chi)$ plane corresponding to the modulation evidence claimed by the Italian collaboration DAMA [28]. I shall just recall here that this modulation evidence can be interpreted in terms of halo neutralino interactions with the most probable values of $M_\chi = 44$ GeV and $\xi\sigma_p = 5.4 \times 10^{-41}$ cm^2. The corresponding 3σ region is reported in figure 8.4.

The CDMS detectors are operated at Stanford beneath only a 16 m water equivalent overburden. A plastic scintillator veto is therefore necessary in order to reject cosmic ray events. The results are based on two data-sets:

- *Two month exposure in 1998*, providing 33 live days collected with one *100 g Si detector* operated with a W–Al film phonon readout (see previous section).

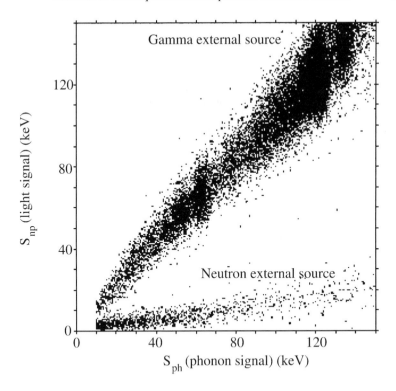

Figure 8.2. Discrimination capability of the CRESST test detector.

In this exposition, the collected statistics correspond to $M_{\text{det}}T = 1.6$ kg day. After the background rejection, only four events survive as slow nuclear recoils.

- *Twelve month exposure in 1999*, providing 96 live days collected with *four 165 g Ge detectors* operated with a Ge ST phonon readout (see previous section). In this exposition, the collected statistics correspond to $M_{\text{det}}T = 10.6$ kg day. After the background rejection, only 17 events survive as slow nuclear recoils. The four Ge detectors are tightly packed in order to increase the neutron multiple scattering probability. Since four recoil events were cut as in coincidence between two detectors (of course the probability of WIMP double scattering is completely negligible), only 13 recoil events attributable to WIMPs survive. (In figure 8.3 the nuclear recoil events are represented by circled points.)

The 13-nuclear-recoil energy spectrum is compatible with the expected WIMP-caused spectrum as deduced by the DAMA neutralino parameters. However, the CDMS collaboration claims that there is clear evidence that these 13 single events are caused by background neutrons. In fact, the background

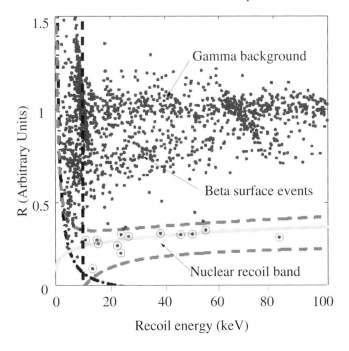

Figure 8.3. Gamma/beta background and nuclear recoils in CDMS results.

neutron spectrum can be estimated by the four Ge multiple events and the four Si nuclear recoils. (Si events cannot be due to WIMPs, otherwise the WIMP rate in Ge would be much higher than observed because of the A^2 term (section 8.1.2) in the cross section.) These safe neutron events, if analysed by means of a Monte Carlo simulation of the neutron background, are fully compatible with 13 background neutron single events in the Ge experiment. In other terms, the Ge multiple events and the Si single events fix the neutron background, that can be subtracted by the Ge single event spectrum, leaving a WIMP signal compatible with zero. Following this analysis, the CDMS collaboration claims to have substantially falsified the DAMA interpretation of the seasonal modulation in terms of the neutralino. In figure 8.4 the CDMS exclusion plot (black thick curve) is compared with the shadow region which represents the DAMA 3σ evidence. The CDMS sensitivity is also reported (as fixed by the *a priori* estimated neutron background). The exclusion plot provided by a powerful conventional technique (Ge semiconductor diodes, no double readout) is shown for comparison.

Anyway, the contrast between DAMA and CDMS looks far from being clarified by the existing data. If it is true that the DAMA results have raised not only excitement but also criticism in part of the DM community [29], it is also clear that the CDMS results would require confirmation with higher statistics and in an environment less affected by the cosmic neutron background. Information

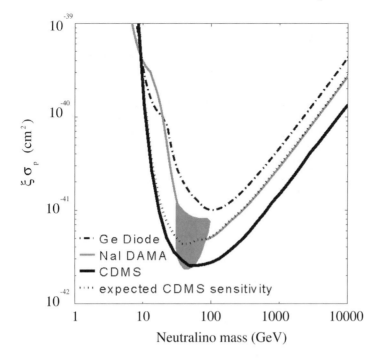

Figure 8.4. Exclusion plots (90% C.L.) and DAMA evidence.

to resolve the dilemma could come from the much larger underground set-up that CDMS is going to operate in the near future and from the DAMA upgrading in terms of total detector mass.

8.4 Other innovative techniques

There are many mid-term projects which are not based on a phonon readout channel but which, however, point to a substantial increase in sensitivity to neutralino interactions. I shall mention here, for lack of space, only three projects. This selection was admittedly also made on the basis of personal taste, besides scientific relevance. For a complete review, I suggest the reader refers to the proceedings of a recent specific conference, for example [30].

The DRIFT experiment [31] represents the only attempt to detect the direction of the nuclear recoil already at a test-phase. It consists of a low-pressure TPC using a 20 Torr Xe–CS_2 gas mixture. Such a device must be able to detect the tiny tracks from nuclear recoils with less than 1 mm track resolution. In addition, large detector masses are necessary. This requirement suggests that the magnetic field should be abandoned, with a consequent deterioration in the space resolution due to enhanced diffusion. In order to solve this problem, the new concept consists of detecting not the drift electrons, but the negative CS_2 ions, with a considerably reduced diffusion because of the large ion mass. This experiment points directly at the most decisive WIMP signature, the diurnal modulation of the recoil average direction (section 8.1.4). By the end of 2001 a 20 m^3 TPC should be in operation.

The ZEPLIN programme [32] is based on double readout (scintillation and charge) in liquid xenon. When an ionizing particle deposits energy in liquid xenon and an electric field is applied, two scintillation pulses are developed. The primary pulse (amplitude S_1) is due to the excitations in Xe atoms produced directly by the particle interaction. The secondary pulse (amplitude S_2) is generated by the drift of the charge created by the primary interaction. Therefore, a low ionizing particle like a slow nuclear recoil will exhibit a secondary pulse depressed in amplitude with respect to the first one, if compared with an electron of the same energy. On a similar footing as in section 8.3.1, S_1/S_2 plays the role of a recoil-specific observable.

The project CUORE [33] ('Cryogenic Underground Observatory for Rare Events') consists of the largest PMD set-up ever conceived. It is based on the experience collected by the Milano group on large mass arrays of low-temperature calorimeters for rare decays [34]. It should consist of a tightly packed array of 1020 TeO$_2$ crystals for a total mass of 0.8 ton, to be cooled down to 10 mK. Each element has a mass of about 800 g and uses a Ge ST (section 8.2.3) as a thermometer. The relevant points of the project are the huge mass (which provides sensitivity to seasonal modulation) and the low background, which can be reduced significantly with respect to the $\sim$1 event/(day kilogram keV) already demonstrated with similar devices. This reduction should be achieved by the operation of the detectors in coincidence, particularly effective in this case due to the minimal amount of inert material among them. A preliminary test of CUORE is in preparation at the underground Gran Sasso Laboratory (Italy).

References

[1] Turner M S and Tyson J A 1999 *Rev. Mod. Phys.* **71** S145
[2] Masiero A and Pascoli S this volume
[3] Dodelson S, Gates E I and Turner M S 1996 *Science* **274** 69D
[4] Sadoulet B 1999 *Rev. Mod. Phys.* **71**
[5] Abbott L and Sikivie P 1983 *Phys. Lett.* B **120** 133
[6] Sikivie P 1983 *Phys. Rev. Lett.* **51** 141
[7] Jungman G *et al* 1996 *Phys. Rev.* **267** 195
[8] Ellis J *et al* 1997 *Phys. Lett.* B **413** 355
[9] Edsjo J and Gondolo P *Phys. Rev.* D **56** 1879
[10] Goodman M W and Witten E 1985 *Phys. Rev.* D **31** 3059
[11] Primack J R *et al* 1988 *Annu. Rev. Nucl. Part. Sci.* **38** 751
[12] Lewin J D and Smith P F 1996 *Astropart. Phys.* **6** 87
[13] Spooner N J C and Kudryavtsev (eds) 1999 *The Identification of Dark Matter* (Singapore: World Scientific)

[14] Freese K *et al* 1988 *Phys. Rev.* D **37** 3388

[15] Bernabei R 1995 *Riv. Nuovo Cimento* **18** 5

[16] Knoll G F 1989 *Radiation Detection and Measurement* (New York: Wiley)

[17] Giuliani A 2000 *Physica* B **280** 501

[18] Alessandrello A *et al* 1997 *Phys. Lett.* B **408** 465

[19] Fiorini E and Niinikoski T O 1984 *Nucl. Instrum. Methods* **224** 83

[20] Moseley S H, Mather J C and McCammon D 1984 *J. Appl. Phys.* **56** 1257

[21] Mott N F 1969 *Phil. Mag.* **19** 835

[22] Giuliani A and Sanguinetti S 1993 *Mater. Sci. Eng.* R **11** 1

[23] Sisti M *et al* 2000 *Nucl. Instrum. Methods* **444** 312

[24] Gaitskell R *et al* 2001 Latest results from CDMS collaboration *Sources and Detection of Dark Matter in the Universe. Proc. 4th Int. Symp. Sources and Detection of Dark Matter/Energy in the Universe, February 23–25, 2000, Marina del Rey, CA* ed D Cline (Berlin: Springer)

[25] Cabrera B *et al* 2000 *Nucl. Instrum. Methods* **444** 304

[26] Chardin G *et al* 2000 *Nucl. Instrum. Methods* **444** 319

[27] Meunier P *et al* 1999 *Appl. Phys. Lett.* **75** 1335

[28] Bernabei R this volume

[29] Gerbier G *et al* 1997 *Preprint* astro-ph/9710181
    Gerbier G *et al* 1999 *Preprint* astro-ph/9902194

[30] Cline D (ed) 2001 *Sources and Detection of Dark Matter in the Universe. Proc. 4th Int. Symp. Sources and Detection of Dark Matter/Energy in the Universe, February 23–25, 2000, Marina del Rey, CA* (Berlin: Springer)

[31] Martoff C J *et al* 2001 DRIFT *Sources and Detection of Dark Matter in the Universe. Proc. 4th Int. Symp. Sources and Detection of Dark Matter/Energy in the Universe, February 23–25, 2000, Marina del Rey, CA* ed D Cline (Berlin: Springer)

[32] Wang H *et al* 2001 Design of the Zeplin II Detector *Sources and Detection of Dark Matter in the Universe. Proc. 4th Int. Symp. Sources and Detection of Dark Matter/Energy in the Universe, February 23–25, 2000, Marina del Rey, CA* ed D Cline (Berlin: Springer)

[33] Fiorini E 1998 *Phys. Rep.* **307** 309

[34] Alessandrello A *et al* 2000 *Phys. Lett.* B **486** 13

Chapter 9

Signature for signals from the dark universe

The DAMA Collaboration
R Bernabei[1], M Amato[2], P Belli[1], R Cerulli[1], C J Dai[3], H L He[3],
G Ignesti[2], A Incicchitti[2], H H Kuang[3], J M Ma[3],
F Montecchia[1], D Prosperi[2]
[1] *Department of Physics, University of Rome 'Tor Vergata' and INFN, Rome, Italy*
[2] *Department of Physics, University of Rome 'La Sapienza' and INFN, Rome, Italy*
[3] *IHEP, Chinese Academy, Beijing, China*

The DAMA experiment is located at the Gran Sasso National Laboratories of the INFN and is searching for dark matter (DM) particles using various scintillators as target-detector systems. In particular the results, presented here, were obtained by analysing, in terms of the WIMP annual modulation signature, the data collected with the highly radiopure ($\sim$100 kg NaI(Tl)) set-up during four annual cycles (total statistics of 57 986 kg day).

9.1 Introduction

In the past few years, the many experimental and theoretical studies have changed the main question on the DM problem from its existence to the nature of its constituents. The stringent limit on the baryonic part (arising from a comparison between the measured relative abundance of light elements with their expectations in the nucleosynthesis scenario) and the results achieved in investigations of the cosmic microwave background (which have ruled out the pure hot DM scenario) support the view that—whatever the DM composition turns out to be (even if a cosmological constant different from zero is definitively demonstrated)—a large amount of CDM is necessary. This can be in the form of WIMPs or axions.

In particular, the WIMPs should be neutral particles in thermal equilibrium in the early stage of the universe, decoupling at the freeze-out temperature, with a cross section for ordinary matter of the order of or lower than the weak one, forming a dissipationless gas trapped in the gravitational field of the galaxy. To be a suitable WIMP candidate a neutral particle should be stable or have a decay time of the order of the age of the universe. The neutralino, which results in stable MSSM and SUGRA models with R-parity conservation, is at present the more studied candidate; it also remains a good candidate in the case of models without R-parity conservation, if the decay time is of the order of the age of the universe. Other candidates can also be considered; moreover, since this type of search requires investigation beyond the SM of particle physics, the possible nature of WIMPs is, in principle, fully open.

WIMPs can be searched for by direct and indirect techniques. However, we have to remark that significant uncertainties exist in every model-dependent analysis and—as can be easily understood—they are even larger in the indirect approach.

In the following we will focus our attention on some of the main points related to the WIMP direct searches by investigating elastic scattering on target nuclei. As regards investigation of WIMP–nucleus inelastic scattering, we only mention them here [1–3], stressing that much lower counting rates for the signal are expected in this case.

The main strategy to search for these processes effectively is based on the use of low radioactive experimental set-ups located deep underground. Significant improvements in the overall radiopurity of the set-up have been reached over several years of work, the ultimate limit remaining as the sea level activation of the materials. This limitation would, however, be significantly overcome if chemical/physical purifications of the used materials could occur just before their storage deep underground and—even more—if all the operations for detector construction were to be performed deep underground.

Another crucial point (as always in experiments which require a very low energy threshold) is the possibility of identifying and effectively rejecting the residual noise above the considered energy threshold. This problem has obviously to be faced with every type of detector. For most of them the rejection is quite uncertain (also affecting the quoted results), because the noise and the 'physical' pulses have indistinguishable features. In contrast, an almost unique effective noise rejection is possible in scintillators:

(i)   when the pulse decay time is relatively long with respect to the fast single photoelectrons from the PMT noise;
(ii) when the number of photoelectrons/keV is really large;
(iii) when the noise contribution from the electronic chain is low; and
(iv) when a sensitive rejection procedure is used.

We note, in addition, that scintillators are unaffected by microphone noise in contrast to ionizing and bolometer detectors.

Although exclusion plots are widely used in practice, many uncertainties arise in comparisons of the results arising from different experiments—even more so when different techniques are used. Furthermore, direct comparison is impossible when different target nuclei are used. To overcome this, it is mandatory to realize experiments with a real signature for the possible signal. If we discard the following possibilities:

(i) a possible comparison between results from different experiments (which can, in principle, be considered since the rate is proportional to A^2 for the spin-independent interactions and to the spin factor for the spin-dependent ones), because e.g. of the relevant role played by the different backgrounds;

(ii) the daily variation of the signal rate [4] (which can, in principle, be considered since the Earth depth crossed by the WIMPs varies during the day inducing a daily variation rate), because this effect is effective only in the case of high cross sections; and

(iii) the correlation of the nuclear recoil track with the Earth's galactic motion (arising from the WIMP velocity distribution), because of the shortness of the induced tracks.

Only the possibility of studying the annual modulation of the WIMP wind [5, 6] remains. This so-called *annual modulation signature* is the annual modulation of the WIMP rate induced by the Earth's motion around the Sun [5–9].

In particular, the DAMA collaboration is performing this investigation with the highly radiopure $\sim$100 kg NaI(Tl) set-up at the Gran Sasso National Laboratory of INFN [7–15].

As has been clearly pointed out by DAMA [7–9, 12, 15], the annual modulation signature is a well-distinguished one, requiring the presence not of a 'generic' rate variation but of a variation according to the following specifications:

(i) the presence of a correlation with the cosine function;
(ii) an appropriate proper period (1 year);
(iii) the proper phase (about 2 June);
(iv) only in a well-defined low-energy region (where WIMP-induced recoils could be significantly present);
(v) for events where only one detector of many actually fires (single 'hit' events) since the probability of a WIMP multi-scattering is negligible (in practice each detector has all the others as a veto);
(vi) with modulated amplitude in the region of maximal sensitivity not exceeding $\lesssim 7\%$.

That all these requirements have been realized by DAMA has been verified by the following actions.

(i) The collection of the whole energy spectrum from single photoelectron to the MeV range;
(ii) the continuous monitoring and control of several parameters; and

(iii) many consistency checks and statistical tests [7–9, 12, 13, 15].

Therefore, to mimic the WIMP annual modulation signature a systematic effect should not only be quantitatively significant, but also able to satisfy the six requirements for a WIMP-induced effect.

In the following, we will summarize only the more recently released results on the WIMP search using the annual modulation signature using the ≃100 kg NaI(Tl) DAMA set-up [12].

However, for the sake of completeness it is worth recalling that the DAMA DM searches are based on the use of

(i) the ~100 kg NaI(Tl) set-up;
(ii) the ~2 l liquid xenon pure scintillator; and
(iii) the CaF$_2$(Eu) prototypes.

Recent references are, for example, [2, 3, 7–10, 12, 13, 16–21]. Moreover, several results on different topics have also been achieved [11, 14, 17, 19, 22–28].

9.2 The highly radiopure ~100 kg NaI(Tl) set-up

A detailed description of the DAMA set-up and of its performances is given in [12], while the stability control of the various parameters, the noise rejection, the efficiency, the calibrations, the higher energy stability, the total hardware rate, etc have been discussed in [8, 9, 12, 13, 15]. Nine 9.70 kg NaI(Tl) detectors have been especially built for the experiment on the WIMP annual modulation signature by means of a joint effort with Crismatec company. The materials used for these detectors have been selected—as well as those for the PMTs—by measuring sample radiopurities with low background germanium detectors deep underground in the low background facility of the Gran Sasso National Laboratory [12]. As regards the samples of powders, their U/Th content was measured in Ispra with a mass spectrometer, while their K content was determined in the chemical department of the University of Rome 'La Sapienza' with an atomic absorption spectrometer. A single growth has been used for all the crystals. The crystals are enclosed in a low radioactive copper box inside a low radioactive shield made from 10 cm of copper and 15 cm of lead; the lead is surrounded by 1.5 mm Cd foils and about 10 cm of polyethylene/paraffin. The copper box is maintained in a nitrogen atmosphere by continuously flushing high-purity nitrogen gas. Each detector is viewed through 10 cm long light guides by two low background EMI9265B53/FL—3 in diameter—PMTs working in coincidences; the hardware threshold for each PMT is at single photoelectron level. The 9.70 kg detectors have tetrasil-B light guides directly coupled to the bare crystals (also acting as windows). Four other crystals of 7.05 kg—originally developed for other purposes—are used as a cut-off for the other detectors and for special triggers; they have tetrasil-B windows and are coupled to the PMTs in one case by tetrasil-B and in the others by noUV-plexiglass light guides. All the crystals

Table 9.1. Released data-sets; the number 1 to 4 refer to different annual cycles.

Period	Statistics (kg day)	Reference
DAMA/NAI-1	4 549	[7]
DAMA/NaI-2	14 962	[8]
DAMA/NaI-3	22 455	[9]
DAMA/NaI-4	16 020	[9]
Total statistics	57 986	[9]
+ DAMA/NaI-0	Limits on recoils fraction by PSD	[10]

have surfaces polished with the same procedure and enveloped in a TETRATEC-4 (Teflon) diffuser such as the light guides.

On the top of the shield a glove-box, maintained in the same nitrogen atmosphere as the Cu box containing the detectors, is directly connected to it through four Cu thimbles in which source holders can be inserted to calibrate all the detectors at the same time without allowing them to enter in direct contact with environmental air. The glove-box is equipped with a compensation chamber. When the source holders are not inserted, the Cu bars completely fill the thimbles. Since this set-up has been realized with the main purpose of studying the annual modulation signature of WIMPs, several parameters are monitored and acquired by CAMAC. A monitoring and alarm system operates continuously by a self-controlled computer processes.

Finally, we recall that the measured low-energy counting rate has been published in various energy intervals [8, 9, 14, 15, 20], while in [26] higher energy regions are shown.

9.3   Investigation of the WIMP annual modulation signature

The present result concerns four years of data-taking for the annual modulation studies, namely DAMA/NaI-1,2,3 and 4 [7–9] for total statistics of 57 986 kg day, the largest statistics ever collected in the field of WIMP search. Moreover, in the final global analysis the constraint, arising from the upper limits on the recoil rate measured in [10] (DAMA/NaI-0), has also been properly included (see table 9.1).

9.3.1   Results of the model-independent approach

In figure 9.1 we show the model-independent residual rate for the cumulative 2–6 keV energy interval as a function of the time [9], which offers immediate evidence for the presence of modulation in the lowest energy region of the experimental data.

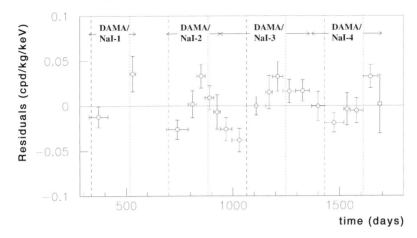

Figure 9.1. Model-independent residual rate in the 2–6 keV cumulative energy interval as a function of the time elapsed since 1 January of the first year of data-taking. The expected behaviour of a WIMP signal is a cosine function with a minimum around the broken vertical lines and with a maximum around the dotted ones.

The χ^2 test of the data in figure 9.1 is not favourable towards the hypothesis of unmodulated behaviour giving a probability of 4×10^{-4}. However, fitting these residuals with the function $A \cos \omega (t - t_0)$ (obviously integrated in each of the considered time bins), one gets for the period $T = 2\pi/\omega = (1.00 \pm 0.01)$ years, when fixing t_0 at 152.5 days and for the phase $t_0 = (144 \pm 13)$ days, when fixing T at 1 year (similar results, but with slightly larger errors, are found when both these parameters are kept free). The modulation amplitude as a free parameter gives $A = (0.022 \pm 0.005)$ cpd kg^{-1} keV^{-1} and $A = (0.023 \pm 0.005)$ cpd kg^{-1} keV^{-1}, respectively. As is evident, the period and the phase fully agree with the ones expected for a WIMP-induced effect.

As we will further comment, this model-independent analysis provides evidence for the possible presence of a WIMP signal independently of the nature of the WIMP and its interaction with ordinary matter. In the following we will briefly summarize the investigation of possible systematics able to mimic such a signature, that is not only quantitatively significant, but also able to satisfy the six requirements given earlier; none has been found. A detailed discussion can be found, for example, in [15].

9.3.2 Main points on the investigation of possible systematics in the new DAMA/NaI-3 and 4 running periods

We have already presented elsewhere the results of the investigations of all the possible known sources of systematics [7–9, 12, 13, 15]; however, in the following we will briefly discuss, in particular, the data from the DAMA/NaI-

3 and DAMA/NaI-4 running periods, which have been recently released [9]; a devoted discussion can be found—as previously mentioned—in [15]. Similar arguments for the DAMA/NaI-1 and DAMA/NaI-2 data have already been discussed elsewhere [7, 8, 13] and at many conferences and seminars.

In our set-up the detectors have been continuously isolated from environmental air for several years; different levels of closures are sealed and maintained in a high-purity nitrogen atmosphere. However, the environmental radon level in the installation is continuously monitored and acquired with the production data; the results of the measurements are at the level of the sensitivity of the used radonmeter. For the sake of completeness, we have examined the behaviour of the environmental radon level with time. When fitting the radon data with a WIMP-like modulation, the amplitudes (0.14 ± 0.25) Bq m^{-3} and (0.12 ± 0.20) Bq m^{-3} are found in the two periods respectively, both consistent with zero. Further arguments are given in [15]. Moreover, we remark that a modulation induced by radon—in every case—would fail some of the six requirements of the annual modulation signature and, therefore, a radon effect can be excluded.

The installation, where the $\sim$100 kg NaI(Tl) set-up operates, is air-conditioned. The operating temperature of the detectors in the Cu box is read by a probe and stored with the production data [12]. In particular, sizeable temperature variations could only induce a light variation in the output, which is negligible considering:

(i)   that around our operating temperature, the average slope of the light output is $\lesssim -0.2\%/^{\circ}$C;
(ii) the energy resolution of these detectors in the keV range; and
(iii) the role of the intrinsic and routine calibrations [12]; see [15].

In addition, every possible effect induced by temperature variations would fail at least some of the six requirements needed to mimic the annual modulation signature; therefore, a temperature effect can be excluded.

In long-term running conditions, knowledge of the energy scale is ensured by periodical calibration with an ^{241}Am source and by continuously monitoring within the same production data (grouping the data approximately into 7 day batches) the position and resolution of the ^{210}Pb peak (46.5 keV) [7–9, 12, 15]. The distribution of the relative variations of the calibration factor (proportionality factor between the area of the recorded pulse and the energy), $tdcal$—without applying any correction—estimated from the position of the ^{210}Pb peak for all the nine detectors during both the DAMA/NaI-3 and the DAMA/NaI-4 running periods, has been investigated. From the measured variation of $tdcal$ an upper limit of $<1\%$ of the modulation amplitude measured at very low energy in [7–9] has been obtained [15].

The only data treatment which is performed on the raw data is to eliminate obvious noise events (which sharply decrease when increasing the number of available photoelectrons) present below approximately 10 keV [12]. The noise in our experiment is given by PMT fast single photoelectrons with decay times of the order of tens of nanoseconds, while the scintillation pulses have decay times of the order of hundreds of nanoseconds. The large difference in decay times and the relatively large number of available photoelectrons response (5.5–7.5 photoelectron/keV depending on the detector) ensures effective noise rejection; see, e.g., [12] for details. To investigate quantitatively the possible role of a noise tail in the data after noise rejection on the annual modulation result, the hardware rate, R_{Hj}, of each detector above a single photoelectron, can be considered. The distribution of $\Sigma_j(R_{Hj} - \langle R_{Hj} \rangle)$ shows a Gaussian behaviour with $\sigma = 0.6\%$ and 0.4% for DAMA/NaI-3 and DAMA/NaI-4, respectively, values well in agreement with those expected on the basis of simple statistical arguments. Moreover, by fitting its time behaviour in both data periods including a WIMP-like modulated term a modulation amplitude compatible with zero $(0.04 \pm 0.12) \times 10^{-2}$ Hz, is obtained. From this value, considering also the typical noise contribution to the hardware rate of the nine detectors, the upper limit on the noise relative modulation amplitude has been derived to be [15] less than

$$\frac{1.6 \times 10^{-3} \text{ Hz}}{9 \times 0.10 \text{ Hz}} \simeq 1.8 \times 10^{-3} \qquad (90\% \text{ C.L.}).$$

This shows that even in the worst hypothetical case of a 10% contamination of the residual noise—after rejection—in the counting rate, the noise contribution to the modulation amplitude in the lowest energy bins would be less than 1.8×10^{-4} of the total counting rate, that is a possible noise modulation could account only for less than 1% of the annual modulation amplitude observed in [9]. In conclusion, there is no evidence that a hypothetical tail of residual noise after rejection plays any role in the results.

The behaviour of the efficiencies during the whole data-taking periods has also been investigated; their possible time variation depends essentially on the stability of the cut efficiencies, which are regularly measured by dedicated calibrations [9, 15]. In this way, the unlikely idea of a possible role played by the efficiency values in the observed effect in [7–9] has also been ruled out [9, 15].

In order to verify the absence of any significant background modulation, the measured energy distribution in energy regions not of interest for the WIMP–nucleus elastic scattering has been investigated [7–9, 13]. For this purpose, we have considered the rate integrated above 90 keV, R_{90}, as a function of time. The distributions of the percentage variations of R_{90} with respect to their mean values for all the crystals during the whole DAMA/NaI-3 and DAMA/NaI-4 running periods show cumulative Gaussian behaviour with $\sigma \simeq 1\%$, well accounted for by the statistical spread expected from the used sampling time [9, 15]. This result excludes any significant background variation. Moreover, including a WIMP-like modulation in the analysis of the time behaviour of R_{90}, an amplitude compatible with zero is found in both the running periods: $-(0.11 \pm 0.33)$ cpd kg^{-1} and $-(0.35 \pm 0.32)$ cpd kg^{-1}. This excludes the presence of a background modulation in the whole energy spectrum at a level much lower than the effect found in the lowest energy region in [7–9]; in fact, if it were otherwise—considering the R_{90} mean values—the modulated term should be of the order of tens of cpd kg^{-1}, that is $\sim 100\sigma$ far away from the measured value. This also accounts for the neutron environmental background; see for further arguments [15]. A similar analysis performed in other energy regions, such as the one just above the first pole of the iodine form factor, leads to the same conclusion.

As regards possible side reactions, the only process which has been found as a hypothetical possibility is the muon flux modulation reported by the MACRO experiment [29]. In fact, MACRO has observed that the muon flux shows a nearly sinusoidal time behaviour with a 1 year period and with a maximum in the summer with amplitude of $\sim 2\%$; this muon flux modulation is correlated with the temperature of the atmosphere. This effect would give, in our set-up, modulation amplitudes much less than 10^{-4} cpd kg^{-1} keV^{-1}, that is much smaller than we observe. Moreover, it will also fail some of the six requirements necessary to mimic the signature. Thus, it can be safely ignored [15]. The search for other possible side reactions able to mimic the signature has so far not offered any other candidate.

For the sake of completeness, we recall that—using pulse shape discrimination—no evidence for the anomalous events with a decay time shorter than the recoils has ever been found in our data [10, 15].

As a result of the model-independent approach and a full investigation of known systematic effects, the presence of an annual modulation compatible with WIMPs in the galactic halo indocates that WIMPs are possible candidates to account for the data, independently of their nature and coupling with ordinary matter.

In the next section a particle candidate will be investigated; for that a model is needed as well as an effective energy and time correlation analysis. We take this occasion to remark that a large scenario exists in the model-dependent analyses not only because various candidates with different couplings can be considered but also because of the large uncertainties affecting several parameters involved in the calculation which are generally neglected, although they should generally play a significant role.

9.3.3   Results of a model-dependent analysis

Properly considering the time occurrence and the energy of each event, a time correlation analysis of the data collected between 2 and 20 keV has been performed, according to the method described in [7–9]. This allows us to test effectively the possible presence in the rate of a contribution having the typical features of a WIMP candidate. In particular we have considered a particle with a dominant spin-independent scalar interaction (which is also possible for the neutralino [30]). A detailed discussion is available in [9]; here the main result is outlined. In the minimization procedure by the standard maximum likelihood method [7–9] the WIMP mass has been varied from 30 GeV up to 10 TeV; the lower bound accounted for results achieved in accelerators. The calculations have been performed according to the same astrophysical, nuclear and particle physics considerations given in [7–9] and to the 90% C.L. recoil limit of [10] (DAMA/NaI-0). Alternative analytical approaches, such as the one based on the χ_{test} variable described in [8] and the Feldman and Cousins method [31], offer substantially the same results.

Since the analysis of each data cycle independently [7–9, 13] gave consistent results, a global analysis has been made properly including both the known uncertainties on astrophysical local velocity, v_0 [21] and the constraint arising from the upper limit on recoils measured in [10] (DAMA/NaI-0). According to [21], the minimization procedure has been repeated by varying v_0 from 170 to 270 km s^{-1} to account for its present uncertainty; moreover, the case of possible bulk halo rotation has also been analysed. The positions of the minima for the log-likelihood function consequently vary [21]; for example, in this model framework for $v_0 = 170$ km s^{-1} the minimum is at $M_W = (72^{+18}_{-15})$ GeV and $\xi\sigma_p = (5.7 \pm 1.1) \times 10^{-6}$ pb, while for $v_0 = 220$ km s^{-1} it is at $M_W = (43^{+12}_{-9})$ GeV and $\xi\sigma_p = (5.4 \pm 1.0) \times 10^{-6}$ pb. The results obtained in this model framework are summarized in figure 9.2, where the regions allowed at 3σ C.L. are shown:

(i)   when $v_0 = 220$ km s^{-1} (dotted contour);
(ii) when the uncertainty on v_0 is taken into account (continuous contour); and
(iii) when possible bulk halo rotation is considered (broken contour).

The latter two calculations have been performed according to [21]. The confidence levels quoted here have also been verified by suitable Monte Carlo calculations; in particular, we note that the Feldman and Cousins analysis [31] of the data gives quite similar results. These regions are well embedded in the Minimal Supersymmetric Standard Model (MSSM) estimates for the neutralino [32]. A quantitative comparison between the results from the model-independent and model-dependent analyses has been discussed in [9].

Finally, many assumptions on the nuclear and particle physics used in these calculation (as well as in those of exclusion plots) are affected by uncertainties, which—when taken into account—would enlarge the regions of figure 9.2 and, as mentioned, consequently vary the positions of the minima for the log-likelihood function. For example, as in [9] we mention the case of the iodine form factor, which depends on the nuclear radius and on the thickness parameter of the nuclear surface; it has been verified that, varying their values with respect to those used in the analysis in [9] by 20%, the locations of the minima will move toward slightly larger M_W and toward lower $\xi\sigma_p$ values, while the calculated 2–6 keV S_m values will increase by about 15%.

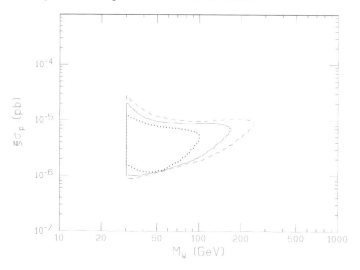

Figure 9.2. Regions allowed at 3σ C.L. in the plane $\xi\sigma_p$ ($\xi = \frac{\rho_{\text{WIMP}}}{0.3\text{ GeV cm}^{-3}}$ and σ_p = WIMP scalar cross section on proton) versus M_W (WIMP mass) by the global analysis: (i) for $v_0 = 220$ km s^{-1} (dotted contour); (ii) when accounting for v_0 uncertainty (170 km s^{-1} $\leq v_0 \leq$ 270 km s^{-1}; continuous contour); and (iii) when considering also a possible bulk halo rotation as in [21] (broken contour). The constraint arising from the measured upper limit on recoils measured in [10] has been properly taken into account. We note that the inclusion of present uncertainties on some nuclear and particle physics parameters would enlarge these regions since the positions of the minima for the log-likelihood function would consequently vary; full estimates are in progress.

9.4 DAMA annual modulation result versus CDMS exclusion plot

As is well known, intrinsic uncertainties exist in the comparison of results achieved by different experiments and, even more, when different techniques are used as in the case of DAMA [7–9] and of CDMS [33]. In fact, DAMA is searching for a distinctive signature by using a large mass NaI(Tl) set-up deep underground, while CDMS is exploiting a widely unknown hybrid bolometer/ionizing technique at a depth of 10 m to reject a huge background. Moreover, always when different target nuclei are used (as is also the case in DAMA and CDMS), no absolute comparison can be pursued at all; only model-dependent comparisons can be considered with further intrinsic uncertainties. In table 9.2 a few numbers are given to offer an immediate view on the two experiments.

The techniques used by CDMS would require several technical papers to be credited at the necessary level (quenching factor values, sensitive volumes, windows for rejection, efficiencies, energy calibrations, etc; the stability of

Table 9.2. Several numbers on the DAMA and CDMS experiments as in [9, 33].

	DAMA	CDMS
Exposure	57 986.0 kg day	10.6 kg day
Depth	1400 m	10 m
Number of events in the observed effect	Total modulated amplitude ~2000 events	13 evt in Ge, 4 evts in Si 4 multiple evts in Ge + Monte Carlo on neutron flux

these quantities during the running period; justification of the performed data selection; quantitative control of systematic uncertainties in the various hardware and software handlings), which have not been made available. Every small deviation from the assumptions used by CDMS in [33] can significantly change their conclusion.

The exclusion plot quoted by CDMS [33] arises from the joint analyses of two different experiments with two different target nuclei (Si and Ge) and, practically, by a neutron Monte Carlo subtraction.

In the Si experiment (used exposure was ~1.5 kg day of the ~3.3 kg day available) a large number of events survived the ionizing/heat discrimination in the whole energy region allowed for recoil candidates. Thereafter, by the so-called athermal pulse shape discrimination, four events remained and were classified as 'mostly neutrons', while all the others as 'surface electrons'. The amount and the Y (ratio between ionizing and heat charges) and energy distributions of the latter ones give a hint that the four 'mostly neutrons' events could indeed be—all or partially—ascribed to the tail of the huge population of 'surface electrons' surviving the ionizing/heat discrimination. Obviously this possibility would significantly change the conclusions in [33].

In the Ge experiment (used exposure was ~10.6 kg day of the ~48 kg day available for three Ge detectors, having already excluded a fourth detector), 13 recoil candidates survive the ionization/heat discrimination. This number of events is largely compatible with the DAMA allowed region estimated in [9] in the framework of a model for a spin-independent candidate with mass above 30 GeV. The interpretation on the real nature of these 13 candidates strongly depends on the Monte Carlo estimates of the neutron background, which is constrained by the hypothesized nature of the four Si candidates and of four multi-hit events. A similar procedure is strongly uncertain since it is based on the previously mentioned assumptions and on the neutron transport code; the latter—as is widely known—is affected by huge uncertainties due to the assumptions on the original neutron energy spectrum and to the transport calculations in all the involved materials. This can be verified by considering that the result of such a calculation gives in [33] about 30 expected neutrons to be compared with the 13 quoted recoil candidates; this, in particular, can suggest an overestimate of the neutron background and, therefore, of the given exclusion plot.

Summarizing we can state that the CDMS result can be expressed by the combination of two quantities: the real number of recoil candidates (when accounting for realistic values of the physical parameters) and the expected number of neutron background. Varying these quantities several different conclusions can be obtained. In every case, a CDMS representative has stated that analysing these data to determine their compatibility with DAMA, the result gives an upper limit for presence of WIMPs in CDMS Ge data of eight events at 90% C.L. [34], evidently compatible with the DAMA allowed region in the model considered in [9]. Moreover, simple calculations assuming again ideal values for the CDMS physical parameters and the values measured for the related quantities in our experiment [7–10, 12] show that in the framework of the model of [9], CDMS should measure from ∼15 events down to less than 1, that is compatibility is still substantially present.

Moreover, we note that the comparison through a model requires, for each considered target nucleus, fixing not only the coupling and the scaling laws, but also several specific different nuclear and particle physics parameters, which are affected instead by uncertainties. The same is for the choice of the astrophysical model, such as the WIMP velocity distribution and the various related parameters. For example if the real WIMP velocity distribution should be such as to enhance, to a certain extent, the modulated part of the signal with respect to the unmodulated one, a comparison in the framework of usual assumptions would fail. The same would hold if the candidate were to have a partial (or total) spin-dependent interaction component (as is also possible for the neutralino) and one of the two experiments is insensitive to spin-dependent interactions (such as practically those using natural Ge). Several other scenarios could also be considered.

For the sake of completeness, we note that in [33] the complete DAMA result has not been considered.

Briefly, many experimental and theoretical reasons do not support the conclusion of [33] to the necessary extent.

9.5   Conclusion

In conclusion, a WIMP contribution to the measured rate is a candidate by the model-independent approach and by the absence of any known systematics able to mimic the signature [7–9, 13, 15] independently of the nature and coupling of the possible particle. The complete global correlation analysis in terms of a spin-independent candidate with a mass greater than 30 GeV favours modulation at approximately 4σ C.L. in the given framework [9]. Moreover, neutralino configurations in the allowed region appear to be of cosmological interest [32].

In [35] a possible heavy neutrino of the fourth family has been considered instead. Further studies on model frameworks are in progress.

The data for a fifth annual cycle are now at hand, while new electronics and data acquisition systems were installed in August 2000. Moreover, after new dedicated R&D for the radiopurification of NaI(Tl) detectors, efforts to increase the experimental sensitivity are in progress; the target mass will become approximately 250 kg.

References

[1] Fushimi K *et al* 1994 *Nucl. Phys.* B *(Proc. Suppl.)* **35** 400
[2] Belli P *et al* 1996 *Phys. Lett.* B **387** 222
[3] Bernabei R *et al* 2000 *New J. Phys.* **2** 15.1–15.7
[4] Collar J I *et al* 1992 *Phys. Lett.* B **275** 181
[5] Drukier K A *et al* 1986 *Phys. Rev.* D **33** 3495
[6] Freese K *et al* 1988 *Phys. Rev.* D **37** 3388
[7] Bernabei R *et al* 1998 *Phys. Lett.* B **424** 195
[8] Bernabei R *et al* 1999 *Phys. Lett.* B **450** 448
[9] Bernabei R *et al* 2000 *Phys. Lett.* B **480** 23
[10] Bernabei R *et al* 1996 *Phys. Lett.* B **389** 757
[11] Bernabei R *et al* 1997 *Phys. Lett.* B **408** 439
[12] Bernabei R *et al* 1999 *Nuovo Cimento* A **112** 545
[13] Belli P *et al* 1999 *3K-Cosmology* (New York: AIP) p 65
[14] Bernabei R *et al* 1999 *Phys. Lett.* B **460** 236
[15] Bernabei R *et al* 2000 *Preprint* ROM2F/2000-26
[16] Belli P *et al* 1996 *Nuovo Cimento* C **19** 537
[17] Bernabei R *et al* 1997 *Astropart. Phys.* **7** 73
[18] Bernabei R *et al* 1998 *Phys. Lett.* B **436** 379
[19] Belli P *et al* 1999 *Nucl. Phys.* B **563** 97
[20] Bernabei R *et al* 1999 *Nuovo Cimento* A **112** 1541
[21] Belli P *et al* 2000 *Phys. Rev.* D **61** 023512
[22] Belli P *et al* 1996 *Astropart. Phys.* **5** 217
[23] Belli P *et al* 1999 *Astropart. Phys.* **10** 115
[24] Belli P *et al* 1999 *Phys. Lett.* B **465** 315
[25] Bernabei R *et al* 1999 *Phys. Rev. Lett.* **83** 4918
[26] Belli P *et al* 1999 *Phys. Rev.* C **60** 065501
[27] Belli P *et al* 2000 *Phys. Rev.* D **61** 117301
[28] Bernabei R *et al* 2000 *Phys. Lett.* B **490** 16
[29] Ambrosio M *et al* 1997 *Astropart. Phys.* **7** 109
[30] Bottino A *et al* 1997 *Phys. Lett.* B **402** 113
[31] Feldman G J and Cousins R D 1998 *Phys. Rev.* D **57** 387
[32] Bottino A *et al* 2000 *Phys. Rev.* D **62** 056006
[33] Abusaidi R *et al* 2000 *Phys. Rev. Lett.* **84** 5699
[34] Shutt T 2000 *Seminar Given at LNGS (March)*
[35] Fargion D *et al* 1998 *Pis. Zh. Eksp. Teor. Fiz.* **68** (*JETP Lett.* **68** 685)

Chapter 10

Neutrino oscillations: a phenomenological overview

GianLuigi Fogli
Dipartimento di Fisica e INFN, Sezione di Bari, Via Amendola 173, 70126 Bari, Italy

The evidence for solar and atmospheric neutrino oscillations is analysed in a three-flavour oscillation framework, including the most recent Super-Kamiokande data, as well as the constraints on ν_e mixing coming from the CHOOZ reactor experiment. The regions of the mass-mixing parameter space compatible with the data are determined and their features discussed. In particular, it is shown that bimaximal mixing (or nearly bimaximal mixing) of atmospheric and solar neutrinos is also possible within the MSW solution to the solar neutrino problem.

10.1   Introduction

The recent atmospheric neutrino data from the Super-Kamiokande (SK) experiment [1] are in excellent agreement with the hypothesis of flavour oscillations in the $\nu_\mu \leftrightarrow \nu_\tau$ channel [2]. Such a hypothesis is consistent with all the SK data, including sub-GeV e-like and μ-like events (SGe, μ), multi-GeV *e*-like and μ-like events (MGe, μ), and upward-going muon events (UPμ), and is also corroborated by independent atmospheric neutrino results from the MACRO [3] and Soudan-2 [4] experiments. Oscillations in the $\nu_\mu \leftrightarrow \nu_\tau$ channel are also compatible with the negative results of the reactor experiment CHOOZ in the $\nu_e \leftrightarrow \nu_e$ channel [5,6].

However, *dominant* $\nu_\mu \leftrightarrow \nu_\tau$ transitions plus *subdominant* $\nu_\mu \leftrightarrow \nu_e$ transitions are also consistent with SK+CHOOZ data, and lead to a much richer three-flavour oscillation phenomenology for atmospheric νs [7]. A three-flavour framework is also needed in order to accommodate, in addition, the evidence for solar ν_e disappearance [8].

In this chapter we analyse atmospheric and solar data in a common 3ν oscillation framework. Concerning atmospheric νs, we include 30 data points from the SK experiment (52 kTy) [1], namely the zenith distributions of sub-GeV events (SG e-like and μ-like, $5 + 5$ bins), multi-GeV events (MGe, μ $5 + 5$ bins) and upward-going muons (UPμ, 10 bins). We also include, when appropriate, the rate of events in the CHOOZ reactor experiment (one bin). Concerning solar neutrinos, we use the total rate information from the Homestake (chlorine), GALLEX+SAGE (gallium), Kamiokande and Super-Kamiokande experiments, as well as the day–night asymmetry and the 18-bin energy spectrum from Super-Kamiokande (825 days) [1], with emphasis on the Mikheyev–Smirnov–Wolfenstein solutions.

10.2   Three-neutrino mixing and oscillations

The combined sources of evidence for neutrino flavour transitions coming from the solar ν problem and from the atmospheric ν anomaly demand an approach in terms of three-flavour oscillations among massive neutrinos (ν_1, ν_2, ν_3) [7–9]. The three-flavour ν parameter space is then spanned by six variables:

$$\delta m^2 = m_2^2 - m_1^2, \tag{10.1}$$

$$m^2 = m_3^2 - m_2^2, \tag{10.2}$$

$$\omega = \theta_{12} \in [0, \pi/2], \tag{10.3}$$

$$\phi = \theta_{13} \in [0, \pi/2], \tag{10.4}$$

$$\psi = \theta_{23} \in [0, \pi/2], \tag{10.5}$$

$$\delta = CP \text{ violation phase}, \tag{10.6}$$

where the θ_{ij} rotations are conventionally ordered as for the quark mixing matrix [10].

In the phenomenologically interesting limit $|\delta m^2| \ll |m^2|$, the two eigenstates closest in mass (ν_1, ν_2) are expected to drive solar ν oscillations, while the 'lone' eigenstate ν_3 drives atmospheric ν oscillations. In such a limit (see [7–10] and references therein) the following occur:

(i)   the phase δ becomes unobservable;
(ii)   the atmospheric parameter space is spanned by (m^2, ψ, ϕ); and
(iii)   the solar ν parameter space is spanned by $(\delta m^2, \omega, \phi)$.

In other words, in the previous limit it can be shown that solar neutrinos probe the composition of ν_e in terms of mass eigenstates

$$\nu_e = U_{e1}\nu_1 + U_{e2}\nu_2 + U_{e3}\nu_3 \tag{10.7}$$

$$= c_\phi(c_\omega\nu_1 + s_\omega\nu_2) + s_\phi\nu_3 \tag{10.8}$$

in the parameter space

$$(\delta m^2, \omega, \phi) \equiv (\delta m^2, U_{e1}^2, U_{e2}^2, U_{e3}^2), \tag{10.9}$$

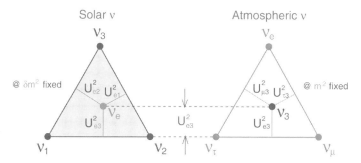

Figure 10.1. Parameter spaces of solar and atmospheric neutrinos in the limit $|\delta m^2| \ll |m^2|$, for assigned δm^2 and m^2. The only common parameter is $U_{e3}^2 = s_\phi^2$.

where $U_{e1}^2 + U_{e2}^2 + U_{e3}^2 = 1$ for unitarity, whereas atmospheric (more generally, 'terrestrial') neutrinos probe the flavour composition of ν_3,

$$\nu_3 = U_{e3}\nu_e + U_{\mu3}\nu_\mu + U_{\tau3}\nu_\tau \qquad (10.10)$$

$$= s_\phi\nu_e + c_\phi(s_\psi\nu_\mu + c_\psi\nu_\tau) \qquad (10.11)$$

in the parameter space

$$(m^2, \psi, \phi) \equiv (m^2, U_{e3}^2, U_{\mu3}^2, U_{\tau3}^2), \qquad (10.12)$$

where $U_{e3}^2 + U_{\mu3}^2 + U_{\tau3}^2 = 1$ for unitarity. The two unitarity constraints can be conveniently embedded [9] in two triangle plots (see figure 10.1), which describe the mixing parameter spaces for given δm^2 and m^2 for solar and atmospheric neutrinos, respectively. The only parameter common to the two triangles is $U_{e3}^2 = s_\phi^2$†.

10.3 Analysis of the atmospheric data

In this section we report an updated analysis of the Super-Kamiokande data, and combine them with the limits coming from the CHOOZ reactor experiment, by assuming the 'standard' three-neutrino framework discussed in the previous section. Details about our calculations can be found in [7]. Constraints on the mass-mixing parameters are obtained through a χ^2 statistics, and are plotted in the atmospheric ν triangle described in figure 10.1.

Figure 10.2 shows the regions favoured at 90% and 99% C.L. in the triangle plots, for representative values of m^2. The CHOOZ data, which exclude a large horizontal strip in the triangle, appear to be crucial in constraining three-flavour mixing. Pure $\nu_\mu \leftrightarrow \nu_e$ oscillations (right-hand side of the triangles) are excluded

† In the special case $\phi = 0$, the atmospheric and solar parameter spaces are decoupled into the two-family oscillation spaces $(\delta m^2, \omega)$ and (m^2, ψ).

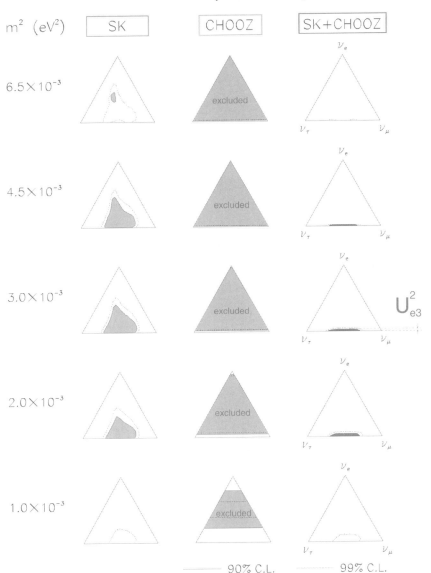

Figure 10.2. Three-flavour analysis in the triangle plot, for five representative values of m^2. Left-hand and middle column: separate analyses of Super-Kamiokande (52 kTy) and CHOOZ data, respectively. Right-hand column: combined SK+CHOOZ allowed regions. Although the SK+CHOOZ solutions are close to pure $\nu_\mu \leftrightarrow \nu_\tau$ oscillations, the allowed values of U_{e3}^2 are not completely negligible, especially in the lower range of m^2.

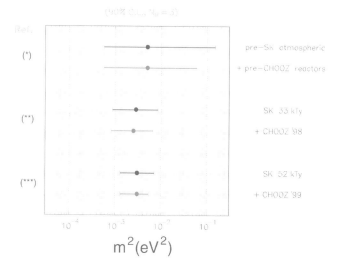

Figure 10.3. 90% C.L. bounds on the mass parameter m^2 from atmospheric data, without and with reactor data. Upper part: pre-SK and pre-CHOOZ bounds. Intermediate part: SK bounds at 32 kTy (+CHOOZ). Lower part: present bounds from SK data at 52 kTy (+CHOOZ).

by SK and CHOOZ independently. The centre of the lower side, corresponding to pure $\nu_\mu \leftrightarrow \nu_\tau$ oscillations with maximal mixing, is allowed in each triangle both by SK and SK+CHOOZ data. However, deviations from maximal ($\nu_\mu \leftrightarrow \nu_\tau$) mixing, as well as subdominant mixing with ν_e, are also allowed to some extent. Such deviations from maximal 2ν mixing are now more constrained than in the previous analysis of the 33 kTy SK data [7], also as a result of tighter constraints from the finalized CHOOZ data [5].

Figure 10.3 shows the progressively tighter constraints on the mass parameter m^2 for unconstrained three-flavour mixing, for pre-SK [11] and post-SK [7] analyses, with and without reactor constraints. The current best-fit value (lower part of figure 10.3) is reached at $m^2 \sim 3 \times 10^{-3}$ eV^2, and is only slightly influenced by the inclusion of CHOOZ data. However, the upper bound on m^2 is significantly improved by including CHOOZ. Note that there is consistency between pre-and post-SK information.

Figures 10.2 and 10.3 clearly show the tremendous impact of the SK experiment in constraining the neutrino oscillation parameter space. Prior to SK, the data could not significantly favour $\nu_\mu \leftrightarrow \nu_\tau$ over $\nu_\mu \leftrightarrow \nu_e$ oscillations, and

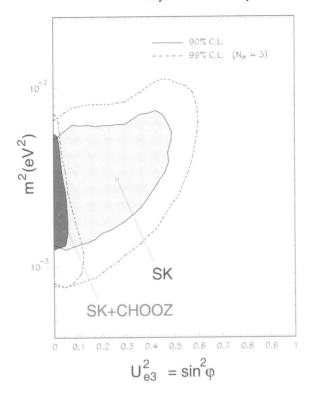

Figure 10.4. Bounds on U_{e3}^2 as a function of m^2 from SK data (52 kTy), with and without the finalized CHOOZ data.

could only put relatively weak bounds on m^2 (see [11]).

The impact of CHOOZ in constraining the mixing matrix element U_{e3}^2 is clearer in figure 10.4, where the 90% and 99% C.L. bounds are shown as a function of m^2, for unconstrained values of the angle ψ. It can be seen that, when CHOOZ data are included, the element U_{e3}^2 cannot be larger than a few percent.

Figure 10.5 shows the best-fit zenith distributions of SGe, μ, MGe, μ and UPμ events, normalized to the no-oscillation rates in each bin, with and without the CHOOZ constraint. The non-zero value of U_{e3}^2 at the best-fit point (SK data only) leads to a slight expected excess in the MGe sample for $\cos\theta \to -1$. A significant reduction in the errors is needed to probe such possible distortions, which would be unmistakable signals of subdominant $\nu_\mu \to \nu_e$ oscillations. Figure 10.5 also shows that, when the results of CHOOZ are included, pure $\nu_\mu \to \nu_\tau$ oscillations represent the best fit to the SK data. In this context, it is useful to show that the pieces of information coming from the *shape* of the zenith distributions (figure 10.5) and from the total event rates are consistent with

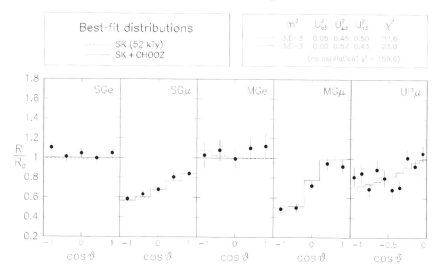

Figure 10.5. SK zenith distributions of leptons at best fit (broken lines), also including CHOOZ (full lines), as compared with the 52 kTy experimental data (dots with error bars). The 3ν mass-mixing values at best fit are indicated in the upper right-hand corner.

each other, contrary to recent claims [12].

To this purpose, figure 10.6 shows the curve of theoretical predictions for maximal 2ν mixing ($U_{\mu3}^2 = U_{\tau3}^2$ and $U_{e3}^2 = 0$) and variable m^2, in the plane of the double ratio of μ-to-e events for SG and MG events, together with the SK data (cross of error bars). The SK data on the double ratio, within one standard deviation, are perfectly consistent with the $\nu_\mu \to \nu_\tau$ oscillation hypothesis at $m^2 \sim 3 \times 10^{-3}$ eV^2.

10.4   Analysis of the solar data

10.4.1   Total rates and expectations

In this section we present an updated phenomenological analysis of the solar neutrino data, assuming oscillations between two and three neutrino families, with emphasis on the MSW [13] solutions.

As far as expectations are concerned, we use the so-called BP98 standard solar model [14] for the electron density in the Sun and for the input neutrino parameters (ν_e fluxes, spectra, and production regions), and compare the predictions to the experimental data for the following observables: total neutrino event rates, SK energy spectrum and SK day–night asymmetry.

The total neutrino event rates are those measured at Homestake [15], Kamiokande [16], SAGE [17], GALLEX [18], and Super-Kamiokande (825 live days) [1]. Since the SAGE and GALLEX detectors measure the same quantity

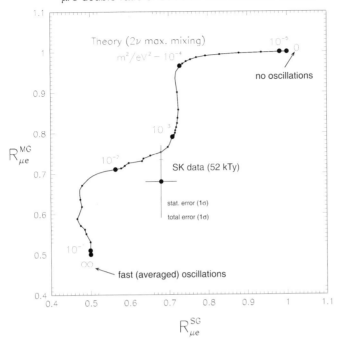

Figure 10.6. Double ratio of μ/e events (data/theory) for SG and MG events in SK: full curve, predictions for maximal $\nu_\mu \to \nu_\tau$ mixing; cross, SK data ($\pm 1\sigma$).

their results are combined in a single (Ga) rate. The Kamiokande and SK data, however, are treated separately (rather than combined in a single datum), since the two experiments, although based on the same ν–e scattering detection technique, have rather different energy thresholds and resolution functions.

The SK electron recoil energy spectrum and its uncertainties (825 lifetime days, $E_e > 5.5$ MeV) are graphically reduced from the 18-bin histograms shown by SK members in recent Summer '99 conferences [1]. Our theoretical calculation of the binned spectrum properly takes into account energy threshold and resolution effects. Standard ^{8}B [19] and *hep* [14] neutrino spectra and fluxes are used, unless otherwise noted. Concerning the SK day–night asymmetry of the event rates, we use the latest measurement [1]: $2(N - D)/(N + D) = 0.065 \pm 0.031 \pm 0.013$.

In the presence of 2ν or 3ν oscillations, the MSW effect in the Sun is computed as in [9]. The additional Earth matter effects are treated as in [20]. The χ^2 analysis basically follows the approach developed in [21, 22], with the necessary updates to take into account the BP98 SSM predictions and the energy spectrum information. Further details can be found in [23].

Solar neutrino problem, 1999

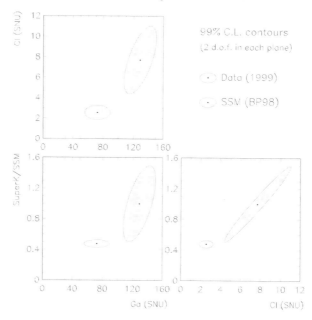

Figure 10.7. The solar neutrino deficit, shown as a discrepancy between data and expectations in the gallium (Ga), chlorine (Cl), and Super-Kamiokande total event rates. In each plane, the error ellipses represent 99% C.L. contours for two degrees of freedom (i.e. $\Delta\chi^2 = 9.21$). The projection of an ellipse onto one of the axis gives approximately the $\pm3\sigma$ range for the corresponding rate.

We start our analysis by comparing the standard (no oscillation) predictions with the experimental data for the Cl, Ga, and SK total rates. Figure 10.7 shows the 99% C.L. error ellipses for data and expectations in the planes charted by the (Cl, Ga), (SK, Ga) and (SK, Cl) total rates. The distance between observations and standard predictions makes the solar ν problem(s) evident. At present, such information is the main evidence for solar neutrino physics beyond the standard electroweak model; however, since the theoretical errors are dominant—as far as total rates are concerned—no substantial improvements can be expected by a reduction in the experimental errors. Conversely, decisive information is expected from the SK spectrum and day–night asymmetry, but no convincing deviation has emerged from such data yet. Therefore, it is not surprising that, in oscillation fits, the total rates mainly determine *allowed* regions, while the SK spectrum and day–night asymmetry determine *excluded* regions.

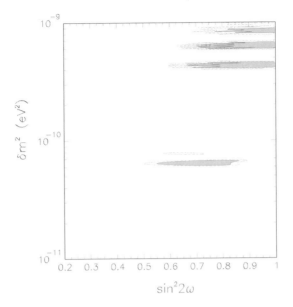

Figure 10.8. Two-generation vacuum solutions to the solar neutrino problem (all data included). Upper solutions fit the SK spectrum better than total rates. Conversely, the solution at lowest δm^2 fits the total rates (Cl + Ga + K + SK) better.

10.4.2 Two-flavour oscillations in vacuum

In figure 10.8 we report our 2ν vacuum oscillation analysis of the solar neutrino data coming from total rates end SK electron energy spectrum. We can see several distinct solutions, allowed at the 90% C.L., with the peculiar behaviour that in general a certain disagreement can be found by a comparison of the total rates and energy spectrum constraints. There are solutions which are preferred by the total rates analysis, but disfavoured by the energy spectrum, and solutions that, conversely, are mainly indicated by the energy spectrum but not by total rates. This behaviour has also been noted in [24]. An interesting feature is that, if one of the vacuum solutions is selected by future data, then we will be able to determine the mass difference δm^2 in a very accurate way.

10.4.3 Two-flavour oscillations in matter

Figure 10.9 shows the results of our 2ν MSW analysis of the solar ν data, shown as C.L. contours in the $(\delta m^2, \sin^2 2\omega / \cos 2\omega)$ plane. The choice of the variable $\sin^2 2\omega / \cos 2\omega$, rather than the usual $\sin^2 2\omega$, allows an expanded view of the large mixing region.

In each of the six panels, we determine the absolute minimum of the χ^2 and then plot the iso-χ^2 contours corresponding to 90%, 95% and 99% C.L. for

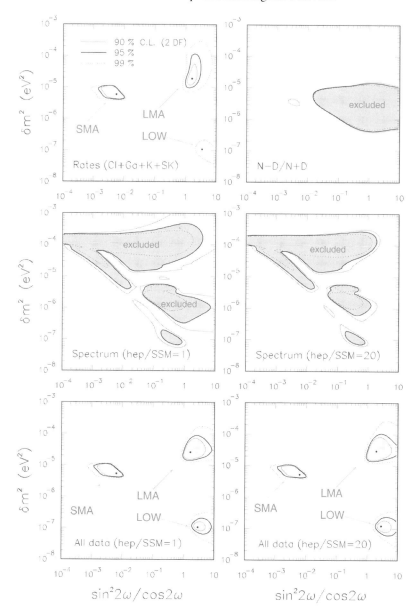

Figure 10.9. Two-generation MSW solutions to the solar neutrino problem. The upper four panels correspond to the following separate fits to data subsets: total rates $(Cl + Ga + K + SK)$; Super-Kamiokande night–day asymmetry $N - D/N + D$; Super-Kamiokande electron energy spectrum with standard *hep* neutrino flux; Super-Kamiokande spectrum with enhanced $(20\times)$ *hep* neutrino flux. The two lower panels show the results of the global fits to all data.

two degrees of freedom (the oscillation parameters). In fits including the total rates, there is a global χ^2 minimum and two local minima; such minima, and the surrounding favoured regions, are usually indicated as MSW solutions at small mixing angle (SMA), large mixing angle (LMA), and low δm^2 (LOW).

The first panel of figure 10.9 refers to the fit to the total rates only. The three χ^2 minima are indicated by dots. The absolute minimum is reached within the SMA solution ($\chi^2_{min} = 1.08$): it represents a very good fit to the data. The LMA solution is also acceptable, while the LOW solution gives a marginal fit.

The SK data on the day–night asymmetry (second panel) and energy spectrum (third panel) exclude large regions in the mass-mixing parameter space; but are unable to (dis)prove any of the three solutions, which in fact are also present in the global fit to all data (fifth panel).

The spectrum information is sensitive to the (uncertain) value of the *hep* neutrino flux; for instance, an enhancement by a factor 20 helps to fit the high-energy part of the SK spectrum [25], and thus it produces a reduction in the excluded regions in the mass-mixing plane (fourth panel in figure 10.9), and a corresponding slight enlargement of the globally allowed regions (sixth panel).

From a careful analysis [23], the following situation emerges for the three MSW solutions SMA, LMA, and LOW. None of them can be excluded at 99% C.L. by the present experimental data. Different pieces of the data give indications that are not as consistent as would be desirable: the total rate information favours the SMA solution, the spectral data favour the LMA and LOW solutions, and the day–night data favour the LMA solution. In a global fit, the three solutions have comparable likelihoods. Although such solutions are subject to change shape and likelihood as more accurate experimental data become available, no dramatic improvement can be really expected in their selection, unless

(1) the theoretical uncertainties on the total rates are reduced to the size of the corresponding experimental uncertainties;
(2) the total errors associated with the SK spectrum and day–night measurement are significantly reduced (by, say, a factor $\sim$2); or
(3) decisive results are found in new generation solar neutrino experiments. Any of these conditions require a time scale of a few years at least; the same time scale should then be expected in order to (patiently) single out one of the three MSW solutions (SMA, LMA, or LOW).

Another aspect of the LMA and LOW solutions emerging from figure 10.9 is their extension to large values of the mixing angle ($\sin^2 2\omega \to 1$), which are often assumed to be realized only through the vacuum oscillation solutions. Since the possibility of nearly maximal (ν_1, ν_2) mixing for solar neutrinos has gained momentum after the SK evidence for maximal (ν_μ, ν_τ) mixing ($\sin^2 2\psi \sim 1$), it is interesting to study it in detail by dropping the usual '2ω' variable and by exploring the full range $\omega \in [0, \pi/2]$, as was done earlier in [9]. The subcase $\omega = \pi/4$ will receive special attention in the next section.

10.4.4   Three-flavour oscillations in matter

As stated in section 10.2, for large values of m^2 ($\gg 10^{-4}$ eV^2) the parameter space relevant for 3ν solar neutrino oscillations is spanned by the variables $(\delta m^2, \omega, \phi)$. As far as ω is taken in its full range $[0, \pi/2]$, one can assume $\delta m^2 > 0$, since the MSW physics is invariant under the substitution $(\delta m^2, \omega) \rightarrow (-\delta m^2, \pi/2 - \omega)$ at any ϕ.

For graphical representations, we prefer to use the mixing variables ($\tan^2 \omega$, $\tan^2 \phi$) introduced in [9], which properly chart both small and large mixing. The case $\tan^2 \phi = 0$ corresponds to the familiar 2ν scenario, except that now we also consider the usually neglected case $\omega > \pi/4$ ($\tan^2 \omega > 1$). For each set of observables (rates, spectrum, day-night difference, and combined data) we compute the corresponding MSW predictions and their uncertainties, identify the absolute minimum of the χ^2 function, and determine the surfaces at $\chi^2 - \chi^2_{\min} = 6.25$, 7.82 and 11.36, which define the volumes constraining the (δm^2, $\tan^2 \omega$, $\tan^2 \phi$) parameter space at 90%, 95% and 99% C.L. Such volumes are graphically presented in (δm^2, $\tan^2 \omega$) slices for representative values of $\tan^2 \phi$.

Figure 10.10 shows the combined fit to all data. The minimum χ^2 is reached within the SMA solution and shows a very weak preference for non-zero values of ϕ ($\tan^2 \phi \simeq 0.1$). It can be seen that the SK spectrum excludes a significant fraction of the solutions at $\delta m^2 \sim 10^{-4}$ eV^2, including the upper part of the LMA solution at small ϕ, and the merging with the SMA solution at large ϕ. In particular, at $\tan^2 \phi = 0.1$ the 95% C.L. upper limit on δm^2 drops from 2×10^{-4} eV^2 (rates only) to 8×10^{-5} eV^2 (all data). This indication tends to disfavour neutrino searches of *CP* violation effects, since such effects decrease with $\delta m^2/m^2$ at $\phi \neq 0$.

The 95% C.L. upper bound on ϕ coming from solar neutrino data alone ($\phi < 55°$–$59°$) is consistent with the one coming from atmospheric neutrino data alone ($\phi < 45°$), as well as with the upper limit coming from the combination of CHOOZ and atmospheric data ($\phi < 15°$) (see figure 10.4). This indication supports the possibility that solar, atmospheric and CHOOZ data can be interpreted in a single three-flavour oscillation framework [7, 23]. In this case, the CHOOZ constraints on ϕ exclude a large part of the 3ν MSW parameter space (basically all but the first two panels in figure 10.9).

However, even small values of ϕ can be interesting for solar ν phenomenology. Figure 10.11 shows the section of the volume allowed in the 3ν MSW parameter space, for $\omega = \pi/4$ (maximal mixing), in the mass-mixing plane (δm^2, $\sin^2 \phi$). All data are included. It can be seen that both the LMA and LOW solutions are consistent with maximal mixing (at 99% C.L.) for $\sin^2 \phi \equiv U^2_{e3} = 0$. Moreover, the consistency of the LOW solution with maximal mixing improves significantly for $U^2_{e3} \simeq 0.1$, while the opposite happens for the LMA solution. This gives the possibility of obtaining nearly bimaximal mixing ($\omega = \psi = \pi/4$ with ϕ small) within the LOW solution to the solar neutrino problem—an interesting possibility for models predicting large mixing angles.

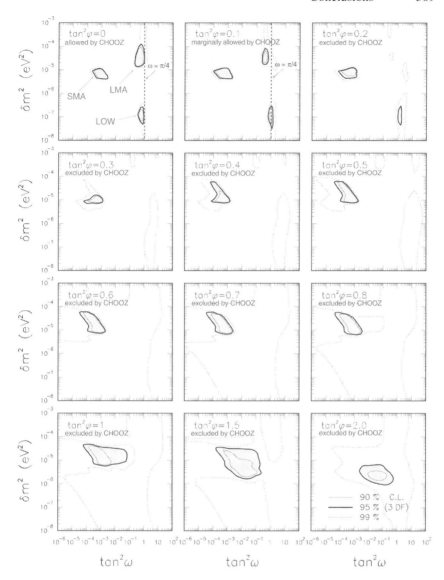

Figure 10.10. Results of the global three-flavour MSW fit to all data. Note that, in the first two panels, the 99% C.L. contours are compatible with maximal mixing ($\tan^2 \omega = 1$) for both the LOW and the LMA solutions. Note that, when the CHOOZ constraints on ϕ are included, only the first two panels are permissible (see figure 10.4).

10.5 Conclusions

We have analysed the most recent experimental evidence for solar and atmospheric ν oscillations in a common theoretical framework including three-

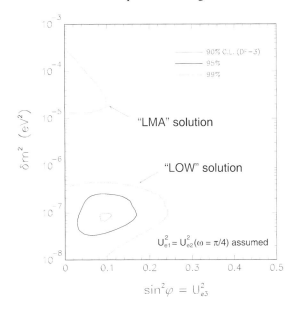

Figure 10.11. Allowed regions in the plane $(\delta m^2, \sin^2 \phi)$, assuming maximal (ν_1, ν_2) mixing $(\omega = \pi/4)$. For $\sin^2 \phi = 0$, both the LMA and LOW solutions are compatible with maximal mixing at 99% C.L. For small values of $\sin^2 \phi$, the maximal mixing case favours the LOW solution.

flavour transitions. We have investigated the regions of the mass-mixing parameter space compatible with the data, with and without the CHOOZ constraints. Such regions are of interest both for model-building and as a guidance for future experimental tests. It turns out that both atmospheric and solar ν data prefer low values of the matrix element U_{e3}^2 even without the inclusion of reactor constraints, which represents a non-trivial consistency check.

The addition of CHOOZ data implies the further restriction $U_{e3}^2 < $ few %. Even within such limits, a novel feature emerges from the 3ν MSW analysis of solar neutrinos [23]: bimaximal mixing of atmospheric and solar νs, usually studied in terms of vacuum solar ν solutions, is possible also within the LMA and LOW MSW solutions.

Acknowledgments

The author would like to thank the organizers of the School in 'Contemporary Relativity and Gravitational Physics' for their kind hospitality. This work is co-financed by the Italian Ministero dell'Università e della Ricerca Scientifica e Tecnologica (MURST) within the 'Astroparticle Physics' project.

References

[1] Kajita T 2000 *Nucl. Phys. B (Proc. Suppl.)* **85** 44

[2] Super-Kamiokande Collaboration 1998 *Phys. Rev. Lett.* **81** 1562

[3] Macro Collaboration 1998 *Phys. Lett.* B **434** 451

[4] Soudan 2 Collaboration 2000 *Nucl. Phys. Proc. Suppl.* **91** 134

[5] CHOOZ Collaboration 1999 *Phys. Lett.* B **466** 415

[6] Mikaelyan L 2000 *Nucl. Phys. B (Proc. Suppl.)* **87** 284

[7] Fogli G L, Lisi E Marrone A and Scioscia G 1999 *Phys. Rev.* D **59** 033001

[8] Fogli G L, Lisi E and Montanino D 1994 *Phys. Rev.* D **49** 3626

[9] Fogli G L, Lisi E and Montanino D 1996 *Phys. Rev.* D **54** 2048

[10] Kuo T K and Pantaleone J 1989 *Rev. Mod. Phys.* **61** 937

[11] Fogli G L, E Lisi, D Montanino and G Scioscia 1997 *Phys. Rev.* D **55** 4385

[12] LoSecco J M *Preprint* hep-ph/9807359
     J M LoSecco *Preprint* hep-ph/9807432

[13] Wolfenstein L 1978 *Phys. Rev.* D **17** 2369
     Mikheyev S P and Smirnov A Yu 1986 *Nuovo Cimento* C **9** 17

[14] Bahcall J N, Basu S and Pinsonneault M 1998 *Phys. Lett.* B **433** 1
     See also J N Bahcall's homepage, www.sns.ias.edu/~jnb

[15] Homestake Collaboration 1998 *Astrophys. J.* **496** 505

[16] Kamiokande Collaboration 1996 *Phys. Rev. Lett.* **77** 1683

[17] SAGE Collaboration 1999 *Phys. Rev.* C **60** 055801

[18] GALLEX Collaboration 1999 *Phys. Lett.* B **447** 127

[19] Bahcall J N *et al* 1996 *Phys. Rev.* C **54** 411

[20] Lisi E and Montanino D 1997 *Phys. Rev.* D **56** 1792

[21] Fogli G L and Lisi E 1994 *Astropart. Phys.* **2** 91

[22] Gonzalez Garcia M C 2000 *Nucl. Phys.* B **573** 3

[23] Fogli G L, Lisi E, Montanino D, and Palazzo A 2000 *Phys. Rev.* D **62** 013002

[24] Bahcall J N, Krastev P I and Smirnov A Yu 2000 *Phys. Lett.* B **477** 401

[25] Bahcall J N and Krastev P I 1998 *Phys. Lett.* B **436** 243

Chapter 11

Highlights in modern observational cosmology

Piero Rosati
European Southern Observatory, Garching b. München,
Germany

11.1 Synopsis

In this chapter, we focus on the fundamental methods of observational cosmology and summarize some of the recent observational results which have deepened our understanding of the structure and evolution of the universe. The chapter is divided into three parts. In the first section, we briefly describe the Friedmann world models, which constitute the theoretical framework, we define the main observables and we illustrate some common applications. In the second section, we describe how galaxy surveys (primarily in the optical band) are utilized to map the structure and evolution of the universe over a large fraction of its age, focusing on observational methodologies and some recent results. In the third section, we describe how surveys of galaxy clusters can be used to constrain cosmological models, and measure the fundamental cosmological parameters. Throughout the chapter, we touch *only* on a few recent highlights in observational cosmology. We refer the reader to fundamental textbooks, such as Longair (1998), Peebles (1993) and Peacock (1999), for a complete overview of the theoretical and observational framework.

11.2 The cosmological framework

This section gives a very brief summary of the basics of Friedmann–Robertson–Walker (FRW) models; only the essentials formulae which are used throughout the chapter and the definition of observable quantities which are often used in cosmology are included.

11.2.1 Friedmann cosmological background

What is generally referred to as the *standard cosmological framework* is the result of the solution of the Einstein equations in the hypothesis that the universe is, *on very large scales, homogeneous* and *isotropic*. There are several pieces of observational evidence which support this *cosmological principle*, such as the distribution of galaxies and clusters of galaxies on large scales and the remarkable isotropy of the cosmic microwave background (CMB).

The FRW models provide the *background* on which the formation and evolution of the large-scale structure in the universe can be studied as the evolution of small perturbations to an otherwise uniform FRW model. The application of the cosmological principle leads to the following FRW spacetime line element (see Landau and Lifshitz (1971) for an elegant and simple derivation):

$$ds^2 = c^2\,dt^2 - R^2(t)\left[\frac{dr_1^2}{1 - kr_1^2} + r_1^2(d\theta^2 + \sin^2\theta\,d\phi^2)\right] \qquad (11.1)$$

$$= c^2\,dt^2 - R^2(t)[dr^2 + S_k^2(r)(d\theta^2 + \sin^2\theta\,d\phi^2)] \qquad (11.2)$$

where two possible definitions of the *comoving coordinate, r,* have been used. This is the coordinate measured by observers at rest with respect to the local matter distribution. The first expression is commonly used in the literature. In the second form, following the notation by Peacock (1999), we have defined:

$$S_k(r) = \begin{cases} \sin(r) & k = 1 \text{ (close)} \\ r & k = 0 \text{ (flat)} \\ \sinh(r) & k = -1 \text{ (open)}. \end{cases} \qquad (11.3)$$

The cases $k = -1, 0, 1$ represent, respectively, an *open universe* (infinite, hyperbolic space), a *flat universe* (infinite, flat space) and a *closed universe* (finite, spherical space).

The solution of the Einstein field equations (with cosmological constant Λ) leads to the following equation for the evolution of the scale factor, $R(t)$:

$$\left(\frac{\dot{R}}{R}\right)^2 = \frac{8\pi G}{3}\rho_{\mathrm{M}} + \frac{1}{3}\Lambda c^2 - \frac{kc^2}{R^2}. \qquad (11.4)$$

This shows three competing terms driving the universal expansion: a matter term, a cosmological constant term and a curvature term. We are neglecting here a radiation term, as appropriate when the universe is dominated by non-relativistic matter ('dust') with density ρ_{M}, i.e. the directly observable universe. The respective fractional contributions to the energy density in the universe at the present epoch are commonly defined as

$$\Omega_{\mathrm{m}} \equiv \frac{8\pi G}{3H_0^2}\rho_{M_0}, \qquad \Omega_\Lambda \equiv \frac{\Lambda c^2}{3H_0^2}, \qquad \Omega_k \equiv -\frac{kc^2}{H_0^2 R_0^2} \qquad (11.5)$$

with

$$\Omega_m + \Omega_\Lambda + \Omega_k = 1, \qquad \Omega_{tot} = \Omega_m + \Omega_\Lambda = 1 - \Omega_k \qquad (11.6)$$

where $H_0 \equiv (\dot{R}/R)_{t=0} = 100 \text{ km s}^{-1} \text{ Mpc}^{-1} h = h(9.78 \times 10^9)^{-1}$ years, is the present value of the *Hubble constant*. The matter density parameter, Ω_m (sometimes denoted as Ω_0), can also be written as $\Omega_m = \rho_0/\rho_{cr}$, where $\rho_{cr} = 3H_0^2/(8\pi G) = 1.9 \times 10^{-29} h^2 \text{ g cm}^{-2}$ is the critical density, which splits open and close models in a matter-dominated universe.

The *deceleration parameter* is also often used:

$$q \equiv -\ddot{R}R/\dot{R}^2 = \Omega_m/2 - \Omega_\Lambda. \qquad (11.7)$$

With these definitions, the equation (11.4) can be written:

$$H^2 = H_0^2 \left[\Omega_m \left(\frac{R_0}{R} \right)^3 + \Omega_k \left(\frac{R_0}{R} \right)^2 + \Omega_\Lambda \right]. \qquad (11.8)$$

11.2.2   Observables in cosmology

Suppose we are at $r = 0$ and observe an object at radial coordinate r_1, when the expansion factor was $R_1 = R(t_1) < R_0$, at some lookback time $t_1 < t_0$. Quantities like r_1, t_1, R_1 are not accessible to measurement. However, there are directly measurable quantities which can be used to test the validity of the FRW metric and to derive its parameters.

First of all, the *redshift*. From the spectrum of a distant source we can easily recognize, say, an emission line whose rest-frame (emitted) wavelength is λ_e. In general, we will measure a redshifted emission line at wavelength λ_0, so that the *redshift z* is defined as

$$1 + z = \frac{\lambda_0}{\lambda_e}. \qquad (11.9)$$

If the expansion factor of the universe was R at redshift z, the following simple relation holds:

$$1 + z = \frac{R_0}{R}. \qquad (11.10)$$

Using this relation, we can now immediately write the *lookback time*, $\tau(z)$, by integrating equation (11.8) after a change of variable, from R to z:

$$\tau(z) = H_0^{-1} \int_0^z (1+z')^{-1} [\Omega_k(1+z')^2 + \Omega_m(1+z')^3 + \Omega_\Lambda]^{-1/2} \, dz'. \quad (11.11)$$

$\tau(z)$ is plotted in figure 11.1 for three different values of $(\Omega_m, \Omega_\Lambda)$. The age of the universe is obtained for $z \to \infty$.

We now examine the other measurable quantities.

11.2.2.1 Angular diameters

Photons from our distant object at radial distance r follow radial, null geodesics ($ds^2 = 0$). Using the FRW metric (11.2), we can then link the angular size ($\Delta\theta$) of an object to its proper length d, perpendicular to the radial coordinate at redshift z:

$$d = R S_k(r)\Delta\theta = R_0 S_k(r)\Delta\theta/(1+z)$$

$$\Delta\theta = \frac{d(1+z)}{d_M} = \frac{d}{D_A} \tag{11.12}$$

where we have defined the *distance measure*, $d_M \equiv R_0 S_k(r)$, and the *angular diameter distance* $D_A = d_M/(1+z)$.

The distance measure out to redshift z, $d_M(z)$, can be derived integrating the equation of motion for a photon, $R\,dr = c\,dt = c\,dR/(RH)$, and using the equations (11.8) and (11.10):

$$
\begin{aligned}
d_M(z) &= \frac{cH_0^{-1}}{|\Omega_k|^{1/2}} S\left\{ |\Omega_k|^{1/2} \int_0^z [\Omega_k(1+z')^2 + \Omega_m(1+z')^3 + \Omega_\Lambda]^{-1/2}\,dz' \right\} \\
&= \frac{cH_0^{-1}}{|\Omega_k|^{1/2}} S\left\{ |\Omega_k|^{1/2} \int_0^z [(1+z')^2(1+\Omega_m z') - z'(2+z')\Omega_\Lambda]^{-1/2}\,dz' \right\}
\end{aligned}
\tag{11.13}
$$

where the multiple function S is defined in (11.3); in the flat case of $\Omega_k = 0$ only the integral remains. Such an integral can easily be evaluated numerically.

For $\Omega_\Lambda = 0$, an analytical solution exists (Mattig 1957):

$$d_M = \frac{2cH_0^{-1}}{\Omega_0^2(1+z)}\{\Omega_0 z + (\Omega_0 - 2)[(\Omega_0 z + 1)^{1/2} - 1]\}. \tag{11.14}$$

Equation (11.12) shows that if a 'standard rod' existed, e.g. a class of objects associated with a fixed physical size with negligible evolutionary effects, then it would be possible to infer cosmological parameters (particularly q_0) by plotting the angular size as a function of redshift (e.g. Kellerman 1993).

11.2.2.2 Apparent intensities

If L is the *rest-frame luminosity* of an object at redshift z (in a given band), then its flux (measured in erg cm^{-2} s^{-1} in cgs units) is

$$S = \frac{L}{4\pi d_M^2(1+z)^2} = \frac{L}{4\pi D_L^2} \tag{11.15}$$

where $D_L = d_M(z)(1+z)$ is the so called *luminosity distance* of the source, which is defined so that the flux assumes the familiar expression in Euclidean geometry (inverse square law). Observations (i.e. fluxes, luminosities) in a given band $[\nu_1, \nu_2]$ can be related to the rest-frame band through the computation of the *K-correction*, K_z, which is essentially the ratio of fluxes in the rest-frame to the observed (redshifted) band $[(1 + z)\nu_1, \nu_2(1 + z)]$. In optical astronomy the magnitude system is used ($m \sim -2.5 \log(S)$) so that (11.15) can be written as a relation between the apparent (m) and absolute magnitude (M) of the object:

$$m = M + 5 \log \left(\frac{D_L}{10 \text{ pc}} \right) + K_z. \qquad (11.16)$$

If the flux spectra density is a power law, i.e. $f_\nu \sim \nu^{-\alpha}$ (like most of the galaxies), then one easily obtains $K_z = 2.5(\alpha - 1) \log(1 + z)$. Such a term can add up to several magnitudes for early type (i.e. red) galaxies at $z \sim 1$.

A low redshift expansion of (11.16) leads to the simple formula (e.g. Sandage 1995):

$$m = 5 \log z + 1.086(1 - q_0)z + 5 \log c H_0^{-1} + M + 25. \qquad (11.17)$$

This shows that if we can recognize a class of astrophysical sources as 'standard candles', by measuring the dimming of these sources over a wide range of redshifts we can measure the deceleration parameter, q_0, and eventually separate Ω_m and Ω_Λ. The application of this fundamental test to high redshifts Type Ia supernovae has lead to spectacular results in recent years (e.g. Perlmutter *et al* 1999, Schmidt *et al* 1998).

11.2.2.3 Number densities

One of the main goal of redshift surveys is to quantify the comoving volume density of objects as a function of redshift. A frequently used quantity is therefore the *comoving volume element* in the redshift interval, z to $z+dz$, in the solid angle $d\Omega$, which follows directly from the FRW metric (11.1), (11.2):

$$dV = \frac{d_M^2}{(1 + \Omega_k c^{-2} H_0^2 d_M^2)^{1/2}} d(d_M) \, d\Omega. \qquad (11.18)$$

Using equation (11.13), and defining the functions $E(z)$ and $A(z)$ as

$$E(z) = \int_0^z [\Omega_k(1 + y)^2 + \Omega_m(1 + y)^3 + \Omega_\Lambda]^{-1/2} \, dy \equiv \int_0^z A(y) \, dy,$$

we have:

$$\frac{dV}{d\Omega \, dz} = (c H_0^{-1})^3 A(z) |\Omega_k|^{-1} S^2 \{|\Omega_k|^{1/2} E(z)\} \qquad (11.19)$$

$$= c H_0^{-1} A(z) d_M^2 \equiv Q(z, \Omega_m, \Omega_\Lambda),$$

where, as usual, we defined S^2 as $\sinh^2$ if $\Omega_k > 0$ (open universe) and $\sin^2$ if $\Omega_k < 0$ (close universe). In the flat case, $S^2 \to E^2(z)$. Remember that Ω_k is not an independent parameter but rather given by $1 - \Omega_m - \Omega_\Lambda$.

For $\Omega_\Lambda = 0$, one finds:

$$\frac{dV}{d\Omega\,dz} = \frac{(cH_0^{-1})^3}{(1+z)^3} \frac{\{q_0 z + (q_0 - 1)[(2q_0 z + 1)^{1/2} - 1]\}^2}{q_0^4(1 + 2q_0 z)^{1/2}} \tag{11.20}$$

$cH_0^{-1} \simeq 3000h^{-1}$ Mpc is the Hubble length.

The volume element (11.19) is plotted in figure 11.1 for three reference models. We will see later that the flat case $(\Omega_m, \Omega_\Lambda) = (0.3, 0.7)$ is currently favoured by measurements. This plot shows that if we peer into a patch of the sky with deep observations, at $z = 2$–3 we have a good chance to explore a large comoving volume (which is ultimately determined by the observational technique).

11.2.2.4 Surface brightnesses

The observed surface brightness Σ_{obs} of an extended object is defined as the flux per unit emitting area. This is the observable that ultimately drives the detection of faint galaxies (rather than its flux), and has the remarkable property of being independent on cosmological parameters. For a FRW model, using equations (11.15), (11.12), it is:

$$\Sigma_{obs} = \frac{S_{obs}(\nu_1, \nu_2)}{\pi\,\Theta^2} = \frac{L_{obs}(\nu_1, \nu_2)K_z}{4\pi\,d_M^2} \frac{d_M^2}{\pi d^2(1+z)^4} = \frac{1}{\pi}\left(\frac{L_{obs}}{4\pi d^2}\right)\frac{K_z}{(1+z)^4}.$$

This is also known as the Tolman law, and can be used as a direct test of the expansion of the universe (e.g. Sandage 1995). $L_{obs}/4\pi d^2$ is the intrinsic surface brightness of the source with physical size d (in units of, e.g., erg s^{-1} kpc^{-2}). Besides the K-correction, this relation shows that the surface brightness of extended objects drops very rapidly with redshifts, making the detection of high-z extended objects difficult.

11.2.3 Applications

One of the most common application of the expressions derived in the previous section is the computation of observed distributions, such as source number counts, or the redshift-dependent volume density of a class of objects, based on known local ($z \simeq 0$) distributions. By comparing these observed distributions, at different redshifts, with those predicted on the basis of observations in the local universe or models of structure formation, one can set constraints on the evolutionary history of a given class of objects, and, *in principle*, on the cosmological model itself (i.e. on Ω_m, Ω_Λ).

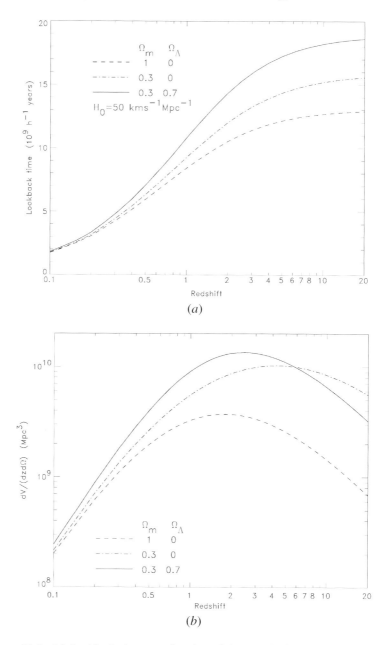

Figure 11.1. (*a*) Lookback time as a function of the redshift for three reference FRW models (Einstein–de Sitter, open, flat). At $z = 20$ the lookback time is approximately 99% of the age of the universe in all models. (*b*) Derivative of the comoving volume element, per unit solid angle, as a function of redshift for the same models.

11.2.3.1 Number counts

By number counts we mean the surface density on the sky of a given class of sources as a function of the limiting flux of the observations (e.g. magnitude, radio flux). This is the simplest observational tool which can be used to study the evolution of a sample of objects, and, to some extent, to test cosmological models. It does not require redshift measurements but only a knowledge of the selection function (indeed, a major challenge in any survey in comology!).

The space density of sources of different intrinsic luminosities, L, is described by the *luminosity function* (LF), $\phi(L)$, so that $dN = \phi(L) \, dL$ is the number of sources per unit volume with luminosity in the range L to $L + dL$. The most common functional form to describe observational data is the one proposed by Schechter (1976):

$$\phi(L) = \frac{\phi_*}{L_*} \left(\frac{L}{L_*}\right)^{-\alpha} e^{-L/L_*}. \tag{11.21}$$

L_* is the *characteristic luminosity* of the population, the normalization ϕ_* determines the volume density of sources, as $n_0 = \int_0^\infty \phi(L) \, dL = \phi_* \Gamma(1 - \alpha)$, where Γ is the gamma-function. The product $\phi_* L_*$ is an estimate of the integrated luminosity of all sources in a given volume, since the *the luminosity density* is defined as $\epsilon_L = \int_0^\infty L\phi(L) \, dL = \phi_* L_* \Gamma(2 - \alpha)$.

The determination of the local LF of galaxies is not completely straighforward since one has to take into account the morphological mix of galaxies (i.e. the existence of a variety of morphological types, from ellipticals to spirals and irregulars) and clustering effects which bias the measurement of the space density. Most of the observations in the nearby universe (e.g. Loveday *et al* 1992) find best-fit parameters:

$$L_* \simeq 10^{10} h^{-2} L_\odot$$

(corresponding to a B band absolute magnitude $M_B \simeq 20 + 5 \log h$);

$$\phi_* \simeq (1.2\text{–}1.5) \times 10^{-2} h^3 \text{ Mpc}^{-3}, \qquad \alpha \simeq 1.$$

Let us consider, for simplicity, the local or nearby Euclidean universe uniformly filled with sources with LF $\phi(L)$. If S is the limiting flux, sources with luminosity L can be observed out to $r = (L/4\pi S)^{1/2}$. The number of sources over the solid angle Ω, observable down to the flux S are:

$$N(> S) = \int \frac{\Omega}{3} r^3 \phi(L) \, dL = \frac{\Omega}{3(4\pi)^{3/2}} S^{-3/2} \int L^{3/2} \phi(L) \, dL.$$

Once the integral over all luminosities is evaluated, the surface density of sources down to the flux S is always $N(> S) \propto S^{-3/2}$. If we use magnitude instead of luminosities, then $\log N(> m) \propto 0.6 \, m$. Therefore, number counts in the nearby universe, where curvature terms can be neglected, are characterized by a *Euclidean slope* of -1.5 (or 0.6 mag). In general, at large distances, curvature effects (cf equations (11.13) and (11.18)) cause number counts to have slopes always shallower than the Euclidean one. However, as we will see in section 11.3.3, evolutionary effects ($\phi = \phi(L, z)$) can counteract such a natural behaviour and produce counts steeper than 1.5.

11.2.3.2 *Redshift distribution and number counts (general case)*

We now have all the ingredients to compute the expected redshift distribution, $n(z)$, and number counts, $n(> S)$, for an evolving population of sources with LF $\phi(L, z)$. Typically, on the basis of the known local LF, $\phi(L, 0)$, one wants to compare the *observed* redshift distribution of sources with the one expected on an empirical evolutionary scenario, or the one predicted by some theory of structure formation. In general, there will be some degree of degeneracy between evolutionary parameters and cosmological paramaters (Ω_m, Ω_Λ) when matching theoretical models with observational data.

With $Q(z, \Omega_m, \Omega_\Lambda)$ given by equation (11.19) (or (11.20)), the number of sources per unit solid angle and redshift, in the luminosity range L to $L + dL$, is:

$$\frac{d^2 N}{d\Omega dz} \phi(L) \, dL = Q(z, \Omega_m, \Omega_\Lambda) \frac{\phi_*}{L_*} \left(\frac{L}{L_*} \right)^{-\alpha} e^{-L/L_*} \, dL. \qquad (11.22)$$

We now change variable, $y = L/L_*$, and call L_1 and L_2 the minimum and maximum luminosity of the source population (for example, a magnitude range within which we want to compute the redshift distribution). Thus, the surface density of sources, per unit redshift, observed down to the flux S can be written as:

$$\frac{dN(> S, z)}{d\Omega \, dz} = \phi_* Q(z, \Omega_m, \Omega_\Lambda) \int_{y_1(z)}^{y_2} y^{-\alpha} e^{-y} \, dy \qquad (11.23)$$
$$= \phi_* \Gamma(1 - \alpha) Q(z, \Omega_m, \Omega_\Lambda)[P(1 - \alpha, y_2) - P(1 - \alpha, y_1)],$$

where P is the generalized Γ-function, $y_2 = L_2/L_*$, and

$$y_1(z) = \max \left(\frac{L_1}{L_*}, \frac{L_{min}(S, z)}{L_*} \right), \qquad L_{min}(S, z) = S 4\pi D_L^2(z) K_z. \qquad (11.24)$$

L_{min} is the rest-frame miminum luminosity detectable at redshift z, at the limiting flux S (equation (11.15)).

The numerical integration of equations (11.23) and (11.24) can also include an evolving LF, e.g. $\phi_* = \phi_*(z)$, $L_* = L_*(z)$. The result can be directly compared with the observed redshift distribution of sources, i.e. the number of sources per deg^2, in each redshift bin. The number counts $n(> S)$ are obtained by integrating (11.23) over all redshifts.

11.3   Galaxy surveys

11.3.1   Overview

Over the last ten years, significant progress has been made in both observational and theoretical studies aimed at understanding the evolutionary history of galaxies, the physical processes driving their evolution and leading to the Hubble sequence of types (ellipticals, spirals, irregulars) that we observe today.

Deep galaxy surveys have had a central role in cosmology back to the pioneering work of Hubble. In the 1960s (see Sandage 1995) several studies used galaxy counts as a tool to test cosmological models; however, it was soon realized that it was difficult to disentangle the effects of *evolution* from those due to the *universal geometry*, as well as the effects of *object selection*, which, if not properly understood, can easily alter the slope of the number counts (see later).

The modern era of observational cosmology began with the advent of CCD detectors in the 1980s and soon after with multi-object spectrographs. Scientific progress has obviously been driven by a series of technological breakthroughs with telescopes and instrumentation, that we can summarize as follows:

- *Mid 1980s*: First deep CCD surveys (Tyson 1988) revealed a large number of faint, blue galaxies in nearly confusion limited images.
- *Early 1990s*: (a) the development of multi-object spectrographs allows the first spectroscopic surveys of distant galaxies (e.g. Ellis *et al* 1996, Lilly *et al* 1995); and (b) central role of Hubble Space Telescope (HST) (resolved images of distant galaxies, morphological information).
- *Mid 1990s*: (a) spectroscopy with the Keck telescope (10 m collecting area) pushed the limit to two magnitudes fainter; (b) significant improvement in near-IR imaging (sensitivity and detector area); and (c) deep imaging in the millimetre wavelength with the SCUBA instrument.
- *Late-1990s*: wide-field optical imaging; (b) high-multiplexing spectroscopy (several hundreds of spectra at once); and (c) 8 m class telescopes with active optics (VLT) (delivering angular resolution of $0.5''$ or better).
- *On-going/upcoming*: (a) next generation of spectrographs + near-IR spectroscopy on 8–10 m class telescopes; (b) Integral-field spectrographs (x, y, λ information); (c) adaptive optics delivering diffraction-limited images ($\sim 0.05''$ resolution); and (d) Advance Camera for Survey on HST (2001).

This rapid technological development has allowed a number of major surveys to be carried out. We can classify those which have had a major impact on the way we understand the structure and evolution of the universe today as follows.

Large area surveys

- APM (Automatic Plate-measuring Machine, e.g. Maddox *et al* 1990)—imaging photographic plates;
- CfA survey (Center for Astrophysics, e.g. Huchra *et al* 1990);
  LCRS (Las Campanas Redshift Survey, e.g. Shectman *et al* 1996)—$\sim 10^4$ galaxy redshifts, over 700 deg^2 out to $z \simeq 0.2$;
- 2dF survey (2 degree field, e.g. Colless 1999)—$\sim 10^5$ redshifts covering 1700 deg^2; and
- SDSS survey (Sloan Digital Sky Survey: http://www.sdss.org)—$\sim 10^6$ redshifts + multicolour imaging (10^4 deg^2, $m_{\lim} \simeq 22$).

The first three surveys have provided the power spectrum of the large-scale structure, by measuring the correlation function over a wide range of scales (see L Guzzo, this volume), and the luminosity functions of different galaxy types in the local universe. The on-going 2dF and SDSS surveys will soon bring these measurements to an unprecedented level of precision.

Deep, small area surveys

- The LDSS autofib survey (Ellis *et al* 1996)—B-band selected redshift survey down to $B \simeq 24$ ($z \lesssim 0.7$).
- The CFRS survey (Lilly *et al* 1995)—I-band selected redshift survey down to $I \simeq 22$ (~ 600 galaxies at $z \lesssim 1$).
- The Keck Survey (Cowie *et al* 1996)—150 galaxy redshifts out to $z \simeq 1.5$ ($22.5 < B < 24$).
- The CNOC2 surveys (Yee *et al* 2000)—6000 galaxy redshifts over 1.5 deg^2 area ($z \lesssim 0.6$).

These surveys have established a clear evolutionary pattern for different galaxy types out to $z \sim 1$ (see section 11.3.3).

Ultra-deep, tiny area surveys

- Hubble Deep Field North and South (e.g. Williams *et al* 1996, Ferguson *et al* 2000)—5 arcmin^2, $m_{\lim} \simeq 29$ (see later).

11.3.2 Survey strategies and selection methods

When planning an imaging survey (not necessarily in optical or near-IR wavelengths which are the primary subject here), the balance between the *depth* and the *solid angle*, as well as the selection of the *observed band* play a central role. These decisions are driven by the nature of the sources under study, as well as their typical volume density and luminosity, i.e. ϕ_* and L_* (see (11.21)). Rare objects, such as quasars or galaxy clusters, require large-area surveys to be found in sizeable numbers. Large surveys also probe the bright end of the LF

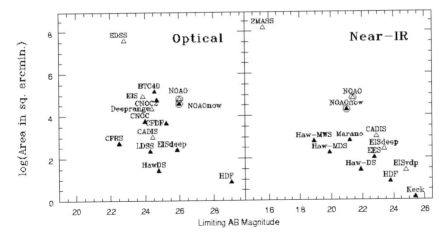

Figure 11.2. Several optical and near-IR surveys (carried out over the last ten years) in the depth–solid angle plane. The AB magnitude system is defined as $m(\mathrm{AB}) = -2.5 \log f_\nu(\mathrm{nJy}) + 31.4$.

of any source population, as opposed to small-area surveys which mostly probe the faint end of the LF ($L \lesssim L_*$). In general, the deeper the survey is the more distant are the L_* objects which can be detected. The combination depth–solid angle will determine the sampled volume at different redshifts, for a given object selection method. Obviously, the product (limiting flux × survey area) is kept approximately constant by observational time constraints. In figure 11.2, we plot several cosmological surveys which have been carried out over the last ten years with the aim of mapping the structure in the universe and understanding its evolution. The Sloan Digital Sky Survey (SDSS) and the Hubble Deep Field (HDF) represent the two complementary extremes, i.e., a shallow survey covering a significant fraction of the sky and a very deep pencil beam survey.

For a given depth and survey area, the probed volume is ultimately determined by the *selection function*, i.e. the set of criteria which lead to the object detection. There are basically three different selection methods:

(1) *Flux-limited selection.* All the sources with a flux greater than a given threshold, S_{lim}, are included in the sample. The simplicity of this method leads to a straightfoward computation of the probed volume (however, see caveats later). If A_S is the survey area, the maximum redshift, z_{max}, at which a source of rest frame luminosity L can be detected, is given implicitily by $L = S_{\mathrm{lim}} 4\pi D_L^2(z_{\mathrm{max}})$ (11.15). Thus, using (11.19), the *survey volume* is:

$$V_{\mathrm{max}}(z, L) = A_S \int_0^{z_{\mathrm{max}}} Q(z, \Omega_m, \Omega_\Lambda) \, dz. \qquad (11.25)$$

Note that the K-correction is also involved in this calculation when converting from observed to rest-frame luminosities. By counting sources in different luminosity–redshift bins one can thus estimate the LF $\phi(z, L)$.

(2) *Colour selection.* Sources are selected on the basis of their flux *and* colour. A relevant case is described in section 11.3.4. The advantage of this method is that it is extremely efficient at isolating objects in a given redshift range, for example a distant volume in the universe. However, the selection function (i.e. the survey volume) critically depends on the knowledge of the spectral energy distribution (SED) of the sources under study.

(3) *Narrow-band filter selection*: This technique consists of selecting sources which have a flux excess when observed through a narrow-band filter, as compared to their broad-band flux. Emission line objects (e.g. starbursts, AGN) are the targets of these surveys. Sources are detected at redshift $1 + z = \lambda_{filter}/\lambda_{em.line}$, within a Δz given by the width of the filter, which needs to be narrow enough ($\lesssim 100$ Å) to boost the contrast of the emitting line object against the background sky. The equivalent width of the emission line ultimately determines the selection function. Several searches for very high redshifts objects have been conducted using the Lyα (1216 Å) as a tracer. Such surveys have had some success (Hu *et al* 1999), but have also underscored the difficulties of this method. First, a very narrow redshift slice is probed, and therefore samples are small and prone to cosmic variance and large-scale structure effects. Second, only a limited portion of the galaxy population (e.g., galaxies with large equivalent width) is selected. These limitations make it difficult to draw statistical conclusions on the volume density, or luminosity density of distant galaxies.

11.3.2.1 Caveats

There are several caveats inherent in the aforementioned selection methods, which if not properly addressed, can lead to a biased view of the evolution of the structure in the universe and underlying cosmological models.

First of all, the flux-limit approach is an idealization of our detection process. Sources are never detected on the basis of their flux, but rather on the basis of their surface brightness (a detection consists of an excess of flux within a given aperture, above a given threshold, which is usually a few times the rms value of the surrounding background). A major concern of any survey is to establish whether the sample is, to a good approximation, *flux-limited* rather than *surface-brightness-limited*. As a result, the flux limit (S_{lim}) should be chosen high enough so to cover the whole range of surface brightness of our sources. Low surface brightness sources will be the first to drop out of the sample if this process is ill defined.

Second, the computation of the *K-correction* requires a knowledge or assumptions on the SED of sources at different redshifts.

Third, the effect of *reddening* especially due to dust enshrouding distant objects (and, to a lesser extent, to intervening neutral hydrogen) can have a

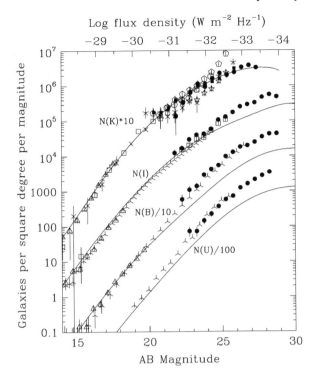

Figure 11.3. A compilation of number counts in the U, B, I, K bands from different surveys (Ferguson *et al* 2000 and references therein). Full symbols are from the HDF North and South, open symbols from several ground-based surveys. Full lines are no-evolution models obtained integrating the observed local luminosity function for $(\Omega_m, \Omega_\Lambda) = (0.3, 0.7)$.

significant impact on the selection function and completeness of the sample by absorbing the UV part of the continuum and selectively suppressing different emission lines.

11.3.3 Galaxy counts and evolution

We show in figure 11.3 a compilation of number counts from ground-based and HST surveys over a 13 magnitude range, as observed in the U, B, I and K passbands (see Ferguson *et al* 2000). Each set is displaced by a factor of 10 for clarity. Full curves represent the theoretical expectations obtained by integrating the local luminosity function assuming *no evolution* and $(\Omega_m, \Omega_\Lambda) = (0.3, 0.7)$, as described in section 11.3.3. These no-evolution (NE) models make reasonable assumptions on the morphological mix of the local galaxy population (relative fraction of irregulars, spirals, ellipticals), their LFs and their SEDs (required to compute the K-corrections). Such assumptions reflect observations of the nearby universe but are still affected by some uncertainty, therefore it is not uncommon to find in the literature NE models which differ by $\sim$50%. This uncertainty will be drastically reduced when the 2dF and SDSS surveys are completed.

A clear trend is apparent in figure 11.3. At blue wavelengths the observed counts exceeds the NE predictions by as much as a factor three, a problem which was recognized in the first deep surveys and which has become known as the *faint blue galaxy excess*. Such an excess progressively disappear at longer wavelengths. Observations in blue filters are sensitive to late type, star-forming galaxies with young stellar populations. Therefore, it had already become evident in the early 1990s (e.g. Ellis *et al* 1996) that this is the galaxy population which has undergone most of the evolution (in luminosity and/or number density) out to $z \sim 1$, i.e. the last 50% of the life of the universe. The first deep redshift surveys (Lilly *et al* 1995) confirmed this scenario directly measuring a significant evolution of the LF for the 'blue population' out to $z \simeq 0.7$, while revealing no significent evolution for the 'red population' consisting of galaxy types earlier than an Sbc (see figure 11.4). Red wavelength observations, particularly in the K-band ($\lambda_0 = 2 \mu$m), collect rest-frame optical light out to $z \sim 3$, thus probing old, long-lasting stellar populations in distant galaxies (i.e. earlier types). All these observations (see also Cowie *et al* 1996) have shown a remarkable increase in the space and/or luminosty density of star-forming galaxies with redshift. However, interpreting these results, and understanding the physical processes responsible for this evolutionary pattern, has remained a difficult task.

In this respect, HST observations have driven us a big step forward by allowing *intrinsic sizes* and *morphologies* of distant galaxies to be measured. The combination of angular resolution (0.05″) and depth has also pushed these studies well beyond $z = 1$. As an example, in figure 11.5 we show number counts for different morphological types as directly determined by the HDF-N images (Driver *et al* 1998). Along with NE model predictions (full lines), *passive evolution models* are also shown. The latter are constructed using spectral synthesis models (e.g. Bruzual and Charlot 1993), assuming a formation redshift (generally varying by type), and a star formation history (with a given initial mass function, IMF). As an example, in figure 11.6 we show the evolution of the SED of a 3 Gyr burst stellar population over approximately a Hubble time. This model well reproduces the evolution of an early type galaxy. The UV luminosity declines rapidly after the end of the burst of star formation, as hot O and B stars burn off the main sequence and the population is more and more dominated by red giants.

In general, passive evolution models are characterized by *luminosity evolution*, which is the result of letting the stellar populations evolve with a pre-defined star formation history, without including any merging. Figure 11.5 confirms that morphologically selected early types show little (simple passive) evolution to faint magnitudes, and hence to relatively high redshifts. Counts of intermediate types (i.e. spiral-like galaxies) are broadly consistent with passive, luminosity evolution models, whereas later types and irregulars are not fitted by

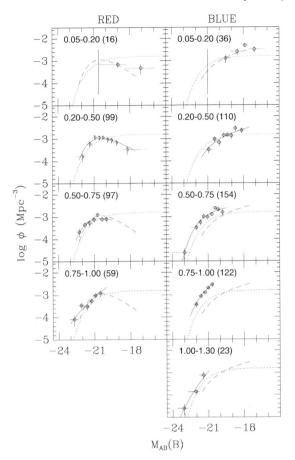

Figure 11.4. Measurement of the LF at different redshifts from the CFRS survey (Lilly *et al* 1995). The redshift bin and number of objects for each LF are given in the label in each panel. The dividing line between 'red' and 'blue' samples corresponds to the rest-frame colour of an Sbc galaxy. A clear evolution is visible in the blue sample, whereas no significant evolution is observed in the red sample out to $z \simeq 0.7$.

any of these models. It is believed that most of the morphological evolution of these irregular and peculiar galaxies occurs at $1 \lesssim z \lesssim 2$, as a result of interactions or merging, to lead to the assembling of the familiar Hubble sequence. In general, fairly complex luminosity evolution models, which also include a prescription for dust obscuration, fail to predict number counts at the faintest magnitudes or the number density of galaxies at $z \gtrsim 2$. This is a clear indication that a much deeper physical understanding of the galaxy formation processes is needed ('active' versus simple passive evolution). Central, unsolved key questions are how the star formation activity is modulated by merging and

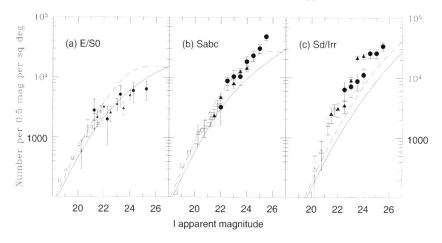

Figure 11.5. Number counts for different morphological types as derived from the HDF-N survey (Driver *et al* 1998). The full and broken curves are predictions from no-evolution and passive evolution models respectively (for $\Omega_m = 1$, $\Omega_\Lambda = 0$).

how the stellar mass is assembled over time in a hierarchical structure formation scenario.

11.3.4 Colour selection techniques

The measurement of the redshift of distant (say $z > 1$), faint galaxies is a time-consuming task and becomes impossible at magnitudes fainter than $\sim$25, even with 8–10 m class telescopes equipped with modern spectrographs. As outlined in section 11.3.3, statistical studies of the nature and evolution of galaxies require an estimate of their SED and their redshift at magnitude selections well beyond the spectroscopic limit. This has stimulated intensive activity over the last few years, aimed at exploiting *colour selection techniques* to *isolate and study* galaxy populations at different redshifts. The basic idea has been to use multi-colour imaging, in as many passbands as possible, to constrain the SED of galaxies by detecting spectral features and measuring the continuum slope, thus estimating the redshift.

The most successful colour selection method in recent years, which has become known as *Lyman break technique*, was devised to detect the ubiquitous Lyman limit discontinuity at 912 Å, which is redshifted into the HST bandpasses at $z \gtrsim 2$ (or at $z \gtrsim 2.5$ for redder ground-based filters) (e.g. Steidel *et al* 1996). This technique is illustrated in figure 11.7 (see the review by Dickinson 1998). A galaxy with an unreddened UV continuum (i.e. a star-forming galaxy or an AGN) has a nearly flat spectrum in f_ν, and a sharp break due to photoelectric absorption of intervening neutral hygrogen (in the galaxy itself and in the intergalctic space along the line of sight) shortward of 912 Å (lyman limit). The integrated effect

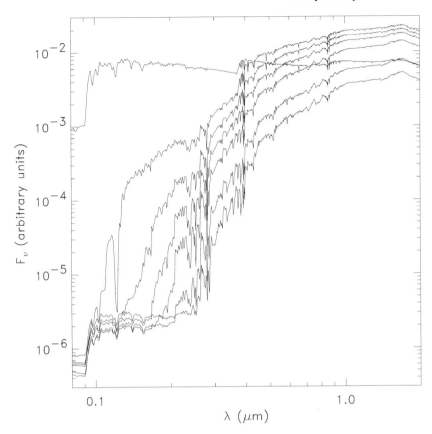

Figure 11.6. Evolution of the spectral energy distribution (SED) of a stellar population modelled as a star formation burst of 3 Gyr, over the lifetime of the universe. From top to bottom, the SEDs are shown at ages: 0.2, 3.2, 3.4, 4, 5, 10, 18 Gyr. The latest Bruzual and Charlot spectral synthesis models have been used.

of neutral hydrogen clouds along the sightline (Lyα forest) produces a further depression blueward of the Lyα, which becomes stronger at higher redshifts. As a result, a star-forming galaxy at $z \simeq 3$ is seen disappearing in the transition from the B to the U band ('U drop-out'). In general, by measuring colours, such as U–B and B–V, one can select a large sample of galaxies around $z \sim 3$, since these sources will stand out in a colour–colour diagram, having very red U–B colours (lyman limit passing through the two filters) and nearly zero B–V colours (flat spectrum). Such a technique was first successfully applied to ground-based imaging data (e.g. Steidel *et al* 1996), which have the advantage of covering much larger solid angles than the HDF, although they cannot match the photometric accuracy of HST, which is critical to measuring colours accurately.

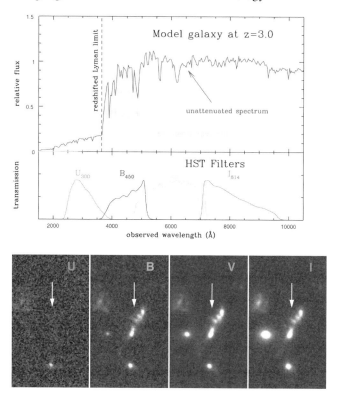

Figure 11.7. Illustration of the Lyman break ('drop-out') technique in the HDF-N from Dickinson (1998). *Top panel*: model spectrum of a star-forming galaxy at $z = 3.0$. Its flat UV continuum (in f_ν units) is truncated by the 912 Å Lyman limit, which is redshifted between the U and B filters of the WFPC2 camera aboard the HST. Intervening neutral hydrogen along the light of sight further suppresses the continuum blueward of Lyα (1216 Å). *Bottom*: HDF-N galaxy, spectroscopically confirmed at $z = 2.8$, as observed in the four WFPC2 bandpasses. Its flux is constant in V and I, it dims in B and completely vanishes in the U-band image.

Follow-up spectroscopy with the Keck telescope has confirmed that objects selected in this fashion were indeed star-forming galaxies at $2 \lesssim z \lesssim 3.5$ (Steidel *et al* 1996). The same technique can be applied to search for higher redshifts galaxies/AGN, for example, objects at $z \gtrsim 4$, the so-called 'B drop-outs' (Steidel *et al* 1999), although it becomes much harder as they become fainter ($R > 24$) and more rare. To date, approximately 900 galaxies have measured with a spectroscopic redshift at $z \simeq 3 \pm 0.5$ and approximately 50 at $4 \lesssim z \lesssim 5$. By exploring relatively large volumes at $z \sim 3$, these studies have taught us much about the star formation density (see section 11.3.5) and large-scale structure (e.g.

Giavalisco *et al* 1998) in the universe back to epochs which represent only 20% of the cosmic time (e.g. Steidel *et al* 1998, 1999).

The Lyman-break technique is just a particular case of a more general method known as *photometric redshifts*. Photometric information from a multi-colour survey can be used as a very low resolution spectrograph to constrain the galaxy SED and thus to estimate the redshift. A good example is shown in figure 11.8 (Giallongo *et al* 1998). A set of SED templates, generally generated with spectral synthesis models (i.e. Bruzual and Charlot models, including UV absorption by the intergalactic medium and dust reddening), is compared with broad photometry data. The best-fit template yields the redshift and the nature of the galaxy.

The photometric redshift technique has been extensively tested in the HDF-N data, since approximately 150 spectroscopic redshifts are available in this field out to $z \simeq 4.5$ and high photometric accuracy can be achieved with the angular resolution and depth of HST images. For example, Benitez (2000) has shown that an accuracy of $\Delta z \leq 0.08(1 + z_{spec})$ can be reached using a Bayesian estimation method (see figure 11.9). With such an accuracy, one can use photometric redshifts to study the evolution of global statistical properties of galaxy populations, such as clustering at $z \lesssim 1$ and the star formation history out to $z \simeq 4$ (see later).

11.3.5 Star formation history in the universe

The UV continuum of a star-forming galaxy probes the emission from young stars and therefore it directly reflects the ongoing star formation rate (SFR). The optimal wavelength range is ~1250–2500 Å, longward of the Lyα forest but at wavelengths short enough that the contribution from older stellar populations can be neglected. In order to establish the relationship between SFR and UV luminosity, evolutionary synthesis models are used. This is a multiparameter exercise though. Basic ingredients include: the metallicity of the stars, the star formation history, the IMF, as well as stellar tracks and atmospheres. A series of these constant SF models, with a range input parameters, is shown in figure 11.10 (lower curves). After ~1 Gyr, the UV luminosity settles around a well defined value which can be used to convert UV luminosities into SFRs. Madau *et al* (1998) used the following relation:

$$\text{SFR}(M_\odot \text{ yr}^{-1}) = 1.4 \times 10^{-28} L_{UV}(\text{erg s}^{-1} \text{ Hz}^{-1}). \qquad (11.26)$$

For models with a short burst of star formation (upper curves) such a simple relation does not exist, although, statistically speaking, (11.26) is still a reasonable approximation, if a sample of galaxies is caught during their first Gyr of life.

Equation (11.26) applies in the wavelength range 1500–2800 Å since the spectrum f_ν of a star-forming galaxy is nearly flat in that region. At $z \gtrsim 1$ optical observations probe this UV rest frame portion of spectrum, therefore the observed luminosity function, or luminosity density, can be directly converted to

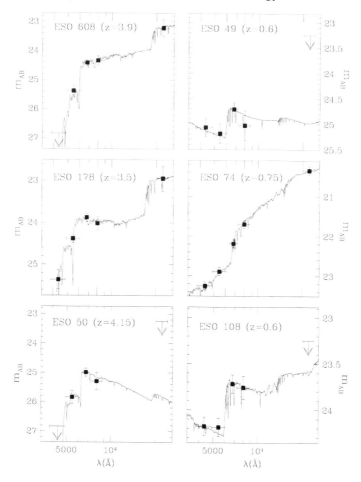

Figure 11.8. Illustration of the photometric redshift technique on a variety of intermediate and high redshift galaxies (Giallongo *et al* 1998). The data points are broad-band photometric meaurements in BVRIK filters used to constrain the spectral energy distribution of galaxies, thus estimating their redshift.

SFR density. By using photometric redshifts (possibly supported by a subset of spectroscopic measurements), one can thus trace the *star formation history* in the distant universe. Madau *et al* (1998) exploited this method to measure the global SFR at $0.5 \lesssim z \lesssim 4$ using HDF and ground-based surveys. This measurement has been repeated by many others in recent years (e.g. Steidel *et al* 1999), and most of the debate has focused on the critical role of *dust* which is surely present in high-z galaxies and is very effective in absorbing UV radiation. To some extent all UV-based SFR measurements are biased low due to dust extinction (e.g. Steidel

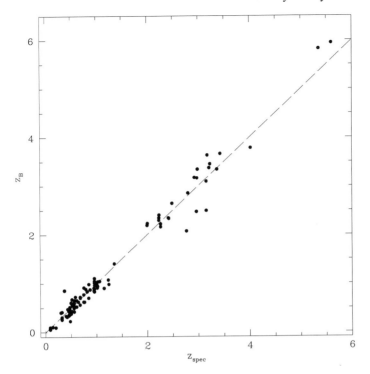

Figure 11.9. Comparison between the spectroscopic redshift (z_{spec}) and the photometric redshift (z_B) in the HDF-N (Benitez 2000).

et al 1999). The standard procedure is to apply statistical corrections, which use empirical correlations of the UV slope β with the extinction derived from the Balmer decrement in nearby starburst galaxies (Calzetti *et al* 1994).

A collection of (mostly dust corrected) estimates of the SFR density over a broad range of redshifts is shown in figure 11.11, which illustrates the great progress made in recent years. This picture seems to suggest that a large fraction of the stars had already been formed by $z \sim 3$. However, global average SFR densities over large cosmic volumes, even in the hypothesis that we can correct for dust extinction, tell us very little about the processes which modulate the star formation (e.g. merging events) and lead to build galaxy masses over time. Future space-based far-infrared (5–30 μm) observations, by providing rest-frame near-IR radiation (which is well correlated with the stellar and dynamical mass) and by measuring the thermally reradiated dust emission in distant galaxy, hold the best promise to shed new light on these issues.

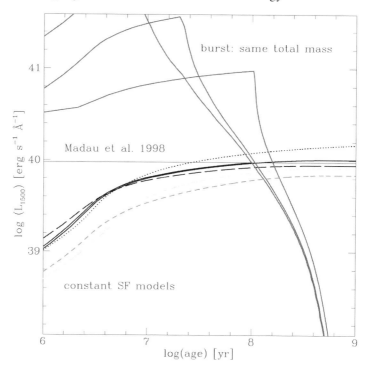

Figure 11.10. Linking the star formation rate (SFR) to the UV luminosity (L_{1500}) using population synthesis models (from Schaerer 1999). Lower curves give the temporal evolution of L_{1500} for models with a constant SFR of $1 M_\odot$ yr^{-1}. Upper curves are models with a burst of SF with duration 5, 20, 100 Myr, forming the same total mass ($10^9 M_\odot$).

11.4　Cluster surveys

11.4.1　Clusters as cosmological probes

The distribution and masses of galaxy clusters are important testing tools for models describing the formation and evolution of cosmic structures. In standard scenarios, clusters form in correspondence with the high peaks (i.e. rare fluctuations) of the primordial density field (e.g. Kaiser 1984). Therefore, both the statistics of their large-scale distribution and their abundance are highly sensitive to the nature of the underlying dark matter density field. Furthermore, their typical scale, $\sim 10h^{-1}$ Mpc relates to fluctuation modes which are just approaching the nonlinear stage of gravitational evolution. Thus, although their internal gravitational and gas dynamics are rather complex, a statistical description of global cluster properties can be obtained by resorting to linear theory or perturbative approaches. By following the redshift evolution of clusters, we have a valuable method to trace the global dynamics of the universe and, therefore, to determine its geometry.

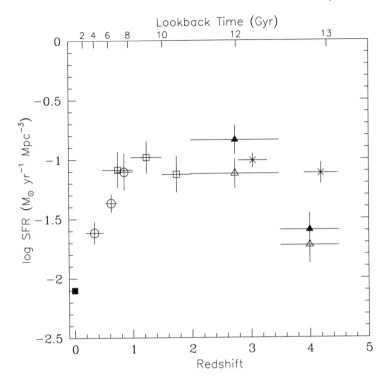

Figure 11.11. History of the star formation rate (SFR) in the universe (~80% of the cosmic time): SFR density versuss. redshift as derived by the UV luminosity density of different distant galaxy samples (see Ferguson *et al* (2000) for a review). Loopback time and distances are computed using $\Omega_m, \Omega_\Lambda, h = 0.3, 0.7, 0.65$.

In this context, the cluster abundance at a given mass has long been recognized as a stringent test for cosmological models. Typical rich clusters have masses of about $5 \times 10^{14}h^{-1}M_\odot$, i.e. similar to the average mass within a sphere of $\sim 8h^{-1}$ Mpc radius in the unperturbed universe. Therefore, the local abundance of clusters is expected to place a constraint on σ_8, the rms mass fluctuation on the $8h^{-1}$ Mpc scale. Analytical arguments based on the approach devised by Press and Schechter (1974) show that the cluster abundance is highly sensitive to σ_8 for a given value of the density parameter Ω_m. Once a model is tuned so as to predict the correct abundance of local ($z \lesssim 0.1$) clusters, its evolution will mainly depend on Ω_m (e.g. Eke *et al* 1996). Therefore, by following the evolution of the cluster abundance with redshift one can constrain the value of the matter density parameter and the fluctuation amplitude level at the cluster scale.

The evolution of cosmic structures, building up in a process of hierarchical clustering, is well illustrated in the VIRGO simulations (Jenkins *et al* 1998) of figure 11.12 (see also the chapter by Anatoly Klypin in this volume). The

z=3   z=1   z=0

ΛCDM

SCDM

OCDM

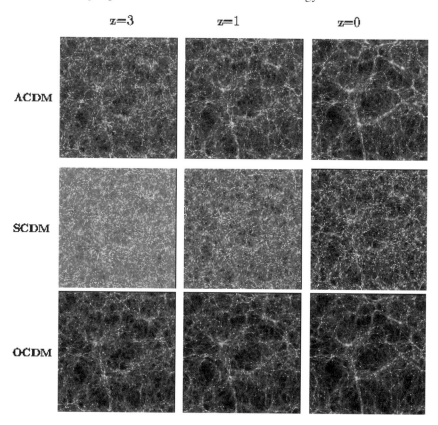

Figure 11.12. Evolution of the cosmic structure (projected mass distribution) from $z = 3$ to the present, as obtained with large N-body simulations by the VIRGO Colloboration (Jenkins *et al* 1998). The three models are Λ-CDM, S(tandard)-CDM and O(pen)-CDM with, respectively, the following parameters $(\Omega_m, \Omega_\Lambda, \Gamma, h) = (0.3, 0.7, 0.21, 0.7), (1, 0, 0.5, 0.5), (0.3, 0, 0.21, 0.7)$. Γ is the shape parameter of the power spectrum. Each box is $240h^{-1}$ Mpc across.

projected mass distribution is shown in three snapshots ($z = 3, 1, 0$), for three different cold dark matter (CDM) models. Model parameters have been chosen to reproduce approximately the same abundance of clusters at $z = 0$ (using a different normalization σ_8). These simulations clearly show that the growth rate of perturbations depends mainly on Ω_m and, to a lesser extent, on Ω_Λ. In low density models, fluctuations start growing in the early universe and stop growing at $1 + z \sim \Omega_m^{-1}$. In SCDM ($\Omega_m = 1$) large structure form much later, and end up evolving rapidly at $z < 1$. The effect of the cosmological constant is to lengthen cosmic time (figure 11.1) and to 'counteract' the effect of gravity, so that perturbations cease to grow at slightly later epochs (a close inspection of figure 11.12 shows indeed less structure at $z = 3$ in the ΛCDM model when compared with OCDM).

One of the fundamental quantities that a CDM model predicts is the *cluster mass function*, $N(M, z)$, i.e. the number of virialized clusters per unit volume and mass, at different epochs. This can be derived by applying cluster-finding algorithms directly on simulations, as in figure 11.12. A very simple and powerful method proposed by Press and Schechter (1974) is, however, often used to compute $N(M, z)$. This analytical approach is found to be in remarkable agreement with N-body simulations, although slight refinements have recently been proposed (Sheth and Tormen 1999). We refer the reader to the original papers or the aforementioned textbooks for a derivation of the Press–Schechter method.

11.4.2   Cluster search methods

The cluster mass is not a direct observable, although several methods exist to estimate the total gravitational mass of clusters. In order to derive the cluster mass function at varying redshifts, one needs three essential tools:

(1) an efficient method to find clusters at least out to $z \simeq 1$;
(2) an estimator (observable), $\hat{M}$, of the cluster mass; and
(3) a simple method to compute the selection function, i.e. the comoving volume within which clusters are found.

We can summarize the methods of finding distant clusters as follows:

- *Galaxy overdensities in optical/IR images*: this is the traditional way which was successfully used by Abell to compile his milestone cluster catalogue. At high redshifts, chance superpositions of unvirialized systems and strong K-corrections for cluster galaxies make optical searches very inefficient. Near-IR searches, supported by some colour information, improve substantially the effectivness of this method. In general, however, the estimate of the survey volume is ill defined and model dependent. In addition, the optical luminosity is poorly correlated with the cluster mass.

- *X-ray selected searches*: arguably, the most efficient method used so far to construct distant cluster samples and to estimate the mass function. The x-ray luminosty is well correlated with the mass and the selection function is straightforward, since it is the one of a (x-ray) flux-limited sample. Possible biases, similar to galaxy searches, are connected to possible surface brightness limits.

- *Search for galaxy overdensities around high-z radio galaxies or AGN*: searches are conducted in near-IR or narrow-band filters. This method has provided so far the only examples of possibly virialized systems at $z > 1.5$ (e.g. Pentericci *et al* 2000).

- *Sunyaev–Zeldovich (SZ) effect*: distortion of the CMB spectrum due to the cluster hot intra-cluster medium. Being a detection in absorption, sensitivity does not depend on redshift. This will possibly be one of the most powerful methods to find distant clusters in the years to come. At present, serendipitous surveys with interferometric techniques (e.g. Carlstrom 1999) cannot cover large areas (i.e. more than ~ 1 deg^2) and their sensitivity is limited to the most x-ray luminous clusters.

- *Clustering of absorption line systems*: this method has lead to a few detections of 'proto-clusters' at $z \gtrsim 2$ (e.g. Francis *et al* 1996). The most serious limitation of this technique is that it is limited to explore small volumes.

To date, the most common procedure used to estimate the cluster mass function has been to exploit x-ray selected samples, for which the survey volume can be computed. Follow-up observations are then used to estimate the cluster mass of a statistical subsample. Most common mass estimators are the temperature of the x-ray emitting gas (directly measured with x-ray spectroscopy), and the galaxy velocity dispersion (virial analysis of galaxy dynamics). We will see later that the x-ray luminosity is also a valid estimator. Gravitational lensing (either in the strong or weak regime) is also a powerful tool to estimate the cluster mass; however, this method is difficult to apply to distant clusters and has some inherent limitations (e.g. mass-sheet degeneracy). For a review of gravitational lensing methods of mass reconstruction, the reader is referred to the chapter by Philippe Jetzer in this volume.

A robust method to quantify the volume density of clusters at different redshifts is to use the x-ray luminosity function (XLF), i.e. the number of clusters per unit volume and per unit x-ray luminosity. By comparing the XLF of an x-ray flux-limited samples of clusters at different redshifts, one can characterize the evolution in luminosity and/or number density. This tool is the exact counterpart of the optical LF used in galaxy surveys (section 11.3.3). Perhaps surprisingly, this standard method applied to cluster surveys has several advantages over galaxy surveys. First, the local XLF is very well determined and no ambiguity exists as from different 'types'. Clusters are basically a single parameter family, the gas temperature, which is also well correlated with the x-ray luminosity. For this reason, K-corrections are also easy to handle as opposed to galaxies in the optical–near-IR. The only point of major concern, as previously discussed, has to do with biases due to surface brightness limits.

In figure 11.13 we show the best determination to date of the XLF from $z \simeq 0$ out to $z \simeq 1.2$, coming from different surveys (Rosati *et al* 1999 and references therein). The most striking result is perhaps the lack of any significant evolution out to $z \simeq 1$, for $L_X \lesssim L_X^* \simeq 5 \times 10^{44}$ erg s^{-1} (i.e. approximately the Coma cluster). This range of luminosities includes the bulk of the cluster population in the universe. However, there is evidence of evolution of the space density of the most luminous, presumably most massive clusters. Using the observed $L_X - T$ relation for clusters and the virial theorem, which links the temperature to the mass, one can show that the XLF can be used as a robust estimator of the cluster

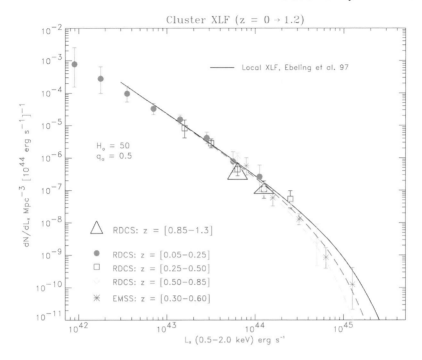

Figure 11.13. The best determination to date of the cluster x-ray luminosity function (i.e. the cluster space density) out to $z \simeq 1.2$. Data points at $z < 0.85$ are derived from a complete RDCS sample of 103 clusters over 47 deg^2, with $F_{X_{lim}} = 3 \times 10^{-14}$ erg s^{-1} cm^{-2} (Rosati *et al* 1999). The triangles represent *a lower limit* (due to incomplete optical identification). to the cluster space density obtained from a fainter and more distant subsample. Long dash curves are Schechter best fits to the XLF $\phi(L_X, z)$, plotted at $z = 0.4$ and $z = 0.6$.

mass function, i.e. $N(L_X, z) \rightarrow N(T, z) \rightarrow N(M, z)$ (e.g. Borgani *et al* 1999). Such a method can be used to set significant constraints on Ω_m (figure 11.14). The fact that a large fraction of relatively massive clusters is already in place at $z \simeq 1$, indicates that the dynamical evolution of structure has proceeded at a relatively slow pace since $z \simeq 1$, a scenario which fits naturally in a low density universe (figure 11.14, see Borgani *et al* 2001, Eke *et al* 1996).

11.4.3 Determining Ω_m and Ω_Λ

Besides the method of the evolution of cluster abundance (which we can call 'universal dynamics'), galaxy clusters, as the largest collapsed objects in the universe, also offer two other independent means to estimate the mean density of matter that participates to gravitational clustering (i.e. Ω_m):

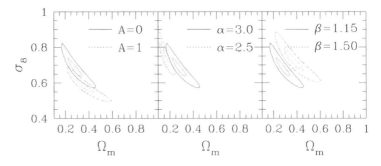

Figure 11.14. Constraints in the plane of the cosmological parameters $\Omega_m - \sigma_8$ derived from the observed evolution of the cluster abundance in the RDCS sample (Borgani *et al* 2001). Contours are 1σ, 2σ and 3σ C.L. The three parameters (A, α, β) describe the uncertainties in converting cluster masses into temperatures ($T \sim M^{2/3}/\beta$), and temperatures into x-ray luminosities ($L_X \sim T^{\alpha}(1 + z)^A$). The two values for each parameter bracket the range which is allowed from current x-ray observations of distant clusters.

(1) $\Omega_b - f_{gas}$ method,
(2) Oort method (M/L) and
(3) universal dynamics.

11.4.3.1 $\Omega_b - f_{gas}$ method (White et al 1993)

A reasonable assumption is that clusters are large enough that they should host a 'fair sample' of the matter in the universe (e.g. there is no special segregation of baryons over the dark matter). In addition, x-ray observations clearly show that most of the baryons in clusters reside in the hot intracluster gas. The gas-to-total-mass ratio, f_{gas}, can be measured using x-ray or SZ observations. The fraction of baryons, $\Omega_b = \rho_B/\rho_{cr}$, is well constrained by the primordial nucleosynthesis theory and the measurement of deuterium abundance from high-z absorption systems. If we know f_{gas} and Ω_b, then we simply have: $\Omega_m = \Omega_b/f_{gas}$.

Deuterium measurements in recent years have settled on the value (Burles and Tytler 1998) $\Omega_b h^2 = 0.02 \pm 0.002$. Ettori and Fabian (1999) have used 36 x-ray clusters to estimate a mean value $\langle f_{gas} \rangle = 0.059 h^{-3/2}$ with a 90% range of $f_{gas} = (0.036\text{--}0.087)h^{-3/2}$. Hence,

$$\Omega_m = \Omega_B/f_{gas} \simeq 0.34 h^{-1/2} \simeq 0.4 \pm 0.2 \qquad \text{(for } H_0 = 65\text{)}, \qquad (11.27)$$

where the error represents an approximate range reflecting the scatter in f_{gas}.

11.4.3.2   Oort method (M/L)

The mean density of the universe is equal to the mass of a large galaxy cluster divided by the equivalent comoving volume in the field from which that mass

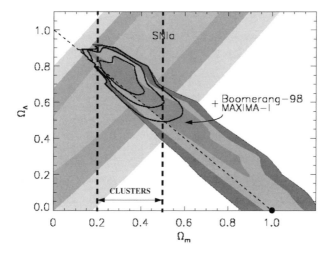

Figure 11.15. Constraints to Ω_m and Ω_Λ from CMB anisotropies (Boomerang: De Bernardis *et al* 2000; Maxima: Hanany *et al* 2000), distant Type Ia supernovae (Perlmutter *et al* 1999; Schmidt *et al* 1998) and several methods based on galaxy clusters.

originated. Such a volume can be evaluated from the ratio of the luminosity of the cluster galaxies, L, with the *field luminosity density*, j_f. Thus,

$$\rho_0 = M_{cl}/V_{cl} = (M/L)_{cl} \times j_f, \quad \text{and} \quad \Omega_m = (M/L)_{cl}/(M/L)_{cr} \quad (11.28)$$

where $(M/L)_{cr} = \rho_{cr}/j_f$.

Important effects which could bias this measurement are luminosity segregation of the cluster versus the field, and differential evolution of the cluster galaxies compared to the field. With enough spectrophotometric data, one can reasonably control these issues. The CNOC survey (e.g. Carlberg *et al* 1996) is the best study to date of cluster dynamics of an x-ray selected sample of 16 clusters *at* $z \lesssim 0.5$. This study lead to a measurement of average mass-to-light ratio $(M/L) = 295 \pm 54\, h\, M_\odot L_\odot^{-1}$, as well as of the luminosity density j_f in the field. Thus, Carlberg *et al* obtain: $\Omega_m = 0.24 \pm 0.05 \pm 0.09$ (the second error is the sytematic one).

Using the constraint on Ω_m derived in the previous section from the application of the third method (universal dynamics), we note a remarkable agreement from *completeley independent techniques* based on galaxy clusters, i.e. $\Omega_m \simeq 0.2$–0.5.

These bounds on the matter density parameter are shown in figure 11.15 together with measurements of $(\Omega_m, \Omega_\Lambda)$ from high redshift supernovae used as standard candles (Perlmutter *et al* 1999, Schmidt *et al* 1998), and from the recent landmark experiments—Boomerang (De Bernardis *et al* 2000) and Maxima (Hanany *et al* 2000) which have measured CMB anisotropies on small scales (see the chapter by Arthur Kosowsky in this volume). The power of these three independent means of measuring $(\Omega_m, \Omega_\Lambda)$ is that they have degeneracies which lie almost orthogonally to each other. The directions of degeneracy in the $(\Omega_m, \Omega_\Lambda)$ plane can be written as

$$\text{SN: } \tfrac{4}{3}\Omega_m - \Omega_\Lambda \simeq \text{constant} \qquad \text{CMB: } \Omega_m + \Omega_\Lambda \simeq \text{constant}$$
$$\text{clusters: } \Omega_m \simeq \text{constant}.$$

These three measurements of the cosmological parameters are well in agreement with each other and define a relatively small allowed region, a circumstance which is sometimes referred to as 'cosmic concordance' (Bahcall *et al* 1999). This explains why by 'standard cosmology' these days one adopts the values $(\Omega_m, \Omega_\Lambda) = (0.3, 0.7)$. Interestingly, the age of the universe for this model is $T_U = 0.965 H_0^{-1}$.

References

Bahcall N, Ostriker J P, Perlmutter S and Steinhardt P J 1999 *Science* **284** 1481
Benitez M 2000 *Astrophys. J.* **536** 571
Borgani S, Rosati P, Tozzi P and Norman C 1999 *Astrophys. J.* **517** 40
Borgani S *et al* 2001 *Astrophys. J.* **559** L71
Bruzual A G and Charlot S 1993 *Astrophys. J.* **405** 538
Burles S and Tytler D 1998 *Astrophys. J.* **507** 732
Calzetti D, Kinney A L and Storchi-Bregmann T 1994 *Astrophys. J.* **429** 582
Carlberg *et al* 1996 *Astrophys. J.* **462** 32
Carlstrom J S 1999 *Phys. Scr.* ed L Bergstrom, P Carlson and C Fransson
Colless M M 1999 *Proc. 'Looking Deep in the Southern Sky' (ESO Astrophysics Symposia)* ed R Morganti and W J Couch (Berlin: Springer) p 9
Cowie L L, Sonfalia A, Hu E M and Cohen J D 1996 *Astron. J.* **112** 839
Dickinson M 1998 *Proc. STScI May 1997 Symposium 'The Hubble Deep Field'* ed M Livio, S M Fall and P Madau *Preprint* astro-ph/9802064
de Bernardis P *et al* 2000 *Nature* **404** 955
Driver S P, Fernandez-Soto A, Couch W J, Odewahn S C, Windhorst R A, Phillipps S, Lanzetta K and Yahil A 1998 *Astrophys. J.* **496** L93
Eke R *et al* 1996 *Mon. Not. R. Astron. Soc.* **282** 263
Ellis R G, Colless M, Broadhurst T J, Heyl J S and Glazebrook K 1996 *Mon. Not. R. Astron. Soc.* **280** 235
Ellis R G 1997 *Annu. Rev. Astron. Astrophys.* **35** 389
Ettori S and Fabian A C 1999 *Mon. Not. R. Astron. Soc.* **305** 834
Ferguson H C, Dickinson M and Williams R 2000 *Annu. Rev. Astron. Astrophys.* **38** 667
Giallongo E, D'Odorico S, Fontana A, Cristiani S, Egami E, Hu E and McMahon R G 1998 *Astron. J.* **115** 2169
Giavalisco M, Steidel C C, Adelberger K L, Dickinson M, Pettini M and Kellogg M 1998 *Astrophys. J.* **503** 543
Hanany S *et al* 2000 *Astrophys. J.* **545** L5
Hu E M, McMahon R G and Cowie L L 1999 *Astrophys. J.* **522** L9

Hubble E 1926 *Astrophys. J.* **64** 321

Huchra J P, Geller M J, de Lapparant V and Corwin H G 1990 *Astrophys. J. Suppl.* **42** 433

Jenkins *et al* 1998 *Astrophys. J.* **499** 20

Kaiser N 1994 *Astrophys. J.* **284** L9

Kellermann K I 1993 *Nature* **361** 134

Landau L D and Lifshitz E M 1971 *The Classical Theory of Fields* (Oxford: Pergamon)

Longair M S 1998 *Galaxy Formation* (Berlin: Springer)

Lilly S J, Tresse L, Hammer F, Crampton D and LeFevre O 1995 *Astrophys. J.* **455** 108

Loveday J, Peterson B A, Efstathiou G and Maddox S J 1992 *Astrophys. J.* **390** 338

Madau P, Pozzetti L and Dickinson M 1998 *Astrophys. J.* **498** 106

Maddox S J, Efstathiou G, Sutherland W J and Loveday J 1990 *Mon. Not. R. Astron. Soc.*
    247 1

Pentericci L *et al* 2000 *Astron. Astrophys.* **361** L25

Peacock J A 1999 *Cosmological Physics* (Cambridge: Cambridge University Press)

Peebles P J E *Principles of Physical Cosmology* (Princeton, NJ: Princeton University Press)

Perlmutter S *et al* 1999 *Astrophys. J.* **517** 565

Press W H and Schechter P 1974 *Astrophys. J.* **187** 425

Rosati P, Della Ceca R, Burg R, Norman C and Giacconi R 1998 *Astrophys. J.* **492** L21

Rosati *et al* 1999 *Proc. 'Large Scale Structure in the X-ray Universe'* ed M Plioniz and
    I Georgantopoulos (Greece: Santorini) (astro-ph/0001119)

Sandage A 1995 *The Deep Universe (Saas-Fee Advanced Course 23)* (Berlin: Springer)

Schaerer D 1999 *Proc. XIXth Moriond Astrophysics Meeting 'Building the Galaxies: from
    the Primordial Universe to the Present'* ed Hammer *et al* (Paris: Editions Frontières)
    (astro-ph/9906014)

Schechter P 1976 *Astrophys. J.* **203** 297

Shectman S A *et al* 1996 *Astrophys. J.* **470** 172

Sheth R K and Tormen G 1999 *Mon. Not. R. Astron. Soc.* **308** 119

Schmidt B P *et al* 1998 *Astrophys. J.* **507** 46

Steidel C C, Giavalisco M, Dickinson M and Adelberger K L 1996 *Astrophys. J.* **462** L17

Steidel C C, Adelberger K L, Dickinson M, Giavalisco M, Pettini M and Kellogg M 1998
    Astrophys. J. **492** 428

Steidel C C, Adelberger K L, Giavalisco M, Dickinson M and Pettini M 1999 *Astrophys. J.*
    519 1

Tyson J A 1988 *Astron. J.* **96** 1

White S D M, Navarro J F, Evrard A E and Frenk C S 1993 *Nature* **366** 429

Yee H K C *et al* 2000 *Astrophys. J. Suppl.* **129** 475 (astro-ph/0004026)

Chapter 12

Clustering in the universe: from highly nonlinear structures to homogeneity

Luigi Guzzo
Osservatorio Astronomico di Brera, Italy

12.1 Introduction

This chapter concentrates on a few specific topics concerning the distribution of galaxies on scales from 0.1 to nearly $1000h^{-1}$ MPc. The main aim is to provide the reader with the information and tools to familiarize him/her with a few basic questions:

(1) What are the scaling laws followed by the clustering of luminous objects over almost four decades of scales?
(2) How do galaxy motions distort the observed maps in redshift space, and how we can correct and use them to our benefit?
(3) Is the observed clustering of galaxies suggestive of a fractal universe? and consequently,
(4) Is our faith in the cosmological principle still well placed? i.e. do we see evidence for a homogeneous distribution of matter on the largest explorable scales, in terms of the correlation function and power spectrum of the distribution of luminous objects?

For some of these questions we have a well-defined answer, but for some others the idea is to indicate the path along which there is still a good deal of exciting work to be done.

12.2 The clustering of galaxies

I believe most of the students reading this book will be familiar with the beautiful *cone diagrams* showing the distribution of galaxies in what have often been called

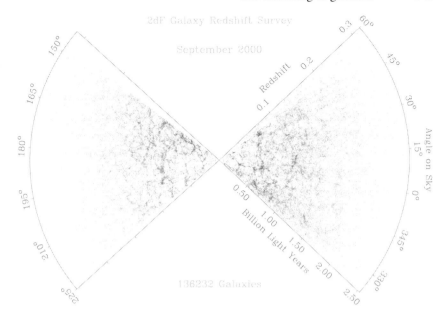

Figure 12.1. The distribution of the nearly 140 000 galaxies observed so far (September 2000) in the 2dF survey (from [3]): compare this picture to that in [2] to see how rapidly this survey is progressing towards its goal of 250 000 redshifts measured (note that this is a projection over a variable depth in declination, due to the survey being still incomplete).

slices of the universe. This has been made possible by the tremendous progress in the efficiency of redshift surveys, i.e. observational campaigns aimed at measuring the distance of large samples of galaxies through the cosmological redshift observed in their spectra. This is one of the very simple, yet fundamental pillars of observational cosmology: reconstructing the three-dimensional positions of galaxies in space to be able to study and characterize statistically their distribution. Figure 12.1 shows the current status of the ongoing 2dF survey and gives an idea of the state of the art, with $\sim$130 000 redshifts measured and a planned final number of 250 000 [1]. From this plot, the main features of the galaxy distribution can be appreciated. One can easily recognize *clusters*, *superclusters* and *voids*, and get the feeling of how the galaxy distribution is extremely inhomogeneous to at least $50h^{-1}$ MPc (see [2] for a more comprehensive review).

The inhomogeneity we clearly see in the galaxy distribution can be quantified at the simplest level by asking what is the *excess* probability over random to find a galaxy at a separation r from another one. This is one way by which one can define the *two-point correlation function*, certainly the most perused statistical estimator in cosmology (see [5] for a more detailed introduction). When we have a catalogue with only galaxy positions on the sky (and usually their magnitudes), however, the first quantity we can compute is the *angular* correlation function

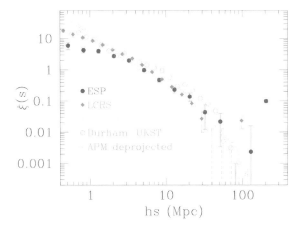

Figure 12.2. The two-point correlation function of galaxies, as measured from a few representative optically-selected surveys (from [2]). The plot shows results from the ESP [9], LCRS [10], APM-Stromlo, [11] and Durham–UKST [12] surveys, plus the real space $\xi(r)$ de-projected from the angular correlation function $w(\theta)$ of the APM survey [13].

$w(\theta)$. This is a projection of the *spatial* correlation function $\xi(r)$ along the redshift path covered by the sample. The relation between the angular and spatial functions is expressed for small angles by the *Limber equation* (see [4] and [5] for definitions and details)

$$w(\theta) = \int_0^\infty dv \, v^4 \phi^2(v) \int_{-\infty}^\infty du \, \xi \left(\sqrt{u^2 + v^2\theta^2} \right) \qquad (12.1)$$

where $\phi(v)$ is the *radial selection function* of the two-dimensional catalogue, that in this version gives the comoving density of objects at a given distance v (which depends, for example, on the magnitude limit of the catalogue and the specific luminosity function of the type of galaxies one is studying). For optically selected galaxies [6, 7] $w(\theta)$ is well described by a power-law shape $\propto \theta^{-0.8}$, corresponding to a spatial correlation function $(r/r_0)^\gamma$, with $r_0 \simeq 5h^{-1}$ Mpc and $\gamma \simeq -1.8$, and a break with a rapid decline to zero around scales corresponding to $r \sim 30h^{-1}$ Mpc.

The advantage of angular catalogues remains the large number of galaxies they include, up to a few millions [6]. Since the beginning of the 1980s (e.g. [8]), redshift surveys have allowed us to compute $\xi(r)$ directly in three-dimensional space, and the most recent samples have pushed these estimates to separations of $\sim 100h^{-1}$ Mpc (e.g. [9]). Figure 12.2 shows the two-point correlation function in *redshift space*,† indicated as $\xi(s)$, for a representative set of published redshift surveys [9–12]. In addition, the dotted lines show the real-space $\xi(r)$ obtained

---

† This means that distances are computed from the redshift in the galaxy spectrum, neglecting the Doppler contribution by its peculiar velocity which adds to the Hubble flow (section 12.3).

through de-projection of the angular $w(\theta)$ from the APM galaxy catalogue [13]. The two different lines correspond to two different assumptions about galaxy clustering evolution, which has to be taken into account in the de-projection, given the depth of the APM survey. This illustrates some of the uncertainties inherent in the use of the angular function. As can be seen from figure 12.2, the shape of $\xi(s)$ below 5–$10h^{-1}$ Mpc is reasonably well described by a power law, but for the four redshift samples the slope is shallower than the canonical ~ -1.8 nicely followed by the APM $\xi(r)$. This is due to the redshift-space smearing of structures that suppresses the true clustering power on small scales, as we shall discuss in the following section. Note how $\xi(s)$ maintains a *low-amplitude*, positive value out to separations of more than $50h^{-1}$ Mpc, showing explicitly why large-size galaxy surveys are important: we need large volumes and good statistics to be able to extract such a weak clustering signal from the noise. Finally, the careful reader might have noticed a small but significant positive change in the slope of the APM $\xi(r)$ (the only one for which we can see the undistorted real-space clustering at small separations), around $r \sim 3$–$4h^{-1}$ Mpc. On scales larger than this, all data show a 'shoulder' before breaking down. This inflection point appears around the scales where $\xi \sim 1$, thus suggesting a relationship with the transition from the linear regime (where each mode of the power spectrum grows by the same amount and the shape is preserved), to fully nonlinear clustering on smaller scales [14]. We shall come back to this in section 12.4.

12.3   Our distorted view of the galaxy distribution

We have just seen an explicit example of how unveiling the true scaling laws describing galaxy clustering from redshift surveys is complicated by the effects of galaxy-peculiar velocities. Separations between galaxies—indicated as s to emphasize this very point—are not measured in real 3D space, but in *redshift space*: what we actually measure when we take the redshift of a galaxy is the quantity $cz = cz_{\text{true}} + v_{\text{pec}//}$, where $v_{\text{pec}//}$ is the component of the galaxy-peculiar velocity along the line of sight. This quantity, while being typically ~ 100 km s^{-1} for 'field' galaxies, can rise above 1000 km s^{-1} in rich clusters of galaxies. As explicitly visible in figure 12.2, the resulting $\xi(s)$ is *flatter* than its real-space counterpart. This is the result of two concurrent effects: on small scales, clustering is suppressed by high velocities in clusters of galaxies, that spread close pairs along the line of sight producing in redshift maps what are sometimes called 'fingers of God'. Many of these are recognizable in figure 12.1 as thin radial structures, particularly in the denser part of the upper cone. The net effect on $\xi(s)$ is, in fact, to suppress its amplitude below ~ 1–$2h^{-1}$ Mpc. However, on larger scales where motions are still coherent, streaming flows towards higher-density structures enhance their apparent contrast when they appear to lie perpendicularly to the line of sight. This, in contrast, amplifies $\xi(s)$ above 10–$20h^{-1}$ Mpc. Both effects can be better appreciated with the help of a computer N-body simulation, for which we have the leisure to see both a real-and a redshift-space snapshot, as in Figure 12.3.

How can we recover the correlation function of the undistorted spatial pattern, i.e. $\xi(r)$? This can be accomplished by computing the two-dimensional correlation function $\xi(r_p, \pi)$, where the radial separation s of a galaxy pair is split into two components, π, parallel to the line of sight, and r_p, perpendicular to it, defined as follows [15]. If d_1 and d_2 are the distances to the two objects (properly computed) and we define the line of sight vector $l \equiv (d_1 + d_2)/2$ and the redshift difference vector $s \equiv d_1 - d_2$, then one defines

$$\pi \equiv \frac{s \cdot l}{|l|} \qquad r_p^2 \equiv s \cdot s - \pi^2. \tag{12.2}$$

The resulting correlation function is a bidimensional map, whose contours at constant correlation look as in the example of figure 12.4. By projecting $\xi(r_p, \pi)$ along the π direction, we obtain a function that is independent of the distortion,

$$w_p(r_p) \equiv 2 \int_0^\infty d\pi\, \xi(r_p, \pi) = 2 \int_0^\infty dy\, \xi_R[(r_p^2 + y^2)^{1/2}] \tag{12.3}$$

and is directly related to the real-space correlation function (here indicated with $\xi_R(r)$ for clarity), as shown. Modelling $\xi_R(r)$ as a power law, $\xi_R(r) = (r/r_0)^{-\gamma}$ we can carry out the integral analytically, yielding

$$w_p(r_p) = r_p \left(\frac{r_0}{r_p}\right)^\gamma \frac{\Gamma(\frac{1}{2})\Gamma(\frac{\gamma-1}{2})}{\Gamma(\frac{\gamma}{2})} \tag{12.4}$$

where Γ is the gamma function. Such a form can then be fitted to the observed $w_p(r_p)$ to recover the parameters describing $\xi(r)$ (e.g. [16]). Alternatively, one can perform a formal Abel inversion of $w_p(r_p)$ [17].

So far, we have treated redshift-space distortions merely as an annoying feature that prevents the true distribution of galaxies from being seen directly. In fact, being a dynamical effect they carry precious direct information on the distribution of mass, independently from the distribution of luminous matter. This information can be extracted, in particular by measuring the value of the *pairwise velocity dispersion* $\sigma_{12}(r)$. This, in practice, is a measure of the small-scale 'temperature' of the galaxy soup, i.e. the amount of kinetic energy produced by the differences in the potential energy created by density fluctuations. Thus, finally, a measure of the mass variance on small scales.

$\xi(r_p, \pi)$ can be modelled as the convolution of the real-space correlation function with the distribution function of pairwise velocities along the line of sight [8, 18], Let $F(w, r)$ be the distribution function of the vectorial velocity differences $w = u_2 - u_1$ for pairs of galaxies separated by a distance r (so it is a function of four variables, w_1, w_2, w_3, r). Let w_3 be the component of w along the direction of the line of sight (that defined by l); we can then consider

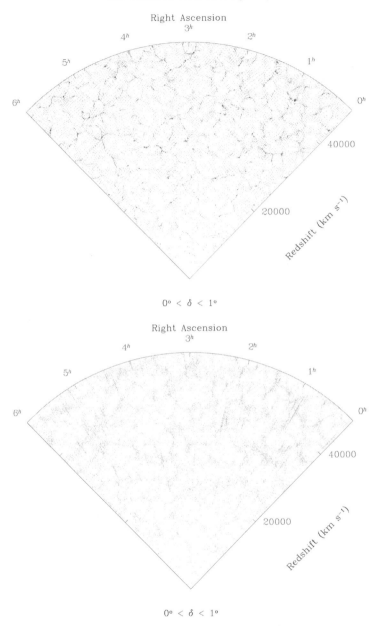

Figure 12.3. Particle distribution from a one-degree thick mock survey through a large-size Open-CDM *N*-body simulation in real (top) and redshift space (bottom). The appearance of the two diagrams gives a clear visual impression of the effect of redshift-space distortions (note that here, unlike in the real survey of figure 12.1, no apparent luminosity selection is applied, i.e. the sample is *volume limited*).

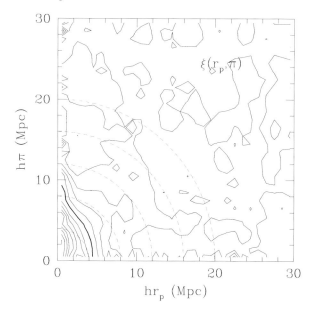

Figure 12.4. The typical appearance of the bidimensional correlation function $\xi(r_p, \pi)$, in this specific case computed for the ESP survey [9]. Note the elongation of the contours along the π direction for small values of r_p, produced by high-velocity pairs in clusters. The broken circles show contours of equal correlation in the absence of distortions.

the corresponding distribution function of w_3,

$$f(w_3, r) = \int dw_1 \, dw_2 \, F(\mathbf{w}, r). \tag{12.5}$$

It is this distribution function that is convolved with $\xi(r)$ to produce the observed $\xi(r_p, \pi)$. If we now call y the component of the separation r along the line of sight, with our convention we have that $w_3 = H_0(\pi - y)$ and the convolution

$$1 + \xi(r_p, \pi) = [1 + \xi(r)] \otimes f(w_3, r), \tag{12.6}$$

can be expressed as

$$1 + \xi(r_p, \pi) = H_0 \int_{-\infty}^{+\infty} dy \, \{1 + \xi[(r_p^2 + y^2)^{\frac{1}{2}}]\} f[H_0(\pi - y)]. \tag{12.7}$$

Note that this expression gives essentially a model description of the *effect* produced by peculiar motions on the observed correlations, but does not take into account the intimate relation between the mass density distribution and the velocity field which is, in fact, a product of mass correlations (see [19] and [20] and references therein). Within this model, therefore, we have no specific physical reason for choosing one or another form for the distribution function f. Peebles [21] first showed that an exponential distribution best fits the observed data, a result subsequently confirmed by N-body models [22]. According to this choice, f can then be parametrized as

$$f(w_3, r) = \frac{1}{\sqrt{2}\sigma_{12}(r)} \exp\left[-\sqrt{2}\left|\frac{w_3(r) - \langle w_3(r)\rangle}{\sigma_{12}(r)}\right|\right] \qquad (12.8)$$

where $\langle w_3(r)\rangle$ and $\sigma_{12}(r)$ are, respectively, the first and second moment of f. The projected *mean streaming* $\langle w_3(r)\rangle$ is usually explicitly expressed in terms of $v_{12}(r)$, the first moment of the distribution F defined earlier, i.e. *the mean relative velocity of galaxy pairs with separation r*, $\langle w_3(r)\rangle = yv_{12}(r)/r$. The final expression for f becomes therefore

$$f(w_3, r) = \frac{1}{\sqrt{2}\sigma_{12}(r)} \exp\left\{-\sqrt{2}H_0\left|\frac{\pi - y\left[1 + \frac{v_{12}(r)}{H_0 r}\right]}{\sigma_{12}(r)}\right|\right\} \qquad (12.9)$$

(see e.g. [18] and [16] for more details).

The practical estimate of $\sigma_{12}(r)$ is typically performed on the data by fitting the model of equation (12.7) to a cut at fixed r_p of the observed $\xi(r_p, \pi)$. To do this, one has first to estimate $\xi(r)$ from the projected function $w_p(r_p)$ and choose a model for the mean streaming $v_{12}(r)$, as e.g. that based on the similarity solution of the BBGKY equations [8]:

$$v_{12}(r) = -H_0 r \frac{F}{1 + (r/r_0)^2}. \qquad (12.10)$$

The traditional approach considers two extreme cases, corresponding to the somewhat idealized situations of *stable clustering* ($F = 1$, a mean infall streaming that compensates exactly the Hubble flow, such that clusters are stable in physical coordinates) and *free expansion* with the Hubble flow ($F = 0$, no mean peculiar streaming). It is instructive to see explicitly what happens to the contours of $\xi(r_p, \pi)$ in these two limiting cases. In figure 12.5, I have used equations (12.7), (12.9) and (12.10) to plot the model for $\xi(r_p, \pi)$, keeping $\sigma_{12}(r)$ fixed and varying the amplitude F of the mean streaming. Here the two competing dynamical effects (small-scale stretching and large-scale compression) are clearly evident. The observational results yield values of σ_{12} at small separations around 300–400 km s^{-1}, with a mild dependence on scale [16, 18, 23]. This value has been shown to be rather sensitive to the survey volume, because of the strong weight the technique puts on galaxy pairs in clusters [23], and the fluctuations in the number of clusters due to their clustering. A different method has been proposed more recently by Landy and collaborators [24] to alleviate this problem. The method is very elegant, and reduces the weight of high-velocity pairs in clusters by working in the Fourier domain where, in addition, the convolution of the two functions becomes a simple product of their transforms. A direct

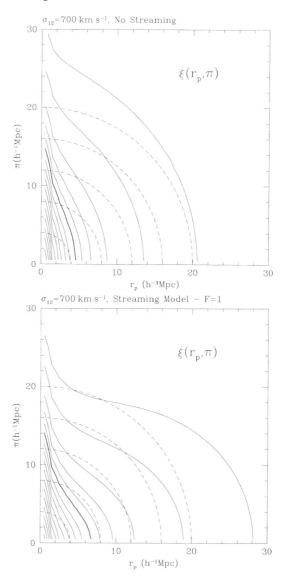

Figure 12.5. The relative effect of the mean streaming $v_{12}(r)$ and pairwise velocity dispersion $\sigma_{12}(r)$ on the shape of the contours of $\xi(r_p, \pi)$, seen through the model of equation (12.7). While a high pairwise dispersion, $\sigma_{12} = 700$ km s^{-1} independent of scale is assumed (a reasonable approximation), the two cases of zero mean streaming ($F = 0$) and stable clustering ($F = 1$) are considered in the infall model of Davis and Peebles [8]. Here the effect of the coherent motions is more evident than in the data plot of figure 12.4: the contours of $\xi(r_p, \pi)$ are clearly compressed along the π direction. This compression is a measure of $\Omega_m^{0.6}/b$.

application to data and N-body simulations under particularly severe survey conditions seems, however, to give results which are not significantly dissimilar to the standard method [25].

Rather than assuming a model for the mean streaming $v_{12}(r)$, one could measure it directly from the compression of the contours of $\xi(r_p, \pi)$, i.e. doing a simultaneous fit to the first and second moment. This quantity also carries important cosmological information, being directly proportional to the parameter $\beta = \Omega_m^{0.6}/b$, where Ω_m is the matter density parameter and b is the *bias parameter* of the class of galaxies one is using (see Peacock, this volume). This has been done, e.g. on the IRAS 1.2 Jy survey [18], but the uncertainty on β is very large due to the weak signal and the need to simultaneously fit both the first and second moments. The situation in this respect will soon improve dramatically thanks to the ongoing 2dF [1] and Sloan (SDSS) surveys [26], that will provide 250 000 and 1 000 000 redshifts respectively.

12.4 Is the universe fractal?

The observation of a power-law shape for the two-point correlation function together with the self-similar aspect of galaxy maps as that of figure 12.1, suggested several years ago a possible description of the large-scale structure of the universe in terms of *fractal objects* [27]. A fractal universe without a cross-over to a homogeneous distribution would imply abandoning the cosmological principle. Also, under such conditions most of our standard statistical descriptions of large-scale structure would be inappropriate [28]: no mean density could be defined and, as a consequence, the whole concept of *density fluctuations* (with respect to a mean density) would make little sense.

It is therefore of significant interest: (1) to compare the scaling properties of galaxy clustering to those expected for a fractal distribution (keeping in mind that on different scales there are different effects at work, as we have seen in the previous section); and (2) to put under serious scrutiny the observational evidences for a convergence of statistical measures to a homogeneus distribution within the boundaries of current samples. Attempts to address these questions using redshift survey data during the last ten years or so have come to different conclusions, mostly because of disagreement on which data can be used and how they should be treated and analysed [29–31]. It is because of the relevance of the issues raised that this subject has been the focus of an intense debate, as also demonstrated by the discussions in this book (see also Montuori, this volume).

12.4.1 Scaling laws

Let us review the arguments for and against the fractal interpretation of the clustering data, by first recalling the basic relations involved.

A fractal set is characterized by a specific *scaling* relation, essentially describing the way the set fills the ambient space. This scaling law can be by itself taken as an heuristic definition of fractal (although it is not strictly equivalent to the formal definition in terms of Hausdorff dimensions, see e.g. [32]): the number of objects counted in spheres of radius r around a randomly chosen object in the set must scale as

$$N(r) \propto r^D \tag{12.11}$$

where D is the *fractal dimension* (or, more correctly, the fractal *correlation* dimension). Analogously, the density within the same sphere will scale as

$$n(r) \propto r^{D-3}. \tag{12.12}$$

Similarly, the expectation value of the density measured within shells of width dr at separation r from an object in the set, the *conditional density* $\Gamma(r)$ [28], will scale in the same way,

$$\Gamma(r) = A \cdot r^{D-3} \tag{12.13}$$

with A being constant for a given fractal set. $\Gamma(r)$ can be directly connected to the standard two-point correlation function $\xi(r)$: suppose for a moment that we can define a mean density $\langle n \rangle$ for this sample (we shall see in a moment what this implies), then it is easy to show that

$$1 + \xi(r) = \frac{\Gamma(r)}{\langle n \rangle} \propto r^{D-3}. \tag{12.14}$$

Therefore, if galaxies are distributed as a fractal, a plot of $1 + \xi(r)$ will have a power-law shape, and in the strong clustering regime (where $\xi(r) \gg 1$) this will also be true for the correlation function itself. This demonstrates the classic argument (see e.g. [5]), that a power-law galaxy correlation function as observed $\xi(r) = (r/r_0)^{-\gamma}$, is consistent with a scale-free, fractal clustering with dimension $D = 3 - \gamma$ (although it does not necessarily imply it: fractals are not the only way to produce power-law correlation functions, see [31]). Note, however, that when $\xi(r) \sim 1$ or smaller, only a plot of $\Gamma(r)$ or $1 + \xi(r)$, and not $\xi(r)$, could properly detect a fractal scaling, if present.

When this happens over a range of scales which is significant with respect to the sample size, the mean density $\langle n \rangle$ becomes an ill-defined quantity which depends on the sample size itself. Considering a spherical sample with radius R_s and the case of a pure fractal for simplicity, the mean density is the integral of equation (12.13)

$$\langle n \rangle = \frac{3A}{D} \cdot R_s^{D-3}, \tag{12.15}$$

and is therefore a function of the sample radius R_s. Under the same conditions, the two-point correlation function becomes

$$\xi(r) = \frac{\Gamma(r)}{\langle n \rangle} - 1 = \frac{D}{3} \cdot \left(\frac{r}{R_s} \right)^{D-3} - 1, \tag{12.16}$$

with a correlation length

$$r_0 = \left(\frac{6}{D}\right)^{\frac{1}{D-3}} \cdot R_s, \qquad (12.17)$$

which also depends on the sample size. Therefore, if the galaxy distribution has a fractal character, with a well-defined dimension D one should observe:

(1) that the number of objects within volumes of increasing radius $N(R)$ grows as R^D;
(2) that analogously, the function $\Gamma(r)$ or, equivalently, $1 + \xi(r)$, is a power law with slope $D - 3$; and
(3) that the correlation length r_0 is a linear function of the sample size.

If the fractal distribution extends only up to a certain scale, the transition to homogeneity would show up first as a flattening of $1+\xi(r)$ and (less rapidly, given that they depend on an integral over r) as a growth $N(r) \propto r^3$ and a convergency of r_0 to a stable value.

12.4.2 Observational evidences

Pietronero [28] originally made the very important point that the use of $\xi(r)$ was not fully justified, given the size (with respect to the clustering scales involved) of the samples available at the time, and the consequent uncertainty on the value of the mean density. In reality, this warning was already clear in the original prescription [5]: one should be confident to have a *fair sample* of the universe before drawing far-reaching conclusions from the correlation function. As often happens, due to the scarcity of data the recommendation was not followed too strictly (see [31] for more discussion on this point).

Although the data available today have increased by an order of magnitude at least, the debate on the scaling properties and homogeneity of the universe is still lively. Given the subject of this book and the extensive use we have made so far of correlation functions, I shall concentrate here on the evidence concerning points 2 and 3 in the previous summary list. In figure 12.6, I have plotted the function $1 + \xi(s)$ for the same surveys of figure 12.2. Taken at face value, the figure shows that the redshift-survey data can be reasonably fitted by a single power law only out to $\sim 5h^{-1}$ Mpc. However, as soon as we compare these to the real space $1 + \xi(r)$ from the APM survey, we realize that what we are seeing here is dominated by the redshift-space distortions. In other words, a fractal dimension on small scales can only be measured from angular or projected correlations, and if the data are interpreted in this way, it is in fact close to $D \simeq 1.2$. Above $\sim 5h^{-1}$ Mpc, a second range follows where D varies between two and three, when moving out to scales approaching $100h^{-1}$ Mpc. The range between $5h^{-1}$ and $\sim 30h^{-1}$ Mpc can, in principle, be described fairly well by a fractal dimension $D \simeq 2$, as originally found in [14], a dimension that could perhaps be *topological* rather than fractal, reflecting a possible sheet-like organization of structures in

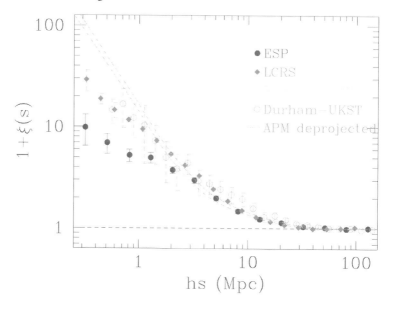

Figure 12.6. The function $1 + \xi(s)$ for the same surveys of figure 12.2. A stable power-law scaling would indicate a fractal range. It is clear how peculiar motions that affect all data plotted but the APM $\xi(r)$ which is computed in projection do significantly distort the overall shape. What would seem to be an almost regular scaling range with $D \sim 2$ from 0.3 to $30h^{-1}$ Mpc, hides in reality a more complex structure, with a clear inflection around $3h^{-1}$ Mpc, which is revealed only when redshift-space effects are eliminated.

this range [33]. Above $100h^{-1}$ Mpc the function $1 + \xi(r)$ seems to be fairly flat, indicating a possible convergence to homogeneity. However, once this is established, this kind of plot does not allow one to deduce evidence of clustering signals of the order of 1%, which can only be seen when the *contrast* with respect to the mean is plotted, i.e. $\xi(s)$. For a similar analysis and more details, see the pedagogical paper by Martìnez [34].

Another way of reading the same statistics and on which I would like to give an update with respect to [31] is the scaling of the correlation length r_0 with the sample size. It is known that for samples which are too small there is indeed a growth of r_0 with the sample size (see e.g. early results in [35]). This is naturally expected: galaxies are indeed clustered with a power-law correlation function, and inevitably samples which are too small will tend statistically to overestimate the mean density, when measuring it in a local volume. When we consider modern samples, however, and we pay attention not to compare apples with pears (galaxies with different morphology and/or different luminosity have different correlation properties, [31]), then the situation is more reassuring: table 12.1 represents an update of that presented in [31], and reports the general properties of the four redshift surveys I have used so far as examples. As the

Table 12.1. The behaviour of the correlation length r_0 for the surveys discussed in previous figures, compared to predictions of a $D = 2$ model. All estimates of r_0 are in *real* space. d is the effective depth of the surveys, while the 'sample radius' R_s has been computed as in [31]. All measures of distance are expressed in h^{-1} Mpc.

Survey	d	R_s	r_0 (predicted)	r_0 (observed)
ESP	~600	5	1.7	$4.50^{+0.22}_{-0.25}$
Durham/UKST	~200	30	10	4.6 ± 0.2
LCRS	~400	32	11	5.0 ± 0.1
Stromlo/APM	~200	83	28	5.1 ± 0.2

survey volumes are not spherical, here the 'sample radius' is defined as that of the maximum sphere contained within the survey boundaries (see [31]). All these are estimates of r_0 in real space. The observed correlation lengths are significantly different from the values predicted by the simple $D = 2$ fractal model. The result would be even worse using $D = 1.2$. The bare evidence from table 12.1 is that the measured values of r_0 are remarkably stable, despite significant changes in the survey volumes and shapes.

The counter-arguments in favour of a fractal interpretation of the available data are instead summarized in the chapter by M Montuori. As the readers can check, the main points of disagreement are related to (a) the use of some samples whose incompleteness is very difficult to assess (as e.g. heterogeneous compilations of data from the literature); and (b) the estimators used for computing the correlation function and the way they take the survey shapes into account. Also on these issues, the 2dF and SDSS surveys will provide data-sets to fully clarify the scene. In fact, preliminary estimates of the correlation function from the 2dF survey provide a result in good agreement with the analyses shown here [1].

12.4.3 Scaling in Fourier space

It is of interest to spend a few words on the complementary, very important view of clustering in Fourier space. The Fourier transform of the correlation function is the power spectrum $P(k)$:

$$P(k) = 4\pi \int_0^\infty \xi(r) \frac{\sin(kr)}{kr} r^2 \, dr, \qquad (12.18)$$

which describes the distribution of power among different wavevectors or *modes* $k = 2\pi/\lambda$ once we decompose the fluctuation field $\delta = \delta\rho/\rho$ over the Fourier basis [4]. The amount of information contained in $P(k)$ is thus formally the same as that yielded by the correlation function, although their estimates are affected

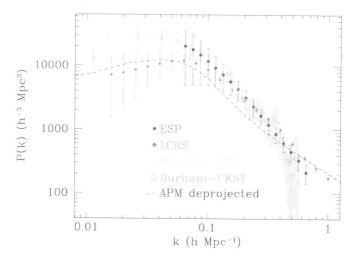

Figure 12.7. The power spectrum of galaxy clustering estimated from the same surveys as in figure 12.2 (also from [2], power spectrum estimates from [36–39]). Also in Fourier space the differences between real- and redshift-space clustering are evident above $k \simeq 0.2h$ Mpc^{-1}.

differently by the uncertainties in the data (e.g. [4, 36]). One practical benefit of the description of clustering in Fourier space through $P(k)$ is that for fluctuations of very long spatial wavelength ($\lambda > 100h^{-1}$ Mpc), where $\xi(r)$ is dangerously close to zero and errors easily make the measured values fluctuate around it (see figure 12.2), $P(k)$ is, in contrast, very large. Around these scales, most models predict a maximum for the power spectrum, the fingerprint of the size of the horizon at the epoch of matter–radiation equivalence. More technical details on power spectra can be found in the chapter by J Peacock in this book.

In figure 12.7, I have plotted the estimates of $P(k)$ for the same surveys of figure 12.2. Here again the projected estimate from the APM survey allows us to disentangle the distortions due to peculiar velocities, which have to be taken properly into account in the comparisons to cosmological models. Here scales are reversed with respect to $\xi(r)$, and the effect manifests itself in the different slopes above $\sim 0.3h$ Mpc^{-1}: an increased slope in real space (broken line) corresponds to a stronger damping by peculiar velocities, diluting the apparent clustering observed in redshift space (all points). Below these strongly nonlinear scales, there is good agreement between the slopes of the different samples (with the exception of the LCRS, see [36] for discussion), with a well-defined k^{-2} power-law range between ~ 0.08 and $\sim 0.3h$ Mpc^{-1}. The APM data show a slope $\sim k^{-1.2}$, corresponding to the $\gamma \simeq -1.8$ range of $\xi(r)$, while at smaller ks (larger scales) they steepen to $\sim k^{-2}$, in agreement with the redshift-space points. It is this change in slope that produces the shoulder observed in $\xi(s)$ (cf.

section 12.2). Peacock [40] showed that such a spectrum is consistent with a steep linear $P(k)$ ($\sim k^{-2.2}$), the same value originally suggested to explain the shoulder when first observed in earlier redshift surveys [14]. A dynamical interpretation of this transition scale has been recently confirmed by a re-analysis of the APM data [41].

At even smaller ks all spectra seem to show an indication for a turnover. However, when errors are checked in detail, they are at most consistent with a flattening, with the Durham–UKST survey providing possibly the cleanest evidence for a maximum around $k \sim 0.03h$ Mpc^{-1} or smaller. A flattening or a turnover to a positive slope would be an indication for a scale over which finally the variance is close to or smaller than that of a random (Poisson) process. But we learn by looking at older data that a turnover can also be an artifact produced when wavelengths comparable to the size of the samples are considered, and here we are close to that case.

12.5 Do we really see homogeneity? Variance on $\sim 1000h^{-1}$ Mpc scales

Wu and collaborators [42] and Lahav [43] nicely reviewed the evidence for a convergence to homogeneity on large scales using several observational tests. On scales corresponding to spatial wavelengths $\lambda \sim 1000h^{-1}$ Mpc, the constraints on the mean-square density fluctuations are provided essentially by the smoothness in the x-ray and microwave backgrounds. Measuring directly the clustering of luminous objects over such enormous volumes, is only now becoming feasible. The 2dF survey will get close to these scales. The SDSS [26] will do even better through a sub-sample of early type galaxies selected as to reach a redshift $z \sim 0.5$. If the goal of a redshift survey is mapping density fluctuations on the larges possible scales a viable alternative to using single galaxies is represented by *clusters of galaxies*. Here I would like to discuss the properties of the largest of such surveys, that is in fact currently producing remarkable results on the amount of inhomogeneity on scales nearing $1000h^{-1}$ Mpc.

12.5.1 The REFLEX cluster survey

With mean separations $>10h^{-1}$ Mpc, clusters of galaxies are ideal objects for sampling efficiently long-wavelength fluctuations over large volumes of the universe. Furthermore, fluctuations in the cluster distribution are amplified with respect to those in galaxies, i.e. they are *biased* tracers of large-scale structure: rich clusters form at the peaks of the large-scale density field, and their variance is amplified by a factor that depends on their mass, as was first shown by Kaiser [44]. X-ray selected clusters have a further major advantage over galaxies or other luminous objects when used to trace and quantify clustering in the universe: their x-ray emission, produced through thermal bremsstrahlung by the thin hot plasma permeating their potential well, is a good measure of their mass and this allows us to directly compare observations to the predictions of cosmological models (see [45] for a review and [46] for a direct application).

The REFLEX (ROSAT-ESO Flux Limited X-ray) cluster survey is the result of the most intensive effort for a homogeneous identification of clusters of galaxies in the ROSAT All Sky Survey (RASS). It combines a thorough analysis of the x-ray data , and extensive optical follow-up with ESO telescopes, to construct a complete flux-limited sample of about 700 clusters with measured redshifts and x-ray luminosities [47, 48]. The survey covers most of the southern celestial hemisphere ($\delta < 2.5°$), at galactic latitude $|b_{\text{II}}| > 20°$ to avoid high absorption and stellar crowding. The present, fully identified version of the REFLEX survey contains 452 clusters and is more than 90% complete to a nominal flux limit of 3×10^{-12} erg s^{-1} cm^{-2} (in the ROSAT band, 0.1–2.4 keV). Mean redshifts for virtually all these have been measured during a long observing campaign with ESO telescopes. Details on the identification procedure and the survey properties can be found in [49], while earlier results are reported in [50,51].

Figure 12.8 shows the spatial distribution of REFLEX clusters, giving evidence for a number of superstructures with sizes $\sim 100h^{-1}$ Mpc. One of the main motivations for this survey was to compute the power spectrum on extremely large scales, benefiting from the efficiency of cluster samples to cover very large volumes of the universe. Figure 12.9 shows the estimates of $P(k)$ from three subsamples of the survey (from [46]).

One of the strong advantages of working with x-ray selected clusters of galaxies is that connection to model predictions is far less ambiguous than with optically selected clusters (e.g. [45, 53]). We have therefore used the specific REFLEX selection function (converted essentially to a selection in mass), to determine that a low-Ω_M model (open or Λ-dominated), best matches *both* the shape and amplitude (i.e. bias value) of the observed power spectrum [46] (broken curve in the figure). In fact, the samples shown here do not reach the maximum spatial wavelengths we can possibly sample with the current data, as the Fourier box could be made to be as large as $1000h^{-1}$ Mpc (the survey reaches $z = 0.3$ with the most luminous objects). In such a case, however, our control over systematic effects becomes poorer, and work is currently undergoing to pin errors down and understand how trustable are our results on ~ 1 Gpc scale, where we do see extra power coming up. At the very least, REFLEX is definitely showing more clustering power on very large scales than any galaxy redshift survey to date. Similar hints for large-scale inhomogeneities seem to be suggested by the most recent analysis of Abell-ACO samples [54].

For $k > 0.05h$ Mpc^{-1}, however, a comparison of REFLEX to galaxy power spectra shows a rather similar shape. This is probably better appreciated by looking at the two-point correlation function $\xi(s)$ [52], compared in figure 12.10 to that of the ESP galaxy redshift survey. The agreement in shape between galaxies and clusters is remarkable on all scales, with a break to zero around 60–$70h^{-1}$ Mpc for both classes of objects. This is, in general, expected in a simple

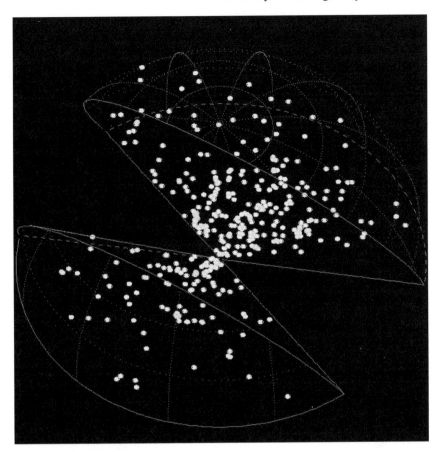

Figure 12.8. The spatial distribution of x-ray clusters in the REFLEX survey, out to $600h^{-1}$ Mpc. Note how, despite the coarser mapping of large-scale structure, filamentary superclusters ('chains' of clusters) are clearly visible.

biasing scenario where clusters represent the high, rare peaks of the mass density distribution. This result strongly corroborates the simpler, reassuring view that at least above $\sim 5h^{-1}$ Mpc the galaxy and mass distributions are linked by a simple constant bias.

12.5.2 'Peaks and valleys' in the power spectrum

Most of the discussion so far has concentrated on the beauty of finding 'smooth' simple shapes for $\xi(r)$ or $P(k)$, as symptoms of an underlying order of Nature. Rather than being a demonstration of Nature's inclination for elegance, however, this smoothness and simplicity might simply indicate our ignorance and lack of data. In fact, while smooth power spectra are predicted in models dominated by

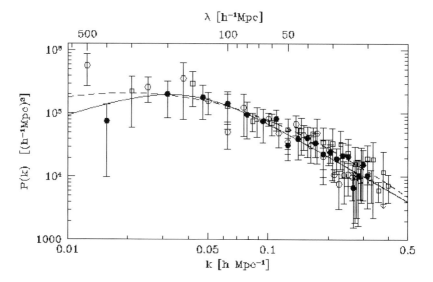

Figure 12.9. Estimates of the power spectrum of x-ray clusters from flux-limited subsamples of the the REFLEX survey, framed within Fourier boxes of 300 (open squares), 400 (filled hexagons), and 500 (open hexagons)h^{-1} Mpc side, containing 133, 188 and 248 clusters, respectively. The two curves correspond to the best-fitting parameters using a phenomenological shape with two power laws (full), or a ΛCDM model, with $\Omega_M = 0.3$ and $\Omega_\Lambda = 0.7$ (broken) (from [46]).

non-interacting dark matter particles, as cold dark matter, a very different situation is expected in cases where ordinary (baryonic) matter plays a more significant role, with wiggles appearing in $P(k)$ that would be difficult to detect with the size and 'Fourier resolution' of our current data-sets.

The possibility that the power spectrum shows a sharp peak (or more peaks) around its maximum has been suggested a few times during the last few years. For example, Einasto and collaborators [55] found evidence for a sharp peak around $k \simeq 0.05h$ Mpc^{-1} in the power spectrum of an earlier sample of Abell clusters, a feature later confirmed with lower significance by a more conservative analysis of the same data [56]. The position of this feature is remarkably close to the $\sim130h^{-1}$ Mpc 'periodicity' revealed by Broadhurst and collaborators in a 'pencil-beam' survey towards the galactic poles [57] and, more recently, in an analysis of the redshift distribution of Lyman-break selected galaxies [58]. Other evidence has been claimed from two-dimensional analyses of redshift 'slices' [59] or QSO superstructures [60].

These observations have stimulated some interesting work on models with high baryonic content. In this case, the power spectrum can exhibit a detectable inprint from 'acoustic' oscillations within the last scattering surface at $z \sim 1000$, the same features observed in the Cosmic Microwave Background (CMB)

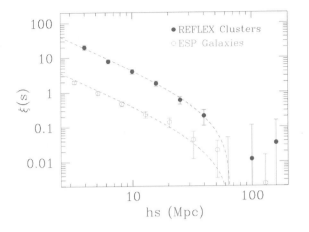

Figure 12.10. The two-point correlation function of the whole flux-limited REFLEX cluster catalogue (filled circles, [52]), compared to that of ESP galaxies (open circles, [9]). The broken curves show the Fourier transform of a phenomenological fit to $P(k)$ which tries to include the large-scale power seen from the largest subsamples (top line). The bottom curve is that obtained after scaling down by an arbitrary bias factor ($b_c^2 = (3.3)^2$ in this specific case).

radiation [61]. While the most recent estimates of the REFLEX power spectrum do not show clear features around the scales of interest to justify 'extreme' high-baryon models (contrary to early indications [62], which shows the importance of the careful assessment of errors), the extra power below $k \sim 0.02$ could still be an indication of an higher-than-conventional baryon fraction [61,63], along the lines that seem to be suggested by the Boomerang CMB results [64].

12.6 Conclusions

At the end of this chapter, a student is possibly more confused than he/she was in the beginning, at least after a first read. I hope, however, that once the dust settles, a few important points emerge. First, that the processes which shaped the large-scale distribution of luminous objects we observe today are different at different scales. At small scales, we observe essentially the outcome of fully nonlinear gravitational evolution that re-shaped the linear power spectrum into a collection of virialized or nearly so structures. Therefore, one cannot naively take the redshift survey data and look for specific patterns or statistical properties without taking into account galaxy peculiar motions. For this reason, one should be careful in over-interpreting things like a single power-law scaling from scales of a tenth of a megaparsecs to hundred megaparsecs, because, again, different phenomena are being compared. However, one can use these distortions to really 'see' how the true mass distribution is, and I have spent a considerable part of this chapter describing some of the techniques in use. Moving to larger and larger scales, we enter a regime where we are lucky enough that we can still see something related to the original scaling law of fluctuations. This is what was originally produced by some generator in the early universe (inflation?) and processed through a matter (dark plus baryons) controlled amplifier. On even larger scales, we hope we are finally entering a regime where the variance in the mass is consistent with a homogeneous distribution, although we have seen that even the largest galaxy and cluster samples are barely sufficient to see hints of that, perhaps suggesting even more inhomogeneity than we expect. Does this mean that we are living in a pure fractal universe? The scaling behaviour of galaxies and the stability of the correlation length seem to imply that this cannot be the case. On top of everything, the smoothness of the cosmic microwave background (treated elsewhere in this book) is probably the most reassuring observation in this respect. What we seem to understand is that our samples still have difficulty in sampling the very largest fluctuations of the density field, properly on scales where this is not fully Poissonian (or sub-Poissonian) yet.

Finally, I hope readers get the message that despite the tremendous progress of the last 25 years which transformed cosmology into a real science, we still have a number of fascinating questions to answer and still feel far away from convincing ourselves that we have understood the universe.

Acknowledgments

Most of the results I have shown here rely upon the work of a number of collaborators in different projects. I would like to thank in particular my colleagues in the REFLEX collaboration, especially C Collins and P Schuecker for the work on correlations and power spectra shown here. Thanks are due to F Governato for providing me with the simulation used for producing Figure 12.3, and to Alberto Fernandez-Soto and Davide Rizzo for a careful reading of the manuscript. Finally, thanks are due to the organizers of the Como School, for their patience in waiting for this chapter and for allowing me extra page space.

References

[1] Colless M 1999 *Proc. II Coral Sea Workshop*
     http://www.mso.anu.edu.au/DunkIsland/Proceedings/
[2] Guzzo L 2000 *Proc. XIX Texas Symposium on Relativistic Astrophysics (Nucl. Phys. Proc. Suppl. 80)* ed E Aubourg *et al*
[3] http://www.mso.anu.edu.au/2dFGRS/
[4] Peacock J A 1999 *Cosmological Physics* (Cambridge: Cambridge University Press)
[5] Peebles P J E 1980 *The Large-Scale Structure of the Universe* (Princeton, NJ: Princeton University Press)
[6] Maddox S J, Efstathiou G, Sutherland W J and Loveday J 1990 *Mon. Not. R. Astron. Soc.* **242** 43p

[7] Heydon-Dumbleton N H, Collins C A and MacGillivray H T 1989 *Mon. Not. R. Astron. Soc.* **238** 379

[8] Davis M and Peebles P J E 1983 *Astrophys. J.* **267** 465

[9] Guzzo L *et al* (ESP Team) 2000 *Astron. Astrophys.* **355** 1

[10] Tucker D L *et al* 1997 *Mon. Not. R. Astron. Soc.* **285** L5

[11] Loveday J, Peterson B A, Efstathiou G and Maddox S J 1992b *Astrophys. J.* **390** 338

[12] Ratcliffe A, Shanks T and Broadbent A *et al* 1996 *Mon. Not. R. Astron. Soc.* **281** L47

[13] Baugh C M 1996 *Mon. Not. R. Astron. Soc.* **280** 267

[14] Guzzo L *et al* 1991 *Astrophys. J.* **382** L5

[15] Fisher K B *et al* 1994a *Mon. Not. R. Astron. Soc.* **266** 50

[16] Guzzo L *et al* 1997 *Astrophys. J.* **489** 37

[17] Ratcliffe A, Shanks T, Parker Q A and Fong D 1998 *Mon. Not. R. Astron. Soc.* **296** 173

[18] Fisher K B *et al* 1994b **267** 927

[19] Fisher K B 1995 *Astrophys. J.* **448** 494

[20] Sheth R K, Hui L, Diaferio A and Scoccimarro R 2000 *Mon. Not. R. Astron. Soc.* **326** 463

[21] Peebles P J E 1976 *Astrophys. Space Sci.* **45** 3

[22] Zurek W, Quinn P J, Warren M S and Salmon J K 1994 *Astrophys. J.* **431** 559

[23] Marzke R O, Geller M J, da Costa L N and Huchra J P 1995 *Astron. J.* **110** 477

[24] Landy S D, Szalay A S and Broadhurst T J 1998 *Astrophys. J.* **494** L133

[25] Quarello S and Guzzo L 2000 *Clustering at High Redshift (ASP Conf. Series 200)* ed A Mazure, O Le Fèvre and V Le Brun (San Francisco, CA: ASP) p 446

[26] Margon B 1998 *Phil. Trans. R. Soc.* A (astro-ph/9805314)

[27] Mandelbrot B B 1982 *The Fractal Geometry of Nature* (San Francisco, CA: Freeman)

[28] Pietronero L 1987 *Physica* A **144** 257

[29] Davis L *Critical Dialogues in Cosmology* ed N Turok (Singapore: World Scientific) p 13

[30] *Critical Dialogues in Cosmology* ed N Turok (Singapore: World Scientific) p 24

[31] Guzzo L 1997 *New Astronomy* **2** 517

[32] Provenzale A 1991 *Applying fractals in Astronomy* ed A Heck and J Perdang (Berlin: Springer)

[33] Provenzale A, Guzzo L and Murante G 1994 *Mon. Not. R. Astron. Soc.* **266** 555

[34] Martìnez V 1999 *Science* **284** 445

[35] Einasto J, Klypin A and Saar E 1986 *Mon. Not. R. Astron. Soc.* **219** 457

[36] Carretti E *et al* 2001 *Mon. Not. R. Astron. Soc.* **324** 1029

[37] Lin H *et al* 1996 *Astrophys. J.* **471** 617

[38] Tadros H and Efstathiou G P 1996 *Mon. Not. R. Astron. Soc.* **282** 138

[39] Hoyle F, Baugh C M, Ratcliffe A and Shanks T 1999 *Mon. Not. R. Astron. Soc.* **309** 659

[40] Peacock J A 1997 *Mon. Not. R. Astron. Soc.* **284** 885

[41] Gaztañaga E and Juszkiewicz R 2000 *Mon. Not. R. Astron. Soc.* submitted (astro-ph/0007087)

[42] Wu K K S, Lahav O and Rees M J 1999 *Nature* **225** 230

[43] Lahav O 2000 *Proc. NATO-ASI Cambridge July 1999* ed R Critenden and N Turok
        (Dordrecht: Kluwer) in press (astro-ph/0001061)
[44] Kaiser N 1984 *Astrophys. J.* **284** L9
[45] Borgani S and Guzzo L 2001 *Nature* **409** 39
[46] Schuecker P *et al* (REFLEX Team) *Astron. Astrophys.* submitted
[47] Böhringer H *et al* (REFLEX Team) 1998 *The Messenger* **94** 21 (astro-ph/9809382)
[48] Guzzo L *et al* (REFLEX Team) 1999 *The Messenger* **95** 27
[49] Böhringer H *et al* (REFLEX Team) 2000 *Astron. Astrophys.* submitted
[50] De Grandi S *et al* (REFLEX Team) 1999 *Astrophys. J.* **513** L17
[51] De Grandi S *et al* (REFLEX Team) 1999b *Astrophys. J.* **514** 148
[52] Collins C A *et al* (REFLEX Team) 2000 *Mon. Not. R. Astron. Soc.* **319** 939
[53] Moscardini L, Matarrese S, Lucchin F and Rosati P 2000 *Mon. Not. R. Astron. Soc.*
        316 283
[54] Miller C J and Batuski D J 2001 *Astrophys. J.* **551** 635
[55] Einasto J *et al* 1997 *Nature* **385** 139
[56] Retzlaff J *et al* 1998 *New Astronomy* **3** 631
[57] Broadhurst T J, Ellis R S, Koo D C and Szalay A S 1990 *Nature* **343** 726
[58] Broadhurst T J and Jaffe A H 1999 *Astrophys. J.* submitted (astro-ph/9904348)
[59] Landy D S *et al* 1996 *Astrophys. J.* **456** L1
[60] Roukema B F and Mamon G 2001 *Astron. Astrophys.* **366** 1
[61] Eisenstein D J, Hu W, Silk J and Szalay A S 1998 *Astrophys. J.* **494** L1
[62] Guzzo L 1999 *Proc. II Coral Sea Workshop*
        http://www.mso.anu.edu.au/DunkIsland/Proceedings/
[63] Guzzo L *et al* (REFLEX Team) 2001 in preparation
[64] De Bernardis P *et al* 2000 *Nature* **404** 955

Chapter 13

The debate on galaxy space distribution: an overview

Marco Montuori and Luciano Pietronero
Deptartment of Physics, University of Rome—'La Sapienza' and
INFM, Rome, Italy

13.1 Introduction

A critical assumption of the hot big bang model of the universe is that matter is homogeneously distributed in space over a certain scale. It is usually assumed that under this condition the Friedmann–Robertson–Walker (FRW) metric correctly describes the dynamics of the universe. Investigating this assumption is then of fundamental importance in cosmology and much current research is devoted to this issue. In this chapter, we will review the current debate on the spatial properties of galaxy distribution.

13.2 The standard approach of clustering correlation

The usual way to investigate the properties of the spatial distribution of glaxies is to measure the two-point autocorrelation function $\xi(r)$ [1]. This is the spatial average of the fluctuations in the galaxy number density at distance r, with respect to a homogeneous distribution of the same number of galaxies. Let $n(r_i)$ the density of galaxies in a small volume δV at the position r_i. The relative fluctuation in δV is

$$\frac{\delta n(r_i)}{\langle n \rangle} = \frac{n(r_i) - \langle n \rangle}{\langle n \rangle} \tag{13.1}$$

where $\langle n \rangle = N/V$ is the density of the sample.

It is clear that the fluctuations are defined with respect to the density of the sample $\langle n \rangle$. The two-point correlation function $\xi(r)$ at scale r is the spatial average of the product of the relative fluctuations in two volumes centred on data points at distance r:

$$\xi(r) = \left\langle \frac{\delta(r_i + r)}{\langle n \rangle} \frac{\delta(r_i)}{\langle n \rangle} \right\rangle_i = \frac{\langle n(r_i)n(r_i + r)\rangle_i}{\langle n \rangle^2} - 1, \qquad (13.2)$$

where the index i means that the average is performed over the all the galaxies in the samples. A set of points is correlated on scale r if $\xi(r) > 0$; it is uncorrelated if $\xi(r) = 0$. In the latter case the points are evenly distributed at scale r or, in another words, they have a homogeneous distribution at scale r. In the definition of $\xi(r)$, the use of the sample density $\langle n \rangle$ as a reference value for the fluctuations of galaxies is the conceptual assumption that the galaxy distribution is homogeneous at the scale of the sample.

In such a framework, a relevant scale r_0 for the correlation properties is usually defined by the condition $\xi(r_0) = 1$. The scale r_0 is called the *correlation length of the distribution*.

13.3 Criticisms of the standard approach

Let us summarize the conclusions of the previous section:

- The $\xi(r)$ analysis assumes homogeneity at the sample size; and
- a characteristic scale for the correlation is defined by the amplitude of $\xi(r)$, i.e. the scale at which $\xi(r)$ is equal to one [1].

These two points raise two main criticisms:

- As the $\xi(r)$ analysis assumes homogeneity, it is not reliable for *testing* homogeneity. In order to use $\xi(r)$ analysis, the density of galaxies in the sample must be a good estimation of the density of the whole distribution of the galaxies. This may either be true or not; in any case, it should be checked *before* $\xi(r)$ analysis is applied [2].
- The correlation length r_0 does not concern the *scale* of fluctuations. In this sense, it is not correct to refer to it as a measure of the characteristic size of correlations and call it the *correlation length*. According to the definition of $\xi(r)$, r_0 simply separates a regime of large fluctuations $\delta n / \langle n \rangle \gg 1$ from a regime of small fluctuations $\delta n / \langle n \rangle \ll 1$ [3,4].

Again the argument is valid if the average density $\langle n \rangle$ of the sample is the average density of the distribution or, in other words, if the distribution is homogeneous on the sample size. In statistical mechanics, the *correlation length* of the distribution is defined by how fast the correlations vanish as a function of the scale, i.e. by the functional form of $\xi(r)$ and not by its amplitude.

In this respect, the first step in a spatial correlation analysis of a data-set should be a study of the density behaviour versus the scale. This should be done without any *a priori* assumptions about the features of the underlying distribution [2].

13.4   Mass–length relation and conditional density

The *mass–length* relation links the average number of points at distance r from any other point of the structure to the scale r. Starting from an ith point occupied by an object of the distribution, we count how many objects $N(< r)_i$ ('mass') are present within a volume of linear size r ('length') [5]. The average over all the points of the structure is:

$$\langle N(< r)_i \rangle = B \cdot r^D. \tag{13.3}$$

The exponent D is called *the fractal dimension* and characterizes in a quantitative way how the system fills the space, while the prefactor B depends on the lower cut-off point of the distribution.

The conditional density $\Gamma(r)$ is the average number of points in a shell of width dr at distance r from any point of the distribution.

According to equation (13.3), $\Gamma(r)$ is:

$$\Gamma(r) = \frac{1}{4\pi r^2 dr} \frac{d\langle N(< r)_i \rangle}{dr} = \frac{BD}{4\pi} \cdot r^{D-3} \tag{13.4}$$

(see [2, 6] for details of the derivation).

13.5   Homogeneous and fractal structure

If the distribution crosses over to a homogeneity distribution at scale r, $\Gamma(r)$ shows a flattening toward a constant value at such a scale. In this case, the fractal dimension in equations (13.3) and (13.4) has the same value as the dimension of the embedding space d, $D = d$ (in three-dimensional space $D = 3$) [2, 5, 6].

If this does not happen, the density of the sample will not correspond to the density of the distribution and it will show correlations up to the sample size. The simplest distribution with such properties is a fractal structure [5]. A fractal consists of a system in which more and more structures appear at smaller and smaller scales and the structures at small scales are similar to those at large scales. The distribution is then self-similar. It has a value of D that is smaller than d, $D < d$. In three-dimensional space $d = 3$, a fractal has $D < 3$ and $\Gamma(r)$ is a *power law*. The value of $N(< r)_i$ largely fluctuates by changing both the starting ith point and the scale r. This is due to the scale-invariant feature of a fractal structure, which does not have a characteristic length [5, 7].

13.6 $\xi(r)$ for a fractal structure

Equation (13.4) shows that $\Gamma(r)$ is a well-defined statistical tool for the generic distribution of points, since it depends only on the intrinsic quantities (B and D). The same is not true for $\xi(r)$ statistics.

Assuming for simplicity a spherical sample volume with radius R_s ($V(R_s) = (4/3)\pi R_s^3$), containing $N(R_s)$ galaxies. The average density of the sample will be

$$\langle n \rangle = \frac{N(R_s)}{V(R_s)} = \frac{3}{4\pi} B R_s^{-(3-D)}. \tag{13.5}$$

For a fractal, $D < 3$ and its average density is a decreasing function of the sample size: $\langle n \rangle \to 0$ for $R_s \to \infty$. Then the average density depends explicitly on the sample size R_s and it is not a meaningful quantity.

From equation (13.2), the expression for $\xi(r)$ for a fractal distribution is [2]:

$$\xi(r) = ((3 - \gamma)/3)(r/R_s)^{-\gamma} - 1. \tag{13.6}$$

From equation (13.6) it follows that, for the *fractal sample* the so-called correlation length r_0 (defined as $\xi(r_0) = 1$) is a linear function of the sample size R_s:

$$r_0 = ((3 - \gamma)/6)^{1/\gamma} R_s. \tag{13.7}$$

It is then a quantity *without* any statistical *significance*, one simply related to the sample size [2].

Neither is $\xi(r)$ *a power law*. For $r \leq r_0$,

$$((3 - \gamma)/3)(r/R_s)^{-\gamma} \gg 1 \tag{13.8}$$

and $\xi(r)$ is well approximated by a *power law* [6].

For larger distances there is clear deviation from power-law behaviour due to the definition of $\xi(r)$. This deviation, however, is just due to the size of the observational sample and does not correspond to any real change in the correlation properties. It is clear that if one estimates the exponent of $\xi(r)$ at distances $r \approx r_0$, one systematically obtains a higher value of the correlation exponent due to the break of $\xi(r)$ in the log–log plot. Only if the sample set has a crossover to homogeneity inside the sample side, is $\xi(r)$ correct. However, this information is given only by the $\Gamma(r)$ analysis which, for this reason, should always come before the $\xi(r)$ investigation.

13.7   Galaxy surveys

Galaxy catalogues are *angular catalogues (three-dimensional)*, which can be computed in *real* or in *redshift space*. The latter defines the galaxy positions by the *redshift distance* s, which is derived by the *galaxy redshift* z, according to *Hubble's law*. s is not the *real* distance, but contains an additional term called the *redshift distortion*, which is small on scales $s > 5h^{-1}$ Mpc [8].

We will report the statistical properties of *redshift surveys*, which contain the large majority of avalaible three-dimensional data.

13.7.1 Angular samples

$\xi(r)$ can be obtained from two-dimensional data, by means of the angular two-point function $w(\theta)$. $\xi(r)$ is reconstructed using the luminosity function, which is derived assuming homogeneneity in the sample [1]. No independent check is usually performed on this assumption. The procedure is currently considered one of the best estimates of three-dimensional clustering properties of galaxies, at least on a small scale ($\leq 20h^{-1}$ Mpc) [9, 10]. Such a claim is considered to be justified by the great quantity of available data in angular catalogues with respect to three-dimensional surveys and by the absence of redshift distortions in the two-dimensional data. The main conclusion obtained by this approach is that the galaxy correlation (more precisely for optical selected galaxies) $\xi_{gg}(r)$ is *quite close to a power law in the range* $10h^{-1}$ kpc–$(10$–$20h^{-1})$ Mpc and more precisely [9, 10]:

$$\xi_{gg}(r) = \left(\frac{r_0}{r}\right)^{-1.77} \tag{13.9}$$

with a *correlation length* $r_0 \approx 4.5 \pm 0.5h^{-1}$ Mpc.

This is considered to give the 'canonical shape and parameter values' of $\xi(r)$ and is a well-established result in cosmology [1, 10–14].

13.7.2 Redshift samples

13.7.2.1 ML samples

An ML sample is simply the whole *redshift catalogue*. By construction, any ML sample is incomplete in the distribution of galaxies. At larger distances, it contains fewer and fewer galaxies, as more and more galaxies fall beyond the threshold of detectability. To account for such an effect, the galaxies in the sample are weighted, according the luminosity function [1].

The value of s_0 in different ML catalogues is found to span from 4.5–$8h^{-1}$ Mpc [10, 13].

$\xi(s)$ does not appear to be *a power law*. According to Guzzo [15], the shape of $\xi(s)$ at very small scales ($<3h^{-1}$ Mpc) is well fitted by a power law with exponent $\gamma = -1$.

13.7.2.2 VL samples

It is possible to extract subsamples from the ML catalogues, which are unaffected by the aforementioned incompleteness. Such samples are called *VL samples* [16]. The main result of $\xi(s)$ analysis is that different VL samples have different values for the correlation length s_0. The general trend is that deeper and brighter samples show larger s_0 (figure 13.1) [6, 15, 19–23]. Again, $\xi(s)$ is *not a power law* in the whole observed range of scale (≈ 1–$50h^{-1}$ Mpc). This has been recognized by several authors, who have performed the fit with the power law in a limited range of scales. The value of the exponent γ (see equation (13.9)) is in the range

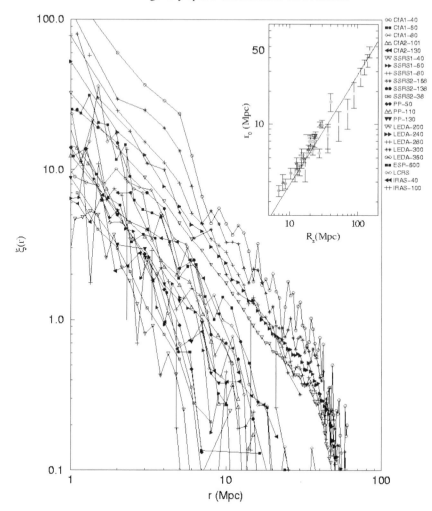

Figure 13.1. $\xi(r)$ measure in various VL galaxy samples. The general trend is an increase of the $\xi(r)$ amplitude for brighter and deeper samples. In the *insert panel* we show the dependence of *correlation length* r_0 on *sample size* R_s for all samples. The linear behaviour is a consequence of the fractal nature of galaxy distribution in these samples.

1.17–2.1, which demonstrates the deviation of $\xi(s)$ from the *canonical power-law* shape [6, 15, 21, 22].

13.8 $\Gamma(r)$ analysis

According to our criticism of the standard analysis, we have measured the galaxy conditional average density $\Gamma(r)$ in all the three-dimensional catalogues

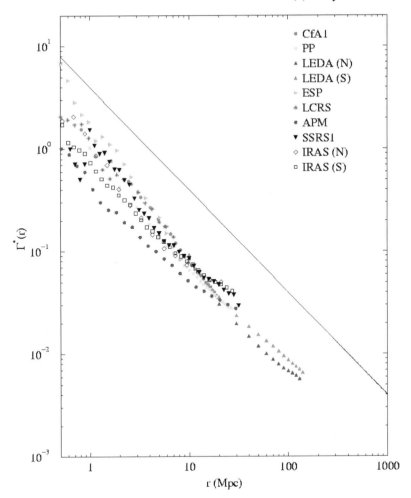

Figure 13.2. Full correlation for the same data of figure 13.1 in the range of distances 0.5–100h^{-1} Mpc. A reference line with a slope −1 is also shown (i.e. fractal dimension $D = 2$.

avalaible. Our analysis was carried out for VL samples; the results are collected in figure 13.2 [6].

We can derive the following conclusions:

- $\Gamma(s)$ measured in different catalogues is a *power law* as a function of the scale s, extending from approximately 1 to 40–50h^{-1} Mpc, without any tendency towards homogenization (flattening) [6]. This implies that all the optical catalogues show well-defined fractal correlations up to their limits, with the fractal dimension $D \simeq 2$ [6].

- Only in a single case, the LEDA database [17, 18], is it possible to reach larger scales of $\sim 100h^{-1}$ Mpc. This data sample has been largely criticized but, to our knowledge, never in a quantitative way. The statistical tests we performed show clearly that up to $50h^{-1}$ Mpc the results are completely consistent with all other data [6]. This agreement also appears to extend to the range $50–100h^{-1}$ Mpc, with the same overall statistical properties found at smaller scales [6].
- We do not detect any difference between the various optical catalogues, as expected if they are simply different parts of the same distribution.
- Such results imply that the $\xi(s)$ analysis is inappropriate as it describes correlations as deviations from an assumed underlaying homogeneity. According to the $\Gamma(s)$ results, the value of s_0 (derived from the $\xi(s)$ approach) has to scale with the sample size R_s. The behaviour observed corresponds to a fractal structure with dimension $D \simeq 2$.

13.9 Interpretation of standard results

Here we attempt a comparison between the different interpretations.

In the standard interpretation, the rough constancy of s_0 for the different ML samples ($s_0 \simeq 4.5–8h^{-1}$ Mpc) and within the angular data is considered evidence for the validity of this approach. Moreover, since the samples have different volumes, these results should discount a *fractal* interpretation, which predicts an increase in s_0 with sample volume [13, 22].

In contrast, in the fractal approach, in our opinion, the analysis of the angular and ML samples is heavily biased by the use of the luminosity function and the corresponding homogeneity assumption. To measure the correlation function of such samples, one has to estimate the number of missing galaxies and their positions in the space. This is done by assuming the existence of a homogeneity scale. As an aside, we stress that the three-dimensional correlation in a fractal structure cannot be reconstructed in such a way from its angular features [6].

Regarding the *shape* of the $\xi(s)$, the difference from a power law is attributed:

(1) in the standard model, to the presence of *redshift distortions* [15]; and
(2) in the fractal model, to the fact that $\xi(s)$ is not in itself *a power law* [6]. If $\Gamma(s)$ is a power law, $\xi(s)$ is expressed by equation (13.6). In particular, it should be close to a power law only on very small scales and with an exponent $\gamma \sim 1$, as in the data reported by Guzzo [15].

With regards to the *VL results*, the increase in s_0 could be due to either of the following cases.

(1) *Luminosity segregation* (standard model). The increase in $\xi(s)$ corresponds to a real change in clustering properties for galaxy distribution, called *luminosity segregation*. This is considered just one aspect of the general

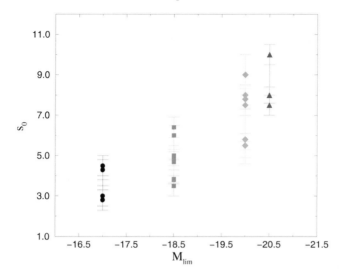

Figure 13.3. *Correlation length* s_0 versus sample luminosity M_{lim}, for several VL samples. VL samples, with the same luminosity M_{lim}, have different *volumes* and very different s_0. This is in contrast to *luminosity segregation* and in agreement with a fractal distribution inside the sample *Volume*.

expected dependence of the clustering features on the internal properties of galaxies, such morphology, colour, surface brightness and internal dynamics [15, 19–23].

(2) In the fractal model, the increase in $\xi(s)$ is just a geometrical effect and it is not related to any variation of the clustering of the corresponding data-set, as shown in equation (13.7). The effect is simply a byproduct of the inappropriate use of a statistical tool for the distribution under analysis [2,6].

The two interpretations seem to be equivalent; this is the reason why the same data-set is considered to confirm both fractality (and no luminosity segregation) and homogeneity (with luminosity segregation).

In our opinion there is a difference between the two interpretations: for the fractal case we have a quantitative prediction of an increase in s_0 within the sample size, while the theoretical expectation for luminosity segregation does not have a general consensus [24].

In principle it is possible to disentangle the two effects. A possible test is presented in figure 13.3. Here, we have reported the value of s_0 for the collection of VL samples versus the luminosity M_{lim} of the corresponding samples. Samples with the same M_{lim} have *different volumes*.

For each value of luminosity M_{lim} there is a range of values of s_0. These appear *in contradiction* to the *luminosity segregation effect*, according to which we should find only a single value for s_0 for samples with the same luminosity

M_{lim}. Experimental uncertainities in the determination of s_0 do not explain such a spread. Conversely, the behaviour seems to be in agreement with a *fractal* distribution of galaxies within the sample size. The spread in the s_0 values for a single M_{lim} is due to the difference in the *volume* between different samples (in agreement with fractal prediction).

Acknowledgments

We warmly thank F Sylos Labini with whom we carried out the large part of this work. We are also grateful to A Gabrielli and M Joyce for many fruitful discussions. We would like to thank the organizers of the Graduate School in Contemporary Relativity and Gravitational Relativistic Cosmology for the invitation to the School. This work has been supported by INFM Forum Project: 'Clustering'.

References

[1] Peebles P J E 1993 *Principles of Physical Cosmology* (Princeton, NJ: Princeton University Press)
[2] Coleman P H and Pietronero L 1992 *Phys. Rep.* **213** 311
[3] Gaite J, Domnguez A and Pérez-Mercader J 1999 *Astrophys. J. Lett.* **522** L5
[4] Gabrielli A, Sylos Labini F and Durrer R 2000 *Astrophys. J. Lett.* **531** L1
[5] Mandelbrot B 1983 *The Fractal Geometry of Nature* (San Francisco, CA: Freeman)
[6] Sylos Labini F, Montuori M and Pietronero L 1998 *Phys. Rep.* **293** 66
[7] Montuori *et al* 1997 *Europhys. Lett.* **39** 103
[8] Hamilton A J S 1998 *The Evolving Universe* ed D Hamilton (Dordrecht: Kluwer Academic) p 185
[9] Peebles J P E 1998 *Les Rencontres de Physique de la Vallee d'Aosta* ed M Greco (Gif-sur-Yvette: Frontières)
[10] Strauss M A and Willick J A 1995 *Phys. Rep.* **261** 271
[11] Kolb E W and Turner M S 1989 *The Early Universe* (Frontiers in Physics) (Boston, MA: Addison-Wesley)
[12] Davis M 1997 *Critical Dialogues in Cosmology* ed N Turok (Singapore: World Scientific) p 13
[13] Guzzo L 1997 *New Astronomy* **2** 517
[14] Wu K K S, Lahav O and Rees M 1999 *Nature* **397** 225
[15] Guzzo L 1998 *Abstracts of the 19th Texas Symposium on Relativistic Astrophysics and Cosmology* ed J Paul and T Montmerle (Paris)
[16] Davis M and Peebles P J E 1983 *Astrophys. J.* **267** 465
[17] Paturel G, Bottinelli L and Gouguenheim L 1994 *Astron. Astrophys.* **286** 768
[18] Di Nella H *et al* 1996 *Astron. Astrophys. Lett.* **308** L33
[19] Davis M *et al* 1988 *Astrophys. J. Lett.* **333** L9
[20] Park C, Vogeley M S, Geller M and Huchra J 1994 *Astrophys. J.* **431** 569

[21]   Benoist C *et al* 1999 *Astrophys. J.* **514** 563
[22]   Willmer C N A, da Costa L N and Pellegrini P S 1998 *Astrophys. J.* **115** 869
[23]   Cappi A *et al* 1998 *Astron. Astrophys.* **335** 779
[24]   Colbert J M *et al* 2000 *Mon. Not. R. Astron. Soc.* **319** 209

Chapter 14

Gravitational lensing

Philippe Jetzer
Laboratory for Astrophysics of PSI and Institute of Theoretical
Physics University of Zürich, Switzerland

14.1   Introduction

Gravitational lensing—i.e. light deflection by gravity—has become, in the last few years, one of the most important fields in present-day astronomy. The enormous activity in this area has mainly been driven by the considerable improvements in observational capabilities. Due to the new wide-field cameras and telescopes which are already in place or will become operational in the near future the rate and quality of the lensing data will increase dramatically. As gravitational lensing is independent of the nature and physical state of the deflecting mass, it is perfectly suited to study dark matter at all scales.

Indeed, the determination of the amount and nature of the matter present in the universe is an important problem for contemporary astrophysics and cosmology. This knowledge is directly related to the question of the fate of the universe: Will it expand forever or, after a phase of expansion, will it collapse again? There are several astrophysical observations which indicate that most of the matter present in the universe is actually dark and, therefore, cannot be detected using telescopes or radiotelescopes. The most recent studies seem to suggest that the total matter density is only about 30% of the 'closure density' of the universe—the amount of mass that would make the universe balance between expanding forever and collapsing. Measurements based on high-redshift supernovae suggest that there is also a non-vanishing cosmological constant, such that the sum of matter density and cosmological constant implies a flat universe [1].

Important evidence for the existence of large quantities of dark matter comes from the measured rotation curves of several hundreds of spiral galaxies [2], which imply the presence of a huge dark halo in which these galaxies are embedded. Typically, a galaxy including its halo contains $\sim$10 times more dark than luminous matter, the latter being in the form of stars and gas. There are also clear indications for the presence of important quantities of dark matter on larger scales, in particular in clusters of galaxies. This was first pointed out in 1933 by Zwicky [3]. Since then, much effort has been put into the search for dark matter, the nature of which is still largely unknown.

The field of gravitational lensing is growing very rapidly and almost daily there are new results. It will not therefore be possible to give here a complete and exhaustive review of the field and of all the results achieved so far. The present chapter is intended more as a way of rapidly acquiring the main ideas and tools of lensing, which will then enable readers to approach the original literature. For more details see the book by Schneider *et al* [4] as well as some reviews [5–7] and the references therein.

Before starting the theory of lensing let us briefly give some historical remarks on the development of the field.

14.1.1 Historical remarks

Nowadays we know that light propagation in a gravitational field has to be described using the theory of general relativity formulated by Einstein in 1915. However, long before then it was argued that gravity might influence the behaviour of light (for a historical account, see, for instance, the book by Schneider *et al* [4]). Indeed, Newton in the first edition of his book on optics which appeared in 1704 discussed the possibility that celestial bodies could deflect the light trajectory. In 1804 the astronomer Soldner published a paper in which he computed the error induced by the light deflection on the determination of the position of stars. To that purpose he used the Newtonian theory of gravity assuming that the light is made of particles. He also estimated that a light ray which just grazes the surface of the sun would be deflected by a value of only 0.85 arcseconds. Within general relativity this value is about twice as much, more precisely 1.75 arcseconds. The first measurement of this effect has been made during the solar eclipse of 29 May 1919 and confirmed the value predicted by general relativity [8].

In 1936 Einstein published a short paper in *Science* in which he computed the deflection of light coming from a distant star by the gravitational field of another star [9]. He mentioned that if the source and the lens are perfectly aligned the image would be a ring. If instead the alignment is not perfect one would see two images with, however, a very small separation angle. Einstein also wrote: 'Of course, there is no hope of observing this phenomenon'. In fact, it has recently been found that Einstein had already made most of the calculations presented in that paper by 1912 as can be seen on some pages of his notebook [10]. The recent developments in microlensing show that Einstein's conclusion, although understandable at that time, was too pessimistic. Indeed, the formulae developed by Einstein in his 1936 paper are still the basis for the description of gravitational lensing.

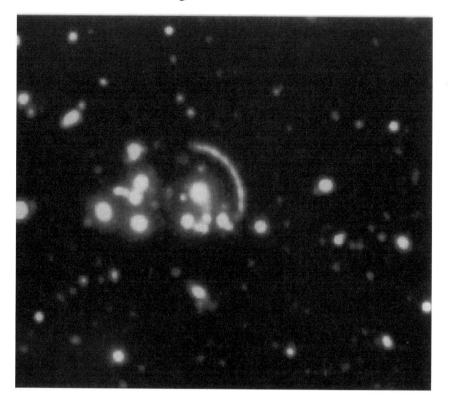

Figure 14.1. Giant arc in Cl2244-02 (image from CFHT). The lensing cluster is at $z = 0.329$ and the source of the arc is a very distant field galaxy at $z = 2.238$. (Courtesy of G Soucail, Observatoire Midi-Pyrénées, ESO Messenger 69, September 1992.)

In the following year (1937) the swiss astronomer Zwicky wrote two short articles in *Physical Review* suggesting that galaxies should be as sources and lenses rather than stars as mentioned by Einstein [11]. He came to the conclusion that such a configuration would have a much higher chance of being seen, since the typical mass of a galaxy is several billion times higher than the mass of a single star. He argued that such configurations must almost certainly be seen. Moreover, he also gave a list of possible applications which included the possibility of determining the total mass of galaxies, including their dark matter content better.

The first gravitational lens was discovered in 1979, when spectra of two point-like quasars which lie only about 6 arcseconds away were obtained. The spectra showed that both objects have the same redshift and are thus at the same distance. Later on a galaxy acting as a lens was also found, making it clear that the two objects are the images of the same quasar, which is lensed. Since then many other examples have been found, and in 1986 the first lensing case with a galaxy acting as a source was discovered. The galaxy then appears distorted as one or more arcs. Many such systems have since then been discovered, with some thanks to the Hubble space telescope. In 1979, Chang and Refsdal [12], and in 1981, Gott [13] noted that even though a point mass in a halo of a distant galaxy would create an unresolvable double image of a background quasar, the time variation of the combined brightness of the two images could be observed. In this way, the effect of non-luminous matter in the form of compact objects could be observed. The term *microlensing* was proposed by Paczyński [14] to describe gravitational lensing which can be detected by measuring the intensity variation of a macro-image made up of any number of unresolved micro-images.

In 1993 the first galactic microlensing events were observed, in which the source is a star in the Large Magellanic Cloud and the galactic bulge. In the former case the lens is a compact object probably located in the galactic halo, whereas in the latter case the lens is a low mass star in the galactic disk or in the bulge itself.

14.2   Lens equation

14.2.1   Point-like lenses

The propagation of light in a curved spacetime is, in general, a complicated problem. However, for almost all relevant applications of gravitational lensing one can assume that the geometry of the universe is described in good approximation by the Friedmann–Lemaître–Robertson–Walker metric. The inhomogeneities in the metric can be considered as local perturbations. Thus the trajectory of the light coming from a distant source can be divided into three distinct parts. In the first, the light coming from a distant source propagates in a flat unperturbed spacetime, near the lens the trajectory is modified due to the gravitational potential of the lens and, afterwards, in the third part the light again travels in an unperturbed spacetime until it reaches to the observer. The region around the lens can be described by a flat Minkowskian spacetime with small perturbations induced by the gravitational potential of the lens. This approximation is valid as long as the Newtonian potential Φ is small, which means $|\Phi| \ll c^2$ (c being the velocity of light), and if the peculiar velocity v of the lens is negligible compared to c. These conditions are almost always fulfilled in all cases of interests for the astrophysical applications. An exception, for instance, is when the light rays get close to a black hole. We will not discuss such cases in the following.

With these simplifying assumptions we can describe the light propagation nearby the lens in a flat spacetime with a perturbation due to the gravitational potential of the lens described in a first-order post-Newtonian approximation. The effect of the spacetime curvature on the light trajectory can be described as an effective refraction index, given by

$$n = 1 - \frac{2}{c^2}\Phi = 1 + \frac{2}{c^2}|\Phi|. \tag{14.1}$$

The Newtonian potential is negative and vanishes asymptotically. As in geometrical optics a refraction index $n > 1$ means that the light travels with a speed which is lower compared with its speed in the vacuum. Thus the effective speed of light in a gravitational field is given by

$$v = \frac{c}{n} \simeq c - \frac{2}{c}|\Phi|. \tag{14.2}$$

Since the effective speed of light is less in a gravitational field, the travel time becomes longer compared to the propagation in empty space. The total time delay Δt is obtained by integrating along the light trajectory from the source until the observer, as follows

$$\Delta t = \int_{\text{source}}^{\text{observer}} \frac{2}{c^3}|\Phi|\, dl. \tag{14.3}$$

This is also called the Shapiro delay.

The deflection angle for the light rays which pass through a gravitational field is given by the integration of the gradient component of n perpendicular to the trajectory itself:

$$\boldsymbol{\alpha} = -\int \boldsymbol{\nabla}_\perp n\, dl = \frac{2}{c^2}\int \boldsymbol{\nabla}_\perp \Phi\, dl. \tag{14.4}$$

For all astrophysical applications of interest the deflection angle is always extremely small, so that the computation can be substantially simplified by integrating $\boldsymbol{\nabla}_\perp n$ along an unperturbed path, rather than the effective perturbed path. The so induced error is of higher order and thus negligible.

As an example let us consider the deflection angle of a point-like lens of mass M. Its Newtonian potential is given by

$$\Phi(b, z) = -\frac{GM}{(b^2 + z^2)^{1/2}}, \tag{14.5}$$

where b is the impact parameter of the unperturbed light ray and z denotes the position along the unperturbed path as measured from the point of minimal distance from the lens. This way we obtain

$$\boldsymbol{\nabla}_\perp \Phi(b, z) = \frac{GM\boldsymbol{b}}{(b^2 + z^2)^{3/2}}, \tag{14.6}$$

where $\boldsymbol{b}$ is orthogonal to the unperturbed light trajectory and is directed towards the point-like lens. Inserting equation (14.6) in equation (14.4) we find, for the the deflection angle,

$$\boldsymbol{\alpha} = \frac{2}{c^2}\int \boldsymbol{\nabla}_\perp \Phi\, dz = \frac{4GM}{c^2 b}\frac{\boldsymbol{b}}{b}. \tag{14.7}$$

The Schwarzschild radius for a body of mass M is given by

$$R_S = \frac{2GM}{c^2},$$ (14.8)

thus the absolute value of the deflection angle can also be written as $\alpha = 2R_S/b$. For the Sun the Schwarzschild radius is 2.95 km, whereas its physical radius is 6.96×10^5 km. Therefore, a light ray which just grazes the solar surface is deflected by an angle corresponding to $1.7''$.

14.2.2 Thin lens approximation

From these considerations one sees that the main contribution to the light deflection comes from the region $\Delta z \sim \pm b$ around the lens. Typically, Δz is much smaller than the distance between the observer and the lens and the lens and the source, respectively. The lens can thus be assumed to be thin compared to the full length of the light trajectory. Thus one considers the mass of the lens, for instance a galaxy cluster, projected onto a plane perpendicular to the line of sight (between the observer and the lens) and going through the centre of the lens. This plane is usually referred to as the lens plane and, similarly, one can define the source plane. The projection of the lens mass on the lens plane is obtained by integrating the mass density ρ along the direction perpendicular to the lens plane:

$$\Sigma(\boldsymbol{\xi}) = \int \rho(\boldsymbol{\xi}, z) \, dz,$$ (14.9)

where $\boldsymbol{\xi}$ is a two-dimensional vector in the lens plane and z is the distance from the plane. The deflection angle at the point $\boldsymbol{\xi}$ is then given by summing over the deflection due to all mass elements in the plane as follows.

$$\boldsymbol{\alpha} = \frac{4G}{c^2} \int \frac{(\boldsymbol{\xi} - \boldsymbol{\xi}')\Sigma(\boldsymbol{\xi}')}{|\boldsymbol{\xi} - \boldsymbol{\xi}'|^2} \, d^2\xi'.$$ (14.10)

In the general case the deflection angle is described by a two-dimensional vector. However, in the special case that the lens has circular symmetry one can reduce the problem to a one-dimensional situation. Then the deflection angle is a vector directed towards the centre of the symmetry with absolute value given by

$$\alpha = \frac{4GM(\xi)}{c^2\xi},$$ (14.11)

where ξ is the distance from the centre of the lens and $M(\xi)$ is the total mass inside a radius ξ from the centre, defined as

$$M(\xi) = 2\pi \int_0^\xi \Sigma(\xi')\xi' \, d\xi'.$$ (14.12)

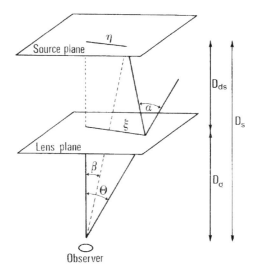

Figure 14.2. Notation for the lens geometry.

14.2.3   Lens equation

The geometry for a typical gravitational lens is given in figure 14.2 A light ray from a source S (in η) is deflected by the lens by an angle α (with impact parameter $|\xi|$) and reaches the observer located in O.

The angle between the optical axis (arbitrarily defined) and the true source position is given by β, whereas the angle between the optical axis and the image position is θ. The distances between the observer and the lens, the lens and the source, and the observer and the source are, respectively, D_d, D_{ds} and D_s. From figure 14.2 one can easily derive (assuming small angles) that $\theta D_s = \beta D_s + \alpha D_{ds}$. Thus the positions of the source and the image are related by the following equation:

$$\beta = \theta - \alpha(\theta)\frac{D_{ds}}{D_s},\qquad(14.13)$$

which is called the *lens equation*. It is a nonlinear equation so that it is possible to have several images θ corresponding to a single source position β.

The lens equation (14.13) can also be derived using the Fermat principle, which is identical to the classical one in geometrical optics but with the refraction index defined as in equation (14.1). The light trajectory is then given by the variational principle

$$\delta \int n \, dl = 0.\qquad(14.14)$$

It expresses the fact that the light trajectory will be such that the travelling time will be extremal. Let us consider a light ray emitted from the source S at time

$t = 0$. It will then proceed straight until it reaches the lens, located at the point I, and where it will be deflected and then proceed again straight to the observer in O. We thus have

$$t = \frac{1}{c} \int \left(1 - \frac{2\phi}{c^2} \right) dl = \frac{l}{c} - \frac{2}{c^3} \int \phi \, dl, \qquad (14.15)$$

where l is the distance SIO (Euclidean distance). The term containing ϕ has to be integrated along the light trajectory. From figure 2.1 we see that

$$l = \sqrt{(\xi - \eta)^2 + D_{ds}^2} + \sqrt{\xi^2 + D_d^2}$$
$$\simeq D_{ds} + D_d + \frac{1}{2D_{ds}} (\xi - \eta)^2 + \frac{1}{2D_d} \xi^2, \qquad (14.16)$$

where η is a two-dimensional vector in the source plane If we take $\phi = -GM/|x|$ (corresponding to a point-like lens of mass M) we get

$$\int_S^I \frac{2\phi}{c^3} \, dl = \frac{2GM}{c^3} \left[\ln \frac{|\xi|}{2D_{ds}} + \frac{\xi \cdot (\eta - \xi)}{|\xi| D_{ds}} + \mathcal{O}\left(\frac{(\eta - \xi)^2}{D_{ds}} \right) \right] \qquad (14.17)$$

and similarly for $\int_I^O 2\phi/c^3 \, dl$.

Only the logarithmic term is relevant for lensing, since the other ones are of higher order. Moreover, instead of a point-like lens we consider a surface mass density $\Sigma(\xi)$ (as defined in equation (14.9)) and so we obtain, for the integral containing the potential term (neglecting higher-order contributions)

$$\frac{2}{c^3} \int \phi \, dl = \frac{4G}{c^3} \int d^2\xi' \, \Sigma(\xi') \ln \frac{|\xi - \xi'|}{\xi_0}, \qquad (14.18)$$

where ξ_0 is a characteristic length in the lens plane and the right-hand side term is defined up to a constant.

The difference in the arrival time between the situation which takes into account the light deflection due to the lens and without the lens, is obtained by summing equation (14.16)–(14.18) and by subtracting the travel time without deflection from S to O. This way one obtains

$$c\Delta t = \hat{\phi}(\xi, \eta) + \text{constant}, \qquad (14.19)$$

where $\hat{\phi}$ is the *Fermat potential* defined as

$$\hat{\phi}(\xi, \eta) = \frac{D_d D_s}{2D_{ds}} \left(\frac{\xi}{D_d} - \frac{\eta}{D_s} \right)^2 - \hat{\psi}(\xi) \qquad (14.20)$$

and

$$\hat{\psi}(\xi) = \frac{4G}{c^2} \int d^2\xi' \, \Sigma(\xi') \ln \left(\frac{|\xi - \xi'|}{\xi_0} \right) \qquad (14.21)$$

is the *deflection potential*, which does not depend on η. The Fermat principle can thus be written as $d\Delta t/d\xi = 0$, and inserting equation (14.19) one once again obtains the lens equation

$$\eta = \frac{D_s}{D_d}\xi - D_{ds}\alpha(\xi), \qquad (14.22)$$

where α is defined in equation (14.10). (If we define $\beta = \eta/D_s$ and $\theta = \xi/D_d$ we obtain equation (14.13). One can also write equation (14.22) as follows.

$$\nabla_\xi \hat{\Phi}(\xi, \eta) = 0, \qquad (14.23)$$

which is an equivalent formulation of the Fermat principle.

The arrival time delay of light rays coming from two different images (due to the same source in η) located in $\xi^{(1)}$ and $\xi^{(2)}$ is given by

$$c(t_1 - t_2) = \hat{\Phi}(\xi^{(1)}, \eta) - \hat{\Phi}(\xi^{(2)}, \eta). \qquad (14.24)$$

14.2.4   Remarks on the lens equation

It is often convenient to write (14.22) in a dimensionless form. Let ξ_0 be a length parameter in the lens plane (whose choice will depend on the specific problem) and let $\eta_0 = (D_s/D_d)\xi_0$ be the corresponding length in the source plane. We set $x = \xi/\xi_0$, $y = \eta/\eta_0$ and

$$\kappa(x) = \frac{\Sigma(\xi_0 x)}{\Sigma_{cr}}, \qquad \alpha(x) = \frac{D_d D_{ds}}{\xi_0 D_s}\hat{\alpha}(\xi_0 x), \qquad (14.25)$$

where we have defined a critical surface mass density

$$\Sigma_{cr} = \frac{c^2}{4\pi G}\frac{D_s}{D_d D_{ds}} = 0.35 \text{ g cm}^{-2}\left(\frac{1 \text{ Gpc}}{D}\right) \qquad (14.26)$$

with $D \equiv \frac{D_d D_{ds}}{D_s}$ (1 Gpc $= 10^9$ pc). Then equation (14.22) reads as follows

$$y = x - \alpha(x), \qquad (14.27)$$

with

$$\alpha(x) = \frac{1}{\pi}\int_{R^2} \frac{x - x'}{|x - x'|^2}\kappa(x')\,d^2x'. \qquad (14.28)$$

In the following we will mainly use the previous notation rather than that in equation (14.28).

An interesting case is a lens with a constant surface mass density Σ. With equation (14.11) one then finds, for the deflection angle,

$$\alpha(\theta) = \frac{4G}{c^2\xi}\Sigma\pi\xi^2 = \frac{4\pi G\Sigma}{c^2}D_d\theta, \qquad (14.29)$$

using $\xi = D_{\rm d}\theta$. In this case the lens equation (14.13) is linear, which means that β is proportional to θ:

$$\beta = \theta - \beta = \theta - \frac{4\pi G}{c^2}\frac{D_{\rm ds}D_{\rm d}}{D_{\rm s}}\Sigma\theta = \theta - \frac{\Sigma}{\Sigma_{\rm cr}}\theta. \tag{14.30}$$

From equation (14.30) we immediately see that for a lens with a critical surface mass density we get for all values of θ: $\beta = 0$. Such a lens would perfectly focus, with a well-defined focal length. Typical gravitational lenses behave, however, quite differently. A lens which has $\Sigma > \Sigma_{\rm cr}$ somewhere in it is defined as *supercritical*, and has, in general, multiple images.

Defining $k(\theta) := \Sigma(\theta D_{\rm d})/\Sigma_{\rm cr}$ we can write the lens equation as

$$\beta = \theta - \tilde{\alpha}(\theta), \tag{14.31}$$

with

$$\tilde{\alpha}(\theta) = \frac{1}{\pi}\int_{R^2} d^2\theta' \, k(\theta')\frac{\theta - \theta'}{|\theta - \theta'|^2}. \tag{14.32}$$

Moreover,

$$\tilde{\alpha}(\theta) = \nabla_\theta \Psi(\theta) \tag{14.33}$$

where

$$\Psi(\theta) = \frac{1}{\pi}\int_{R^2} d^2\theta' \, k(\theta') \ln|\theta - \theta'|. \tag{14.34}$$

The Fermat potential is given by

$$\Phi(\theta, \beta) = \tfrac{1}{2}(\theta - \beta)^2 - \Psi(\theta) \tag{14.35}$$

and we then obtain the lens equation from

$$\nabla_\theta \Phi(\theta, \beta) = 0. \tag{14.36}$$

Note that

$$\Delta\Psi = 2k \geq 0 \tag{14.37}$$

(using $\Delta \ln|\theta| = 2\pi\delta^2(\theta)$), since k as a surface mass density is always positive (or vanishes).

The flux of a source, located in β, in the solid angle $d\Omega(\beta)$ is given by

$$S(\beta) = I_\nu \, d\Omega(\beta). \tag{14.38}$$

I_ν is the intensity of the source in the frequency ν. $S(\beta)$ is the flux one would see if there were no lensing. However, the observed flux from the image located in θ is

$$S(\theta) = I_\nu \, d\Omega(\theta). \tag{14.39}$$

I_ν does not change, since the total number of photons stays constant as well as their frequency. The amplification factor μ is thus given by the ratio

$$\mu = \frac{d\Omega(\boldsymbol{\theta})}{d\Omega(\boldsymbol{\beta})} = \frac{1}{\det A(\boldsymbol{\theta})}, \tag{14.40}$$

with

$$A(\boldsymbol{\theta}) = \frac{d\boldsymbol{\beta}}{d\boldsymbol{\theta}} \left(A_{ij} = \frac{d\beta_i}{d\theta_j} = \delta_{ij} - \Psi_{,ij} \right), \tag{14.41}$$

(where $\Psi_{,ij} = \partial_i \partial_j \Psi$) which is the Jacobi matrix of the corresponding lens mapping given by equation (14.31). Notice that the amplification factor μ can be positive or negative. The corresponding image will then have *positive or negative parity*, respectively.

For some values of $\boldsymbol{\theta}$, $\det A(\boldsymbol{\theta}) = 0$ and thus $\mu \to \infty$. The points (or the curve) $\boldsymbol{\theta}$ in the lens plane for which $\det A(\boldsymbol{\theta}) = 0$ are defined as *critical points (or critical curve)*. At these points the geometrical optics approximation used so far breaks down. The corresponding points (or curve) of the critical points in the source plane are the so called *caustics*.

The matrix A_{ij} is often parametrized as follows.

$$A_{ij} = \begin{pmatrix} 1 - k - \gamma_1 & -\gamma_2 \\ -\gamma_2 & 1 - k + \gamma_1 \end{pmatrix} \tag{14.42}$$

with $\gamma_1 = \frac{1}{2}(\Psi_{,11} - \Psi_{,22})$, $\gamma_2 = \Psi_{,12} = \Psi_{,21}$ and $\boldsymbol{\gamma} = (\gamma_1, \gamma_2)$. We have therefore

$$\det A_{ij} = (1 - k)^2 - \gamma^2 \tag{14.43}$$

and $\gamma = \sqrt{\gamma_1^2 + \gamma_2^2}$,

$$\operatorname{tr} A_{ij} = 2(1 - k). \tag{14.44}$$

The eigenvalues of A_{ij} are $a_{1,2} = 1 - k \pm \gamma$.

In the next paragraphs, we study how small circles in the source plane are deformed. Consider a small circular source with radius R at $\boldsymbol{y}$, bounded by a curve described by

$$c(t) = \boldsymbol{y} + \begin{pmatrix} R \cos t \\ R \sin t \end{pmatrix} \qquad (0 \leq t \leq 2\pi). \tag{14.45}$$

The corresponding boundary curve of the image is

$$d(t) = \boldsymbol{x} + A^{-1} \begin{pmatrix} R \cos t \\ R \sin t \end{pmatrix}. \tag{14.46}$$

Inserting the parametrization (14.42) one finds that the image is an ellipse centred on $\boldsymbol{x}$ with semi-axes parallel to the main axes of A, with magnitudes

$$\frac{R}{|1 - \kappa \pm \gamma|}, \tag{14.47}$$

and the position angles $\varphi_\pm$ for the axes are

$$\tan \varphi_\pm = \frac{\gamma_1}{\gamma_2} \mp \sqrt{\left(\frac{\gamma_1}{\gamma_2}\right)^2 + 1} \quad \text{or} \quad \tan 2\varphi_\pm = -\frac{\gamma_2}{\gamma_1}. \tag{14.48}$$

The ellipticity of the image is defined as follows.

$$\epsilon = \epsilon_1 + i\epsilon_2 = \frac{1-r}{1+r} e^{2i\varphi}, \quad r \equiv \frac{b}{a}, \tag{14.49}$$

where φ is the position angle of the ellipse and a and b are the major and minor semi-axes, respectively. a and b are given by the inverse of the eigenvalues of the matrix A_{ij} defined in equation (14.42), thus $a = (1 - k - \gamma)^{-1}$ and $b = (1 - k + \gamma)^{-1}$. ϵ describes the orientation and the shape of the ellipse and is thus observable. Let us denote $g = |\epsilon|$ with

$$g = \frac{\gamma}{1-\kappa} \quad \left(g = \frac{\gamma}{1-\kappa}\right), \tag{14.50}$$

which is called the *reduced shear*. One often uses a complex notation with $\gamma = \gamma_1 + i\gamma_2$ and then accordingly one defines a complex reduced shear.

14.2.4.1 *Classification ordinary images*

If we consider a fixed value for β, then $\Phi(\theta, \beta)$ defines a (two-dimensional) surface for the arrival time of the light. Ordinary images, for which $\det A(\theta) \neq 0$, are formed at the points θ, where $\nabla_\theta \Phi(\theta, \beta) = 0$. Thus the images are localized at extremal or saddle points of the surface $\Phi(\theta, \beta)$ and are classified as follows.

- *Images of type I*: These correspond to minima of Φ, with $\det A > 0$, $\operatorname{tr} A > 0$ (and thus $\gamma < 1 - k \leq 1$, $a_i > 0$, $\mu \geq \frac{1}{1-\gamma^2} \geq 1$).
- *Images of type II*: These correspond to saddle points of Φ, with $\det A < 0$ (then $(1 - k)^2 < \gamma^2$, $a_2 > 0 > a_1$).
- *Images of type III*: These correspond to maxima of Φ, with $\det A > 0$, $\operatorname{tr} A < 0$ (with $(1 - k)^2 > \gamma^2$, $k > 1$, $a_i < 0$).

Consider a thin lens with a smooth surface mass density $k(\theta)$, which decreases faster than $|\theta|^{-2}$ for $|\theta| \to \infty$. For such a lens the total mass is finite and the deflection angle $\alpha(\theta)$ is continuous and tends to zero for $|\theta| \to \infty$, therefore α is bounded: $|\alpha| \leq \alpha_0$. Moreover, let us denote by n_I the number of images of type I for a source located in β, similarly for n_II and n_III and define $n_\text{tot} = n_\text{I} + n_\text{II} + n_\text{III}$. If these conditions are fulfilled then the following theorems hold.

Theorem 14.1. *If the previous conditions hold and β is not situated on a caustic, the following conditions apply:*

(a) $n_I \geq 1$
(b) $n_{tot} < \infty$
(c) $n_I + n_{III} = 1 + n_{II}$
(d) for $|\boldsymbol{\beta}|$ sufficiently large $n_{tot} = n_I = 1$.

It thus follows from (c) that the total number of images $n_{tot} = 1 + 2n_{II}$ is odd. The number of images with positive parity ($n_I + n_{III}$) exceeds by one those with negative parity (n_{II}); $n_{II} \geq n_{III}$ and $n_{tot} > 1$ if and only if $n_{II} \geq 1$. The number of images is odd; however, in practice some images may be very faint or be covered by the lens itself and are thus not observable.

Theorem 14.2. *The image of the source which will appear first to the observer is of type I and it is at least as bright as the unlensed source would appear ($\mu(\boldsymbol{\theta}_1) \geq 1$).*

For a proof of the two theorems we refer to [4]. The second theorem is a consequence of the fact that the surface mass density k is a positive quantity.

14.3   Simple lens models

14.3.1   Axially symmetric lenses

Let us consider a lens with an axially symmetric surface mass density, that is $\Sigma(\boldsymbol{\xi}) = \Sigma(|\boldsymbol{\xi}|)$, in which case the lens equation reduces to a one-dimensional equation. By symmetry we can restrict the impact vector $\boldsymbol{\theta}$ to be on the positive θ_1-axis, thus we have $\boldsymbol{\theta} = (\theta, 0)$ with $\theta > 0$. We can then use polar coordinates: $\boldsymbol{\theta}' = \theta'(\cos\phi, \sin\phi)$ (thus $d^2\theta' = \theta' d\theta' d\phi$). With $k(\boldsymbol{\theta}) = k(\theta)$ we get for equation (14.32)

$$\alpha_1(\theta) = \frac{1}{\pi} \int_0^\infty \theta' \, d\theta' \, k(\theta') \int_0^{2\pi} d\phi \, \frac{\theta - \theta' \cos\phi}{\theta^2 + \theta'^2 - 2\theta\theta' \cos\phi}, \quad (14.51)$$

$$\alpha_2(\theta) = \frac{1}{\pi} \int_0^\infty \theta' \, d\theta' \, k(\theta') \int_0^{2\pi} d\phi \, \frac{-\theta' \sin\phi}{\theta^2 + \theta'^2 - 2\theta\theta' \cos\phi}. \quad (14.52)$$

Due to symmetry, $\boldsymbol{\alpha}$ is parallel to $\boldsymbol{\theta}$ and with equation (14.52) we get $\alpha_2(\theta) = 0$. Only the mass inside the disc of radius θ around the centre of the lens contributes to the light deflection, therefore from equation (14.51) one finds

$$\alpha(\theta) \equiv \alpha_1(\theta) = \frac{2}{\theta} \int_0^\theta \theta' \, d\theta' k(\theta') \equiv \frac{m(\theta)}{\theta}. \quad (14.53)$$

This way we can write the lens equation as

$$\beta = \theta - \alpha(\theta) = \theta - \frac{m(\theta)}{\theta} \quad (14.54)$$

for $\theta \geq 0$. Due to the axial symmetry it is enough to consider $\beta \geq 0$. Since $m(\theta) \geq 0$ it follows that $\theta \geq \beta$ (for $\theta \geq 0$). Instead of equation (14.34) we get

$$\Psi(\theta) = 2 \int_0^\theta \theta' \, d\theta' \, k(\theta') \ln\left(\frac{\theta}{\theta'}\right),$$ (14.55)

whereas the Fermat potential can be written as

$$\Phi(\theta, \beta) = \tfrac{1}{2}(\theta - \beta)^2 - \Psi(\theta).$$ (14.56)

This way we get the lens equation (14.54) from

$$\frac{\partial \Phi(\theta, \beta)}{\partial \theta} = 0.$$ (14.57)

To get the Jacobi matrix we write:

$$\boldsymbol{\alpha}(\boldsymbol{\theta}) = \frac{m(\theta)}{\theta^2} \boldsymbol{\theta} \qquad (\text{with } \boldsymbol{\theta} = (\theta_1, \theta_2) \text{ and } \theta = |\boldsymbol{\theta}|)$$

and thus

$$A = \begin{pmatrix} 1 & 0 \\ 0 & 1 \end{pmatrix} - \frac{m(\theta)}{\theta^4} \begin{pmatrix} \theta_2^2 - \theta_1^2 & -2\theta_1\theta_2 \\ -2\theta_1\theta_2 & \theta_1^2 - \theta_2^2 \end{pmatrix} - \frac{2k(\theta)}{\theta^2} \begin{pmatrix} \theta_1^2 & \theta_1\theta_2 \\ \theta_1\theta_2 & \theta_2^2 \end{pmatrix}, \quad (14.58)$$

where we made use of $m'(\theta) = 2\theta k(\theta)$. The determinant of the Jacobi matrix is given by

$$\det A = \left(1 - \frac{m}{\theta^2}\right)\left(1 - \frac{d}{d\theta}\left(\frac{m}{\theta}\right)\right) = \left(1 - \frac{m}{\theta^2}\right)\left(1 + \frac{m}{\theta^2} - 2k\right). \quad (14.59)$$

14.3.1.1   Tangential and radial critical curves

The critical curves (the points for which $\det A(\theta) = 0$) are then circles of radius θ. From equation (14.59) we see that there are two possible cases:

(1) $\frac{m}{\theta^2} = 1$: defined as *tangential critical curve*; and
(2) $\frac{d}{d\theta}\left(\frac{m}{\theta}\right) = 1$: defined as *radial critical curve*.

For case (1) one gets $m/\theta = \theta$ and thus from the lens equation (14.54) we see that $\beta = 0$ is the corresponding caustic, which reduces to a point. If the axial symmetric gets only slightly perturbed this degeneracy is lifted.

We can look at the critical points on the θ_1-axis with $\boldsymbol{\theta} = (\theta, 0)$, $\theta > 0$. Then

$$A = 1 - \frac{m(\theta)}{\theta^2} \begin{pmatrix} -1 & 0 \\ 0 & +1 \end{pmatrix} - \frac{m'}{\theta} \begin{pmatrix} 1 & 0 \\ 0 & 0 \end{pmatrix} \quad (14.60)$$

and this matrix must have an eigenvector X with eigenvalue zero. For symmetry reasons, the vector must be either tangential, $X = (0, 1)$, or normal, $X = (1, 0)$, to the critical curve (which must be a circle). We see readily that the first case occurs for a tangential critical curve, and the second for a radial critical curve. The image of a circle (in the source plane) which lies close to a tangential critical curve will be deformed to an ellipse with major axis tangential to the critical curve. However, if the image of a circle gets close to a radial critical curve it will be deformed to an ellipse with major axis radial to the critical curve.

For a tangential critical curve ($|\boldsymbol{\theta}| = \theta_t$) we get

$$m(\theta_t) = \int_0^{\theta_t} 2\theta \kappa(\theta) \, d\theta = \theta_t^2. \tag{14.61}$$

With the definition of κ this translates to

$$\int_0^{\xi_t} 2\xi \, \Sigma(\xi) \, d\xi = \xi_t^2 \Sigma_{cr}. \tag{14.62}$$

The total mass $M(\xi_t)$ inside the critical curve is thus

$$M(\xi_t) = \pi \xi_t^2 \Sigma_{cr}. \tag{14.63}$$

This shows that the average density $\langle \Sigma \rangle_t$ inside the tangential critical curve is equal to the critical density Σ_{cr}. This can be used to estimate the mass of a deflector if the lens is sufficiently strong and the geometry is such that almost complete Einstein rings are formed.

14.3.1.2 Einstein radius

For a lens with axial symmetry we get, with (14.11), the following equation:

$$\beta(\theta) = \theta - \frac{D_{ds}}{D_s D_d} \frac{4GM(\theta)}{c^2 \theta}, \tag{14.64}$$

from which we see that the image of a source, which is perfectly aligned (that means $\beta = 0$), is a ring if the lens is supercritical. By setting $\beta = 0$ in equation (14.64) we get the radius of the ring

$$\theta_E = \left(\frac{4GM(\theta_E)}{c^2} \frac{D_{ds}}{D_d D_s} \right)^{1/2}, \tag{14.65}$$

which is called *Einstein radius*. The Einstein radius depends not only on the characteristics of the lens but also on the various distances.

The Einstein radius sets a natural scale for the angles entering the description of the lens. Indeed, for multiple images the typical angular separation between the different images turns out to be of order $2\theta_E$. Moreover, sources with angular distances smaller than θ_E from the optical axis of the system are magnified quite substantially whereas sources which are at a distance much greater than θ_E are only weakly magnified.

In several lens models the Einstein radius delimits the region within which multiple images occur, whereas outside this region there is a single image. By comparing equation (14.26) with equation (14.65) we see that the surface mass density inside the Einstein radius precisely corresponds to the critical density. For a point-like lens with mass M the Einstein radius is given by

$$\theta_E = \left(\frac{4GM}{c^2} \frac{D_{ds}}{D_d D_s} \right)^{1/2},$$ (14.66)

or instead of an angle one often also uses

$$R_E = \theta_E D_d = \left(\frac{4GM}{c^2} \frac{D_{ds} D_d}{D_s} \right)^{1/2}.$$ (14.67)

To get some typical values we can consider the following two cases: a lens of mass M located in the galactic halo at a distance of $D_d \sim 10$ kpc and a source in the Magellanic Cloud, in which case

$$\theta_E = (0.9'' \times 10^{-3}) \left(\frac{M}{M_\odot} \right)^{1/2} \left(\frac{D}{10 \text{ kpc}} \right)^{-1/2}$$ (14.68)

and a lens with the mass of galaxy (including its halo) $M \sim 10^{12} M_\odot$ located at a distance of $D_d \sim 1$ Gpc

$$\theta_E = 0.9'' \left(\frac{M}{10^{12} M_\odot} \right)^{1/2} \left(\frac{D}{\text{Gpc}} \right)^{-1/2},$$ (14.69)

where $D = D_d D_s / D_{ds}$.

14.3.2 Schwarzschild lens

A particular case of a lens with axial symmetry is the Schwarzschild lens, for which $\Sigma(\xi) = M\delta^2(\xi)$ and thus $m(\theta) = \theta_E^2$. The source is also considered as point-like, this way we get, for lens equation (14.13), the following expression

$$\beta = \theta - \frac{\theta_E^2}{\theta},$$ (14.70)

where θ_E is given by equation (14.66). This equation has two solutions:

$$\theta_\pm = \frac{1}{2} \left(\beta \pm \sqrt{\beta^2 + 4\theta_E^2} \right).$$ (14.71)

Therefore, there will be two images of the source located one inside the Einstein radius and the other outside. For a lens with axial symmetry the amplification is given by

$$\mu = \frac{\theta}{\beta} \frac{d\theta}{d\beta}.$$ (14.72)

For the Schwarzschild lens, which is a limiting case of an axial symmetric one, we can substitute β using equation (14.71) and obtain this way the amplification for the two images

$$\mu_{\pm} = \left[1 - \left(\frac{\theta_E}{\theta_{\pm}}\right)^4\right]^{-1} = \frac{u^2 + 2}{2u\sqrt{u^2 + 4}} \pm \frac{1}{2}. \tag{14.73}$$

$u = r/R_E$ is the ratio between the impact parameter r, that is the distance between the lens and the line of sight connecting the observer and the source and the Einstein radius R_E defined in equation (14.67). u can also be expressed as β/θ_E. Since $\theta_- < \theta_E$ we have that $\mu_- < 0$. The negative sign for the amplification indicates that the parity of the image is inverted with respect to the source. The total amplification is given by the sum of the absolute values of the amplifications for each image

$$\mu = |\mu_+| + |\mu_-| = \frac{u^2 + 2}{u\sqrt{u^2 + 4}}. \tag{14.74}$$

If $r = R_E$ then we get $u = 1$ and $\mu = 1.34$, which corresponds to an increase of the apparent magnitude of the source of $\Delta m = -2.5 \log \mu = -0.32$. For lenses with a mass of the order of a solar mass and which are located in the halo of our galaxy the angular separation between the two images is far too small to be observable. Instead, one observes a time-dependent change in the brightness of the the the source star. This situation is also referred to as *microlensing*.

Much research activity is devoted to studying microlensing in the context of quasar lensing. Today, several cases of quasars which are lensed by foreground galaxies, producing multiple observable images are known. The stars contained in the lensing galaxy can act as microlenses on the quasar and, as a result, induce time-dependent changes in the quasar brightness, but in a rather complicated way, since here the magnification is a coherent effect of many stars at the same time. This is an interesting field of research, which will lead to important results on the problem of the dark matter in galaxies [15]. However, we will not discuss *extragalactic* microlensing in detail (see, for instance, [16]), whereas we will report in some depth on *galactic* microlensing (see section 14.4).

The time delay between the two images of a Schwarzschild lens is given by

$$c\Delta t = \frac{4GM}{c^2}\left(\frac{1}{2}u\sqrt{u^2 + 4} + \ln\frac{\sqrt{u^2 + 4} + u}{\sqrt{u^2 + 4} - u}\right). \tag{14.75}$$

The two images have a comparable luminosity only if $u \leq 1$ (otherwise the difference is such that one image is no longer observable since it gets too faint). For $u = 1$ one obtains $\Delta t \sim 4R_S/c$ (typically for a galaxy with mass $M = 10^{12} M_\odot$ one finds $\Delta t \sim 1.3$ years). Such measurements are important since they allow to determine the value H_0 of the Hubble constant (see section 14.5.1).

14.3.3 Singular isothermal sphere

A simple model for describing the matter distribution in a galaxy is to assume that the stars forming the galaxy behave like the particles in an ideal gas, confined by the total gravitational potential, which we assume to have spherical shape. The equation of state of the 'particles' (stars) has the form

$$p = \frac{\rho k_B T}{m}, \qquad (14.76)$$

where ρ and m are the matter density and the mass of a star, respectively. In the equilibrium case the temperature T is defined via the one-dimensional dispersion velocity σ_v of the stars as obtained from

$$m\sigma_v^2 = k_B T. \qquad (14.77)$$

In principle the temperature could depend on the radius; however, in the simplest model, of the isothermal spherical model, one assumes that the temperature is constant and hence also σ_v. The equation for hydrostatic equilibrium is given by

$$\frac{p'}{\rho} = -\frac{GM(r)}{r^2}, \qquad (14.78)$$

with

$$M'(r) = 4\pi r^2 \rho, \qquad (14.79)$$

where $M(r)$ is the mass inside the sphere of radius r. A solution of the previous equations is

$$\rho(r) = \frac{\sigma_v^2}{2\pi G} \frac{1}{r^2}. \qquad (14.80)$$

This mass distribution is called *singular isothermal sphere* (it is indeed singular for $r \to 0$). Since $\rho(r) \sim r^2$, $M(r) \sim r$, the velocity of the stars in the gravitational field of an isothermal sphere is given by

$$v_{rot}^2(r) = \frac{GM(r)}{r} = 2\sigma_v^2, \qquad (14.81)$$

which is constant. Such a mass distribution can (at least in a qualitative way) describe the flat rotation curves of the galaxies, as measured beyond a certain galactic radius. Thus the dark matter in the halo can, in a first approximation, be described by a singular isothermal sphere model.

The projected mass density on the lens plane perpendicular to the line of sight is:

$$\Sigma(\xi) = \frac{\sigma_v^2}{2G} \frac{1}{\xi}, \qquad (14.82)$$

where ξ is the distance (in the lens plane) from the the centre of mass. For the light deflection angle we get

$$\hat{\alpha} = 4\pi \frac{\sigma_v^2}{c^2} = 1.4'' \left(\frac{\sigma_v}{220 \text{ km s}^{-1}} \right)^2 \tag{14.83}$$

independent of the position ξ (220 km s^{-1} is a typical value for the rotation velocity in spiral galaxies).

The Einstein radius R_E is given by

$$R_E = 4\pi \frac{\sigma_v^2}{c^2} \frac{D_{ds} D_d}{D_s} = \hat{\alpha} \frac{D_{ds} D_d}{D_s} = \alpha D_d. \tag{14.84}$$

Multiple images occur only if the source is located within the Einstein radius. Let be $\xi_0 = R_E$, then $\Sigma(\xi) = \Sigma(x\xi_0)$ where $x = \xi/\xi_0$. This way the lens equation becomes

$$y = x - \frac{x}{|x|}. \tag{14.85}$$

For $0 < y < 1$ we have two solutions: $x = y + 1$ and $x = y - 1$. For $y > 1$ (the source is located outside the Einstein radius) there is only one image: $x = y + 1$. The images with $x > 0$ are of type I, whereas the ones with $x < 0$ are of type II. If the singularity in $\xi = 0$ is removed then there will be a third image in the centre.

The amplification of an image in x is given by

$$\mu = \frac{|x|}{|x| - 1} \tag{14.86}$$

(the circle $|x| = 1$ corresponds to a tangential critical curve). For $y \to 1$ the second image (corresponding to the solution $x = y - 1$) becomes very faint.

The potential is given by $\psi(x) = |x|$ and the time delay between the images is

$$c\Delta t = \left(4\pi \left(\frac{\sigma_v}{c} \right)^2 \right)^2 \frac{D_d D_{ds}}{D_s} 2y. \tag{14.87}$$

14.3.4   Generalization of the singular isothermal sphere

The singular isothermal sphere model can, for instance, be generalized by adopting for the projected mass density Σ the following expression

$$\Sigma(\xi) = \Sigma_0 \frac{1 + p(\xi/\xi_c)^2}{(1 + (\xi/\xi_c)^2)^{2-p}}, \tag{14.88}$$

with $0 \le p \le 1/2$ and Σ_0 is the central density. ξ_c is a typical distance of the order of the scale on which the matter decreases, often one can take it as the core radius of a galaxy. $p = 0$ corresponds to the Plummer distribution, whereas for $p = 1/2$ we get the isothermal sphere for large values of ξ.

Defining $x = \xi/\xi_0$ and $k_0 = \Sigma_0/\Sigma_{cr}$ we can write equation (14.88) as

$$k(x) = k_0 \frac{1 + px^2}{(1 + x^2)^{2-p}}. \tag{14.89}$$

The deflection potential is given by

$$\Psi(x) = \frac{k_0}{2p}[(1 + x^2)^p - 1], \tag{14.90}$$

which is valid for $p \neq 0$, whereas for $p = 0$ we get

$$\Psi(x) = \frac{k_0}{2} \ln(1 + x^2). \tag{14.91}$$

Thus the lens equation is

$$y = x - \alpha(x) = x - \frac{k_0 x}{(1 + x^2)^{1-p}}. \tag{14.92}$$

If $k_0 > 1$ there is one tangential critical curve for $x = x_t$, where $x_t = \sqrt{k_0^{1/1-p} - 1}$, and a radial critical curve for $x = x_r$, which is defined by the equation

$$1 - k_0(1 + x_r^2)^{p-2}[1 + (2p - 1)x_r^2] = 0. \tag{14.93}$$

The corresponding caustics are given by $y_t \equiv y(x_t) = 0$, whereas

$$y_r \equiv |y(x_r)| = \frac{2(1 - p)x_r^3}{1 - (1 - 2p)x_r^2}. \tag{14.94}$$

Sources with $|y| < y_r$ lead to the formation of three images, whereas for $|y| > y_r$ there is only one image. The three images are at: $x > x_t$ (image of type I), $-x_t < x < r_r$ (image of type II) and $-x_r < x < 0$ (image of type III).

14.3.5 Extended source

The magnification for an extended source with surface brightness profile $I(y)$ is given by

$$\mu_e = \frac{\int I(y)\mu_p(y)\,d^2y}{\int I(y)\,d^2y}, \tag{14.95}$$

where $\mu_p(y)$ is the magnification of a point source at position y. As an example let us consider a disk-like source with radius R centred in y with a brightness profile $I(r/R)$, where r is the distance of a source point from the centre of the source. Adopting polar coordinates centred on the circular source, we obtain

$$
\mu_e(y) = \left[2\pi \int_0^\infty I(r/R) r \, dr \right]^{-1} \int_0^\infty I(r/R) r \, dr
$$
$$
\times \int_0^{2\pi} d\phi \, \frac{\mu_p(y) y}{\sqrt{y^2 + r^2 + 2ry \cos \phi}}. \tag{14.96}
$$

For a uniform brightness profile the maximum of μ_e is at $y = 0$ (with $\mu_e(0) = 2/R$ if μ_p is the magnification of a point source, since then $\mu_p(y)y \to 0$ for $y \to 0$). Indeed, for a Schwarzschild lens with $\mu_p = (y^2 + 2)/(y\sqrt{y^2 + 4})$ one finds

$$
\mu_e^{max} = \frac{\sqrt{4 + R^2}}{R}. \tag{14.97}
$$

14.3.6 Two point-mass lens

A natural generalization of the Schwarzschild lens is to consider a lens with two point masses. This case is also of relevance for the applications, since many binary microlensing events have been observed. For several point masses M_i located at transversal positions ξ_i the general formula equation (14.10) for the deflection angle gives

$$
\alpha(\xi) = \sum_{i=1}^N \frac{4GM_i}{c^2} \frac{\xi - \xi_i}{|\xi - \xi_i|^2}. \tag{14.98}
$$

Let $M = \sum_i^N M_i$ be the total mass and $M_i = \eta_i M$. For the typical length scale ξ_0 we choose the Einstein radius equation (14.67) for the total mass. Then the lens map becomes

$$
y = x - \sum_{1=1}^N \frac{\eta_i}{|x - x_i|^2} (x - x_i), \tag{14.99}
$$

where $x_i = \xi_i/\xi_0$. For a detailed discussion see [4].

14.4 Galactic microlensing

14.4.1 Introduction

There are cases in which the deflection angles are tiny, of the order of milliarcseconds or smaller, such that the multiple images are not observable. However, lensing magnifies the affected source, and since the lens and the source are moving relative to each other, this can be detected as a time-variable brightness. This behaviour is referred to as gravitational microlensing, a powerful method to search for dark matter in the halo of our own galaxy, if it consists of massive astrophysical compact halo objects (MACHOs), and to study the content of low-mass stars in the galactic disk.

The idea to use gravitational light deflection to detect MACHOs in the halo of our galaxy by monitoring the light variability of millions of stars in the Large Magellanic Cloud (LMC) was first proposed by Paczyński in 1986 [17] and then further developed—from a theoretical point of view—in a series of papers by De Rújula *et al* [18, 19], Griest [20] and Nemiroff [21]. Following these first studies, the field has grown very rapidly, especially since the discovery of the first microlensing events at the end of 1993 and many new applications have been suggested, including the detection of Earth-like planets around stars in our galaxy. (For reviews on microlensing see, for instance, [22–25].)

Since the discovery of the first microlensing events in September 1993 by monitoring millions of stars in the Large Magellanic Cloud (LMC) and in the direction of the galactic centre, several hundreds of events have been found. The still few observed events towards the LMC indicate that the halo dark matter fraction in the form of MACHOs is of the order of 20%, assuming a standard spherical halo model.

The best evidence for dark matter in galaxies comes from the observed rotation curves in spiral galaxies. Measurements of the rotation velocity v_{rot} of stars up to the visible edge of the spiral galaxies (of about 10 kpc) and of atomic hydrogen gas in the disk beyond the optical radius (by measuring the Doppler shift in the characteristic 21-cm radio line emitted by neutral hydrogen gas) imply that v_{rot} remains constant out to very large distances, rather than showing a Keplerian fall-off, as expected if there is no more matter beyond the visible edge.

There are also measurements of the rotation velocity for our own galaxy. However, these observations turn out to be rather difficult, and the rotation curve has been measured accurately only up to a distance of about 20 kpc. Without any doubt, our own galaxy has a typical flat rotation curve and thus it is possible to search directly for dark matter characteristic of spiral galaxies in the Milky Way.

The question which naturally arises is the nature of dark matter in galactic halos. A possibility is that the dark matter is comprised of baryons, which have been processed into compact objects (MACHOs), such as stellar remnants (for a detailed discussion see [26]). If their mass is below $\sim 0.08 M_\odot$, they are too light to ignite hydrogen-burning reactions. Otherwise, MACHOs might be either low-mass (~ 0.1–$0.3 M_\odot$) hydrogen burning stars (also called M-dwarfs) or white dwarfs. As a matter of fact, a deeper analysis makes the M-dwarf option look problematic. The null result of several searches for low-mass stars both in the disk and in the halo of our galaxy suggests that the halo cannot be mainly in the form of hydrogen-burning main-sequence M-dwarfs. Optical imaging of high-latitude fields taken with the Wide Field Planetary Camera of the Hubble Space Telescope indicates that less than $\sim 6\%$ of the halo can be in this form [27]. However, this result is derived under the assumption of a smooth spatial distribution of M-dwarfs, and the problem becomes considerably less severe in the case of a clumpy distribution [28]. Recent observations of four nearby spiral galaxies carried out with the satellite Infrared Space Observatory (ISO) also seem to exclude M-dwarfs as significantly contributing to halo dark matter [29].

A scenario with white dwarfs as a major constituent of the galactic halo dark matter has been explored [30]. However, it requires a rather *ad hoc* initial mass function sharply peaked around 2–6$M_\odot$. Future Hubble Deep Field exposures could either find the white dwarfs or put constraints on their fraction in the halo [31]. A substantial component of neutron stars and black holes with masses higher than $\sim 1 M_\odot$ is also excluded, for otherwise they would lead to an overproduction of heavy elements relative to the observed abundances.

A further possibility is that the hydrogen gas is in molecular form, clumped into very cold clouds, as we proposed some years ago [32, 33]. Indeed, the observation of such clouds is very difficult and, therefore, at present there are no stringent limits on their contribution to the halo dark matter [34].

14.4.1.1   Microlensing probability

When a MACHO of mass M is sufficiently close to the line of sight between us and a more distant star, the light from the source star suffers a gravitational deflection and we see two images of the source (figure 14.3). For most applications we can consider the lens and the source as point-like and thus use the Schwarzschild lens approximation previously discussed. R_E is then defined in equation (14.67).

For a cosmological situation, where the lens is a galaxy or even a cluster of galaxies and the source is a very distant quasar, one indeed sees two or more images which are typically separated by an angle of some arcseconds. However, in the situation being considered here, namely of a MACHO of typically $\sim 0.1 M_\odot$ and a source star located in the LMC at about 50 kpc from us, the separation angle turns out to be of the order of some milli-arcseconds. Thus, the images cannot be seen separately. However, the measured brightness of the source star varies with time. It increases until the MACHO reaches the shortest distance from the line of sight between the observer on Earth and the source star. Afterwards, the brightness decreases and eventually returns to its usual unlensed value. The magnification of the original star brightness turns out to be typically of the order of 30% or even more, corresponding to an increase of at least 0.3 magnitudes of the source star (see figures 14.4 and 14.5). Such an increase is easily observable.

An important quantity is the optical depth τ due to gravitational microlensing, which is the probability that a source is found within a circle of radius $r \le R_E$ around a MACHO. It is defined as follows

$$\tau = \int_0^1 \mathrm{d}x \, \frac{4\pi G}{c^2} \rho(x) D_s^2 x (1-x) \qquad (14.100)$$

with $\rho(x)$ being the mass density along the line of sight at distance $s = x D_s$ from the observer.

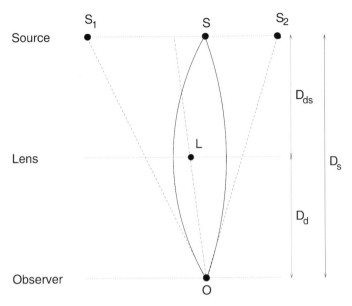

Figure 14.3. The set-up of a gravitational lens situation: The lens L located between source S and observer O produces two images S_1 and S_2 of the background source. D_d is the distance between the observer and the lens, D_s between the observer and the source and D_{ds} between the lens and the source.

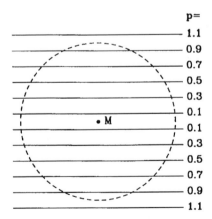

Figure 14.4. Einstein ring (broken curve) and some possible relative orbits of a background star with projected minimal distances $p = r/R_E = 0.1, 0.3, \ldots, 1.1$ from a MACHO M (from [22]).

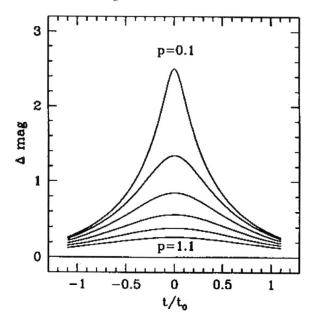

Figure 14.5. Light curves for the different cases of figure 4.2. The maximal magnification is $\Delta m = 0.32$ mag, if the star just touches the Einstein radius ($p = 1.0$). For smaller values of p the maximum magnification gets larger. t is the time in units of t_0 (from [22]).

We can easily compute τ assuming that the mass distribution in the galactic halo is of the following form

$$\rho(r) = \frac{\rho_0(a^2 + R_{GC}^2)}{a^2 + r^2}, \tag{14.101}$$

which is consistent with a flat rotation curve. $|r|$ is the distance from Earth, a is the core radius, ρ_0 the local density nearby the solar system of dark matter and R_{GC} the distance to the galactic centre. Standard values for these parameters are: $\rho_0 = 0.3 \text{ GeV cm}^{-3} = 7.9 \times 10^{-3} M_\odot \text{ pc}^{-3}$, $a = 5.6$ kpc and $R_{GC} = 8.5$ kpc.

Assuming a spherical halo made entirely of MACHOs, one finds an optical depth towards the LMC of $\tau = 5 \times 10^{-7}$. This means that at any one moment out of 2 million stars, one is being lensed. From this number it can be seen that in order to obtain a reasonable number of microlensing events, an experiment has to monitor several million stars in the LMC or in other targets such as the galactic centre region (also referred to as the galactic bulge).

The magnification of the brightness of a star by a MACHO is a time-dependent effect, since the MACHO, which acts as a lens, changes its location relative to the line of sight to the source as it moves along its orbit around the galaxy. Typically, the velocity transverse to the line of sight for a MACHO

Table 14.1. The expected number of events N_{ev} is obtained for a halo made entirely of MACHOs of a given mass.

MACHO mass ($M_\odot$)	Mean R_E (km)	Mean microlensing duration	N_{ev}
10^{-1}	0.3×10^9	1 month	4.5
10^{-2}	10^8	9 days	15
10^{-4}	10^7	1 day	165
10^{-6}	10^6	2 h	1662

in the galactic halo is $v_T \approx 200$ km s^{-1}, which can be inferred from the measured rotation curve of our galaxy. Clearly, the duration of the microlensing phenomenon and thus of the brightness increase of the source star depends on the MACHO mass, its distance and transverse velocity (see table 14.1).

Since the light deflection does not depend on the frequency of the light, the change in luminosity of the source star will be achromatic. For this reason, the observations are done in different wavelengths in order to check that. Moreover, the light curve will be symmetric with respect to the maximum value, since the transverse velocity of the MACHO is in excellent approximation constant during the period in which the lensing occurs. The probability that a given star is lensed twice is practically zero. Therefore, the achromaticity, symmetry and uniqueness of the signal are distinctive features that allow a microlensing event to be discriminated from background events such as variable stars (some of which are periodic, others show chromaticity and most often the light curve is not symmetric).

14.4.1.2 Microlensing towards the LMC

Another important quantity is the microlensing rate, which depends on the mass and velocity distributions of MACHOs. To determine this one has to model the galaxy and its halo. For simplicity one usually assumes a spherically symmetric shape for the halo with matter density decreasing as $1/r^2$ with distance as in equation (14.101), to obtain naturally a flat rotation curve. The velocity distribution is assumed to be Maxwellian. The least known quantity is the mass distribution of the MACHOs. For that, one makes the simplifying assumption that all MACHOs have the same mass. The number N_{ev} of microlensing events (such that the increase in magnitude is at least 30%) can then be computed. Table 14.1 shows some values for N_{ev} assuming monitoring of a million stars for 1 year in the LMC.

Microlensing allows the detection of MACHOs located in the galactic halo in the mass range $10^{-7} < M/M_\odot < 1$ [19], as well as MACHOs in the disk or bulge of our galaxy [35, 36].

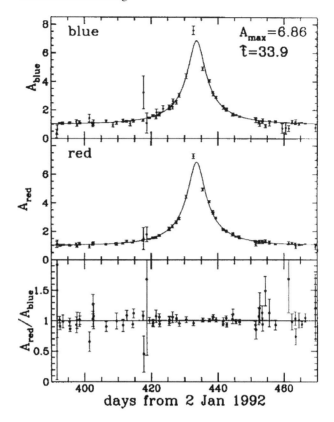

Figure 14.6. Microlensing event observed by the MACHO collaboration in their first year of data towards the LMC. The event lasted about 33 days. The data are shown for blue light, red light and the ratio red light to blue light, which for perfect achromaticity should be equal to one (from [38]).

In September 1993, the French collaboration EROS (Expérience de Recherche d'Objets Sombres) [37] announced the discovery of two microlensing candidates, and the American–Australian collaboration MACHO (for the collaboration they use the same acronym as for the compact objects) of one candidate [38] by monitoring several millions of stars in the LMC (figure 14.6).

The MACHO team went on to report the observation of 13 to 17 events (one being a binary lensing event; see figure 4.5) by analysing their 5.7 year of LMC data [39]. The inferred optical depth is $\tau = 1.2^{+0.4}_{-0.3} \times 10^{-7}$ with an additional 20% to 30% of systematic error. Correspondingly, this implies that about 20% of the halo dark matter is in the form of MACHOs with a most likely mass in the range 0.15–$0.9 M_{\odot}$ depending on the halo model. Moreover, it might well be that not all the MACHOs are in the halo: some could be stars in the LMC itself or

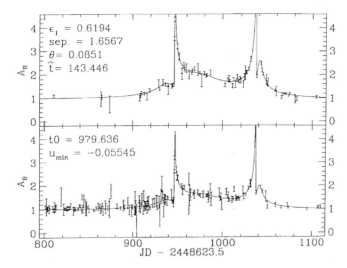

Figure 14.7. Binary microlensing event towards the LMC by the MACHO collaboration (taken from the web page http://darkstar.astro.washington.edu). The two light curves correspond to observations in different colours taken in order to test achromaticity.

located in an extended disk of our galaxy, in which case an average mass value including all events would produce an incorrect value. These considerations show that, at present, the values for the average mass as well as the fraction of halo dark matter in the form of MACHOs have to be treated with care.

As mentioned, one of the events discovered was due to a lens made from two objects, namely a binary system. Such events are more rare, but their observation is not surprising; since almost 50% of the stars are double systems, it is quite plausible that MACHOs also form binary systems. The light curve is then more complicated than for a single MACHO.

EROS has also searched for very-low-mass MACHOs by looking for microlensing events with time scales ranging from 30 min to 7 days [40]. The lack of candidates in this range places significant constraints on any model for the halo that relies on objects in the range $5 \times 10^{-8} < M/M_\odot < 2 \times 10^{-2}$. Indeed, such objects may make up at most 20% of the halo dark matter (in the range between $5 \times 10^{-7} < M/M_\odot < 2 \times 10^{-3}$ at most 10%). Similar conclusions have also been reached by the MACHO group [39].

A few events have also been discovered towards the Small Magellanic Cloud [41, 42].

14.4.1.3 *Microlensing towards other targets*

To date, the MACHO [43] and OGLE collaborations have found several hundred microlensing events towards the galactic bulge, most of which are listed among the alert events, which are constantly updated†. During their first season, the MACHO team found 45 events towards the bulge, which led to an estimated optical depth of $\tau \simeq 2.43^{+0.54}_{-0.45} \times 10^{-6}$, which is roughly in agreement with the OGLE result [44], and also implies the presence of a bar in the galactic centre. They also found three events by monitoring the spiral arms of our galaxy in the region of Gamma Scutum. Meanwhile, the EROS II collaboration also found some events towards the spiral arm regions. These results are important for studying the structure of our galaxy [45].

Microlensing searches towards the Andromeda galaxy (M31) have also been proposed [46–48]. In this case, however, one has to use the so-called 'pixel-lensing' method. Since the source stars are, in general, no longer resolvable, one has to consider the luminosity variation of a whole group of stars, which are, for instance, registered on a single pixel element of a CCD camera. This makes the subsequent analysis more difficult; however, if successful it allows M31 and other objects to be used as targets, which would otherwise not be possible to use. For information on the shape of the dark halo, which is presently unknown, it is important to observe microlensing in different directions. Two groups have started to perform searches: the French AGAPE (Andromeda Gravitational Amplification Pixel Experiment) [49, 50] and the American VATT/COLUMBIA [51] [52] which uses the 1.8-m VATT-telescope (Vatican Advanced Technology Telescope). Both teams showed that the pixel-lensing method works; however, the small number of observations so far does not allow firm conclusions to be drawn. Both the AGAPE and VATT/COLUMBIA teams found some candidate events which are consistent with microlensing; however, additional observations are needed to confirm them.

There are also networks involving different observatories with the aim of performing accurate photometry on alert microlensing events and in particular with the goal to find planets [53–55].

Although a rather young observational technique, microlensing has already enabled us to make substantial progress and the prospects for further contributions to solve important astrophysical problems look very bright.

14.5   The lens equation in cosmology

Until now, we have considered only almost static, weak localized perturbations of Minkowski spacetime. In cosmology the unperturbed spacetime background is given by a Robertson–Walker metric and this induces various changes in the previous discussions. It turns out that the final result for the lens map and the time delay looks practically unchanged, essentially we only have to insert some obvious redshift factors and interpret all distances as *angular diameter distances*.

---

† Current information on the MACHO collaboration's alert events is maintained at the WWW site http://darkstar.astro.washington.edu.

We recall that the expression for the time delay in an almost Newtonian situation is given by equation (14.19) with equations (14.20), (14.21):

$$c\Delta t = \frac{D_d D_s}{2 D_{ds}} \left(\frac{\xi}{D_d} - \frac{\eta}{D_s} \right)^2 - \hat{\psi}(\xi) + \text{constant.} \tag{14.102}$$

Note that

$$\left(\frac{\xi}{D_d} - \frac{\eta}{D_s} \right) = (\theta - \beta).$$

If the distances involved are cosmological, we must multiply the whole expression by $(1 + z_d)$, where z_d is the redshift of the lens. In addition all distances must be interpreted as angular diameter distances. (For a detailed derivation we refer to the book by Schneider *et al* [4] or [56]). With these modifications we obtain for the time delay,

$$c\Delta t = (1 + z_d) \left[\frac{D_d D_s}{2 D_{ds}} (\theta - \beta)^2 - \hat{\Psi}(\xi) \right] + \text{constant,} \tag{14.103}$$

where the prefactor of the first term is proportional to $1/H_0$ (H_0 is the present Hubble parameter).

For cosmological applications, it is convenient to rewrite the potential term using the length scale $\xi_0 = D_d$ as defined in equation (14.18) and $\theta = \xi/D_d$. This way we get

$$\hat{\psi}(\xi) = 4G \int d^2\theta' \, D_d^2 \Sigma(D_d\theta') \ln |\theta - \theta'| = 2 R_S \tilde{\psi}(\theta), \tag{14.104}$$

where $R_S = 2GM$ is the Schwarzschild radius of the total mass M of the lens, and

$$\tilde{\psi}(\theta) = \int d^2\theta' \, \tilde{\Sigma}(\theta') \ln |\theta - \theta'|, \tag{14.105}$$

with

$$\tilde{\Sigma}(\theta) := \frac{\Sigma(D_d\theta)}{M} D_d^2. \tag{14.106}$$

This quantity gives the fraction of the total mass M per unit solid angle as seen by the observer. We can now write equation (14.103) in the form

$$c\Delta t = \hat{\phi}(\theta, \beta) + \text{constant,} \tag{14.107}$$

where $\hat{\phi}$ is the *cosmological Fermat potential*:

$$\hat{\phi}(\theta, \beta) = \frac{1}{2}(1 + z_d)\frac{D_d D_s}{D_{ds}}(\theta - \beta)^2 - 2 R_S (1 + z_d)\tilde{\psi}(\theta). \tag{14.108}$$

For a Friedmann–Lemaitre model with density parameter Ω_0 and vanishing cosmological constant Λ, the angular diameter distance $D(z_1, z_2)$ between two events at redshifts z_1 and $z_2(z_1 < z_2)$, is given by

$$D(z_1, z_2) = 2c \frac{\sqrt{1 + \Omega_0 z_1}(2 - \Omega_0 + \Omega_0 z_2) - \sqrt{1 + \Omega_0 z_2}(2 - \Omega_0 + \Omega_0 z_1)}{H_0 \Omega_0^2 (1 + z_2)^2 (1 + z_1)}.$$

(14.109)

Equations (14.107)–(14.108) provide the basis for the determination of the Hubble parameter with gravitational lensing. One should also take into account that the universe might have a clumpy structure, which then affects the light propagation (for details on this problem see [57, 58]).

From equation (14.108) we obtain the cosmological lens mapping using Fermat's principle, which implies that $\nabla_\theta \hat{\phi}(\theta, \beta) = 0$ and gives an equation identical to equation (14.22), but, with the present meaning of the symbols, it holds for arbitrary redshifts.

Consider two images at the (observed) positions θ_1, θ_2, with separation $\theta_{12} \equiv \theta_1 - \theta_2$ and time delay Δt_{12}. Using the lens equation we obtain

$$\theta_{12} = 2R_S \frac{D_{ds}}{D_d D_s} \left[\frac{\partial \tilde{\psi}}{\partial \theta} \bigg|_{\theta_1} - \frac{\partial \tilde{\psi}}{\partial \theta} \bigg|_{\theta_2} \right].$$

(14.110)

The time delay $\Delta t_{12} = \hat{\phi}(\theta_1, \beta) - \hat{\phi}(\theta_2, \beta)$ contains the unobservable angle β, but this can be eliminated with the lens equation and equation (14.110):

$$\Delta t_{12} = 2R_S(1 + z_d) \left\{ \frac{1}{2} \left(\frac{\partial \tilde{\psi}}{\partial \theta} \bigg|_{\theta_1} + \frac{\partial \tilde{\psi}}{\partial \theta} \bigg|_{\theta_2} \right) \cdot \theta_{12} - \left(\tilde{\psi}(\theta_1) - \tilde{\psi}(\theta_2) \right) \right\}.$$

(14.111)

Given a lens model (i.e. $\tilde{\Sigma}(\theta)$), then equations (14.110) and (14.111) give a relation between the observables θ_{12}, Δt_{12} and H_0, provided that Ω_0, z_d, z_s are also known. Fortunately, the dependence on Ω_0 is, in practice, not strong.

Consider as an example a point source lensed by a point mass (Schwarzschild lens). Then $\tilde{\psi}(\theta) = \ln|\theta|$ and equation (14.110) gives

$$\theta_{12} = 2R_S \frac{D_{ds}}{D_d D_s} \left(\frac{1}{\theta_1} - \frac{1}{\theta_2} \right),$$

(14.112)

However, equation (14.111) becomes

$$\Delta t_{12} = 2R_S(1 + z_d) \left\{ \frac{\theta_2^2 - \theta_1^2}{2|\theta_1 \theta_2|} + \ln \left| \frac{\theta_2}{\theta_1} \right| \right\}.$$

(14.113)

We write this in terms of the ratio v of the magnifications. Using equation (14.74) one finds $v = \ln(\theta_2/\theta_1)^2$ and thus

$$\Delta t_{12} = R_S(1 + z_d)\{v^{1/2} - v^{-1/2} + \ln v\}.$$

(14.114)

14.5.1 Hubble constant from time delays

As first noted by Refsdal in 1964 [59], time delay measurements can yield, in principle, the Hubble parameter. Unfortunately, the use of this method requires a reliable lens model. This introduces systematic uncertainties. Moreover, the cosmological Fermat potential involves the density parameter Ω_0 and Λ (set equal to zero in equation (14.109)). The dependence on Ω_0 and Λ is, however, not strong, at least in some redshift domains ($z_s \leq 2$, $z_d \leq 0.5$).

Measuring the time delay is not an easy task as the history of the famous double QSO0957+561 demonstrates. Fortunately, the time delay for QSO0957+561 is now well known: $\Delta t = 417 \pm 3$ days [60]. Modellings lead to a best estimate of $H_0 \simeq 61$ km s^{-1} Mpc^{-1}. For this example there are constraints for modelling the lens; nevertheless, it is difficult to assess an error for the value of H_0.

Another example is the Einstein ring system B0218+357. A single galaxy is responsible for the small image splitting of $0.3''$. The time delay was reported to be 12 ± 3 days and the value $H_0 \sim 70$ km s^{-1} Mpc^{-1} was deduced. The ongoing surveys will hopefully find new lenses that possess the desirable characteristics for a reliable determination of H_0.

Besides having the above mentioned problems, the determination of H_0 through gravitational lensing offers also some advantages compared to the other methods. It can be directly used for large redshifts (~ 0.5) and it is independent of any other method. Moreover, it is based on fundamental physics, while other methods rely on models for variable stars (Cepheids), or supernova explosions (type II) or empirical calibrations of standard candles (Tully–Fisher distances, type I supernovae).

Finally, we note that lensing can also lead to bounds on the cosmological constant. The volume per unit redshift of the universe at high redshifts increases for a large Λ. This implies that the relative number of lensed sources for a given comoving number density of galaxies increases rapidly with Λ. This can be used to constrain Λ by making use of the observed probability of lensing. Various authors have used this method and came up with a limit $\Omega_\Lambda \leq 0.6$ for a universe with $\Omega_0 + \Omega_\Lambda = 1$. It remains to be seen whether such bounds, based on lensing statistics, can be improved.

14.6 Galaxy clusters as lenses

Galaxy clusters similarly to galaxies can act as gravitational lenses for more distant galaxies. One classifies the observed lensing effects due to clusters into two types:

(1) rich centrally condensed clusters produce sometimes giant arc when a background galaxy turns out to be almost aligned with one of the cluster caustics (*strong lensing*) (see, for instance, figure 14.1); and

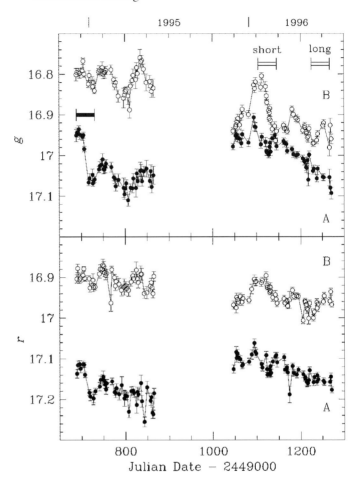

Figure 14.8. Light curves of the two images of the gravitationally lensed quasar Q0957+561. Note the sudden decrease in image A at the beginning in the 1995 season (taken from T Kundić *et al* 1997 [60]).

(2) every cluster produces weakly distorted images of a large number of background galaxies (*weak lensing*) (A nice example is in figure 14.11).

Both of these cases have been observed and have provided important information on the distribution of the matter in galaxy clusters. For the analysis of giant arcs, we have to use parametrized lens models which are fitted to the observational data. The situation is much better for weak lensing, because there now exist several parameter-free reconstruction methods of projected mass distributions from weak lensing data now exist.

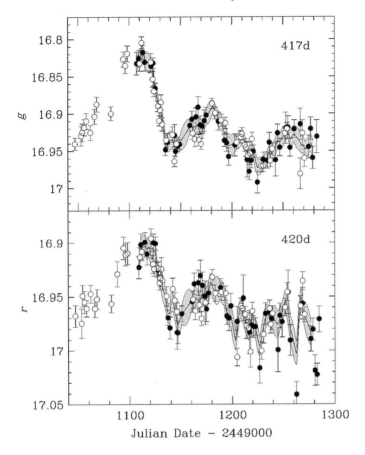

Figure 14.9. The light curve of image A of figure 14.3 is advanced by the optimal value of the time delay, 417 days (taken from T Kundić *et al* 1997 [60]).

Strong lensing requires that the mass density per surface Σ has to be in some parts of the lens bigger than the critical mass density given by

$$\Sigma \geq \Sigma_{\text{cr}} = \frac{c^2 D_{\text{s}}}{4\pi G D_{\text{d}} D_{\text{ds}}}. \tag{14.115}$$

Indeed, if this condition is satisfied there will be one or more caustics. The observation of an arc in a cluster of galaxies allows the projected cluster mass which lies inside a circle traced by the arc, even if no ring-shaped image is produced to be easily estimated. For an axisymmetric lens, the average surface mass density within the tangential critical curve is given by Σ_{cr}. Tangentially oriented large arcs occur close to the tangential critical curves, and thus the radius θ_{arc} of the circle traced by the arc gives an estimate of the Einstein radius θ_{E}.

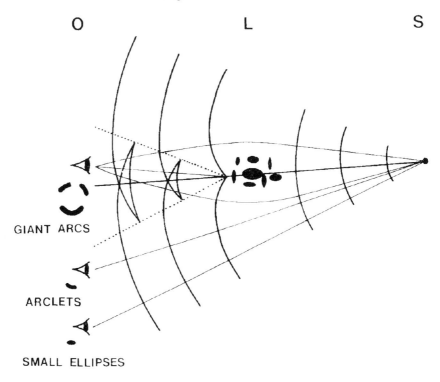

O   L   S

GIANT ARCS

ARCLETS

SMALL ELLIPSES

Figure 14.10. Wavefronts in the presence of a cluster perturbation.

Inside the so defined circle the surface mass is Σ_{cr}, and this way, knowing the redshifts of the lens and the source, one finds the total mass enclosed by $\theta = \theta_{arc}$

$$M(<\theta) = \Sigma_{cr}\pi(D_d\theta)^2 \simeq 1.1 \times 10^{14} M_\odot \left(\frac{\theta}{30''}\right)^2 \left(\frac{D_d}{1\,\text{Gpc}}\right), \qquad (14.116)$$

A mass estimate with this procedure is useful and often quite accurate.

   If we assume that the cluster can, at least as a first approximation, be described as a singular isothermal sphere, then using equation (14.84) we obtain for the dispersion velocity in the cluster

$$\sigma_v \simeq 10^3\,\text{km s}^{-1} \left(\frac{\theta}{28''}\right)^{1/2} \left(\frac{D_s}{D_{ds}}\right)^{1/2}. \qquad (14.117)$$

   A limitation of strong lensing is that it is model-dependent and, moreover, one can only determine the mass inside a cylinder of the inner part of a lensing cluster. The fact that the observed giant arcs never have a counter-arc of comparable brightness and even small counter-arcs are rare, implies that the lensing geometry has to be non-spherical.

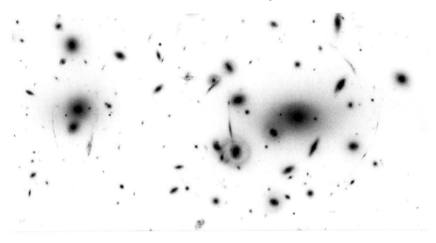

Figure 14.11. Hubble Space Telescope image of the cluster Abell 2218. Beside arcs around the two centres of the cluster, many arclets can be seen (NASA HST Archive).

A remarkable phenomenon is the occurrence of so-called *radial arcs* in galaxy clusters. These are *radially* rather than tangentially elongated, as most luminous arcs are. They are much less numerous (examples: MS 2137, Abell 370). Their position has been interpreted in terms of the turnover of the mass profile and a core radius $\sim 20h^{-1}$ kpc has been deduced, quite independent of any details of the lens model. There are, however, other mass profiles which can produce radial arcs, and have no flat core; even singular density profiles can explain radial arcs [61]. Such singular profiles of the dark matter are consistent with the large core radii inferred from x-ray emission.

14.6.1 Weak lensing

There is a population of distant blue galaxies in the universe whose spatial density reaches 50–100 galaxies per square arc minute at faint magnitudes. The images of these distant galaxies are coherently distorted by any foreground cluster of galaxies. Since they cover the sky so densely, the distortions can be determined statistically (individual weak distortions cannot be determined, since galaxies are not intrinsically round). Typical separations between arclets are $\sim (5\text{--}10)''$ and this is much smaller than the scale over which the gravitational cluster potential changes appreciably.

Starting with a paper by Kaiser and Squires [62], a considerable amount of theoretical work on various parameter-free reconstruction methods has recently been carried out. The main problem consists in making an optimal use of limited noisy data, without modelling the lens. For reviews see [63, 64]. The derivation of most of the relevant equations becomes much easier when using a complex formulation of lensing theory (see, for instance, [65]). In the following we will, however, not use it.

The reduced shear g is, in principle, observable over a large region. What we are really interested in, however, is the mean curvature κ, which is related to the surface mass density. Since

$$g = \frac{\gamma}{1 - \kappa} \tag{14.118}$$

we first look for relations between the shear $\boldsymbol{\gamma} = (\gamma_1, \gamma_2)$ and κ.

From equation (14.37) we get that

$$\Delta \Psi = 2k \tag{14.119}$$

or if, instead, we use the notation $\boldsymbol{\theta} = (\theta_1, \theta_2)$ for the image position equation (14.119) can be explicitly written as

$$k(\boldsymbol{\theta}) = \frac{1}{2} \left(\frac{\partial^2 \Psi(\boldsymbol{\theta})}{\partial \theta_1^2} + \frac{\partial^2 \Psi(\boldsymbol{\theta})}{\partial \theta_2^2} \right). \tag{14.120}$$

Using the definition for γ_i as given in equation (14.42) we find

$$\gamma_1(\boldsymbol{\theta}) = \frac{1}{2} \left(\frac{\partial^2 \Psi(\boldsymbol{\theta})}{\partial \theta_1^2} - \frac{\partial^2 \Psi(\boldsymbol{\theta})}{\partial \theta_2^2} \right) \equiv D_1 \Psi \tag{14.121}$$

and

$$\gamma_2(\boldsymbol{\theta}) = \frac{\partial^2 \Psi(\boldsymbol{\theta})}{\partial \theta_1 \partial \theta_2} \equiv D_2 \Psi. \tag{14.122}$$

where

$$D_1 := \frac{1}{2} \left(\partial_1^2 - \partial_2^2 \right), \qquad D_2 := \partial_1 \partial_2. \tag{14.123}$$

Note the identity

$$D_1^2 + D_2^2 = \tfrac{1}{4} \Delta^2. \tag{14.124}$$

Hence

$$\Delta \kappa = 2 \sum_{i=1,2} D_i \gamma_i. \tag{14.125}$$

Here, we can substitute the reduced shear, given by equation (14.118), on the right-hand side for γ_i. This gives the important equation

$$\Delta \kappa = 2 \sum_i D_i [g_i (1 - \kappa)]. \tag{14.126}$$

For a given (measured) $\boldsymbol{g}$ this equation does not determine uniquely κ, indeed equation (14.126) remains invariant under the substitution

$$\kappa \to \lambda \kappa + (1 - \lambda) \tag{14.127}$$

where λ is a real constant. This is the so-called *mass-sheet degeneracy* (a homogeneous mass sheet does not produce any shear).

Equation (14.126) can be turned into an integral equation, by making use of the fundamental solution

$$\mathcal{G} = \frac{1}{2\pi} \ln |\boldsymbol{\theta}| \tag{14.128}$$

for which $\Delta \mathcal{G} = \delta^2$ (δ^2 is the two-dimensional delta function). Then we get

$$k(\boldsymbol{\theta}) = 2 \int_{R^2} d^2\theta' \mathcal{G}(\boldsymbol{\theta} - \boldsymbol{\theta}') \sum_{i=1,2} (D_i \gamma_i)(\boldsymbol{\theta}') + k_0. \tag{14.129}$$

After some manipulations we can bring equation (14.129) into the following form

$$k(\boldsymbol{\theta}) = \frac{1}{\pi} \sum_{i=1,2} \int_{R^2} d^2\theta' \, [\tilde{D}_i(\boldsymbol{\theta} - \boldsymbol{\theta}') \gamma_i(\boldsymbol{\theta}')] + k_0, \tag{14.130}$$

or, in terms of the reduced shear,

$$k(\boldsymbol{\theta}) = k_0 + \frac{1}{\pi} \sum_{i=1,2} \int_{R^2} d^2\theta' \, [\tilde{D}_i(\boldsymbol{\theta} - \boldsymbol{\theta}')(g_i(1 - k))(\boldsymbol{\theta}')], \tag{14.131}$$

where

$$D_1 \ln |\boldsymbol{\theta}| = \frac{\theta_2^2 - \theta_1^2}{|\boldsymbol{\theta}|^4} \equiv \tilde{D}_1, \qquad D_2 \ln |\boldsymbol{\theta}| = -\frac{2\theta_1\theta_2}{|\boldsymbol{\theta}|^4} \equiv \tilde{D}_2. \tag{14.132}$$

The crucial fact is that $\gamma(\boldsymbol{\theta})$ is an observable quantity and thus using equation (14.130) one can infer the matter distribution of the considered galaxy cluster. This result is, however, fixed up to an overall constant k_0 (problem of the mass-sheet degeneracy).

As discussed in section 14.2.4 we can define the ellipticity ϵ of an image of a galaxy as

$$\epsilon = \epsilon_1 + i\epsilon_2 = \frac{1 - r}{1 + r} e^{2i\varphi}, \qquad r \equiv \frac{b}{a} \tag{14.133}$$

where φ is the position angle of the ellipse and a and b are the major and minor semi axis, respectively. a and b are given by the inverse of the eigenvalues of the matrix defined in equation (14.42). If we take the average on the ellipticity due to lensing and make use of equation (14.133) as well as of the expressions for a and b we find the relation

$$\langle \epsilon \rangle = \left\langle \frac{\gamma}{1 - k} \right\rangle. \tag{14.134}$$

The angle bracket means average over a finite sky area. In the weak lensing limit $k \ll 1$ and $|\gamma| \ll 1$ the mean ellipticity directly relates to the shear: $\langle \gamma_1(\boldsymbol{\theta}) \rangle \simeq \langle \epsilon_1(\boldsymbol{\theta}) \rangle$ and $\langle \gamma_2(\boldsymbol{\theta}) \rangle \simeq \langle \epsilon_2(\boldsymbol{\theta}) \rangle$. Thus a measurement of the average ellipticity allows γ, to be determined and, making use of equation (14.130) one can get the surface mass density k of the lens. Recently, several groups have reported the detection of cosmic shear, which clearly demonstrates the technical feasibility of using weak lensing surweys to measure dark matter clustering and the potential for cosmological measurements, in particular with the upcoming wide-field CCD cameras [67,68].

14.6.2   Comparison with results from x-ray observations

Beside the lensing technique, there are two other methods for determining mass distributions of clusters:

(1) *the observed velocity dispersion*, combined with the Jeans–equation from stellar dynamics gives the total mass distribution, if it is assumed that light traces mass; and

(2) *x-ray observations of the intracluster gas*, combined with the condition of hydrostatic equilibrium and spherical symmetry also lead to the total mass distribution as well as to the baryonic distribution.

If the hydrostatic equilibrium equation for the hot gas

$$\frac{\mathrm{d}P_g}{\mathrm{d}r} = -\rho_g \frac{GM_t(r)}{r^2} \tag{14.135}$$

is combined with the ideal equation of state $P_g = (k_B T_g/\mu m_H)\rho_g$ and assuming spherical symmetry, one easily finds for the total mass profile

$$M_t(r) = -\frac{k_B T_g}{G \mu m_H}\left(\frac{\mathrm{d}\ln\rho_g}{\mathrm{d}\ln r} + \frac{\mathrm{d}\ln T_g}{\mathrm{d}\ln r}\right)r. \tag{14.136}$$

The right-hand side can be determined from the intensity distribution and some spectral information. (At present, the latter is not yet good enough, because of relatively poor resolution which, however, will change with the XMM survey.)

Weak lensing, together with an analysis of x-ray observations, offers a unique possibility for probing the relative distributions of the gas and the dark matter, and for studying the dynamical relationship between the two. As an example consider the cluster of galaxies A2163 (z=0.201) which is one of the two most massive clusters known so far.

ROSAT measurements reach out to $2.3h^{-1}$ Mpc ($\sim$15 core radii)(h being the Hubble constant in units of 100). The total mass is 2.6 times greater than that of COMA, but the gas mass fraction, $\sim 0.1h^{-3/2}$ is typical for rich clusters. The data together suggest that there was a recent merger of two large clusters. The optical observations of the distorted images of background galaxies were made with the CFHT telescope. The resulting lensing and x-ray mass profiles are compared in figure 14.12. The data-sets only overlap out to a radius of $200'' \simeq 500h^{-1}$ kpc to which the lensing studies were limited. It is evident

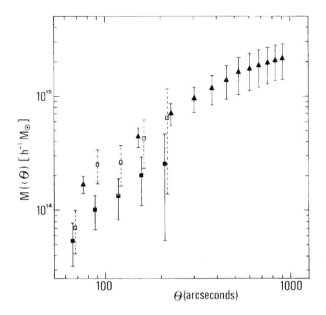

Figure 14.12. The radial mass profiles determined from the x-ray and lensing analysis for Abell 2163. The triangles display the total mass profile determined from the x-ray data. The filled squares are the weak lensing estimates 'corrected' for the mean surface density in the control annulus determined from the x-ray data. The conversion from angular to physical units is $60'' = 0.127h^{-1}$ Mpc (taken from Squires *et al* 1997 [66]).

that the lensing mass estimates are systematically lower by a factor of ~ 2 than the x-ray results, but generally the results are consistent with each other, given the substantial uncertainties. There are reasons that the lensing estimate may be biased downward. Correcting for this gives the results displayed by open squares. The agreement between the lensing and x-ray results then becomes quite impressive. The rate and quality of such data will increase dramatically during the coming years. With weak lensing one can also test the dynamical state of clusters. By selecting the relaxed ones one can then determine, with some confidence, the relative distributions of gas and dark matter.

In addition, it will become possible to extend the investigations to supercluster scales, with the aim of determining the power spectrum and obtain information on the cosmological parameters [63, 64].

References

[1] Perlmutter S *et al* 1999 *Astrophys. J.* **517** 565
[2] Trimble V 1987 *Annu. Rev. Astron. Astrophys.* **25** 425
[3] Zwicky F 1933 *Helv. Phys. Acta* **6** 110

[4]　Schneider P, Ehlers J and Falco E E 1992 *Gravitational Lensing* (Berlin: Springer)

[5]　Refsdal S and Surdej J 1994 Gravitational lenses *Rep. Prog. Phys.* **56** 117

[6]　Narayan R and Bartelmann M 1999 Lectures on gravitational lensing *Formation of Structure in the Universe* ed A Dekel and J P Ostriker (Cambridge: Cambridge University Press)

[7]　Straumann N, Jetzer Ph and Kaplan J 1998 *Topics on Gravitational Lensing (Napoli Series on Physics and Astrophysics)* (Naples: Bibliopolis)

[8]　Dyson F W, Eddington A S and Davidson C R 1920 *Mem. R. Astron. Soc.* **62** 291

[9]　Einstein A 1936 *Science* **84** 506

[10]　Renn J, Sauer T and Stachel J 1997 *Science* **275** 184

[11]　Zwicky F 1937 *Phys. Rev.* **51** 290
　　　Zwicky F 1937 *Phys. Rev.* **51** 679

[12]　Chang K and Refsdal S 1979 *Nature* **282** 561

[13]　Gott R J 1981 *Astrophys. J.* **243** 140

[14]　Paczyński B 1986 *Astrophys. J.* **301** 503

[15]　Schmidt R and Wambsganss J 1998 *Astron. Astrophys.* **335** 379

[16]　Wambsganss J 2001 *Microlensing 2000: A New Era of Microlensing Astrophysics* ed J W Menzies (San Francisco, CA: ASP)

[17]　Paczyński B 1986 *Astrophys. J.* **304** 1

[18]　De Rújula A, Jetzer Ph and Massó 1991 *Mon. Not. R. Astron. Soc.* **250** 348

[19]　De Rújula A, Jetzer Ph and E Massó 1992 *Astron. Astrophys.* **254** 99

[20]　Griest K 1991 *Astrophys. J.* **366** 412

[21]　Nemiroff R J 1991 *Astron. Astrophys.* **247** 73

[22]　Paczyński B 1996 *Annu. Rev. Astron. Astrophys.* **34** 419

[23]　Roulet E and Mollerach S 1997 *Phys. Rep.* **279** 67

[24]　Zakharov A F and Sazhin M V 1998 *Phys. Usp.* **41** 945

[25]　Jetzer Ph 1999 *Naturwissenschaften* **86** 201

[26]　Carr B 1994 *Annu. Rev. Astron. Astrophys.* **32** 531

[27]　Bahcall J, Flynn C, Gould A and Kirhakos S 1994 *Astrophys. J.* **435** L51

[28]　Kerins E J 1997 *Astron. Astrophys.* **322** 709

[29]　Gilmore G and Unavane M 1998 *Mon. Not. R. Astron. Soc.* **301** 813

[30]　Tamanaha C M, Silk J, Wood M A and Winget D E 1990 *Astrophys. J.* **358** 164

[31]　S D Kawaler 1996 *Astrophys. J.* **467** L61

[32]　Paolis F De, Ingrosso G, Jetzer Ph and Roncadelli M 1995 *Phys. Rev. Lett.* **74** 14

[33]　Paolis F De, Ingrosso G, Jetzer Ph and Roncadelli M 1995 *Astron. Astrophys.* **295** 567

[34]　Paolis F De, Ingrosso G, Jetzer Ph and Roncadelli M 1999 *Astrophys. J.* **510** L103

[35]　Paczyński B 1991 *Astrophys. J.* **371** L63

[36]　Griest K *et al* 1991 *Astrophys. J.* **372** L79

[37]　Aubourg E *et al* 1993 *Nature* **365** 623

[38]　Alcock C *et al* 1993 *Nature* **365** 621
　　　Alcock C *et al* 1995 *Astrophys. J.* **445** 133

[39]　Alcock C *et al* 2000 *Astrophys. J.* **542** 281

[40]　Renault C *et al* 1997 *Astron. Astrophys.* **324** L69

[41]　Alcock C *et al* 1997 *Astrophys. J.* **491** L11

[42]　Palanque-Delabrouille N *et al* 1999 *Astron. Astrophys.* **332** 1

[43]　Alcock C *et al* 1997 *Astrophys. J.* **479** 119

[44]　Udalski A *et al* 1994 *Acta Astron.* **44** 165

[45] Grenacher L, Jetzer Ph, Strässle M and De Paolis F 1999 *Astron. Astrophys.* **351** 775
[46] Crotts A P 1992 *Astrophys. J.* **399** L43
[47] Baillon P, Bouquet A, Giraud-Héraud Y and Kaplan J 1993 *Astron. Astrophys.* **277** 1
[48] Jetzer Ph 1994 *Astron. Astrophys.* **286** 426
[49] Ansari R *et al* 1997 *Astron. Astrophys.* **324** 843
[50] Ansari R *et al* 1999 *Astron. Astrophys.* **344** L49
[51] Crotts A P S and Tomaney A B 1996 *Astrophys. J.* **473** L87
[52] Crotts A and Uglesich R 2001 *Microlensing 2000: A New Era of Microlensing Astrophysics* ed J W Menzies (San Francisco, CA: ASP)
[53] Mao S and Paczyński B 1991 *Astrophys. J.* **374** L37
[54] Gould A and Loeb A 1992 *Astrophys. J.* **396** 104
[55] Bennett D and Rhie S H 1996 *Astrophys. J.* **472** 660
[56] Straumann N 1999 *Lectures on Gravitational Lensing* Troisième Cycle de la Physique en Suisse Romande
[57] Sachs R K 1961 *Proc. R. Soc.* A **264** 309
[58] Dyer C C and Roeder R C 1973 *Astrophys. J.* **180** L31
[59] Refsdal S 1966 *Mon. Not. R. Astron. Soc.* **134** 315
[60] Kundić T *et al* 1997 *Astrophys. J.* **482** 648
[61] Bartelmann M 1996 *Astron. Astrophys.* **313** 697
[62] Kaiser N and Squires G 1993 *Astrophys. J.* **404** 441
[63] Mellier Y 1999 *Annu. Rev. Astron. Astrophys.* **37** 127
[64] Bartelmann M and Schneider P 1999 Weak gravitational lensing *Preprint* astro-ph/9912508
[65] Straumann N 1997 *Helv. Phys. Acta* **70** 894
[66] Squires G *et al* 1997 *Astrophys. J.* **482** 648
[67] Van Waerbeke L *et al* 2000 *Astron. Astrophys.* **358** 30
[68] Wittman D M *et al* 2000 *Nature* **405** 143

Chapter 15

Numerical simulations in cosmology

Anatoly Klypin
Astronomy Department, New Mexico State University, Las Cruces, USA

15.1   Synopsis

In section 15.2 we give a short description of different methods used in cosmology. The focus is on the major features of N-body simulations: equations, main numerical techniques, the effects of resolution and methods of halo identification.

In section 15.3 we give a summary of recent results on spatial and velocity biases in cosmological models. Progress in numerical techniques made it possible to simulate halos in large volumes with such an accuracy that halos survive in dense environments of groups and clusters of galaxies. Halos in simulations look like real galaxies, and, thus, can be used to study the biases—differences between galaxies and the dark matter. The biases depend on scale, redshift and circular velocities of selected halos. Two processes seem to define the evolution of the spatial bias: (1) statistical bias and (2) merger bias (merging of galaxies, which happens preferentially in groups, reduces the number of galaxies, but does not affect the clustering of the dark matter). There are two kinds of velocity bias. The pair-wise velocity bias is $b_{12} = 0.6$–0.8 at $r < 5h^{-1}$ Mpc, $z = 0$. This bias mostly reflects the spatial bias and provides almost no information on the relative velocities of the galaxies and the dark matter. One-point velocity bias is a better measure of the velocities. Inside clusters the galaxies should move slightly faster ($b_v = 1.1$–1.3) than the dark matter. Qualitatively this result can be understood using the Jeans equations of stellar dynamics. For the standard LCDM model we find that the correlation function and the power spectrum of galaxy-size halos at $z = 0$ are antibiased on scales $r < 5h^{-1}$ Mpc and $k \approx (0.15$–$30)h$ Mpc^{-1}. In section 15.4 we give a review of the different properties of dark matter halos. Taken from different publications, we present results on (1) the mass and velocity

420

functions, (2) density and velocity profiles and (3) concentration of halos. The results are not sensitive to the parameters of cosmological models, but formally most of them were derived for popular flat ΛCDM model. In the range of radii $r = (0.005–1)r_{vir}$ the density profile for a quiet isolated halo is very accurately approximated by a fit suggested by Moore *et al* (1997): $\rho \propto 1/x^{1.5}(1 + x^{1.5})$, where $x = r/r_s$ and r_s is a characteristic radius. The fit suggested by Navarro *et al* (1995), $\rho \propto 1/x(1 + x)^2$, also gives a very satisfactory approximation with relative errors of about 10% for radii not smaller than 1% of the virial radius. The mass function of $z = 0$ halos with mass below $\approx 10^{13}h^{-1}M_\odot$ is approximated by a power law with slope $\alpha = -1.85$. The slope increases with the redshift. The velocity function of halos with $V_{max} < 500$ km s^{-1} is also a power law with the slope $\beta = -3.8–4$. The power law extends to halos at least down to 10 km s^{-1}. It is also valid for halos inside larger virialized halos. The concentration of halos depends on mass (more massive halos are less concentrated) and environment, with isolated halos being less concentrated than halos of the same mass inside clusters. Halos have intrinsic scatter of concentration: at 1σ level halos with the same mass have $\Delta(\log c_{vir}) = 0.18$ or, equivalently, $\Delta V_{max}/V_{max} = 0.12$. Velocity anisotropy for both sub-halos and the dark matter is approximated by $\beta(r) = 0.15 + 2x/[x^2 + 4]$, where x is the radius in units of the virial radius.

15.2 Methods

15.2.1 Introduction

Numerical simulations in cosmology have a long history and numerous important applications. The different aspects of the simulations including the history of the subject were reviewed recently by Bertschinger (1998); see also Sellwood (1987) for an older review. More detailed aspects of simulations were discussed by Gelb (1992), Gross (1997) and Kravtsov (1999). Numerical simulations play a very significant role in cosmology. It all started in the 1960s (Aarseth 1963) and 1970s (Peebles 1970, Press and Schechter 1974) with simple N-body problems solved using N-body codes with a few hundred particles. Later the Particle–Particle code (direct summation of all two-body forces) was polished and brought to the state of art (Aarseth 1985). Already those early efforts brought some very valuable fruits. Peebles (1970) studied the collapse of a cloud of particles as a model of cluster formation. The model had 300 points initially distributed within a sphere with no initial velocities. After the collapse and virialization the system looked like a cluster of galaxies. Those early simulations of cluster formation, though producing cluster-like objects, signalled the first problem—a simple model of an initially isolated cloud (top-hat model) results in a density profile for the cluster which is way too steep (power-law slope -4) as compared with real clusters (slope -3). The problem was addressed by Gunn and Gott (1972), who introduced the notion of secondary infall in an effort to solve the problem. Another keystone work of those times is the paper by White (1976), who studied the collapse of 700 particles with different masses. It was shown that if one distributes the mass of a cluster to individual galaxies, two-body scattering will result in mass segregation not compatible with observed clusters. This was another manifestation of the dark matter in clusters. This time it was shown that inside a cluster the dark matter cannot reside inside individual galaxies.

The survival of substructures in galaxy clusters was another problem addressed in that paper. It was found that halos of dark matter, which in real life may represent galaxies, do not survive in the dense environment of galaxy clusters. White and Rees (1978) argued that the real galaxies survive inside clusters because of energy dissipation by the baryonic component. That point of view was accepted for almost 20 years. Only recently was it shown that the energy dissipation probably does not play a dominant role in the survival of galaxies and the dark matter halos are not destroyed by tidal stripping and galaxy–galaxy collisions inside clusters (Klypin *et al* 1999a (KGKK), Ghigna *et al* 2000). The reason why early simulations came to a wrong result was purely numerical: they did not have enough resolution. But 20 years ago it was impossible to make a simulation with sufficient resolution. Even if at that time we had present-day codes, it would have taken about 600 years to make one run.

The generation of initial conditions with a given amplitude and spectrum of fluctuations was a problem for some time. The only correctly simulated spectrum was the flat spectrum which was generated by randomly distributing particles. In order to generate fluctuations with a power spectrum, say $P(k) \propto k^{-1}$, Aarseth *et al* (1979) placed particles along rods. Formally, it generates the spectrum, but the distribution has nothing to do with cosmological fluctuations, which have random phases. Doroshkevich *et al* (1980) and Klypin and Shandarin (1983) were the first to use the Zeldovich (1970) approximation to set the initial conditions. Since then this method has been used to generate initial conditions for arbitrary initial spectrum of perturbations.

Starting in the mid-1980s the field of numerical simulations has blossomed: new numerical techniques have been invented, old ones perfected. The number of publications based on numerical modelling has skyrocketed. To a large extent, this has changed our way of doing cosmology. Instead of questionable assumptions and waving-hands arguments, we have tools for testing our hypotheses and models. As an example, I mention two analytical approximations which were validated by numerical simulations. The importance of both approximations is difficult to overestimate. The first is the Zeldovich approximation, which paved the way for understanding the large-scale structure of the galaxy distribution. The second is the Press and Schechter (1974) approximation, which gives the number of objects formed at different scales at different epochs. Both approximations cannot be formally proved. The Zeldovich approximation is not formally applicable for hierarchical clustering. It must start with smooth perturbations (a truncated spectrum). Nevertheless, numerical simulations have shown that even for the hierarchical clustering the approximation can be used with appropriate filtering of the initial spectrum (see Sahni and Coles (1995) and references therein). The Press–Schechter approximation is also difficult to justify without numerical simulations. It operates with an initial spectrum and a linear theory, but then (a very long jump) it predicts the number of objects at very nonlinear stage. Because it is not based on any realistic theory of nonlinear evolution, it was an ingenious but wild guess. If anything, the approximation is based on a simple spherical top-hat model. But simulations show that objects do not form in this way——they are formed in a complicated fashion through multiple mergers and accretion along filaments. Still this very simple and very useful prescription gives quite accurate predictions.

This chapter is organized in the following way. Section 15.2 gives the equations which we solve to follow the evolution of initially small fluctuations. Initial conditions are discussed in section 15.3. A brief discussion of different methods is given in section 15.4. The effects of the resolution and some other technical details are also discussed in section 15.5. Identification of halos ('galaxies') is discussed in section 15.6.

15.2.2 Equations of evolution of fluctuations in an expanding universe

Usually the problem of the formation and dynamics of cosmological objects is formulated as an N-body problem: for N point-like objects with given initial positions and velocities, find their positions and velocities at any later moment. It should be remembered that this is just a short-cut in our formulation—to make things simple. While it is still mathematically correct in many cases, it does not give a correct explanation for what we do. If we are literally to take this approach, we should follow the motion of zillions of axions, baryons, neutrinos and whatever else our universe is made of. So, what has it to do with the motion of those few millions of particles in our simulations? The correct approach is to start with the Vlasov equation coupled with the Poisson equation and with appropriate initial and boundary conditions. If we neglect the baryonic component, which of course is very interesting, but would complicate our situation even more, the system is described by distribution functions $f_i(x, \dot{x}, t)$ which should include all different clustered components i. For a simple CDM model we have only one component (axions or whatever it is). For more complicated Cold plus Hot Dark Matter (CHDM) with several different types of neutrinos the system includes one distribution function for the cold component and one distribution function for each type of neutrino (Klypin *et al* 1993). In the comoving coordinates x, the equations for the evolution of f_i are:

$$\frac{\partial f_i}{\partial t} + \dot{x}\frac{\partial f_i}{\partial x} - \nabla\phi\frac{\partial f_i}{\partial p} = 0, \qquad p = a^2\dot{x}, \qquad (15.1)$$

$$\nabla^2\phi = 4\pi Ga^2(\rho_{dm}(x, t) - \langle\rho_{dm}(t)\rangle) = 4\pi Ga^2\Omega_{dm}\delta_{dm}\rho_{cr}, \qquad (15.2)$$

$$\delta_{dm}(x, t) = (\rho_{dm} - \langle\rho_{dm}\rangle)/\langle\rho_{dm}\rangle), \qquad (15.3)$$

$$\rho_{dm}(x, t) = a^{-3}\sum_i m_i \int d^3p \, f_i(x, \dot{x}, t). \qquad (15.4)$$

Here $a = (1 + z)^{-1}$ is the expansion parameter, $\boldsymbol{p} = a^2\dot{\boldsymbol{x}}$ is the momentum, Ω_{dm} is the contribution of the clustered dark matter to the mean density of the universe, m_i is the mass of a particle of the ith component of the dark matter. The solution of the Vlasov equation can be written in terms of equations for the characteristics, which *look* like equations of particle motion:

$$\frac{d\boldsymbol{p}}{da} = -\frac{\nabla\phi}{\dot{a}}, \qquad \frac{d\boldsymbol{v}}{dt} + 2\frac{\dot{a}}{a}\boldsymbol{v} = -\frac{\nabla\phi'}{a^3}, \tag{15.5}$$

$$\frac{d\boldsymbol{x}}{da} = \frac{\boldsymbol{p}}{\dot{a}a^2}, \qquad \frac{d\boldsymbol{x}}{dt} = \boldsymbol{v}, \tag{15.6}$$

$$\nabla^2\phi = 4\pi G\Omega_0\delta_{dm}\rho_{cr,0}/a, \qquad \phi' = a\phi, \tag{15.7}$$

$$\dot{a} = H_0\sqrt{1 + \Omega_0\left(\frac{1}{a} - 1\right) + \Omega_\Lambda(a^2 - 1)}. \tag{15.8}$$

In these equations $\rho_{cr,0}$ is the critical density at $z = 0$; Ω_0, and $\Omega_{\Lambda,0}$, are the density of the matter and of the cosmological constant in units of the critical density at $z = 0$.

The distribution function f_i is constant along each characteristic. This property should be preserved by numerical simulations. The complete set of characteristics coming through every point in the phase space is equivalent to the Vlasov equation. We cannot have the complete (infinite) set, but we can follow the evolution of the system (with some accuracy), if we select a representative sample of characteristics. One way of doing this would be to split the initial phase space into small domains, to take only one characteristic as being representative of each volume element, and to follow the evolution of the system of 'particles' in a self-consistent way. In models with one 'cold' component of clustering dark matter (like the CDM or ΛCDM models) the initial velocity is a unique function of the coordinates (only the 'Zeldovich' part is present, no thermal velocities). This means that we need only to split the coordinate space, not the velocity space. For complicated models with a significant thermal component, the distribution in the full phase space should be taken into account. Depending on what we are interested in, we might split the initial space into equal-size boxes (a typical set-up for PM or P³M simulations) or we could divide some area of interest (say, where a cluster will form) into smaller boxes, and use much bigger boxes outside the area (to mimic the gravitational forces of the outside material). In any case, the mass assigned to a 'particle' is equal to the mass of the domain it represents. Now we can think of the 'particle' either as a small box, which moves with the flow but does not change its original shape, or as a point-like particle. Both presentations are used in simulations. None is superior to another.

There are different forms of final equations. Mathematically they are all equivalent but computationally there are very significant differences. There are considerations, which may affect the choice of a particular form of the equations. Any numerical method gives more accurate results for a variable, which changes slowly with time. For example, for the gravitational potential we can choose either

ϕ or ϕ'. At early stages of evolution perturbations still grow almost linearly. In this case we expect that $\delta_{dm} \propto a$, $\phi \approx$ constant and $\phi' \approx a$. Thus, ϕ can be a better choice because it does not change. This is especially helpful, if the code uses the gravitational potential from a previous moment of time as an initial 'guess' for the current moment, as happens in the case of the ART code. In any case, it is better to have a variable which does not change much. For equations of motion we can choose, for example, either the first equations in (15.5)–(15.6) or the second equations. If we choose the 'momentum' $p = a^2 \dot{x}$ as the effective velocity and take the expansion parameter a as the time variable, then for linear growth we expect the change of coordinates per each step to be constant: $\Delta x \propto \Delta a$. Numerical integration schemes should not have a problem with this type of growth. For the t and v variables, the rate of change is more complicated: $\Delta x \propto a^{-1/2} \Delta t$, which may produce some errors at small expansion parameters. The choice of variables may affect the accuracy of the solution even at a very nonlinear stage of the evolution as was argued by Quinn *et al* (1997).

15.2.3 Initial conditions

15.2.3.1 The Zeldovich approximation

The Zeldovich approximation is commonly used to set initial conditions. The approximation is valid in mildly nonlinear regimes and is much superior to the linear approximation. We slightly rewrite the original version of the approximation to incorporate cases (like CHDM) when the growth rates $g(t)$ depends on the wavelength of the perturbation $|\mathbf{k}|$. In the Zeldovich approximation the comoving and Lagrangian coordinates are related in the following way:

$$x = q - \alpha \sum_k g_{|k|}(t) S_{|k|}(q), \qquad p = -\alpha a^2 \sum_k g_{|k|}(t) \left(\frac{\dot{g}_{|k|}}{g_{|k|}} \right) S_{|k|}(q), \quad (15.9)$$

where the displacement vector S is related to the velocity potential Φ and the power spectrum of fluctuations $P(|\mathbf{k}|)$:

$$S_{|k|}(q) = \nabla_q \Phi_{|k|}(q), \qquad \Phi_{|k|} = \sum_k a_k \cos(kq) + b_k \sin(kq), \qquad (15.10)$$

where a and b are Gaussian random numbers with mean zero and dispersion $\sigma^2 = P(k)/k^4$:

$$a_k = \sqrt{P(|k|)} \frac{\text{Gauss}(0, 1)}{|k|^2}, \qquad b_k = \sqrt{P(|k|)} \frac{\text{Gauss}(0, 1)}{|k|^2}. \qquad (15.11)$$

The parameter α, together with the power spectrum $P(k)$, define the normalization of the fluctuations.

In order to set the initial conditions, we choose the size of the computational box L and the number of particles N^3. The phase space is divided into small equal cubes of size $2\pi/L$. Each cube is centred on a harmonic $\boldsymbol{k} = 2\pi/L \times \{i, j, k\}$, where $\{i, j, k\}$ are integer numbers with limits from zero to $N/2$. We realize the spectrum of perturbations $a_{\boldsymbol{k}}$ and $b_{\boldsymbol{k}}$, and find the displacement and the momenta of particles with $\boldsymbol{q} = L/N \times \{i, j, k\}$ using equation (15.9). Here $i, j, k = 1, N$.

15.2.3.2 Power spectrum

There are approximations of the power spectrum $P(k)$ for a wide range of cosmological models. The publicly available COSMICS code (Bertschinger 1996) gives accurate approximations for the power spectrum. Here we follow Klypin and Holtzman (1997) who give the following fitting formula:

$$P(k) = \frac{k^n}{(1 + P_2 k^{1/2} + P_3 k + P_4 k^{3/2} + P_5 k^2)^{2P_6}}. \tag{15.12}$$

The coefficients P_i are presented by Klypin and Holtzman (1997) for a variety of models. A comparison of some of the power spectra with the results from COSMICS (Bertschinger 1996) indicate that the errors of the fits are smaller than 5%. Table 15.1 gives the parameters of the fits for some popular models. The power spectrum of cosmological models is often approximated using a fitting formula given by Bardeen *et al* (1986, BBKS):

$$P(k) = k^n T^2(k),$$
$$T(k) = \frac{\ln(1 + 2.34q)}{2.34q}[1 + 3.89q + (16.1q)^2 + (5.4q)^3 + (6.71q)^4]^{-1/4},$$
$$\tag{15.13}$$

where $q = k/(\Omega_0 h^2 \text{ Mpc}^{-1})$. Unfortunately, the accuracy of this approximation is not great and it should not be used for accurate simulations. We find that the following approximation, which is a combination of a slightly modified BBKS fit and the Hu and Sugiyama (1996) scaling with the amount of baryons, provides errors in the power spectrum which are less than 5% for the range of wavenumbers $k = (10^{-4}$–$40)h$ Mpc^{-1} and for $\Omega_b/\Omega_0 < 0.1$:

$$P(k) = k^n T^2(k),$$
$$T(k) = \frac{\ln(1 + 2.34q)}{2.34q}[1 + 13q + (10.5q)^2 + (10.4q)^3 + (6.51q)^4]^{-1/4},$$
$$q = \frac{k(T_{\text{CMB}}/2.7 \text{ K})^2}{\Omega_0 h^2 \alpha^{1/2}(1 - \Omega_b/\Omega_0)^{0.60}}, \qquad \alpha = a_1^{-\Omega_b/\Omega_0} a_2^{-(\Omega_b/\Omega_0)^3},$$
$$a_1 = (46.9\Omega_0 h^2)^{0.670}[1 + (32.1\Omega_0 h^2)^{-0.532}],$$
$$a_2 = (12\Omega_0 h^2)^{0.424}[1 + (45\Omega_0 h^2)^{-0.582}]. \tag{15.14}$$

Table 15.1. Approximations of the power spectra.

Ω_0	Ω_{bar}	h	P_2	P_3	P_4	P_5	P_6
0.3	0.035	0.60	−1.7550E+00	6.0379E+01	2.2603E+02	5.6423E+02	9.3801E-01
0.3	0.030	0.65	−1.6481E+00	5.3669E+01	1.6171E+02	4.1616E+02	9.3493E-01
0.3	0.026	0.70	−1.5598E+00	4.7986E+01	1.1777E+02	3.2192E+02	9.3030E-01
1.0	0.050	0.50	−1.1420E+00	2.9507E+01	4.1674E+01	1.1704E+02	9.2110E-01
1.0	0.100	0.50	−1.3275E+00	3.0152E+01	5.5515E+01	1.2193E+02	9.2847E-01

15.2.3.3 *Multiple masses: high resolution for a small region*

In many cases we would like to set initial conditions in such a way that inside some specific region(s) there are more particles and the spectrum is better resolved. A rigorous but complicated approach for the problem is described by Bertschinger (2001). Here I give a simplified prescription. The procedure has two steps. First, we run a low-resolution simulation which has a sufficiently large volume to include the effects of the environment. For this run all the particles have the same mass. A halo is picked for rerunning with high resolution. Second, using particles of the halo, we identify a region in the Lagrangian (initial) space, where the resolution should be increased. We add high-frequency harmonics, which are not present in the low-resolution run. We then add the contributions from all the harmonics and get initial displacements and momenta (equation (15.9)). Let us be more specific. In order to add the new harmonics, we must specify (1) how we divide the phase space and place the harmonics and (2) how we sum the contributions of the harmonics.

The simplest way is to divide the phase space into many small boxes of size $2\pi/L$, where L is the box size. This is the same division, which we use to set the low-resolution run. But now we extend it to very high frequencies up to $2\pi/L \times N/2$, where N is the new effective number of particles. For example, we used $N = 64$ for the low-resolution run. For a high-resolution run we may choose $N = 1024$. Simply replace the value and run the code again. Of course, we really cannot do it because it would generate too many particles. Instead, in some regions, where the resolution should not be high, we combine particles together (by taking average coordinates and average velocities) and replace many small-mass particles with fewer larger ones. The top panel in figure 15.1 gives an example of mass refinement. Note that we try to avoid jumps that are too large in the mass resolution by creating layers of particles of increasing mass.

This approach is correct and relatively simple. It may seem that it takes too much CPU time to obtain the initial conditions. In practice, CPU time is not much of an issue because initial conditions are generated only once and it takes only a few CPU hours even for a 1024^3 mesh. For most applications 1024^3 particles is more than enough. The problem arises when we want to have more

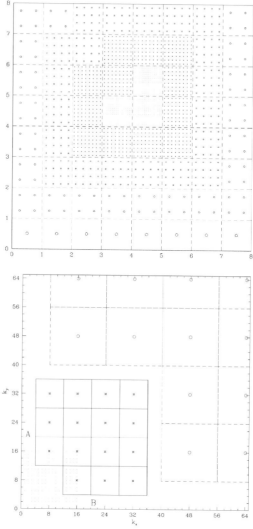

Figure 15.1. An example of the construction of mass refinement in real space (top) and in phase space (bottom). In real space (*top panel*) three central blocks of particles were marked for highest mass resolution. Each block produces 16^2 of smallest particles. Adjacent blocks get one step lower resolution and produce 8^2 particles each. The procedure is repeated recursively. In phase space (*bottom panel*) small points in the left-hand bottom corner represent the harmonics used for the low-resolution simulation. For the high-resolution run with box ratios 1:1/8:1/16 the phase space is sampled more coarsely, but high frequencies are included. Each harmonic (different markers) represents a small cube of the phase space indicated by squares. In this case the matching of the harmonics is not perfect: there are overlapping blocks and gaps. In any case, the waves inside domains A and B are missed in the simulation.

then 1024^3 particles. We simply do not have enough computer memory to store the information for all the harmonics. In this case we must decrease the resolution in the phase space. It is a bit easier to understand the procedure, if we consider phase-space diagrams like the one presented in figure 15.3. The low-resolution run in this case was done for 32^3 particles with harmonics up to $16 \times 2\pi/L$ (small points). For the high-resolution run we choose a region of size 1/8 of the original large box. Inside the small box we place another box, which is twice as small. Thus, we will have three levels of mass refinement. For each level we have the corresponding size of the phase-space block. The size is defined by the size of real-space box and is equal to $2\pi/L \times K$, $K = 1, 8, 16$. Harmonics from different refinements should not overlap: if a region in the phase space is represented on a lower level of resolution, it should not appear in the higher resolution level. This is why the rows of the highest resolution harmonics (circles) with $K_x = 16$ and $K_y = 16$ are absent in figure 15.3: they have already been covered by the lower resolution blocks marked by stars. Figure 15.3 clearly illustrates that matching harmonics is a complicated process: we failed to do the match because there are partially overlapping blocks and there are gaps. We can get much better results, if we assume different ratios of the sizes of the boxes. For example, if instead of box ratios 1:1/8:1/16, we choose ratios 1:3/32:5/96, the coverage of the phase space is almost perfect as shown in figure 15.2.

15.2.4 Codes

There are many different numerical techniques to follow the evolution of a system of many particles. For earlier reviews see Hockney and Eastwood (1981), Sellwood (1987) and Bertschinger (1998). Most of the methods for cosmological applications take some ideas from three techniques: the Particle–Mesh (PM) code, direct summation or the Particle–Particle code and the TREE code. For example, the Adaptive Particle–Particle/Particle–Mesh (AP^3M) code (Couchman 1991) is a combination of the PM code and the Particle–Particle code. The Adaptive-Refinement-Tree code (ART) (Kravtsov *et al* 1997, Kravtsov 1999) is an extension of the PM code with the organization of meshes in the form of a tree. All methods have their advantages and disadvantages.

15.2.4.1 The PM code

This uses a mesh to produce the density and potential. As a result, its resolution is limited by the size of the mesh. There are two advantages of the method: (i) it is fast (the smallest number of operations per particle per time step of all the other methods); and (ii) it typically uses a very large number of particles. The latter can be crucial for some applications. There are several modifications of the code. 'Plain-vanilla' PM was described by Hockney and Eastwood (1981). It includes a cloud-in-cell density assignment and a seven-point discrete analogue of the Laplacian operator. Higher-order approximations improve the accuracy on

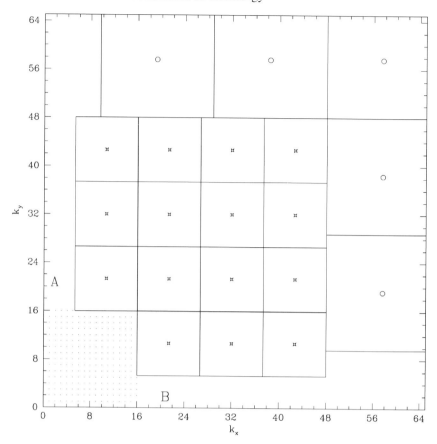

Figure 15.2. Another example of construction of mass refinement in phase space. For the high-resolution run with box ratios 1:3/3:5/96 the phase space is sampled without overlapping blocks or gaps.

large distances but degrades the resolution (e.g. Gelb 1992). The PM code is available (Klypin and Holtzman 1997).

15.2.4.2   The P^3M code

The P^3M code is described in detail in Hockney and Eastwood (1981) and Efstathiou *et al* (1985). It has two parts: the PM part, which takes care of the large-scale forces; and the PP part, which adds the small-scale particle–particle contribution. Because of strong clustering at late stages in the evolution, the PP part becomes prohibitively expensive once large objects start to form in large numbers. A significant speed is achieved in a modified version of the code, which introduces sub-grids (the next levels of PM) in areas with high density

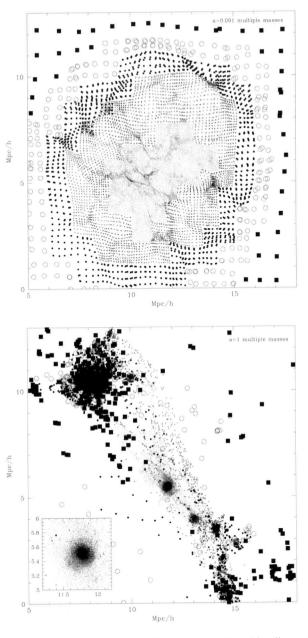

Figure 15.3. Distribution of particles of different masses in a thin slice going through the centre of halo A_1 at redshift 10 (top panel) and at redshift zero (bottom panel). To avoid crowding of points the thickness of the slice is made smaller in the centre (about $30h^{-1}$ kpc) and larger ($1h^{-1}$ Mpc) in the outer parts of the forming halo. Particles of different mass are shown with different symbols.

(Couchman 1991). With modification the code is as fast as the TREE code even for heavily clustered configurations. The code expresses the inter-particle force as a sum of a short-range force (computed by a direct particle–particle pair force summation) and the smoothly varying part (approximated by the particle–mesh force calculation). One of the major problems for these codes is the correct splitting of the force into a short-range and a long-range part. The grid method (PM) is only able to produce reliable inter-particle forces down to a minimum of at least two grid cells. For smaller separations the force can no longer be represented on the grid and therefore one must introduce a cut-off radius r_e (larger than two grid cells), where for $r < r_e$ the force should smoothly go to zero. The parameter r_e defines the chaining-mesh and for distances smaller than this cut-off radius r_e a contribution from the direct particle–particle (PP) summation needs to be added to the total force acting on each particle. Again this PP force should smoothly go to zero for very small distances in order to avoid unphysical particle–particle scattering. This cut-off of the PP force determines the overall force resolution of a P^3M code.

The most widely used version of this algorithm is currently the adaptive P^3M (AP^3M) code of Couchman (1991), which is available publicly. The smoothing of the force in this code is connected to an S_2 sphere, as described in Hockney and Eastwood (1981).

15.2.4.3   *The TREE code*

The TREE code is the most flexible code in the sense of the choice of boundary conditions (Appel 1985, Barnes and Hut 1986, Hernquist 1987). It is also more expensive than PM: it takes 10–50 times more operations. Bouchet and Hernquist (1988) and Hernquist *et al* (1991) extended the code for periodical boundary conditions, which is important for simulating large-scale fluctuations. Some variants of TREE are publicly available. A very useful example is the GADGET code available at http://www.mpa-garching.mpg.de/gadget/right.html. There are variants of the code modified for massively parallel computers and there are variants with variable time stepping, which is vital for extremely high-resolution simulations.

15.2.4.4   *The ART code*

Multi-grid methods were introduced long ago, but only recently have they started to produce important results. Examples of adaptive multi-grid codes are the Adaptive Refinement Tree code (ART; Kravtsov *et al* 1997), the AMR code written by Bryan and Norman and MLAPM (Knebe *et al* 2001). The ART code reaches high-force resolution by refining all high-density regions with an automated refinement algorithm. The refinements are recursive: the refined regions can also be refined, each subsequent refinement having half of the previous level's cell size. This creates a hierarchy of refinement meshes with different resolutions covering the regions of interest. The refinement is done cell-by-cell (individual cells can be refined or de-refined) and meshes are not constrained to have a rectangular (or any other) shape. This allows the code to refine the required regions in an efficient manner. The criterion for refinement is the *local overdensity* of particles: the code refines an individual cell only if the density of particles (smoothed with the cloud-in-cell scheme; Hockney and Eastwood 1981) is higher than n_{TH} particles, with typical values $n_{TH} = 2–5$. The Poisson equation on the hierarchy of meshes is solved first on the base grid using FFT techniques and then on the subsequent refinement levels. On each refinement level the code obtains the potential by solving the Dirichlet boundary problem with boundary conditions provided by the already existing solution at the previous level or from the previous moment of time.

Figure 15.4 (courtesy of A Kravtsov) gives an example of mesh refinement for the hydro-dynamical version of the ART code. The code produced this refinement mesh for a spherical strong explosion (Sedov solution).

The refinement of the time integration mimics the spatial refinement and the time step for each subsequent refinement level is twice as small as the step on the previous level. Note, however, that particles on the same refinement level move with the same step. When a particle moves from one level to another, the time step changes and its position and velocity are interpolated to appropriate time moments. This interpolation is first-order accurate in time, whereas the rest of the integration is done with the second-order accurate-time centred leap-frog scheme. All equations are integrated with the expansion factor a as a time variable and the global time step hierarchy is thus set by the step Δa_0 at the zeroth level (uniform base grid). The step on level L is then $\Delta a_L = \Delta a_0/2^L$.

What code is the best? Which one to choose? There is no unique answer—everything depends on the problem, which we are addressing. If you intend to study the structure of individual galaxies in the large-scale environment, you must have a code with very high resolution, variable time stepping and multiple masses. In this case the TREE or ART codes should be the choice.

15.2.5   Effects of resolution

As the resolution of the simulations improves and the range of their applications broaden, it becomes increasingly important to understand their limits. The effects of resolution and convergence studies were studied in a number of publications (e.g. Moore *et al* 1998, Frenk *et al* 1999, Knebe *et al* 2000, Ghigna *et al* 2000, Klypin *et al* 2001). Knebe *et al* (2000) made a detailed comparison of realistic simulations done with three codes: ART, AP^{3}M and PM. Here we present some of their results and main conclusions. The simulations were done for the standard CDM model with the dimensionless Hubble constant $h = 0.5$ and $\Omega_0 = 1$. The simulation box of $15h^{-1}$ Mpc had 64^3 equal-mass particles, which gives the mass resolution (mass per particle) of $3.55 \times 10^9 h^{-1} M_\odot$. Because of the low resolution of the PM runs, we show results only for the other two codes. For the ART code

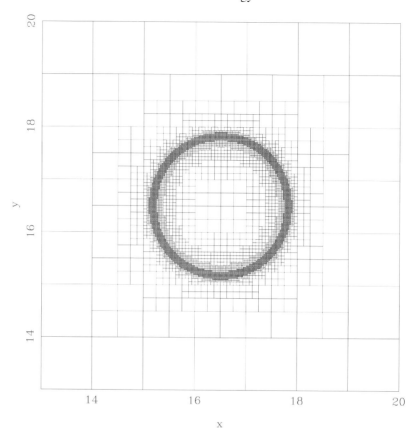

Figure 15.4. An example of a refinement structure constructed by the (hydro)ART code for spherical strong explosion (courtesy of A Kravtsov).

the force resolution is practically fixed by the number of particles. The only free parameter is the number of steps on the lowest (zero) level of resolution. In the case of the AP^3M, besides the number of steps, one can also request the force resolution. Parameters from two runs with the ART code and five simulations with the AP^3M are given in table 15.2.

Figure 15.5 shows the correlation function for the dark matter down to the scale of $5h^{-1}$ kpc, which is close to the force resolution of all our high-resolution simulations. The correlation function in the AP^3M_1 and ART_2 runs are similar to those of AP^3M_5 and ART_1 respectively and are not shown for clarity. We can see that the AP^3M_5 and the ART_1 runs agree to $\lesssim 10\%$ over the whole range of scales. The correlation amplitudes of runs AP^3M_{2-4}, however, are systematically lower at $r \lesssim 50$–$60h^{-1}$ kpc (i.e. the scale corresponding to ≈ 15–20 resolutions), with the AP^3M_3 run exhibiting the lowest amplitude. The fact that the AP^3M_2

Table 15.2. Parameters of the numerical simulations.

Simulation	Softening (h^{-1} kpc)	Dyn. range	Steps (min–max)	N_{steps}/dyn. range
AP^3M_1	3.5	4267	8000	1.87
AP^3M_2	2.3	6400	6000	0.94
AP^3M_3	1.8	8544	6000	0.70
AP^3M_4	3.5	4267	2000	0.47
AP^3M_5	7.0	2133	8000	3.75
ART_1	3.7	4096	660–21 120	2.58
ART_2	3.7	4096	330–10 560	5.16

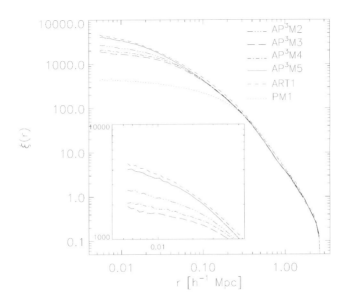

Figure 15.5. The correlation function of dark matter particles. Note that the range of correlation amplitudes is different in the inset panel.

correlation amplitude deviates less than that of the AP^3M_3 run indicates that the effect is very sensitive to the force resolution.

Note that the AP^3M_3 run has formally the best force resolution. Thus, one would naively expect that it would give the largest correlation function. At scales $\lesssim 30h^{-1}$ kpc the deviations of the AP^3M_3 from the ART_1 or the AP^3M_5 runs are ≈ 100–200%. We attribute these deviations to the numerical effects: the high force resolution in AP^3M_3 was not adequately supported by the time integration. In other words, the AP^3M_3 had too few time steps. Note that it had quite a large

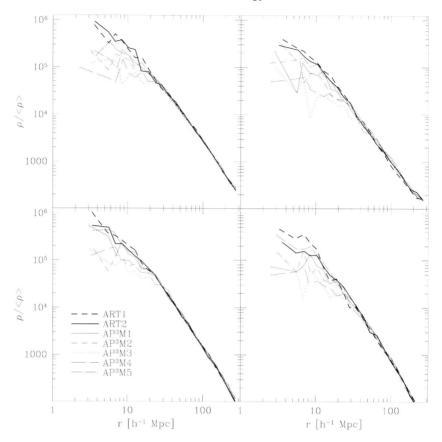

Figure 15.6. Density profiles of four largest halos in simulations of Knebe *et al* (1999). Note that the AP^{3}M$_3$ run has formally the best force resolution, but its actual resolution was much lower because of an insufficient number of steps.

number of steps (6000), not much smaller than the AP^{3}M$_5$ (8000). But for its force resolution, it should have many more steps. The lack of the number of steps was devastating.

Figure 15.6 presents the density profiles of four of the most massive halos in our simulations. We have not shown the profile of the most massive halo because it appears to have undergone a recent major merger and is not very relaxed. In this figure, we present only profiles of halos in the high-resolution runs. Not surprisingly, the inner density of the PM halos is much smaller than in the high-resolution runs and their profiles deviate strongly from the profiles of high-resolution halos at the scales shown in figure 15.6. A glance at figure 15.6 shows that all profiles agree well at $r \gtrsim 30h^{-1}$ kpc. This scale is about eight times smaller than the mean inter-particle separation. Thus, despite the very different resolutions, time steps and numerical techniques used for the simulations, the convergence is observed at a scale much lower than the mean inter-particle separation, argued by Splinter *et al* (1998) to be the smallest trustworthy scale.

Nevertheless, there are systematic differences between the runs. The profiles in two ART runs are identical within the errors indicating convergence (we have run an additional simulation with time steps twice as small as those in the ART_1 finding no difference in the density profiles). Among the AP^3M runs, the profiles of the AP^3M_1 and AP^3M_5 are closer to the density profiles of the ART halos than the rest. The AP^3M_2, AP^3M_3 and AP^3M_4, despite the higher force resolution, exhibit lower densities in the halo cores, the AP^3M_3 and AP^3M_4 runs being the most deviant.

These results can be interpreted, if we examine the trend of the central density, as a function of the ratio of the number of time steps to the dynamic range of the simulations (see table 15.2). The ratio is smaller when either the number of steps is smaller or the force resolution is higher. The agreement in the density profiles is observed when this ratio is $\gtrsim 2$. This suggests that for a fixed number of time steps, there should be a limit on the force resolution. Conversely, for a given force resolution, there is a lower limit on the required number of time steps. The exact requirements would probably depend on the code type and the integration scheme. For the AP^3M code our results suggest that the ratio of the number of time steps to the dynamic range should be no less than one. It is interesting that the deviations in the density profiles are similar to and are observed at the same scales as the deviations in the DM correlation function (figure 15.5), suggesting that the correlation function is sensitive to the central density distribution of dark matter halos.

15.2.6   Halo identification

Finding halos in dense environments is a challenge. Some of the problems that any halo-finding algorithm faces are not numerical. They exist in the real universe. We select a few typical difficult situations.

(1) *A large galaxy with a small satellite.* Examples: LMC and the Milky Way or the M51 system. Assuming that the satellite is bound, do we have to include the mass of the satellite in the mass of the large galaxy? If we do, then we count the mass of the satellite twice: once when we find the satellite and then when we find the large galaxy. This does not seem reasonable. If we do not include the satellite, then the mass of the large galaxy is underestimated. For example, the binding energy of a particle at the distance of the satellite will be wrong. The problem arises when we try to assign particles to different halos in an effort to find the masses of halos. This is very difficult to do for particles moving between halos. Even if a particle at some moment has negative energy relative to one of the halos, it is not guaranteed that it belongs to the halo. The gravitational potential changes with time, and the particle may end up falling onto another halo. This is not just a precaution. This

actually was found very often in real halos when we compared the contents of halos at different redshifts. Interacting halos exchange mass and lose mass. We try to avoid the situation: instead of assigning mass to halos, we find the maximum of the 'rotational velocity', $\sqrt{GM/R}$, which, observationally, is a more meaningful quantity.

(2) *A satellite of a large galaxy.* The previous situation is now viewed from a different angle. How can we estimate the mass or the rotational velocity of the satellite? The formal virial radius of the satellite is large: the big galaxy is within the radius. The rotational velocity may rise all the way to the centre of the large galaxy. In order to find the outer radius of the satellite, we analyse the density profile. At small distances from the centre of the satellite the density steeply declines, but then it flattens out and may even increase. This means that we have reached the outer border of the satellite. We use the radius at which the density starts to flatten out as the first approximation for the radius of the halo. This approximation can be improved by removing unbound particles and checking the steepness of the density profile in the outer part.

(3) *Tidal stripping.* Peripheral parts of galaxies, responsible for extended flat rotation curves outside of clusters, are very likely tidally stripped and lost when the galaxies fall into a cluster. The same happens with halos: a large fraction of the halo mass may be lost due to stripping in dense cluster environments. Thus, if an algorithm finds that 90% of the mass of a halo identified at an early epoch is lost, it does not mean that the halo was destroyed. This is not a numerical effect and is not due to 'lack of physics'. This is a normal situation. What is left of the halo, given that it still has a large enough mass and radius, is a 'galaxy'.

There are different methods of identifying collapsed objects (halos) in numerical simulations.

The *Friends-Of-Friends (FOF)* algorithm was used a lot and still has its adepts. If we imagine that each particle is surrounded by a sphere of radius $bd/2$, then every connected group of particles is identified as a halo. Here d is the mean distance between particles, and b is called the *linking parameter*, which typically is 0.2. The dependence of groups on b is extremely strong. The method stems from an old idea of using percolation theory to discriminate between cosmological models. Because of this, FOF is also called the percolation method, which is wrong because the percolation is about groups spanning the whole box, not collapsed and compact objects. FOF was criticized for failing to find separate groups in cases when those groups were obviously present (Gelb 1992). The problem originates from the tendency of FOF to 'percolate' through bridges connecting interacting galaxies or galaxies in high-density backgrounds.

DENMAX tried to overcome the problems of FOF by dealing with density maxima (Gelb 1992, Bertschinger and Gelb 1991). It finds the maxima of density and then tries to identify particles, which belong to each maximum (halo). The procedure is quite complicated. First, the density field is constructed. Second, the density (with a negative sign) is treated as a potential in which particles start to move as in a viscous fluid. Eventually, particles sink to the bottom of the potentials (which are also maxima density). Third, only particles with negative energy (relative to their group) are retained. Just as in the case of FOF, we can easily imagine situations when (this time) DENMAX should fail; for example, two colliding galaxies in a cluster of galaxies. They should just pass each other because of large relative velocity. In the moment of collision DENMAX ceases to 'see' both galaxies because all particles have positive energies. This is probably a quite unlikely situation. The method is definitely one of the best at present. The only problem is that it seems to be too complicated for the present state of simulations. DENMAX has two siblings—SKID (Stadel *et al*) and BDM (Klypin and Holtzman 1997)—which are frequently used.

'Overdensity 200'. There is no name for this method, but it is often used. Find the density maximum, place a sphere and find the radius, within which the sphere has the mean overdensity 200 (or 177 if you really want to follow the top-hat model of nonlinear collapse).

15.3   Spatial and velocity biases

15.3.1   Introduction

The distribution of galaxies is probably biased with respect to the dark matter. Therefore, galaxies can be used to probe the matter distribution only if we understand the bias. Although the problem of bias has been studied extensively in the past (e.g. Kaiser 1984, Davis *et al* 1985, Dekel and Silk 1986), new data on high redshift clustering and the anticipation of coming measurements have recently generated substantial theoretical progress in the field. The breakthrough in an analytical treatment of the bias was the paper by Mo and White (1996), who showed how bias can be predicted in the framework of the extended Press–Schechter approximation. A more elaborate analytical treatment has been developed by Catelan *et al* (1998a, b), Porciani *et al* (1998) and Sheth and Lemson (1999). The effects of nonlinearity and stochasticity were considered in Dekel and Lahav (1999) (see also Taruya and Suto 2000).

Valuable results are produced by 'hybrid' numerical methods in which low-resolution N-body simulations (typical resolution $\sim$20 kpc) are combined with semi-analytical models of galaxy formation (e.g. Diaferio *et al* 1999, Benson *et al* 2000, Somerville *et al* 2001). Typically, the results of these studies are very close to those obtained with brute-force approach of high-resolution ($\lesssim$2 kpc) N-body simulations (e.g. Colín *et al* 1999, Ghigna *et al* 1998). This agreement is quite remarkable because the methods are very different. It may indicate that the biases of galaxy-size objects are controlled by the random nature of the clustering and merging of galaxies and by dynamical effects, which cause the merging, because those are the only common effects in those two approaches.

Direct N-body simulations can be used for studies of the biases only if they have very high mass and force resolution. Because of numerous numerical effects, halos in low-resolution simulations do not survive in dense environments of clusters and groups (e.g. Moore *et al* 1996, Tormen *et al* 1998, Klypin *et al* 1999a). Estimates of the necessary resolution are given in Klypin *et al* (1999a). Indeed, recent simulations, which have sufficient resolution, have found hundreds of galaxy-size halos moving inside clusters (Ghigna *et al* 1998, Colín *et al* 1999, Moore *et al* 1999, Okamoto and Habe 1999).

It is very difficult to make accurate and trustworthy predictions of luminosities for galaxies, which should be hosted by dark matter halos. Instead of luminosities or virial masses we suggest using circular velocities V_c for both numerical and observational data. For a real galaxy its luminosity tightly correlates with the circular velocity. So, one has a good idea what the circular velocity of the galaxy is. Nevertheless, direct measurements of circular velocities of a large complete sample of galaxies are extremely important because it will provide a direct way of comparing theory and observations. This chapter is mostly based on results presented in Colín *et al* (1999, 2000) and Kravtsov and Klypin (1999).

15.3.2 Oh, bias, bias

There are numerous aspects and notions related with the bias. One should be really careful to understand what type of bias is used. Results can be dramatically different. We start by introducing the overdensity field. If $\bar{\rho}$ is the mean density of some component (e.g. the dark matter or halos), then for each point x in space we have $\delta(x) \equiv [\rho(x) - \bar{\rho}]/\bar{\rho}$. The overdensity can be decomposed into the Fourier spectrum, for which we can find the power spectrum $P(k) = \langle |\delta_k|^2 \rangle$. We can then find the correlation function $\xi(r)$ and the rms fluctuation of $\delta(R)$ smoothed on a given scale R. We can construct the statistics for each component: dark matter, galaxies or halos with given properties. Each statistics gives its own definition of bias b:

$$P_h(k) = b_P^2 P_h(k), \qquad \xi_h(r) = b_\xi^2 \xi_{dm}(r), \qquad \delta_h(R) = b_\delta \delta_{dm}(R). \quad (15.15)$$

The three estimates of the bias b are related. In a special case, when the bias is linear, local, and scale independent all three forms of bias are all equal. In general case they are different and they are complicated nonlinear functions of scale, mass of the halos or galaxies and redshift. The dependence on the scale is not local in the sense that the bias in a given position in space may depend on environment (e.g. density and velocity dispersion) on a larger scale. Bias has memory: it depends on the local history of the fluctuations. There is another complication: bias very likely is not a deterministic function. One source of this stochasticity is that it is non-local. Dependence on the history of clustering may also introduce some random effect.

There are some processes which we know create and affect the bias. At high redshifts there is statistical bias: in a Gaussian correlated field, high-density regions are more clustered than the field itself (Kaiser 1984). Mo and White (1996) showed how the extended Press–Schechter formalism can be used to derive of the bias of the dark matter halos. In the limit of small perturbations on large scales the bias is (Catelan *et al* 1998b, Taruya and Suto 2000)

$$b(M, z, z_{\text{f}}) = 1 + \frac{\nu^2 - 1}{\delta_{\text{c}}(z, z_{\text{f}})}. \tag{15.16}$$

Here $\nu = \delta_{\text{c}}(z, z_{\text{f}})/\sigma(M, z)$ is the relative amplitude of a fluctuation on scale M in units of the rms fluctuation $\sigma(M, z)$ of the density field at redshift z. The parameter z_{f} is the redshift of halo formation. The critical threshold of the top-hat model is $\delta_{\text{c}}(z, z_{\text{f}}) = \delta_{c,0} D(z)/D(z_{\text{f}})$, where D is the growth factor of perturbations and $\delta_{c,0} = 1.69$. At high redshifts, parameter ν for galaxy-size fluctuations is very large and δ_{c} is small. As a result, galaxy-size halos are expected to be more clustered (strongly biases) compared to the dark matter. The bias is larger for more massive objects. As the fluctuations grow, newly formed galaxy-size halos do not have such high peaks as at large redshifts and the bias tends to decrease. It also loses its sensitivity.

At later stages another process starts to change the bias. In group and cluster progenitors the merging and destruction of halos reduces the number of halos. This does not happen in the field where the number of halos of given mass may only increase with time. As a result, the number of halos inside groups and cluster progenitors is reduced relative to the field. This produces (anti)bias: there is a relatively smaller number of halos compared with the dark matter. This merging bias does not depend on the mass of halos and it has a tendency to slow down once a group becomes a cluster with a large relative velocity of halos (Kravtsov and Klypin 1999).

Here is a list of different types of bias. We classify them into three groups: (1) measures of bias, (2) terms related with the description of biases and (3) physical processes, which produce or change the bias.

15.3.2.1 *Measures of bias*

(i)   Bias measured in a statistical sense (e.g. ratio of correlation functions $\xi_{\text{h}}(r) = b^2 \xi_{\text{dm}}(r)$).
(ii) Bias measured point-by-point (e.g. $\delta_{\text{h}}(x) - \delta_{\text{dm}}(x)$ diagrams).

15.3.2.2 *Description of biases*

(i)   Local and non-local bias. For example, $b(R) = \sigma_{\text{h}}(R)/\sigma_{\text{m}}(R)$ is the local bias. If $b = b(R; \tilde{R})$, the bias is non-local, where $\tilde{R}$ is some other scale or scales.

(ii) Linear and nonlinear bias. If in $\xi_h(r) = b^2 \xi_{dm}(r)$ the bias b does not depend on ξ_{dm}, it is the linear bias.

(iii) Scale-dependent and scale-independent bias. If b does not depend on the scale at which the bias is estimated, the bias is scale independent. Note that, in general, the bias can be nonlinear and scale independent, but this highly unlikely.

(iv) Stochastic and deterministic.

15.3.2.3   Physical processes, which produce or change the bias

(i)   Statistical bias. This arises when a specific subset of points is selected from a Gaussian field.

(ii) Merging bias. This is produced due to merging and destruction of halos.

(iii) Physical bias.   This includes any bias due to physical processes inside forming galaxies.

15.3.3   Spatial bias

Colín *et al* (1999) have simulated different cosmological models and, using the simulations, have studied halo biases. Most of the results presented here are for the currently favoured ΛCDM model with the following parameters: $\Omega_0 = 1 - \Omega_\Lambda = 0.3$, $h = 0.7$, $\Omega_b = 0.032$, $\sigma_8 = 1$. The model was simulated with 256^3 particles in a $60h^{-1}$ Mpc box. The formal mass and force resolutions are $m_1 = 1.1 \times 10^9 h^{-1} M_\odot$ and $2h^{-1}$ kpc. The Bound Density Maximum halo finder was used to identify halos with at least 30 bound particles. For each halo we find the maximum circular velocity $V_c = \sqrt{GM(<r)/r}$.

In figure 15.7 we compare the evolution of the correlation functions of the dark matter and halos. There are remarkable differences between the halos and the dark matter. The correlation functions of the dark matter always increases with time (but the rate is different on different scales) and it is never a power law. The correlation function of the halos at redshifts decreases and then starts to increase again. It is accurately described by a power law with slope $\gamma = (1.5\text{--}1.7)$. Figure 15.8 presents a comparison of the theoretical and observational data on correlation functions and power spectra. The dark matter clearly predicts much too high a clustering amplitude. The halos are much closer to the observational points and predict antibias. For the correlation function the antibias appears on scales $r < 5h^{-1}$ Mpc; for the power spectrum the scales are $k > 0.2h$ Mpc^{-1}. One may get an impression that the antibias starts at longer waves in the power spectrum $\lambda = 2\pi/k \approx 30h^{-1}$ Mpc compared with $r \approx 5h^{-1}$ Mpc in the correlation function. There is no contradiction: sharp bias at small distances in the correlation function when Fourier transformed to the power spectrum produces antibias at very small wavenumbers. Thus, the bias should be taken into account at long waves when dealing with the power spectra. There is an inflection point in the power spectrum where the nonlinear power spectrum start to go upward (if

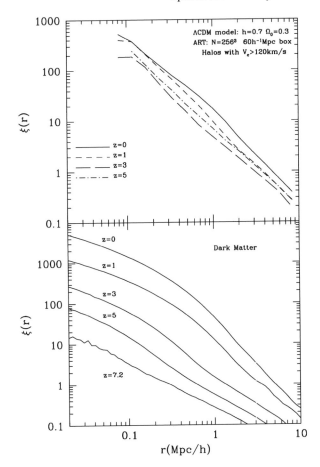

Figure 15.7. Evolution of the correlation function of the dark matter and halos. The correlation function of the dark matter increases monotonically with time. At any given moment it is not a power law. The correlation function of halos is a power law, but it is not monotonic in time.

one moves from low to high k) compared with the prediction of the linear theory. The exact position of this point may have been affected by the finite size of the simulation box $k_{min} = 0.105h^{-1}$ Mpc, but the effect is expected to be small.

At $z = 0$ the bias hardly depends on the mass limit of the halos. There is a tendency of more massive halos to be more clustered at very small distances $r < 200h^{-1}$ kpc, but at this stage it is not clear that this is not due to residual numerical effects around centres of clusters. The situation is different at high redshift. At very high redshifts $z > 3$ galaxy-size halos are very strongly (positively) biased. For example, at $z = 5$ the correlation function of halos

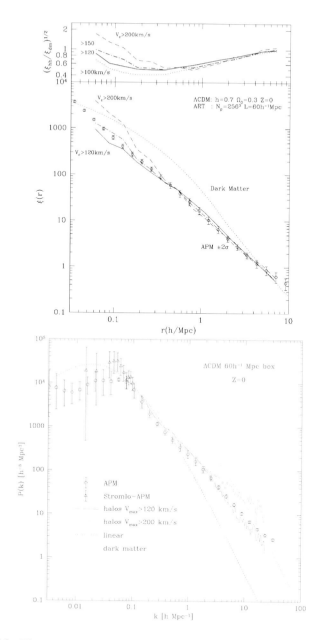

Figure 15.8. The correlation function and the power spectrum of halos with different limiting circular velocities in the ΛCDM model. The results are compared with the observational data from the APM and Stromlo–APM surveys. The bias is scale dependent but it does not depend much on the halo mass.

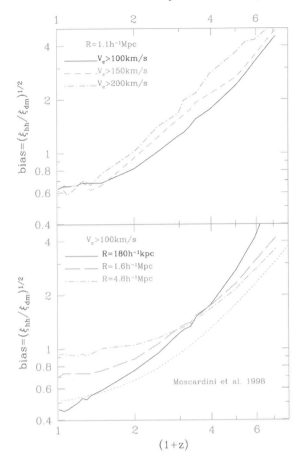

Figure 15.9. *Top panel*: The evolution of bias at comoving scale of $0.54h^{-1}$ Mpc for halos with different circular velocities. *Bottom panel*: Dependence of the bias on the scale for halos with the same circular velocity.

with $v_c > 150$ km s^{-1} was 15 times larger than that of the dark matter at $r = 0.5h^{-1}$ Mpc (see figure 8 in Colín *et al* (1999). The bias was also very strongly mass-dependent with more massive halos being more clustered. At smaller redshifts the bias was declining quickly. Around $z = 1$–2 (the exact value depends on the halo circular velocity) the bias crossed unity and became less than unity (antibias) at later redshifts.

The evolution of bias is illustrated by figure 15.10. The figure shows that, at all epochs, the overdensity of halos tightly correlates with the overdensity of the dark matter. The slope of the relation depends on the dark matter density and evolves with time. At $z > 1$ halos are biased ($\delta_h > \delta_{dm}$) in overdense regions with

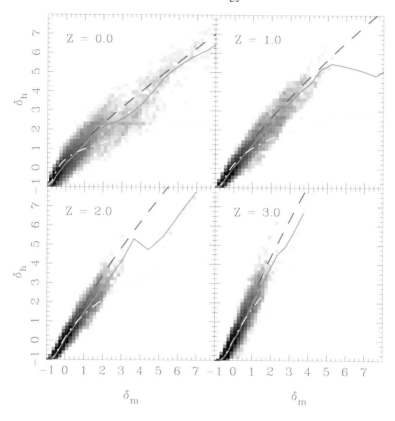

Figure 15.10. Overdensity of halos δ_h versus the overdensity of the dark matter δ_{dm}. The overdensities are estimated in spheres of radius $R_{TH} = 5h^{-1}$ Mpc. The intensity of the grey shade corresponds to the natural logarithm of the number of spheres in a two-dimensional grid in δ_h–δ_{dm} space. The full curves show the average relation. The chain curve is a prediction of an analytical model, which assumes that formation redshift z_f of halos coincides with observation redshift (typical assumption for the Press–Schechter approximation). The long-dashed curve is for a model, which assumes that the substructure survives for some time after it falls into a larger object: $z_f = z + 1$.

$\delta_{dm} > 1$ and antibiased in underdense regions with $\delta_{dm} < -0.5$ At low redshifts there is an antibias at large overdensities and almost no bias at low densities.

Figure 15.11 shows the density profiles for a cluster with mass $2.5 \times 10^{14}h^{-1}M_\odot$. There is antibias on scales below $300h^{-1}$ kpc. This is an example of the merging and destruction bias. Some of the halos have merged or were destroyed by the central cD halo of the cluster. As the result, there is a smaller number of halos in the central part compared with what we would expect if the number density of halos had followed the density of the dark matter (the full curve

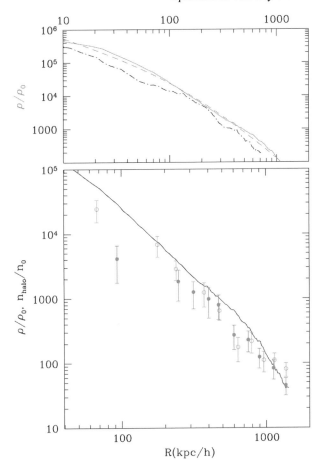

Figure 15.11. Density profiles for a cluster with mass $2.5 \times 10^{14} h^{-1} M_\odot$. *Top panel:* Dark matter density in units of the mean matter density at $z = 0$ (full curve) and at $z = 1$ (chain curve). The Navarro–Frenk–White profile (broken curve) provides a very good fit at $z = 0$. The $z = 1$ profile is given in proper (not comoving) units. *Bottom panel:* Number density profiles of halos in the cluster at $z = 0$ (full circles) and at $z = 1$ (open circles) compared with the $z = 0$ dark matter profile (full curve). There is antibias on scales below $300 h^{-1}$ kpc.

in the bottom panel). Note that, in the outer parts of the cluster, the halos closely follow the dark matter.

15.3.4   Velocity bias

There are two statistics, which measure velocity biases—differences in velocities of the galaxies (halos) and the dark matter. For a review of the results and

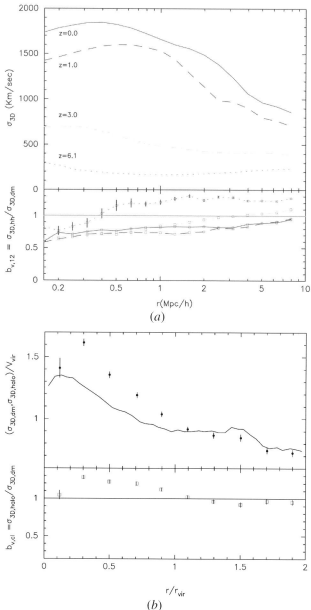

Figure 15.12. (*a*) Two-point velocity bias. (*b*) *Top panel*: 3D rms velocity for halos (circles) and for dark matter (full curve) in the 12 largest clusters. *Bottom panel*: velocity bias in the clusters. The bias in the first point increases to 1.2 if the central cD halos are excluded from analysis. Errors correspond to 1-sigma errors of the mean obtained by averaging over 12 clusters at two moments of time. Fluctuations for individual clusters are larger.

references see Colín *et al* (2000). Two-particle or pairwise velocity bias (PVB) measures the relative velocity dispersion in pairs of objects with given separation r: $b_{12} = \sigma_{h-h}(r)/\sigma_{dm-dm}(r)$. Figure 15.12 (left-hand panel) shows this bias. It is very sensitive to the number of pairs inside clusters of galaxies, where relative velocities are largest. Removal of a few pairs can substantially change the value of the bias. This 'removal' happens when halos merge or are destroyed by central cluster halos.

The one-point velocity bias is estimated as a ratio of the rms velocity of halos to that of the dark matter: $b_1 = \sigma_h/\sigma_{dm}$. It is typically applied to clusters of galaxies where it is measured at different distances from the cluster centre. For an analysis of the velocity bias in clusters, Colín *et al* (2000) have selected the 12 most massive clusters in a simulation of the ΛCDM model. The most massive cluster had virial mass $6.5 \times 10^{14} h^{-1} M_\odot$ comparable to that of the Coma cluster. The cluster had 246 halos with circular velocities larger than 90 km s^{-1}. There were three Virgo-type clusters with virial masses in the range $(1.6\text{–}2.4) \times 10^{14} h^{-1} M_\odot$ and with approximately 100 halos in each cluster. Just like the spatial bias, the PVB is positive at large redshifts (except for the very small scales) and decreases with the redshift. At lower redshifts it does not evolve much and stays below unity (antibias) at scales below $5h^{-1}$ Mpc on level $b_{12} \approx (0.6\text{–}0.8)$.

Figure 15.13 shows the one-point velocity bias in clusters at $z = 0$. Note that the sign of the bias is now different: the halos move slightly faster than the dark matter. The bias is stronger in the central parts ($b_1 = 1.2\text{–}1.3$) and goes to almost no bias ($b_1 \approx 1$) at the virial radius and above. Both the antibias in the pairwise velocities and positive one-point bias are produced by the same physical process—merging and destruction of halos in the central parts of groups and clusters. The difference is in the different weighting of halos in these two statistics. A smaller number of high-velocity pairs significantly changes the PVB, but it only slightly affects the one-point bias because it is normalized to the number of halos at a given distance from the cluster centre. At the same time, merging preferentially happens for halos, which move with a smaller velocity at a given distance from the cluster centre. Slower halos have shorter dynamical times and have smaller apocentres. Thus, they have a better chance to be destroyed and merge with the central cD halo. Because low-velocity halos are eaten up by the central cD, the velocity dispersion of those which survive is larger. Another way of addressing the issue of velocity bias is to use the Jeans equations. If we have a tracer population, which is in equilibrium in a potential produced by mass $M(<r)$, then

$$-r\sigma_r^2(r) \left[\frac{d \ln \sigma_r^2(r)}{d \ln r} + \frac{d \ln \rho(r)}{d \ln r} + 2\beta(r) \right] = GM(<r), \qquad (15.17)$$

where ρ is the number density of the tracer, β is the velocity anisotropy, and σ_r is the rms radial velocity. The right-hand side of the equation is the same for

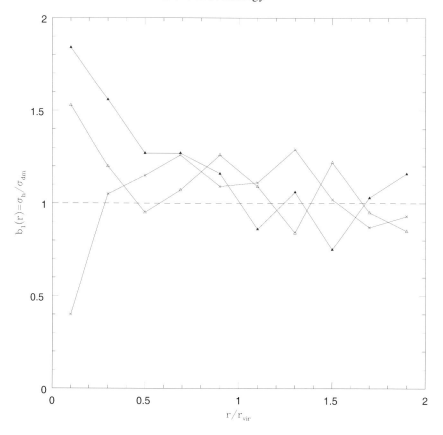

Figure 15.13. One-point velocity bias for three Virgo-type clusters in the simulation. Central cD halos are not included. Fluctuations in the bias are very large because each cluster has only $\sim$100 halos with $V_c > 90$ km s^{-1} and because of substantial substructure in the clusters.

the dark matter and the halos. If the term in the brackets were to be the same, there would be no velocity bias. But there is systematic difference between the halos and the dark matter: the slope of the distribution halos in a cluster $\frac{d \ln \rho(r)}{d \ln r}$ is smaller than that of the dark matter (see Colín *et al* 1999, Ghigna *et al* 2000). The reason for the difference in the slopes is the same—merging with the central cD. Other terms in the equation also have small differences but the main contribution comes from the slope of the density. Thus, as long as we have spatial antibias of the halos, there should be a small positive one-point velocity bias in the clusters and a very strong antibias in the pairwise velocity. The exact values of the biases are still under debate, but one thing seems to be certain: the two biases go hand in hand.

The velocity bias in clusters is difficult to measure because it is small. Figure 15.12 may be misleading because it shows the average trend but it does not give the level of fluctuations for a single cluster. Note that the errors in the plots correspond to the error of the mean obtained by averaging over 12 clusters and two close moments of time. The fluctuations for a single cluster are much larger. Figure 15.12 shows results for three Virgo-type clusters in the simulation. The noise is very large both because of poor statistics (small number of halos) and the noise produced by residual non-equilibrium effects (substructure). A comparable (but slightly smaller) value of b_v was recently found in simulations by Ghigna *et al* (1999) for a cluster in the same mass range as that in figure 15.12. Unfortunately, it is difficult to make a detailed comparison with their results because Ghigna *et al* (1999) use only one hand-picked cluster for a different cosmological model. Very likely their results are dominated by the noise due to residual substructure. The results of another high-resolution simulation by Okamoto and Habe (1999) are consistent with our results.

15.3.5 Conclusions

There are a number of physical processes which can contribute to the biases. In this contribution we explore the dynamical effects in the dark matter itself, which result in differences in the spatial and velocity distribution of the halos and the dark matter. Other effects related to the formation of the luminous parts of galaxies can also produce or change biases. At this stage it is not clear how strong these biases are. Because there is a tight correlation between the luminosity and circular velocity of galaxies, any additional biases are limited by the fact that galaxies 'know' how much dark matter they have.

Biases in the halos are reasonably well understood and can be approximated on a few megaparsec scales by analytical models. We find that the biases in the distribution of the halos are sufficient to explain within the framework of standard cosmological models the clustering properties of galaxies on a vast ranges of scales from 100 kpc to dozens of megaparsecs. Thus, there is neither need nor much room for additional biases in the standard cosmological model.

In any case, biases in the halos should be treated as benchmarks for more complicated models, which include non-gravitational physics. If a model cannot reproduce the biases of halos or it does not have enough halos, it should be rejected, because it fails to give the correct dynamics for the main component of the universe—the dark matter.

15.4 Dark matter halos

15.4.1 Introduction

During the last decade there has been an increasing interest in testing the predictions of variants of the cold dark matter (CDM) models at sub-galactic

($\lesssim$100 kpc) scales. This interest was first induced by indications that the observed rotation curves in the central regions of dark-matter-dominated dwarf galaxies are at odds with predictions of hierarchical models. Specifically, it was argued (Moore 1994, Flores and Primack 1994) that the circular velocities, $v_c(r) \equiv [GM(< r)/r]^{1/2}$, at small galactocentric radii predicted by the models are too high and increase too rapidly with increasing radius compared to the observed rotation curves. The steeper than expected rise in $v_c(r)$ implies that the *shape* of the predicted halo density distribution is incorrect and/or that the DM halos formed in CDM models are too concentrated (i.e. have too much of their mass concentrated in the inner regions).

In addition to the density profiles, there is an alarming mismatch in the predicted abundance of small-mass ($\lesssim 10^8$–$10^9 h^{-1} M_\odot$) galactic satellites and the observed number of satellites in the Local Group (Kauffmann *et al* 1993, Klypin *et al* 1999b, Moore *et al* 1999). Although this discrepancy may well be due to feedback processes such as photoionization that prevent gas collapse and star formation in the majority of the small-mass satellites (e.g. Bullock *et al* 2000), the mass scale at which the problem sets in is similar to the scale in the spectrum of primordial fluctuations that may be responsible for the problems with density profiles. In the age of precision cosmology that the forthcoming MAP and Planck cosmic microwave background anisotropy satellite missions are expected to bring about, tests of the cosmological models at small scales may prove to be the final frontier and the ultimate challenge to our understanding of the cosmology and structure formation in the universe. However, this obviously requires detailed predictions and checks from the theoretical side and higher resolution/quality observations and thorough understanding of their implications and associated caveats from the observational side. In this section we focus on the theoretical predictions of the density distribution of DM halos and some problems with comparing these predictions to observations.

A systematic study of halo density profiles for a wide range of halo masses and cosmologies was carried out by Navarro *et al* (1996, 1997; hereafter NFW), who argued that an analytical profile of the form $\rho(r) = \rho_s(r/r_s)^{-1}(1 + r/r_s)^{-2}$ provides a good description of halo profiles in their simulations for all halo masses and in all cosmologies. Here, r_s is the scale radius which, for this profile corresponds to the scale at which d$\log \rho(r)$/d$\log r|_{r=r_s} = -2$. The parameters of the profile are determined by the halo's virial mass M_{vir} and *concentration* defined as $c \equiv r_{vir}/r_s$. NFW argued that there is a tight correlation between c and M_{vir}, which implies that the density distributions of halos of different masses can, in fact, be described by a one-parameter family of analytical profiles. Further studies by Kravtsov *et al* (1997, 1999), Jing (2000) and Bullock *et al* (2001), although confirming the $c(M_{vir})$ correlation, indicated that there is a significant scatter in the density profiles and concentrations for DM halos of a given mass.

Following the initial studies by Moore (1994) and Flores and Primack (1994), Kravtsov *et al* (1999) presented a systematic comparison of the results of numerical simulations with rotation curves of a sample of 17 DM-dominated dwarf and low-surface-brightness (LSB) galaxies. Based on these comparisons, we argued that there does not seem to be a significant discrepancy in the *shape* of the density profiles at the scales probed by the numerical simulations ($\gtrsim 0.02$–$0.03 r_{vir}$, where r_{vir} is the halo's virial radius). However, these conclusions were subject to several caveats and had to be tested. First, the observed galactic rotation curves had to be re-examined more carefully and with higher resolution. The fact that all of the observed rotation curves used in earlier analyses were obtained using relatively low-resolution HI observations, required checks of the possible beam smearing effects. Also, the possibility of non-circular random motions in the central regions that could modify the rotation velocity of the gas (e.g. Binney and Tremain 1987, p 198) had to be considered. Second, the theoretical predictions had to be tested for convergence and extended to scales $\lesssim 0.01 r_{vir}$.

Moore *et al* (1998; see also a more recent convergence study by Ghigna *et al* 2000) presented a convergence study and argued that the mass resolution has a significant impact on the central density distribution of halos. They argued that at least several million particles per halo are required to model the density profiles at scales $\lesssim 0.01 r_{vir}$ reliably. Based on these results, Moore *et al* (1999) advocated a density profile of the form $\rho(r) \propto (r/r_0)^{-1.5}[1 + (r/r_0)^{1.5}]^{-1}$, that behaves similarly ($\rho \propto r^{-3}$) to the NFW profile at large radii, but is steeper at small r: $\rho \propto r^{-1.5}$. Most recently, Jing and Suto (2000) presented a systematic study of density profiles for halo masses ranging from $2 \times 10^{12} h^{-1} M_\odot$ to $5 \times 10^{14} h^{-1} M_\odot$. The study was uniform in mass and force resolution featuring $\sim (5$–$10) \times 10^5$ particles per halo and a force resolution of $\sim 0.004 r_{vir}$. They found that the galaxy-mass halos in their simulations are well fitted by profile† $\rho(r) \propto (r/r_0)^{-1.5}[1 + r/r_0]^{-1.5}$, but that cluster-mass halos are well described by the NFW profile, with a logarithmic slope of the density profiles at $r = 0.01 r_{vir}$ changing from ≈ -1.5 for $M_{vir} \sim 10^{12} h^{-1} M_\odot$ to ≈ -1.1 for $M_{vir} \sim 5 \times 10^{14} h^{-1} M_\odot$. Jing and Suto interpreted these results as evidence that the profiles of DM halos are not universal.

The rotation curves of a number of dwarf and LSB galaxies have recently been reconsidered using Hα observations (e.g. Swaters *et al* 2000, van den Bosch *et al* 2000). The results show that, for some galaxies, Hα rotation curves are significantly different in their central regions than the rotation curves derived from HI observations. This indicates that the HI rotation curves are affected by beam smearing (Swaters *et al* 2000). It is also possible that some of the difference may be due to real differences in the kinematics of the two tracer gas components (ionized and neutral hydrogen). Preliminary comparisons of the new Hα rotation curves with model predictions show that the NFW density profiles are consistent with the observed *shapes* of the rotation curves (van den Bosch *et al* 2000). Moreover, cusp density profiles with inner logarithmic slopes as steep as ~ -1.5 also seem to be consistent with the data (van den Bosch *et al* 2000). Nevertheless,

---

† Note that their profile is somewhat different from the profile advocated by Moore *et al*, but behaves similarly to the latter at small radii.

CDM halos appear to be too concentrated (Navarro and Swaters 2000, McGaugh *et al* 2000) compared to galactic halos and therefore the problem remains.

New observational and theoretical developments show that a comparison of model predictions to the data is not straightforward. Decisive comparisons require the convergence of theoretical predictions and understanding the kinematics of the gas in the central regions of the observed galaxies. In this section we present convergence tests designed to test the effects of mass resolution on the density profiles of halos formed in the currently popular CDM model with cosmological constant (ΛCDM) and simulated using the multiple mass resolution version of the Adaptive Refinement Tree code (ART). We also discuss some caveats in drawing conclusions about the density profiles from the fits of analytical functions to numerical results and their comparisons to the data.

15.4.2   Dark matter halos: the NFW and the Moore *et al* profiles

Before we fit the analytical profiles to real dark matter halos or compare them with observed rotational curves, it is instructive to compare different analytical approximations. Although the NFW and Moore *et al* profiles predict different behaviour for $\rho(r)$ in the central regions of a halo, the scale where this difference becomes significant depends on the specific values of the halo's characteristic density and radius. Table 15.3 presents the different parameters and statistics associated with the two analytical profiles. For the NFW profile more information can be found in Klypin *et al* (1999a, b, 2001), Lokas and Mamon (2000) and in Widrow (2000).

Each profile is set by two independent parameters. We choose these to be the characteristic density ρ_0 and radius r_s. In this case all expressions describing the different properties of the profiles have a simple form and do not depend on the concentration. The concentration or the virial mass appears only in the normalization of the expressions. The choice of the virial radius (e.g. Lokas and Mamon 2000) as a scale unit results in more complicated expressions with an explicit dependence on the concentration. In this case, one also has to be careful about the definition of the virial radius, as there are several different definitions in the literature. For example, it is often defined as the radius, r_{200}, within which the average density is 200 times the *critical density*. In this section the virial radius is defined as the radius within which the average density is equal to the density predicted by the top-hat model: it is δ_{TH} times the *average matter density* in the universe. For the $\Omega_0 = 1$ case the two existing definitions are equivalent. In the case of $\Omega_0 = 0.3$ models, however, the virial radius is about 30% larger than r_{200}.

There is no unique way of defining a consistent concentration for the different analytical profiles. Again, it is natural to use the characteristic radius r_s to define the concentration: $c \equiv r_{vir}/r_s$. This simplifies the expressions. At the same time, if we fit the same dark matter halo with the two profiles, we will get different concentrations because the values of the corresponding r_s will be different. Alternatively, if we choose to match the outer parts of the profiles

Table 15.3. Comparison of the NFW and Moore *et al* profiles.

Parameter	NFW	Moore *et al*
Density $x = r/r_s$	$\rho = \dfrac{\rho_0}{x(1+x)^2}$ $\rho \propto x^{-3}$ for $x \gg 1$ $\rho \propto x^{-1}$ for $x \ll 1$ $\rho/\rho_0 = 1/4.00$ at $x = 1$ $\rho/\rho_0 = 1/21.3$ at $x = 2.15$	$\rho = \dfrac{\rho_0}{x^{1.5}(1+x)^{1.5}}$ $\rho \propto x^{-3}$ for $x \gg 1$ $\rho \propto x^{-1.5}$ for $x \ll 1$ $\rho/\rho_0 = 1/2.00$ at $x = 1$ $\rho/\rho_0 = 1/3.35$ at $x = 1.25$
Mass $M = 4\pi\rho_0 r_s^3 f(x)$ $M = M_{\text{vir}} f(x)/f(C)$ $M_{\text{vir}} = \frac{4}{3}\pi\rho_{\text{cr}}\Omega_0\delta_{\text{TH}} r_{\text{vir}}^3$	$f(x) = \ln(1+x) - \dfrac{x}{1+x}$	$f(x) = \frac{2}{3}\ln(1 + x^{3/2})$
Concentration $C = r_{\text{vir}}/r_s$	$C_{\text{NFW}} = 1.72 C_{\text{Moore}}$ (for the same M_{vir} and r_{max}) $C_{1/5} \approx \dfrac{C_{\text{NFW}}}{0.86 f(C_{\text{NFW}})+0.1363}$ (error $<3\%$ for $C_{\text{NFW}} = 5 - 30$) $C_{\gamma=-2} = C_{\text{NFW}}$	$C_{\text{Moore}} = C_{\text{NFW}}/1.72$ $C_{1/5} = \dfrac{C_{\text{Moore}}}{[(1+C_{\text{Moore}}^{3/2})^{1/5}-1]^{2/3}}$ $C_{0.0} \approx \dfrac{C_{\text{Moore}}}{[C_{\text{Moore}}^{3/10}-1]^{2/3}}$ $C_{\gamma=-2} = 2^{3/2} C_{\text{Moore}}$ $C_{\gamma=-2} = \approx 2.83 C_{\text{Moore}}$
Circular velocity $v_c^2 = \dfrac{GM_{\text{vir}}}{r_{\text{vir}}} \dfrac{C}{x} \dfrac{f(x)}{f(C)}$ $v_c^2 = v_{\text{max}}^2 \dfrac{x_{\text{max}}}{x} \dfrac{f(x)}{f(x_{\text{max}})}$ $v_{\text{vir}}^2 = \dfrac{GM_{\text{vir}}}{r_{\text{vir}}}$	$x_{\text{max}} \approx 2.15$ $v_{\text{max}}^2 \approx 0.216 v_{\text{vir}}^2 \dfrac{C}{f(C)}$	$x_{\text{max}} \approx 1.25$ $v_{\text{max}}^2 \approx 0.466 v_{\text{vir}}^2 \dfrac{C}{f(C)}$

(say, $r > r_s$) as closely as possible, we may choose to change the ratio of the characteristic radii $r_{s,\text{NFW}}/r_{s,\text{Moore}}$ in such a way that both profiles reach the maximum circular velocity v_c at the same physical radius r_{max}. In this case, the formal concentration of the Moore *et al* profile is 1.72 times smaller than that of the NFW profile. Indeed, with this normalization the profiles look very similar in the outer parts as one finds in figure 15.14. Table 15.3 also gives two other 'concentrations'. The concentration $C_{1/5}$ is defined as the ratio of virial radius to the radius, which encompasses one-fifth of the virial mass (Avila-Reese *et al* 1999). For halos with $C_{\text{NFW}} \approx 5.5$ this one-fifth mass concentration is equal to C_{NFW}. One can also define the concentration as the ratio of the virial radius to the radius at which the logarithmic slope of the density profile is equal to -2. This scale corresponds to r_s for the NFW profile and $\approx 0.35 r_s$ for the Moore *et al* profile.

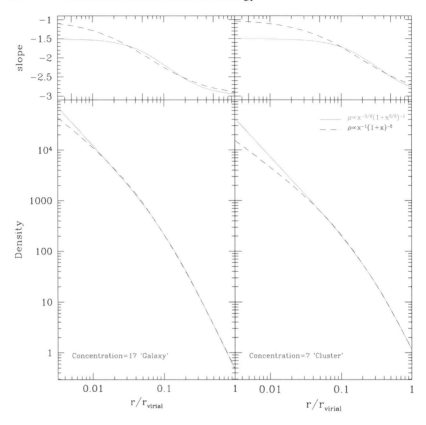

Figure 15.14. Comparison of the Moore *et al* and NFW profiles. Each profile is normalized to have the same virial mass and the same radius of the maximum circular velocity. *Left panels*: High-concentration halo with concentrations typical for small galaxies $C_{NFW} = 17$. *Right panels*: Low-concentration halo with concentrations typical for clusters of galaxies. The deviations are very small ($<3\%$) for radii $r > 1/2r_s$. The top panels show the local logarithmic slope of the profiles. Note that for the high concentration halo the slope of the profile is significantly larger than the asymptotic value -1 even at very small radii $r \approx 0.01/r_{vir}$.

Figure 15.14 presents a comparison of the analytic profiles normalized to have the same virial mass and the same radius r_{max}. We show the results for halos with low and high concentration values which are representative of cluster- and low-mass galaxy halos, respectively. The bottom panels show the profiles, while the top panels show the corresponding logarithmic slope as a function of the radius. The figure shows that the two profiles are very similar throughout the main body of the halos. Only in the very central region do the differences become significant. The difference is more apparent in the logarithmic slope than in the actual density profiles. Moreover, for galaxy-mass halos the difference sets in at a rather small radius ($\lesssim 0.01 r_{\text{vir}}$), which would correspond to scales less than 1 kpc for the typical DM-dominated dwarf and LSB galaxies. In most analyses involving galaxy-size halos, the differences between the NFW and Moore *et al* profiles are irrelevant, and the NFW profile should provide an accurate description of the density distribution.

Note also that for galaxy-size (e.g. high-concentration) halos the logarithmic slope of the NFW profile does not reach its asymptotic inner value of -1 at scales as small as $0.01 r_{\text{vir}}$. For $\sim 10^{12} h^{-1} M_{\odot}$ halos the logarithmic slope of the NFW profile is ≈ -1.4–1.5, while for cluster-size halos this slope is ≈ -1.2. This dependence of slope at a given fraction of the virial radius on the virial mass of the halo is very similar to the results plotted in figure 3 of Jing and Suto (2000). They interpreted it as evidence that the halo profiles are not universal. It is obvious, however, that their results are consistent with the NFW profiles and the dependence of the slope on mass can simply be a manifestation of the well-studied $c_{\text{vir}}(M)$ relation.

To summarize, we find that the differences between the NFW and Moore *et al* profiles are very small ($\Delta \rho / \rho < 10\%$) for radii above 1% of the virial radius. The differences are larger for halos with smaller concentrations. For the NFW profile, the asymptotic value of the central slope $\gamma = -1$ is not achieved even at radii as small as 1–2% of the virial radius.

15.4.3 Properties of dark matter halos

Some properties of halos depend on the large-scale environment in which the halos are found. We will call a halo *distinct* if it is not inside a virial radius of another (larger) halo. A halo is called a *sub-halo* if it is inside another halo. The number of sub-halos depends on the mass resolution—the deeper we go, the more sub-halos we will find. Most of the results given here are based on a simulation, which was complete to masses down to $10^{11} h^{-1} M_{\odot}$ or, equivalently, to the maximum circular velocity of 100 km s^{-1}.

15.4.3.1 Mass and velocity distribution functions

The halo mass and velocity function has been extensively analysed by Sigad *et al* (2000) for halos in the ΛCDM model. Additional results can also be found in Ghigna *et al* (1999), Moore *et al* (1999), Klypin *et al* (1999b) and Gottlöber *et al* (1998). Figure 15.15 compares the mass function of sub-halos and distinct halos. The Press–Schechter approximation overestimates the mass function by a factor of twofor $M < 5 \times 10^{12} h^{-1} M_{\odot}$ and it somewhat underestimates it at larger masses. A more advanced approximation given by Sheth and Tormen is more accurate. On scales below $10^{14} h^{-1} M_{\odot}$ the mass function is close to a power law with slope $\alpha \approx -1.8$. There is no visible difference in the slope for sub-halos and for the distinct halos.

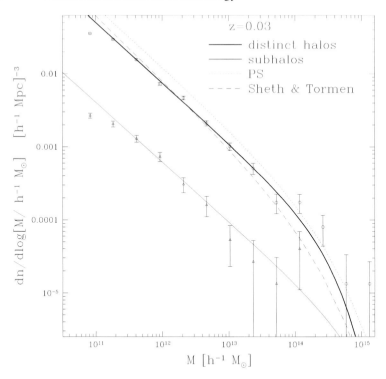

Figure 15.15. The mass function for distinct halos (top) and for sub-halos bottom). Raw counts are marked by symbols with error bars. The curves are Schechter-function fits. The Press–Schechter (dotted) and Sheth–Tormen (dashes) predictions for distinct halos are also shown. On scales below $10^{14}h^{-1}M_\odot$ the mass function is close to a power law with slope $\alpha \approx -1.8$. There is no visible difference in the slope for sub-halos and that for distinct halos. (After Sigad *et al* 2000.)

For each halo one can measure the maximum circular velocity V_{max}. In many cases (especially for sub-halos) V_{max} is a better measure of the size of the halo. It is also related more closely with the observed properties of galaxies hosted by halos. Figure 15.16 presents the velocity distribution functions of different types of halo. In addition to distinct halos and sub-halos, we also show isolated halos and halos in groups and clusters. Here isolated halos are defined as halos with a mass less than $10^{13}h^{-1}M_\odot$, which are not inside a larger halo and which do not have sub-halos more massive than $10^{11}h^{-1}M_\odot$. The velocity function is approximated by a power law $dn = \Phi_* V_{max}^\beta \, dV_{max}$ with slope $\beta \approx -3.8$ for distinct halos. The slope depends on the environment: $\beta \approx -3.1$ for halos in groups and $\beta \approx -4$ for isolated halos. Klypin *et al* (1999b) and Ghigna *et al* (1999) found that the slope $\beta \approx -3.8$–4 of the velocity function extends to much smaller halos with velocities down to 10 km s^{-1}.

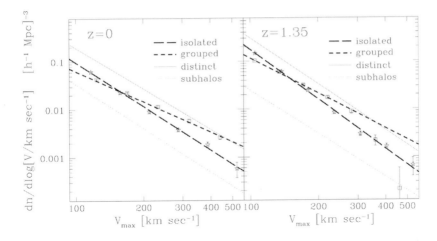

Figure 15.16. Velocity functions for isolated halos (squares) and for halos in groups and clusters. Halos with mass less than $10^{13}h^{-1}M_\odot$ are used for the plots. (After Sigad *et al* 2000.)

15.4.3.2 *Correlation between characteristic density and radius*

The halo density profiles are approximated by the NFW profile:

$$\rho = \frac{\rho_0}{(r/r_0)[1 + r/r_0]^2}. \tag{15.18}$$

Kravtsov *et al* (1999) found the correlation between the two parameters of halos: ρ_0 and r_s. Figure 15.17 compares the results for the DM halos with those for DM-dominated, LSB galaxies and dwarf galaxies. The halos are consistent with observational data: smaller halos are denser.

15.4.3.3 *Correlations between mass, concentration and redshift*

Navarro *et al* (1997) argued that the halo profiles have a universal shape in the sense that the profile is uniquely defined by the virial mass of the halo. Bullock *et al* (2001) analysed concentrations of thousands of halos at different redshifts. To some degree they confirm the conclusions of Navarro *et al* (1997): halo concentration correlates with its mass. However, some significant deviations were also found. There is no one-to-one relation between concentration and mass. It appears the the universal profile should only be treated as a trend: the halo concentration does increase as the halo mass decreases, but there are large deviations for individual halos from that 'universal' shape. Halos have an intrinsic scatter of concentration: at the 1σ level halos with the same mass have $\Delta(\log c_{\mathrm{vir}}) = 0.18$ or, equivalently, $\Delta V_{\mathrm{max}}/V_{\mathrm{max}} = 0.12$.

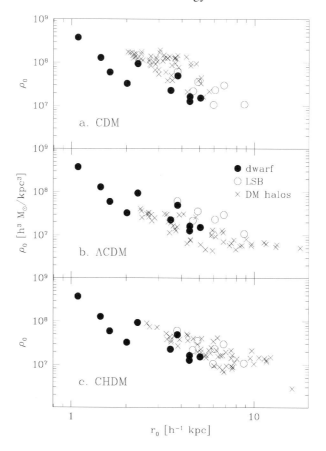

Figure 15.17. Correlation of the characteristic density ρ_0 and radius r_0 for the dwarf and LSB galaxies (full and open circles) and for DM halos (crosses) in different cosmological models. The halos are consistent with observational data: smaller halos are denser. (After Kravtsov *et al* 1999.)

15.4.3.4 *Velocity anisotropy*

Inside a large halo, sub-halos or DM particles do not move on either circular or radial orbits. A velocity ellipsoid can be measured at each position inside a halo. It can be characterized by an anisotropy parameter defined as $\beta(r) = 1 - V_\perp^2/2V_r^2$. Here $V_\perp^2$ is the velocity dispersion perpendicular to the radial direction and V_r^2 is the radial velocity dispersion. For pure radial motions $\beta = 1$. For isotropic velocities $\beta = 0$. The function $\beta(r)$ was estimated for halos in different cosmological models (see Colín *et al* 1999 for references). By studying 12 rich clusters with many sub-halos inside each of them, Colín *et al* (1999) found that both the sub-halos and DM particles can be described by the same anisotropy

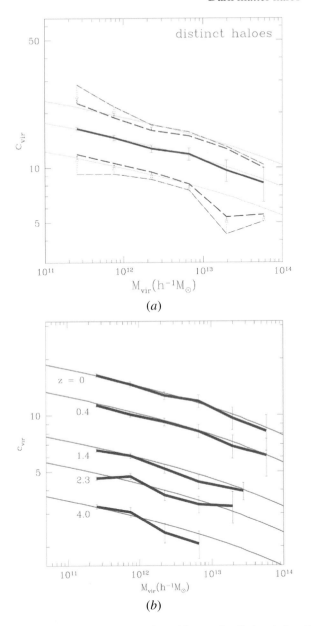

Figure 15.18. (*a*) Dependence of concentration with mass for distinct halos. The bold full curve is the median value. The errors are errors of the mean due to sampling. The outer chain curves encompass 68% of halos in the simulations. The broken curves and arrows indicate values corrected for the noise in halo profiles. Thin curves are different analytical models. (*b*) Median halo concentration as a function of mass for different redshifts. The thin lines show the predictions of an analytical model. (After Bullock *et al* 2001.)

parameter

$$\beta(r) = 0.15 + \frac{2x}{x^2 + 4}, \qquad x = r/r_{\text{vir}}. \qquad (15.19)$$

15.4.4   Halo profiles: convergence study

The following results are based on Klypin *et al* (2001).

15.4.4.1   Numerical simulations

Using the ART code (Kravtsov *et al* 1997, Kravtsov 1999), we simulate a flat low-density cosmological model (ΛCDM) with $\Omega_0 = 1 - \Omega_\Lambda = 0.3$, the Hubble parameter (in units of 100 km s^{-1} Mpc^{-1}) $h = 0.7$, and the spectrum normalization $\sigma_8 = 0.9$. We run two sets of simulations with $30h^{-1}$ Mpc and $25h^{-1}$ Mpc computational box. The first simulations were run to the present moment $z = 0$. The second set of simulations had higher mass resolution and therefore produced more halos but were run only to $z = 1$.

In all our simulations the step in the expansion parameter was chosen to be $\Delta a_0 = 2 \times 10^{-3}$ on the zero level of resolution. This gives about 500 steps for an entire run to $z = 0$. A test run was done with a time step twice as small as that for a halo of comparable mass (but with a smaller number of particles) as studied in this chapter. We did not find any visible deviations in the halo profile. In the first set of simulations, the highest refinement level was ten, which corresponds to $500 \times 2^{10} \approx 500\,000$ time steps at the tenth level. For the second set of simulations, nine levels of refinement were reached which corresponds to $128\,000$ steps at the ninth level.

In the following sections we present the results for four halos. The first halo (A) was the only halo selected for re-simulation in the first set of simulations. In this case the selected halo was relatively quiescent at $z = 0$ and had no massive neighbours. The halo was located in a long filament bordering a large void. It was about 10 Mpc away from the nearest cluster-size halo. After the high-resolution simulation was completed we found that the nearest galaxy-size halo was about 5 Mpc away. The halo had a fairly typical merging history with an $M(t)$ track slightly lower than the average mass growth predicted using extended Press–Schechter model. The last major merger event occurred at $z \approx 2.5$; at lower redshifts the mass growth (the mass in this time interval has grown by a factor of three) was due to slow and steady mass accretion.

The second set of simulations was done in a different way. In the low-resolution run we selected three halos in a well-pronounced filament. Two of the halos were neighbours located at about 0.5 Mpc from each other. The third halo was 2 Mpc away from this pair. Thus, the halos were not selected to be too isolated as was the case in the first set of runs. Moreover, the simulation was analysed at an earlier moment ($z = 1$) where halos are more likely to be unrelaxed. Therefore, we consider the halo A from the first set as an example

Table 15.4. Parameters of halos.

z	M_{vir} $M_\odot/h$	R_{vir} kpc h^{-1}	V_{max} km s^{-1}	N_{part}	m_{part} $M_\odot/h$	Form. res. kpc h^{-1}	C_{NFW}	RelEr NFW	RelEr Moore
(1) (2)	(3)	(4)	(5)	(6)	(7)	(8)	(9)	(10)	(11)
A$_1$ 0	1.97×10^{12}	257	247.0	1.2×10^5	1.6×10^7 0.23		17.4	0.17	0.20
A$_2$ 0	2.05×10^{12}	261	248.5	1.5×10^4	1.3×10^8 0.91		16.0	0.13	0.16
A$_3$ 0	1.98×10^{12}	256	250.5	1.9×10^3	1.1×10^9 3.66		16.6	0.16	0.10
B 1	8.5×10^{11}	241	195.4	7.1×10^5	1.2×10^6 0.19		12.3	0.23	0.16
C 1	6.8×10^{11}	208	165.7	5.0×10^5	1.2×10^6 0.19		11.9	0.37	0.20
D 1	9.6×10^{11}	245	202.4	7.9×10^5	1.2×10^6 0.19		9.5	0.25	0.60

of a rather isolated well-relaxed halo. In many respects, this halo is similar to halos simulated by other research groups that used multiple mass resolution techniques. The three halos from the second set of simulations can be viewed as being representative of more typical halos, not necessarily well relaxed and located in more crowded environments.

The parameters of the simulated DM halos are listed in table 15.4. Columns represent:

(1) the halo 'name' (halos A$_1$, A$_2$, A$_3$ are halo A re-simulated with different resolutions);
(2) the redshift at which the halo was analysed;
(3)–(5) the virial mass, comoving virial radius and maximum circular velocity. At $z = 0$ ($z = 1$) the virial radius was estimated as the radius within which the average overdensity of matter is 340 (180) times larger than the mean cosmological density of matter at that redshift;
(6) the number of particles within the virial radius;
(7) the smallest particle mass in the simulation;
(8) formal force resolution achieved in the simulation. As we will show later, convergent results are expected at scales larger than four times the formal resolution;
(9) the halo concentration as estimated from NFW profile fits to halo density profiles;
(10) the maximum relative error of the NFW fit: $\rho_{NFW}/\rho_h - 1$ (the error was estimated inside $50h^{-1}$ kpc radius);
(11) the same as in the previous column, but for the fits of profile advocated by Moore *et al.*

Halo A in the first set of simulations was re-simulated three times with increasing mass resolution. For each simulation, we considered outputs at four moments in the interval to $z = 0$–0.03. The parameters of the halos in these simulations averaged over the four moments are presented in the first three rows of table 15.4. We did not find any systematic change with resolution in the values of the halo parameters either on the virial radius scale or around the maximum of the circular velocity ($r = (30-40)h^{-1}$ kpc).

The top panel in figure 15.19 shows the central region of halo A_1 (see table 15.4). This plot is similar to figure 1(a) in Moore *et al* (1998) in that all profiles are drawn to the formal force resolution. The straight lines indicate the slopes of two power laws: $\gamma = -1$ and $\gamma = -1.4$. The figure indeed shows that, at around 1% of the virial radius, the slope is steeper than -1 and the central slope increases as we increase the mass resolution. Moore *et al* (1998) interpreted this behaviour as evidence that the profiles are steeper than those predicted by the NFW profile. We also note that the results of our highest resolution run A_1 are qualitatively consistent with the results from Kravtsov *et al* (1999). Indeed, if the profiles are considered down to the scale of *two* formal resolutions, the density profile slope in the very central part of the profile $r \lesssim 0.01r_{\mathrm{vir}}$ is close to $\gamma = -0.5$.

The profiles in figure 15.19 reflect the density distribution in the cores of simulated halos. However, the interpretation of these profiles is not straightforward because it requires an assessment of the numerical effects. The formal resolution does not usually even correspond to the scale where the numerical force is fully Newtonian (usually it is still considerably 'softer' than the Newtonian value). In the ART code, the inter-particle force reaches (on average) the Newtonian value at about two formal resolutions (see Kravtsov *et al* 1997). The effects of force resolution can be studied by re-simulating the same objects with higher force resolution and comparing the density profiles. Such a convergence study was done in Kravtsov *et al* (1998) where it was found that *for a fixed mass resolution* the halo density profiles converge at scales above two formal resolutions. Second, the local dynamical time for particles moving in the core of a halo is very short. For example, particles on the circular orbit of the radius $1h^{-1}$ kpc from the centre of halo A makes about 200 revolutions over the Hubble time. Therefore, if the time step is insufficiently small, numerical errors in these regions will tend to grow especially fast. The third possible source of numerical error is the mass resolution. Poor mass resolution in simulations with good force resolution may, for example, lead to two-body effects (e.g. Knebe *et al* 2000). An insufficient number of particles may also result in a 'grainy' potential in halo cores and thereby affect the accuracy of the orbit integration. In these effects, the mass resolution may be closely inter-related with the force resolution.

It is clear thus that, in order to draw conclusions unaffected by numerical errors, one has to determine the range of trustworthy scales using convergence analysis. The bottom panel in figure 15.19 shows that, for the halo A simulations, the convergence for vastly different mass and force resolution is reached for scales greater than or approximately equal to four formal force resolutions (all profiles in this figure are plotted down to the radius of four formal force resolutions). For all resolutions, there are more than 200 particles within the radius of four resolutions from the halo centre. For the highest resolution simulation (halo A_1) convergence is reached at scales $\gtrsim 0.005r_{\mathrm{vir}}$.

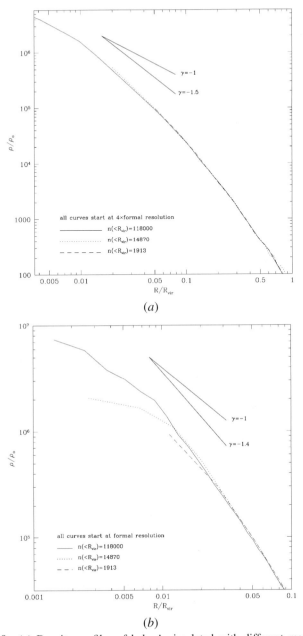

Figure 15.19. (*a*) Density profiles of halo A simulated with different mass and force resolutions. The profiles are plotted down to the formal force resolution of each simulation. (*b*) The profiles plotted down to *four formal resolutions*. It is clear that for vastly different mass (from 2000 to 120 000 particles in the halo) and force (from $3.66h^{-1}$ kpc to $0.23h^{-1}$ kpc) resolutions, convergence is reached at these scales.

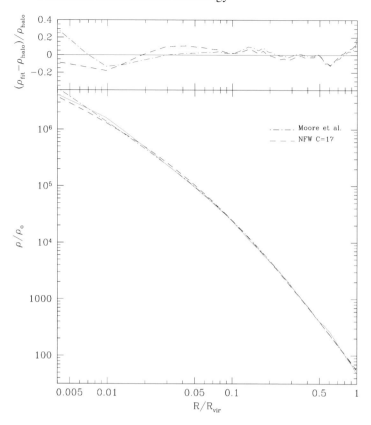

Figure 15.20. Fits of the NFW and Moore *et al* halo profiles to the profile of halo A₁ (*bottom panel*). The *top panel* shows the fractional deviations of the analytic fits from the numerical profile. Note that both analytical profiles fit the numerical profile equally well: fractional deviations are smaller than 20% over almost three decades in the radius.

In order to judge which profile provides a better description of the simulated profiles we fitted the NFW and Moore *et al* analytical profiles. Figure 15.20 presents the results of the fits and shows that both profiles fit the numerical profile equally well: fractional deviations of the fitted profiles from the numerical one are smaller than 20% over almost three decades in the radius. It is thus clear that the fact that the numerical profile has a slope steeper than -1 at the scale of $\sim 0.01 r_{\mathrm{vir}}$ does not mean that a good fit of the NFW profile (or even analytical profiles with shallower asymptotic slopes) cannot be obtained.

There is certainly a certain degree of degeneracy in fitting various analytic profiles to the numerical results. Figure 15.21 illustrates this further by showing results of fitting profiles (full curves) of the form $\rho(r) \propto (r/r_0)^{-\gamma}[1 + (r/r_0)^{\alpha}]^{-(\beta-\alpha)/\gamma}$ to *the same* (halo A₁) simulated halo profile shown as full

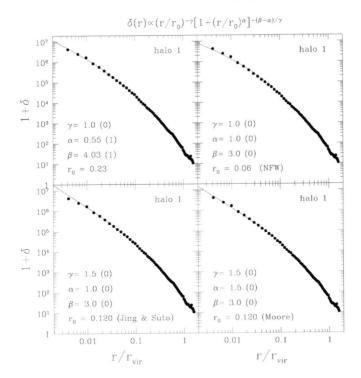

$$\delta(r) \propto (r/r_0)^{-\gamma}[1+(r/r_0)^\alpha]^{-(\beta-\alpha)/\gamma}$$

Figure 15.21. Analytical fits to the density profile of halo A_1 (see table 15.4) from our set of simulations. The fits are of the form $\rho(r) \propto (r/r_0)^{-\gamma}[1 + (r/r_0)^\alpha]^{-(\beta-\alpha)/\gamma}$. The legend in each panel indicates the corresponding values of α, β and γ of the fit; the digit in parentheses indicates whether the parameter was kept fixed (0) or not (1) during the fit. Note that various sets of parameters, α, β, γ, provide equally good fits to the simulated halo profile in the whole resolved range of scales $\approx(0.005-1)r_{\rm vir}$. This indicates a large degree of degeneracy in the parameters α, β and γ.

circles. The legend in each panel indicates the corresponding values of α, β and γ of the fit; the digit in parentheses indicates whether the parameter was kept fixed (0) or not (1) during the fit. The two right-hand panels show the fits of the NFW and Moore *et al* profiles; the bottom left-hand panel shows fit of the profiles used by Jing and Suto (2000). The top left-hand panel shows a fit in which the inner slope was fixed but α and β were fitted. The figure shows that all four analytic profiles can provide a nice fit to the numerical profile in the whole range $(0.005-1)r_{\rm vir}$.

15.4.4.2   Halo profiles at $z = 1$

As we have mentioned, the halo A analysed in the previous section is somewhat special because it was selected as an isolated relaxed halo. In order to reach unbiased conclusions, in this section we will present an analysis of halos from the second set of simulations (halos B, C and D in table 15.4) which were not selected to be relaxed or isolated. Based on the results of the convergence study presented in the previous section, we will consider profiles of these halos only at scales above four formal resolutions using results starting only from four formal resolutions and not less than 200 particles. Note that these conditions are probably more stringent than necessary because these halos were simulated with five to seven times more particles per halo. There is an advantage in analysing halos at a relatively high redshift. Halos of a given mass will have a lower concentration (see Bullock *et al* 2001). A lower concentration implies a large scale at which the asymptotic inner slope is reached. Profiles of the high-redshift halos should, therefore, be more useful in discriminating between the analytic models with different inner slopes.

We found that a substantial substructure is present inside the virial radius in all three halos. Figure 15.22 shows the profiles of these halos at $z = 1$. There profiles are not as smooth as that of halo A_1 due to their substructure. Note that bumps and depressions visible in the profiles cannot have a significantly larger amplitude than the shot noise. Halo C appeared to be the most relaxed of the three halos. It also had its last major merger somewhat earlier than the other two. Halo D had a major merger event at $z \approx 2$. Remnants of the merger are still visible as a hump at radii around $100h^{-1}$ kpc. Non-uniformities in the profiles caused by the substructure may substantially bias the analytic fits to the entire range of scales below the virial radius. Therefore, we used only the central, presumably more relaxed, regions in the analytic fits: $r < 50h^{-1}$ kpc for halo D and $r < 100h^{-1}$ kpc for halos B and C (fits using only central $50h^{-1}$ kpc did not change the results).

The best-fit parameters were obtained by minimizing the maximum fractional deviation of the fit: $\max(\mathrm{abs}(\log \rho_{\mathrm{fit}}) - \log \rho_{\mathrm{h}})$. Minimizing the sum of the squares of deviations (χ^2), as is often done, can result in larger errors at small radii with the false impression that the fit fails because it has a wrong central slope. The fit that minimizes the maximum deviations improves the NFW fit for points in the range of radii $(5–20)h^{-1}$ kpc, where the NFW fit would appear to be below the data points if the fit was done by χ^2 minimization. This improvement comes at the expense of a few points around $1h^{-1}$ kpc. For example, if we fit halo B by using χ^2 minimization, the concentration decreases from 12.3 (see table 15.4) to 11.8. We also made a fit for halo B assuming even more stringent limits on the effects of numerical resolution. By minimizing the maximum deviation we fitted the halo starting at six times the formal resolution. Inside this radius there were about 900 particles. The resulting parameters of the fit were close to those in table 15.4: $C_{\mathrm{NFW}} = 11.8$, and the maximum error of the NFW fit was 17%.

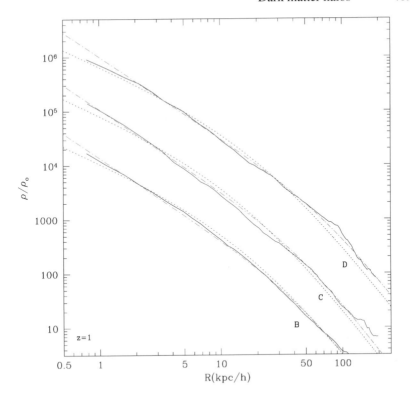

Figure 15.22. Profiles of halos B, C and D at $z = 1$. The profiles of halos C and D were offset downwards by factors of 10 and 100 for clarity. The full curves show simulated profiles, while the dotted and chain curves show the NFW and Moore *et al* fits, respectively. The halo profiles in the simulations are plotted down to four formal resolutions. Each halo had more than 200 particles inside the smallest plotted scale.

We found that the errors in the Moore *et al* fits were systematically smaller than those of the NFW fits, though the differences were not dramatic. The Moore *et al* fit failed for halo D. It formally gave very small errors, but this was done for a fit with an unreasonably small concentration ($C = 2$). When we constrained the approximation to have a concentration twice as large compared with the best NFW fit, we were able to obtain a reasonable fit (this fit is shown in figure 15.22). Nevertheless, the central part was fitted poorly in this case.

Our analysis therefore failed to determine which analytic profile provides a better description of the density distribution in simulated halos. Despite the larger number of particles per halo and lower concentrations of halos, the results are still inconclusive. The Moore *et al* profile is a better fit to the profile of halo C; the NFW profile is a better fit to the central part of halo D. Halo B represents an intermediate case where both profiles provide equally good fits (similar to the analysis of halo A).

Note that there seems to be real deviations in the parameters of halos of the same mass. Halos B and D have the same virial radii and nearly the same circular velocities, yet their concentrations are different by 30%. We find the same differences in estimates of $C_{1/5}$ concentrations, which do not depend on the specifics of an analytic fit. The central slope at around 1 kpc also changes from halo to halo.

15.4.4.3 *Summary*

In this section we have given a review of some of the internal properties of DM halos focusing mostly on their profiles and concentrations. Our results are mostly based on simulations done with the ART code, which is capable of handling particles with different masses, variable force and time resolution. In runs with the highest resolution, the code achieved (formal) dynamical range of $2^{17} = 131\,072$ with 500 000 steps for particles at the highest level of resolution.

Our conclusions regarding the convergence of the profiles differ from those of Moore *et al* (1998). If we take into account only the radii, at which we believe the numerical effects (the force resolution, the resolution of initial perturbations and two-body scattering) to be small, then we find that the slope and amplitude of the density do not change when we change the force and mass resolution. This result is consistent with what was found in simulations of the 'Santa Barbara' cluster (Frenk *et al* 1999): at a fixed *resolved* scale the results do not change as the resolution increases. For the ART code the results converged at four times the formal force resolution and more than 200 particles. These convergence limits very likely depend on the particular code used and on the duration of the integration.

We reproduce Moore *et al*'s results regarding convergence and the results from Kravtsov *et al* (1998) regarding shallow central profiles, but only when we considered points inside unresolved scales. We conclude that those results followed from an overly optimistic interpretation of the numerical accuracy of the simulations.

For the galaxy-size halos considered in this section with masses $M_{\rm vir} = 7 \times 10^{11} h^{-1} M_\odot$ to $2 \times 10^{12} h^{-1} M_\odot$ and concentrations $C = 9\text{--}17$ both the NFW profile, $\rho \propto r^{-1}(1+r)^{-2}$, and the Moore *et al* profile, $\rho \propto r^{-1.5}(1+r^{1.5})^{-1}$, give good fits with an accuracy of about 10% for radii not smaller than 1% of the virial radius. None of the profiles is significantly better than the other.

Halos with the same mass may have different profiles. No matter what profile is used—NFW or Moore *et al*—there is no universal profile: halo mass does not yet define the density profile. Nevertheless, the universal profile is an extremely useful notion which should be interpreted as the general trend $C(M)$ of halos with a larger mass to have a lower concentration. Deviations from the general $C(M)$ are real and significant (Bullock *et al* 2001). It is not yet clear but it seems very likely that the central slopes of halos also have real fluctuations. The fluctuations in the concentration and central slopes are important for interpreting the central parts of rotation curves.

References

Aarseth S J 1963 *Mon. Not. R. Astron. Soc.* **126** 223
——1985 *Multiple Time Scales* ed J W Brackbill and B J Cohen (New York: Academic Press) p 377
Aarseth S J, Gott J R and Turner E L 1979 *Astrophys. J.* **228** 664
Appel A 1985 *SIAM J. Sci. Stat. Comput.* **6** 85
Avila-Reese *et al* 1999 *Mon. Not. R. Astron. Soc.* **310** 527
Barnes J and Hut P 1986 *Nature* **324** 446
Benson A J, Cole S, Frenk C S, Baugh C M and Lacey C D 2000 *Mon. Not. R. Astron. Soc.* **311** 793 (astro-ph/9903343)
Bertschinger E 1998 *Annu. Rev. Astron. Astrophys.* **36** 599
——2001 *Preprint* astro-ph/0103301
Bertschinger E and Gelb J 1991 *Comput. Phys.* **5** 164
Binney J and Tremaine S 1987 *Galactic Dynamics* (Princeton, NJ: Princeton University Press)
Bouchet F R and Hernquist L 1988 *Astrophys. J. Suppl.* **68** 521
Bullock J S, Kolatt T S, Sigad Y, Somerville R S, Kravtsov A V, Klypin A, Primack J P and Dekel A 2001 *Mon. Not. R. Astron. Soc.* **321** 559 (astro-ph/9908159)
Bullock J S, Kravtsov A V and Weinberg D H 2000 *Astrophys. J.* **539** 517 (astro-ph/0002214)
Catelan P, Lucchin F, Matarrese S and Porciani C 1998a *Mon. Not. R. Astron. Soc.* **297** 692
Catelan P, Matarrese S and Porciani C 1998b *Astrophys. J. Lett.* **502** 1
Colín P, Klypin A and Kravtsov A 2000 *Astrophys. J.* **539** 561 (astro-ph/9907337)
Colín P, Klypin A A, Kravtsov A V and Khokhlov A M 1999 *Astrophys. J.* **523** 32 (astro-ph/9809202)
Couchman H M P 1991 *Astrophys. J.* **368** 23
Davis M, Efstathiou G, Frenk C S and White S D M 1985 *Astrophys. J.* **292** 371
Dekel A and Lahav O 1999 *Astrophys. J.* **520** 24 (astro-ph/9806193)
Dekel A and Silk J 1986 *Astrophys. J.* **303** 39
Diaferio A, Kauffmann G, Colberg J M and White S D M 1999 *Mon. Not. R. Astron. Soc.* **307** 537 (astro-ph/9812009)
Doroshkevich A G, Kotok E V, Novikov I D, Polyudov A N and Sigov Yu S 1980 *Mon. Not. R. Astron. Soc.* **192** 321
Efstathiou G, Davis M, Frenk C S and White S D M 1985 *Astrophys. J. Suppl.* **57** 241
Flores R A and Primack J R 1994 *Astrophys. J.* **427** L1
Frenk C *et al* 1999 *Astrophys. J.* **525** 630
Gelb J 1992 *PhD Thesis* MIT
Ghigna S, Moore B, Governato F, Lake G, Quinn T and Stadel J 1998 *Mon. Not. R. Astron. Soc.* **300** 146
——*Observational Cosmology: The Development of Galaxy Systems* ed G Giuricin, M Mezzetti and P Solucci (San Francisco, CA: ASP) p 140
——2000 *Astrophys. J.* **544** 616

Gottlöber S, Klypin A and Kravtsov A V 1998 *Observational Cosmology: The Development of Galaxy Systems (ASP Conf. Series 176, 1999)* ed G Giuricin, M Mezetti and P Salucci (San Francisco, CA: ASP) p 418

Gross M 1997 *PhD Thesis* University of California, Santa Cruz

Gunn J E and Gott J R 1972 *Astrophys. J.* **176** 1

Hernquist L 1987 *Astrophys. J. Suppl.* **64** 715

Hernquist L, Bouchet F R and Suto Y 1991 *Astrophys. J. Suppl.* **75** 231

Hockney R W and Eastwood J W 1981 *Numerical Simulations Using Particles* (New York: McGraw-Hill)

Hu W and Sugiyama 1996 *Astrophys. J.* **471** 542

Jing Y P 2000 *Astrophys. J.* **535** 30 (astro-ph/9901340)

Jing Y P and Suto Y 2000 *Astrophys. J.* **529** L69

Kaiser N 1984 *Astrophys. J.* **284** L9

Kauffmann G, White S D M and Guiderdoni B 1993 *Mon. Not. R. Astron. Soc.* **264** 201

Klypin A, Gotlöber S, Kravtsov A V and Khokhlov A 1999a *Astrophys. J.* **516** 530 (KGKK)

Klypin A and Holtzman J 1997 *Preprint* astro-ph/9712217

Klypin A, Holtzman J, Primack J and Regos E 1993 *Astrophys. J.* **416** 1

Klypin A, Kravtsov A V, Bullock J S and Primack J P 2001 *Astrophys. J.* **554** 903

Klypin A, Kravtsov A V, Valenzuela O and Prada F 1999b *Astrophys. J.* **522** 82

Klypin A and Shandarin S F 1983 *Mon. Not. R. Astron. Soc.* **204** 891

Knebe A, Green A and Binney J 2001 *Mon. Not. R. Astron. Soc.* **325** 845

Knebe A, Kravtsov A V, Gottlöber S and Klypin A 2000 *Mon. Not. R. Astron. Soc.* **317** 630

Kravtsov A V 1999 *PhD Thesis* New Mexico State University

Kravtsov A V and Klypin A 1999 *Astrophys. J.* **520** 437

Kravtsov A V, Klypin A, Bullock J S and Primack J P 1999 *Astrophys. J.* **502** 48

Kravtsov A V, Klypin A and Khokhlov A 1997 *Astrophys. J. Suppl.* **111** 73

Lokas Elanol Mammon G A 2001 *Mon. Not. R. Astron. Soc.* **321** 155

Mo H J and White S D M 1996 *Mon. Not. R. Astron. Soc.* **282** 347

Moore B 1994 *Nature* **370** 629

Moore B, Ghigna S, Governato F, Lake G, Quinn T, Stadel J and Tozzi P 1999 *Astrophys. J. Lett.* **524** L19 (astro-ph/9907411)

Moore B, Governato F, Quinn T, Stadel J and Lake G 1998, *Astrophys. J.* **499** L5

Moore B, Katz N and Lake G 1996 *Astrophys. J.* **456** 455

Moore B, Quinn T, Governato F, Stadel J and Lake G 1999 *Mon. Not. R. Astron. Soc.* **310** 1147

Navarro J F, Frenk C S and White S D M 1996 *Astrophys. J.* **462** 563

——1997 *Astrophys. J.* **490** 493

Okamoto T and Habe A 1999 *Astrophys. J.* **516** 591

Peebles P J E 1970 *Astron. J.* **75** 13

Porciani C, Matarrese S, Lucchin F and Catelan P 1998 *Mon. Not. R. Astron. Soc.* **298** 1097 (astro-ph/9801290)

Press W H and Schechter P L 1974 *Astrophys. J.* **187** 425

Quinn T, Katz N and Stadel J and Lake G 1997 *Preprint* astro-ph/9710043

Sahni V and Coles P 1995 *Phys. Rep.* **262** 2

Sellwood J A 1987 *Annu. Rev. Astron. Astrophys.* **25** 151

Sheth R K and Lemson G 1999 *Mon. Not. R. Astron. Soc.* **305** 946 (astro-ph/9808138)

Sigad Y, Kolatt T S, Bullock J S, Kravtsov A V, Klypin A, Primack J R and Dekel A 2000
    in preparation

Somerville R, Lemson G, Sigad Y, Dekel A, Kauffmann G and White S D M 2001 *Mon.*
    Not. R. Astron. Soc. **320** 289 (astro-ph/9912073)

Splinter R *et al* 1998 *Astrophys. J.* **498** 38

Swaters R A, Madore B F and Trewhella M 2000 *Astrophys. J.* **531** L107

Taruya A and Suto Y 2000 *Astrophys. J.* **542** 559 (astro-ph/0004288)

Tormen G, Diaferio A and Syer D 1998 *Mon. Not. R. Astron. Soc.* **299** 728

van den Bosch F C, Robertson B E, Dalcanton J J and de Blok W J G 2000 *Astrophys. J.*
    119 1579

White S D M 1976 *Mon. Not. R. Astron. Soc.* **177** 717

White S D M and Rees M J 1978 *Mon. Not. R. Astron. Soc.* **183** 341

Widrow L 2000 *Astrophys. J.* **131** 39

Zeldovich Ya B 1970 *Astron. Astrophys.* **5** 84

Index